Universitext

Universitext

Universitext is a series of textbooks that presents material from a wide variety of mathematical disciplines at master's level and beyond. The books, often well class-tested by their author, may have an informal, personal, even experimental approach to their subject matter. Some of the most successful and established books in the series have evolved through several editions, always following the evolution of teaching curricula, into very polished texts.

Thus as research topics trickle down into graduate-level teaching, first textbooks written for new, cutting-edge courses may make their way into *Universitext*.

More information about this series at http://www.springer.com/series/223

Leonor Godinho · José Natário

An Introduction to Riemannian Geometry

With Applications to Mechanics and Relativity

Leonor Godinho
Departamento de Matemática
Instituto Superior Técnico
Lisbon
Portugal

José Natário
Departamento de Matemática
Instituto Superior Técnico
Lisbon
Portugal

ISSN 0172-5939
ISBN 978-3-319-08665-1
DOI 10.1007/978-3-319-08666-8

ISSN 2191-6675 (electronic)
ISBN 978-3-319-08666-8 (eBook)

Library of Congress Control Number: 2014943752

Mathematics Subject Classification: 53-01, 70-01, 83-01

Springer Cham Heidelberg New York Dordrecht London

Printed on acid-free paper

Springer is part of Springer Science+Business Media (www.springer.com)

Preface

This book is based on a one-semester course taught since 2002 at Instituto Superior Técnico (Lisbon) to mathematics, physics and engineering students. Its aim is to provide a quick introduction to differential geometry, including differential forms, followed by the main ideas of Riemannian geometry (minimizing properties of geodesics, completeness and curvature). Possible applications are given in the final two chapters, which have themselves been independently used for one-semester courses on geometric mechanics and general relativity. We hope that these will give mathematics students a chance to appreciate the usefulness of Riemannian geometry, and physics and engineering students an extra motivation to learn the mathematical background.

It is assumed that readers have basic knowledge of linear algebra, multivariable calculus and differential equations, as well as elementary notions of topology and algebra. For their convenience (especially physics and engineering students), we have summarized the main definitions and results from this background material at the end of each chapter as needed.

To help readers test and consolidate their understanding, and also to introduce important ideas and examples not treated in the main text, we have included more than 330 exercises, of which around 140 are solved in Chap. 7 (the solutions to the full set are available for instructors). We hope that this will make this book suitable for self-study, while retaining a sufficient number of unsolved exercises to pose a challenge.

We now give a short description of the contents of each chapter.

Chapter 1 discusses the basic concepts of differential geometry: differentiable manifolds and maps, vector fields and the Lie bracket. In addition, we give a brief overview of Lie groups and Lie group actions.

Chapter 2 is devoted to differential forms, covering the standard topics: wedge product, pull-back, exterior derivative, integration and the Stokes theorem.

Riemannian manifolds are introduced in Chap. 3, where we treat the Levi–Civita connection, minimizing properties of geodesics and the Hopf–Rinow theorem.

Chapter 4 addresses the notion of curvature. In particular, we use the powerful computational method given by the Cartan structure equations to prove the Gauss–Bonnet theorem. Constant curvature and isometric embeddings are also discussed.

Chapter 5 gives an overview of geometric mechanics, including holonomic and non-holonomic systems, Lagrangian and Hamiltonian mechanics, completely integrable systems and reduction.

Chapter 6 treats general relativity, starting with a geometric introduction to special relativity. The Einstein equation is motivated via the Cartan connection formulation of Newtonian gravity, and the basic examples of the Schwarzschild solution (including black holes) and cosmology are studied. We conclude with a discussion of causality and the celebrated Hawking and Penrose singularity theorems, which, although unusual in introductory texts, are very interesting applications of Riemannian geometry.

Finally, we want to thank the many colleagues and students who read this text, or parts of it, for their valuable comments and suggestions. Special thanks are due to our colleague and friend Pedro Girão.

Contents

Chapter 1
Differentiable Manifolds

In pure and applied mathematics, one often encounters spaces that locally look like $\mathbb{R}^n$, in the sense that they can be locally parameterized by n coordinates: for example, the n-dimensional sphere $S^n \subset \mathbb{R}^{n+1}$, or the set $\mathbb{R}^3 \times SO(3)$ of configurations of a rigid body. It may be expected that the basic tools of calculus can still be used in such spaces; however, since there is, in general, no canonical choice of local coordinates, special care must be taken when discussing concepts such as derivatives or integrals whose definitions in $\mathbb{R}^n$ rely on the preferred Cartesian coordinates.

The precise definition of these spaces, called **differentiable manifolds**, and the associated notions of differentiation, are the subject of this chapter. Although the intuitive idea seems simple enough, and in fact dates back to Gauss and Riemann, the formal definition was not given until 1936 (by Whitney).

The concept of spaces that locally look like $\mathbb{R}^n$ is formalized by the definition of **topological manifolds**: topological spaces that are locally homeomorphic to $\mathbb{R}^n$. These are studied in Sect. 1.1, where several examples are discussed, particularly in dimension 2 (surfaces).

Differentiable manifolds are defined in Sect. 1.2 as topological manifolds whose changes of coordinates (maps from $\mathbb{R}^n$ to $\mathbb{R}^n$) are smooth (C^∞). This enables the definition of **differentiable functions** as functions whose expressions in local coordinates are smooth (Sect. 1.3), and **tangent vectors** as directional derivative operators acting on real-valued differentiable functions (Sect. 1.4). Important examples of differentiable maps, namely **immersions** and **embeddings**, are examined in Sect. 1.5.

Vector fields and their **flows** are the main topic of Sect. 1.6. A natural differential operation between vector fields, called the **Lie bracket**, is defined; it measures the non-commutativity of their flows and plays a central role in differential geometry.

Section 1.7 is devoted to the important class of differentiable manifolds which are also groups, the so-called **Lie groups**. It is shown that to each Lie group one can associate a **Lie algebra**, i.e. a vector space equipped with a Lie bracket. **Quotients** of manifolds by actions of Lie groups are also treated.

Orientability of a manifold (closely related to the intuitive notion of a surface "having two sides") and **manifolds with boundary** (generalizing the concept of a surface bounded by a closed curve, or a volume bounded by a closed surface)

© Springer International Publishing Switzerland 2014
L. Godinho and J. Natário, *An Introduction to Riemannian Geometry*, Universitext,
DOI 10.1007/978-3-319-08666-8_1

are studied in Sects. 1.8 and 1.9. Both these notions are necessary to formulate the celebrated Stokes theorem, which will be proved in Chap. 2.

1.1 Topological Manifolds

We will begin this section by studying spaces that are locally like $\mathbb{R}^n$, meaning that there exists a neighborhood around each point which is homeomorphic to an open subset of $\mathbb{R}^n$.

Definition 1.1 A **topological manifold** M of dimension n is a topological space with the following properties:

 (i) M is **Hausdorff**, that is, for each pair p_1, p_2 of distinct points of M there exist neighborhoods V_1, V_2 of p_1 and p_2 such that $V_1 \cap V_2 = \varnothing$.
 (ii) Each point $p \in M$ possesses a neighborhood V homeomorphic to an open subset U of $\mathbb{R}^n$.
 (iii) M satisfies the **second countability axiom**, that is, M has a countable basis for its topology.

Conditions (i) and (iii) are included in the definition to prevent the topology of these spaces from being too strange. In particular, the Hausdorff axiom ensures that the limit of a convergent sequence is unique. This, along with the second countability axiom, guarantees the existence of partitions of unity (cf. Sect. 7.2 of Chap. 2), which, as we will see, are a fundamental tool in differential geometry.

Remark 1.2 If the dimension of M is zero then M is a countable set equipped with the discrete topology (every subset of M is an open set). If dim $M = 1$, then M is locally homeomorphic to an open interval; if dim $M = 2$, then it is locally homeomorphic to an open disk etc.

Example 1.3

(1) Every open subset M of $\mathbb{R}^n$ with the subspace topology (that is, $U \subset M$ is an open set if and only if $U = M \cap V$ with V an open set of $\mathbb{R}^n$) is a topological manifold.
(2) (*Circle*) The **circle**

$$S^1 = \left\{ (x, y) \in \mathbb{R}^2 \mid x^2 + y^2 = 1 \right\}$$

with the subspace topology is a topological manifold of dimension 1. Conditions (i) and (iii) are inherited from the ambient space. Moreover, for each point $p \in S^1$ there is at least one coordinate axis which is not parallel to the vector n_p normal to S^1 at p. The projection on this axis is then a homeomorphism between a (sufficiently small) neighborhood V of p and an interval in $\mathbb{R}$.

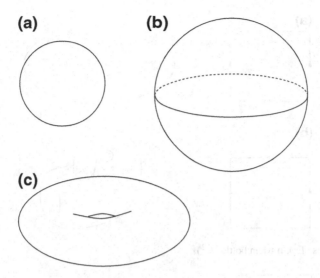

Fig. 1.1 **a** S^1; **b** S^2; **c** Torus of revolution

(3) (*2-sphere*) The previous example can be easily generalized to show that the
2-sphere
$$S^2 = \left\{(x, y, z) \in \mathbb{R}^3 \mid x^2 + y^2 + z^2 = 1\right\}$$

with the subspace topology is a topological manifold of dimension 2.

(4) (*Torus of revolution*) Again, as in the previous examples, we can show that the
surface of revolution obtained by revolving a circle around an axis that does not
intersect it is a topological manifold of dimension 2 (Fig. 1.1).

(5) The surface of a cube is a topological manifold (homeomorphic to S^2).

Example 1.4 We can also obtain topological manifolds by identifying edges of certain polygons by means of homeomorphisms. The edges of a square, for instance,
can be identified in several ways (see Figs. 1.2 and 1.3):

(1) (*Torus*) The **torus** T^2 is the quotient of the unit square $Q = [0, 1]^2 \subset \mathbb{R}^2$ by the
equivalence relation

$$(x, y) \sim (x + 1, y) \sim (x, y + 1),$$

equipped with the quotient topology (cf. Sect. 1.10.1).

(2) (*Klein bottle*) The **Klein bottle** K^2 is the quotient of Q by the equivalence
relation
$$(x, y) \sim (x + 1, y) \sim (1 - x, y + 1).$$

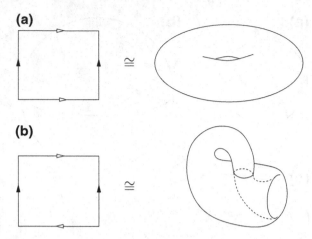

Fig. 1.2 **a** Torus (T^2); **b** Klein bottle (K^2)

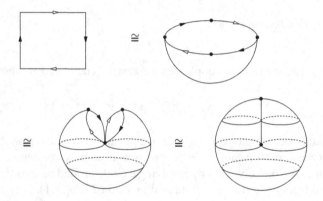

Fig. 1.3 Projective plane ($\mathbb{R}P^2$)

(3) (*Projective plane*) The **projective plane** $\mathbb{R}P^2$ is the quotient of Q by the equivalence relation

$$(x, y) \sim (x + 1, 1 - y) \sim (1 - x, y + 1).$$

Remark 1.5

(1) The only compact connected 1-dimensional topological manifold is the circle S^1 (see [Mil97]).
(2) The **connected sum** of two topological manifolds M and N is the topological manifold $M\#N$ obtained by deleting an open set homeomorphic to a ball on each manifold and gluing the boundaries, which must be homeomorphic to spheres, by a homeomorphism (cf. Fig. 1.4). It can be shown that any com-

Fig. 1.4 Connected sum of two tori

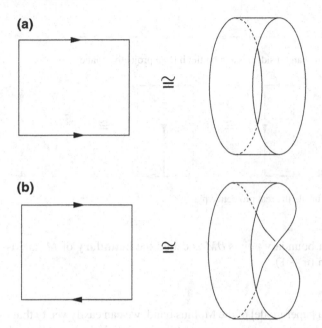

Fig. 1.5 a Cylinder; **b** Möbius band

pact connected 2-dimensional topological manifold is homeomorphic either to S^2 or to connected sums of manifolds from Example 1.4 (see [Blo96, Mun00]).

If we do not identify all the edges of the square, we obtain a cylinder or a **Möbius band** (cf. Fig. 1.5). These topological spaces are examples of **manifolds with boundary**.

Definition 1.6 Consider the **closed half space**

$$\mathbb{H}^n = \left\{ \left(x^1, \ldots, x^n\right) \in \mathbb{R}^n \mid x^n \geq 0 \right\}.$$

A **topological manifold with boundary** is a Hausdorff space M, with a countable basis of open sets, such that each point $p \in M$ possesses a neighborhood V which is homeomorphic either to an open subset U of $\mathbb{H}^n \setminus \partial \mathbb{H}^n$, or to an open subset U of $\mathbb{H}^n$, with the point p identified to a point in $\partial \mathbb{H}^n$. The points of the first type are called **interior points**, and the remaining are called **boundary points**.

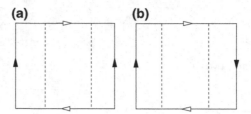

Fig. 1.6 Möbius band inside **a** Klein bottle; **b** Real projective plane

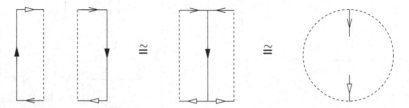

Fig. 1.7 Disk inside the real projective plane

The set of boundary points ∂M is called the **boundary** of M, and is a manifold of dimension $(n-1)$.

Remark 1.7

1. Making a paper model of the Möbius band, we can easily verify that its boundary is homeomorphic to a circle (not to two disjoint circles), and that it has only one side (cf. Fig. 1.5).
2. Both the Klein bottle and the real projective plane contain Möbius bands (cf. Fig. 1.6). Deleting this band on the projective plane, we obtain a disk (cf. Fig. 1.7). In other words, we can glue a Möbius band to a disk along their boundaries and obtain $\mathbb{R}P^2$.

Two topological manifolds are considered the same if they are homeomorphic. For example, spheres of different radii in $\mathbb{R}^3$ are homeomorphic, and so are the two surfaces in Fig. 1.8. Indeed, the knotted torus can be obtained by cutting the torus along a circle, knotting it and gluing it back again. An obvious homeomorphism is then the one which takes each point on the initial torus to its final position after cutting and gluing (however, this homeomorphism cannot be extended to a homeomorphism of the ambient space $\mathbb{R}^3$).

Exercise 1.8

(1) Which of the following sets (with the subspace topology) are topological manifolds?

    (a) $D^2 = \{(x, y) \in \mathbb{R}^2 \mid x^2 + y^2 < 1\}$;

    (b) $S^2 \setminus \{p\}$ $(p \in S^2)$;

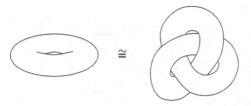

Fig. 1.8 Two homeomorphic topological manifolds

    (c) $S^2 \setminus \{p, q\}$ $(p, q \in S^2, p \neq q)$;
    (d) $\{(x, y, z) \in \mathbb{R}^3 \mid x^2 + y^2 = 1\}$;
    (e) $\{(x, y, z) \in \mathbb{R}^3 \mid x^2 + y^2 = z^2\}$;

(2) Which of the manifolds above are homeomorphic?
(3) Show that the Klein bottle K^2 can be obtained by gluing two Möbius bands together through a homeomorphism of the boundary.
(4) Show that:

    (a) $M \# S^2 = M$ for any 2-dimensional topological manifold M;
    (b) $\mathbb{R}P^2 \# \mathbb{R}P^2 = K^2$;
    (c) $\mathbb{R}P^2 \# T^2 = \mathbb{R}P^2 \# K^2$.

(5) A **triangulation** of a 2-dimensional topological manifold M is a decomposition of M in a finite number of triangles (i.e. subsets homeomorphic to triangles in $\mathbb{R}^2$) such that the intersection of any two triangles is either a common edge, a common vertex or empty (it is possible to prove that such a triangulation always exists). The **Euler characteristic** of M is

$$\chi(M) := V - E + F,$$

where V, E and F are the number of vertices, edges and faces of a given triangulation (it can be shown that this is well defined, i.e. does not depend on the choice of triangulation). Show that:

    (a) adding a vertex to a triangulation does not change $\chi(M)$;
    (b) $\chi\left(S^2\right) = 2$;
    (c) $\chi\left(T^2\right) = 0$;
    (d) $\chi\left(K^2\right) = 0$;
    (e) $\chi\left(\mathbb{R}P^2\right) = 1$;
    (f) $\chi(M \# N) = \chi(M) + \chi(N) - 2$.

1.2 Differentiable Manifolds

Recall that an n-dimensional topological manifold is a Hausdorff space with a countable basis of open sets such that each point possesses a neighborhood homeomorphic

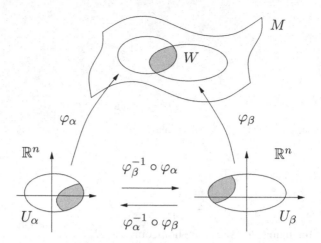

Fig. 1.9 Parameterizations and overlap maps

to an open subset of $\mathbb{R}^n$. Each pair (U, φ), where U is an open subset of $\mathbb{R}^n$ and $\varphi : U \to \varphi(U) \subset M$ is a homeomorphism of U to an open subset of M, is called a **parameterization**. The inverse φ^{-1} is called a **coordinate system** or **chart**, and the set $\varphi(U) \subset M$ is called a **coordinate neighborhood**. When two coordinate neighborhoods overlap, we have formulas for the associated coordinate change (cf. Fig. 1.9). The idea to obtain differentiable manifolds will be to choose a sub-collection of parameterizations so that the coordinate changes are differentiable maps.

Definition 2.1 An n-dimensional **differentiable** or **smooth manifold** is a topological manifold of dimension n and a family of parameterizations $\varphi_\alpha : U_\alpha \to M$ defined on open sets $U_\alpha \subset \mathbb{R}^n$, such that:

(i) the coordinate neighborhoods cover M, that is, $\bigcup_\alpha \varphi_\alpha(U_\alpha) = M$;
(ii) for each pair of indices α, β such that

$$W := \varphi_\alpha(U_\alpha) \cap \varphi_\beta(U_\beta) \neq \varnothing,$$

the overlap maps

$$\varphi_\beta^{-1} \circ \varphi_\alpha : \varphi_\alpha^{-1}(W) \to \varphi_\beta^{-1}(W)$$
$$\varphi_\alpha^{-1} \circ \varphi_\beta : \varphi_\beta^{-1}(W) \to \varphi_\alpha^{-1}(W)$$

are C^∞;
(iii) the family $\mathcal{A} = \{(U_\alpha, \varphi_\alpha)\}$ is maximal with respect to (i) and (ii), meaning that if $\varphi_0 : U_0 \to M$ is a parameterization such that $\varphi_0^{-1} \circ \varphi$ and $\varphi^{-1} \circ \varphi_0$ are C^∞ for all φ in $\mathcal{A}$, then (U_0, φ_0) is in $\mathcal{A}$.

Remark 2.2

(1) Any family $\mathcal{A} = \{(U_\alpha, \varphi_\alpha)\}$ that satisfies (i) and (ii) is called a C^∞-**atlas** for M. If $\mathcal{A}$ also satisfies (iii) it is called a **maximal atlas** or a **differentiable structure**.

(2) Condition (iii) is purely technical. Given any atlas $\mathcal{A} = \{(U_\alpha, \varphi_\alpha)\}$ on M, there is a unique maximal atlas $\widetilde{\mathcal{A}}$ containing it. In fact, we can take the set $\widetilde{\mathcal{A}}$ of all parameterizations that satisfy (ii) with every parameterization on $\mathcal{A}$. Clearly $\mathcal{A} \subset \widetilde{\mathcal{A}}$, and one can easily check that $\widetilde{\mathcal{A}}$ satisfies (i) and (ii). Also, by construction, $\widetilde{\mathcal{A}}$ is maximal with respect to (i) and (ii). Two atlases are said to be **equivalent** if they define the same differentiable structure.

(3) We could also have defined C^k-manifolds by requiring the coordinate changes to be C^k-maps (a C^0-manifold would then denote a topological manifold).

Example 2.3

(1) The space $\mathbb{R}^n$ with the usual topology defined by the Euclidean metric is a Hausdorff space and has a countable basis of open sets. If, for instance, we consider a single parameterization $(\mathbb{R}^n, id)$, conditions (i) and (ii) of Definition 2.1 are trivially satisfied and we have an atlas for $\mathbb{R}^n$. The maximal atlas that contains this parameterization is usually called the **standard differentiable structure** on $\mathbb{R}^n$. We can of course consider other atlases. Take, for instance, the atlas defined by the parameterization $(\mathbb{R}^n, \varphi)$ with $\varphi(x) = Ax$ for a nonsingular $(n \times n)$-matrix A. It is an easy exercise to show that these two atlases are equivalent.

(2) It is possible for a manifold to possess non-equivalent atlases: consider the two atlases $\{(\mathbb{R}, \varphi_1)\}$ and $\{(\mathbb{R}, \varphi_2)\}$ on $\mathbb{R}$, where $\varphi_1(x) = x$ and $\varphi_2(x) = x^3$. As the map $\varphi_2^{-1} \circ \varphi_1$ is not differentiable at the origin, these two atlases define different (though, as we shall see, diffeomorphic) differentiable structures [cf. Exercises 2.5(4) and 3.2(6)].

(3) Every open subset V of a smooth manifold is a manifold of the same dimension. Indeed, as V is a subset of M, its subspace topology is Hausdorff and admits a countable basis of open sets. Moreover, if $\mathcal{A} = \{(U_\alpha, \varphi_\alpha)\}$ is an atlas for M and we take the U_α for which $\varphi_\alpha(U_\alpha) \cap V \neq \varnothing$, it is easy to check that the family of parameterizations $\widetilde{\mathcal{A}} = \{(\widetilde{U}_\alpha, \varphi_\alpha|_{\widetilde{U}_\alpha})\}$, where $\widetilde{U}_\alpha = \varphi_\alpha^{-1}(V)$, is an atlas for V.

(4) Let $M_{n \times n}$ be the set of $n \times n$ matrices with real coefficients. Rearranging the entries along one line, we see that this space is just $\mathbb{R}^{n^2}$, and so it is a manifold. By Example 3, we have that $GL(n) = \{A \in M_{n \times n} \mid \det A \neq 0\}$ is also a manifold of dimension n^2. In fact, the determinant is a continuous map from $M_{n \times n}$ to $\mathbb{R}$, and $GL(n)$ is the preimage of the open set $\mathbb{R} \backslash \{0\}$.

(5) Let us consider the n-**sphere**

$$S^n = \left\{ \left(x^1, \ldots, x^{n+1}\right) \in \mathbb{R}^{n+1} \mid \left(x^1\right)^2 + \cdots + \left(x^{n+1}\right)^2 = 1 \right\}$$

and the maps

$$\varphi_i^+ : U \subset \mathbb{R}^n \to S^n$$

$$\left(x^1, \ldots, x^n\right) \mapsto \left(x^1, \ldots, x^{i-1}, g\left(x^1, \ldots, x^n\right), x^i, \ldots, x^n\right),$$

$$\varphi_i^- : U \subset \mathbb{R}^n \to S^n$$

$$\left(x^1, \ldots, x^n\right) \mapsto \left(x^1, \ldots, x^{i-1}, -g\left(x^1, \ldots, x^n\right), x^i, \ldots, x^n\right),$$

where

$$U = \left\{\left(x^1, \ldots, x^n\right) \in \mathbb{R}^n \mid \left(x^1\right)^2 + \cdots + \left(x^n\right)^2 < 1\right\}$$

and

$$g\left(x^1, \ldots, x^n\right) = \left(1 - \left(x^1\right)^2 - \cdots - \left(x^n\right)^2\right)^{\frac{1}{2}}.$$

Being a subset of $\mathbb{R}^{n+1}$, the sphere (equipped with the subspace topology) is a Hausdorff space and admits a countable basis of open sets. It is also easy to check that the family $\left\{\left(U, \varphi_i^+\right), \left(U, \varphi_i^-\right)\right\}_{i=1}^{n+1}$ is an atlas for S^n, and so this space is a manifold of dimension n (the corresponding charts are just the projections on the hyperplanes $x^i = 0$).

(6) We can define an atlas for the surface of a cube $Q \subset \mathbb{R}^3$ making it a smooth manifold: Suppose the cube is centered at the origin and consider the map $f : Q \to S^2$ defined by $f(x) = x/\|x\|$. Then, considering an atlas $\{(U_\alpha, \varphi_\alpha)\}$ for S^2, the family $\{(U_\alpha, f^{-1} \circ \varphi_\alpha)\}$ defines an atlas for Q.

Remark 2.4 There exist topological manifolds which admit no differentiable structures at all. Indeed, Kervaire presented the first example (a 10-dimensional manifold) in 1960 [Ker60], and Smale constructed another one (of dimension 12) soon after [Sma60]. In 1956 Milnor [Mil07] had already given an example of a 8-manifold which he believed not to admit a differentiable structure, but that was not proved until 1965 (see [Nov65]).

Exercise 2.5

(1) Show that two atlases $\mathcal{A}_1$ and $\mathcal{A}_2$ for a smooth manifold are equivalent if and only if $\mathcal{A}_1 \cup \mathcal{A}_2$ is an atlas.

(2) Let M be a differentiable manifold. Show that a set $V \subset M$ is open if and only if $\varphi_\alpha^{-1}(V)$ is an open subset of $\mathbb{R}^n$ for every parameterization $(U_\alpha, \varphi_\alpha)$ of a C^∞ atlas.

(3) Show that the two atlases on $\mathbb{R}^n$ from Example 2.3(1) are equivalent.

(4) Consider the two atlases on $\mathbb{R}$ from Example 2.3(2), $\{(\mathbb{R}, \varphi_1)\}$ and $\{(\mathbb{R}, \varphi_2)\}$, where $\varphi_1(x) = x$ and $\varphi_2(x) = x^3$. Show that $\varphi_2^{-1} \circ \varphi_1$ is not differentiable at the origin. Conclude that the two atlases are not equivalent.

(5) Recall from elementary vector calculus that a **surface** $S \subset \mathbb{R}^3$ is a set such that, for each $p \in S$, there is a neighborhood V_p of p in $\mathbb{R}^3$ and a C^∞ map

$f_p : U_p \to \mathbb{R}$ (where U_p is an open subset of $\mathbb{R}^2$) such that $S \cap V_p$ is the graph of $z = f_p(x, y)$, or $x = f_p(y, z)$, or $y = f_p(x, z)$. Show that S is a smooth manifold of dimension 2.

(6) (Product manifold) Let $\{(U_\alpha, \varphi_\alpha)\}$, $\{(V_\beta, \psi_\beta)\}$ be two atlases for two smooth manifolds M and N. Show that the family $\{(U_\alpha \times V_\beta, \varphi_\alpha \times \psi_\beta)\}$ is an atlas for the product $M \times N$. With the differentiable structure generated by this atlas, $M \times N$ is called the **product manifold** of M and N.

(7) (Stereographic projection) Consider the n-sphere S^n with the subspace topology and let $N = (0, \ldots, 0, 1)$ and $S = (0, \ldots, 0, -1)$ be the north and south poles. The **stereographic projection** from N is the map $\pi_N : S^n\backslash\{N\} \to \mathbb{R}^n$ which takes a point $p \in S^n\backslash\{N\}$ to the intersection point of the line through N and p with the hyperplane $x^{n+1} = 0$ (cf. Fig. 1.10). Similarly, the stereographic projection from S is the map $\pi_S : S^n\backslash\{S\} \to \mathbb{R}^n$ which takes a point p on $S^n\backslash\{S\}$ to the intersection point of the line through S and p with the same hyperplane. Check that $\{(\mathbb{R}^n, \pi_N^{-1}), (\mathbb{R}^n, \pi_S^{-1})\}$ is an atlas for S^n. Show that this atlas is equivalent to the atlas on Example 2.3(5). The maximal atlas obtained from these is called the **standard differentiable structure** on S^n.

(8) (*Real projective space*) The **real projective space** $\mathbb{R}P^n$ is the set of lines through the origin in $\mathbb{R}^{n+1}$. This space can be defined as the quotient space of S^n by the equivalence relation $x \sim -x$ that identifies a point to its antipodal point.

(a) Show that the quotient space $\mathbb{R}P^n = S^n/\sim$ with the quotient topology is a Hausdorff space and admits a countable basis of open sets. (Hint: Use Proposition 10.2).

(b) Considering the atlas on S^n defined in Example 2.3(5) and the canonical projection $\pi : S^n \to \mathbb{R}P^n$ given by $\pi(x) = [x]$, define an atlas for $\mathbb{R}P^n$.

(9) We can define an atlas on $\mathbb{R}P^n$ in a different way by identifying it with the quotient space of $\mathbb{R}^{n+1}\backslash\{0\}$ by the equivalence relation $x \sim \lambda x$, with $\lambda \in \mathbb{R}\backslash\{0\}$. For that, consider the sets $V_i = \{[x^1, \ldots, x^{n+1}] \mid x^i \neq 0\}$ (corresponding to the

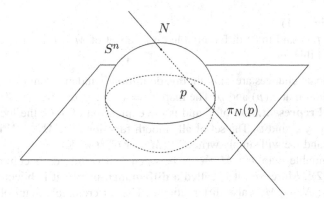

Fig. 1.10 Stereographic projection

set of lines through the origin in $\mathbb{R}^{n+1}$ that are not contained on the hyperplane $x^i = 0$) and the maps $\varphi_i : \mathbb{R}^n \to V_i$ defined by

$$\varphi_i\left(x^1, \ldots, x^n\right) = \left[x^1, \ldots, x^{i-1}, 1, x^i, \ldots, x^n\right].$$

Show that:

(a) the family $\{(\mathbb{R}^n, \varphi_i)\}$ is an atlas for $\mathbb{R}P^n$;
(b) this atlas defines the same differentiable structure as the atlas on Exercise 2.5(8).

(10) (*A non-Hausdorff manifold*) Let M be the disjoint union of $\mathbb{R}$ with a point p and consider the maps $f_i : \mathbb{R} \to M$ ($i = 1, 2$) defined by $f_i(x) = x$ if $x \in \mathbb{R}\backslash\{0\}$, $f_1(0) = 0$ and $f_2(0) = p$. Show that:

(a) the maps $f_i^{-1} \circ f_j$ are differentiable on their domains;
(b) if we consider an atlas formed by $\{(\mathbb{R}, f_1), (\mathbb{R}, f_2)\}$, the corresponding topology will not satisfy the Hausdorff axiom.

1.3 Differentiable Maps

In this book the words **differentiable** and **smooth** will be used to mean **infinitely differentiable** (C^∞).

Definition 3.1 Let M and N be two differentiable manifolds of dimension m and n, respectively. A map $f : M \to N$ is said to be **differentiable** (or **smooth**, or C^∞) at a point $p \in M$ if there exist parameterizations (U, φ) of M at p (i.e. $p \in \varphi(U)$) and (V, ψ) of N at $f(p)$, with $f(\varphi(U)) \subset \psi(V)$, such that the map

$$\hat{f} := \psi^{-1} \circ f \circ \varphi : U \subset \mathbb{R}^m \to \mathbb{R}^n$$

is smooth (Fig. 1.11).

The map f is said to be differentiable on a subset of M if it is differentiable at every point of this set.

As coordinate changes are smooth, this definition is independent of the parameterizations chosen at $f(p)$ and p. The map $\hat{f} := \psi^{-1} \circ f \circ \varphi : U \subset \mathbb{R}^m \to \mathbb{R}^n$ is called a **local representation** of f and is the expression of f on the local coordinates defined by φ and ψ. The set of all smooth functions $f : M \to N$ is denoted $C^\infty(M, N)$, and we will simply write $C^\infty(M)$ for $C^\infty(M, \mathbb{R})$.

A differentiable map $f : M \to N$ between two manifolds is continuous [cf. Exercise 3.2(2)]. Moreover, it is called a **diffeomorphism** if it is bijective and its inverse $f^{-1} : N \to M$ is also differentiable. The differentiable manifolds M and N will be considered the same if they are **diffeomorphic**, i.e. if there exists a diffeomorphism $f : M \to N$. A map f is called a **local diffeomorphism** at a point

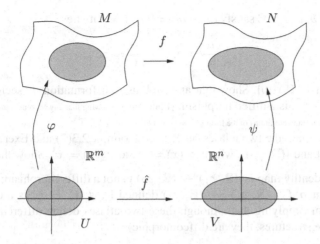

Fig. 1.11 Local representation of a map between manifolds

$p \in M$ if there are neighborhoods V of p and W of $f(p)$ such that $f|_V : V \to W$ is a diffeomorphism.

For a long time it was thought that, up to a diffeomorphism, there was only one differentiable structure for each topological manifold (the two different differentiable structures in Exercises 2.5(4) and 3.2(6) are diffeomorphic—cf. Exercise 3.2(6). However, in 1956, Milnor [Mil56] presented examples of manifolds that were homeomorphic but not diffeomorphic to S^7. Later, Milnor and Kervaire [Mil59, KM63] showed that more spheres of dimension greater than 7 admitted several differentiable structures. For instance, S^{19} has 73 distinct smooth structures and S^{31} has $16,931,177$. More recently, in 1982 and 1983, Freedman [Fre82] and Gompf [Gom83] constructed examples of non-standard differentiable structures on $\mathbb{R}^4$.

Exercise 3.2

(1) Prove that Definition 3.1 does not depend on the choice of parameterizations.
(2) Show that a differentiable map $f : M \to N$ between two smooth manifolds is continuous.
(3) Show that if $f : M_1 \to M_2$ and $g : M_2 \to M_3$ are differentiable maps between smooth manifolds M_1, M_2 and M_3, then $g \circ f : M_1 \to M_3$ is also differentiable.
(4) Show that the **antipodal map** $f : S^n \to S^n$, defined by $f(x) = -x$, is differentiable.
(5) Using the stereographic projection from the north pole $\pi_N : S^2 \setminus \{N\} \to \mathbb{R}^2$ and identifying $\mathbb{R}^2$ with the complex plane $\mathbb{C}$, we can identify S^2 with $\mathbb{C} \cup \{\infty\}$, where ∞ is the so-called **point at infinity**. A **Möbius transformation** is a map $f : \mathbb{C} \cup \{\infty\} \to \mathbb{C} \cup \{\infty\}$ of the form

$$f(z) = \frac{az + b}{cz + d},$$

where $a, b, c, d \in \mathbb{C}$ satisfy $ad - bc \neq 0$ and ∞ satisfies

$$\frac{\alpha}{\infty} = 0, \qquad \frac{\alpha}{0} = \infty$$

for any $\alpha \in \mathbb{C} \setminus \{0\}$. Show that any Möbius transformation f, seen as a map $f : S^2 \to S^2$, is a diffeomorphism. (**Hint:** Start by showing that any Möbius transformation is a composition of transformations of the form $g(z) = \frac{1}{z}$ and $h(z) = az + b$).

(6) Consider again the two atlases on $\mathbb{R}$ from Example 2.3(2) and Exercise 2.5(4), $\{(\mathbb{R}, \varphi_1)\}$ and $\{(\mathbb{R}, \varphi_2)\}$, where $\varphi_1(x) = x$ and $\varphi_2(x) = x^3$. Show that:

(a) the identity map $i : (\mathbb{R}, \varphi_1) \to (\mathbb{R}, \varphi_2)$ is not a diffeomorphism;
(b) the map $f : (\mathbb{R}, \varphi_1) \to (\mathbb{R}, \varphi_2)$ defined by $f(x) = x^3$ is a diffeomorphism (implying that although these two atlases define different differentiable structures, they are diffeomorphic).

1.4 Tangent Space

Recall from elementary vector calculus that a vector $v \in \mathbb{R}^3$ is said to be **tangent** to a surface $S \subset \mathbb{R}^3$ at a point $p \in S$ if there exists a differentiable curve $c : (-\varepsilon, \varepsilon) \to S \subset \mathbb{R}^3$ such that $c(0) = p$ and $\dot{c}(0) = v$ [cf. Exercise 2.5(5)]. The set T_pS of all these vectors is a 2-dimensional vector space, called the **tangent space** to S at p, and can be identified with the plane in $\mathbb{R}^3$ which is tangent to S at p (Fig. 1.12).

To generalize this to an abstract n-dimensional manifold we need to find a description of v which does not involve the ambient Euclidean space $\mathbb{R}^3$. To do so, we notice that the components of v are

$$v^i = \frac{d(x^i \circ c)}{dt}(0),$$

where $x^i : \mathbb{R}^3 \to \mathbb{R}$ is the ith coordinate function. If we ignore the ambient space, $x^i : S \to \mathbb{R}$ is just a differentiable function, and

$$v^i = v(x^i),$$

where, for any differentiable function $f : S \to \mathbb{R}$, we define

$$v(f) := \frac{d(f \circ c)}{dt}(0).$$

This allows us to see v as an operator $v : C^\infty(S) \to \mathbb{R}$, and it is clear that this operator completely determines the vector v. It is this new interpretation of tangent vector that will be used to define tangent spaces for manifolds.

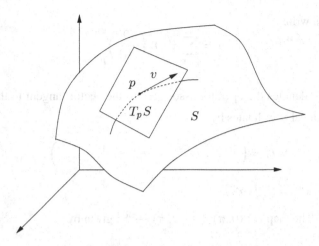

Fig. 1.12 Tangent vector to a surface

Definition 4.1 Let $c : (-\varepsilon, \varepsilon) \to M$ be a differentiable curve on a smooth manifold M. Consider the set $C^\infty(p)$ of all functions $f : M \to \mathbb{R}$ that are differentiable at $c(0) = p$. The **tangent vector to the curve** c at p is the operator $\dot{c}(0) : C^\infty(p) \to \mathbb{R}$ given by

$$\dot{c}(0)(f) = \frac{d(f \circ c)}{dt}(0).$$

A **tangent vector** to M at p is a tangent vector to some differentiable curve $c : (-\varepsilon, \varepsilon) \to M$ with $c(0) = p$. The **tangent space** at p is the space $T_p M$ of all tangent vectors at p.

Choosing a parameterization $\varphi : U \subset \mathbb{R}^n \to M$ around p, the curve c is given in local coordinates by the curve in U

$$\hat{c}(t) := \left(\varphi^{-1} \circ c \right)(t) = \left(x^1(t), \ldots, x^n(t) \right),$$

and

$$\dot{c}(0)(f) = \frac{d(f \circ c)}{dt}(0) = \frac{d}{dt} \left(\overbrace{(f \circ \varphi)}^{\hat{f}} \circ \overbrace{(\varphi^{-1} \circ c)}^{\hat{c}} \right)_{|t=0} =$$

$$= \frac{d}{dt} \left(\hat{f} \left(x^1(t), \ldots, x^n(t) \right) \right)_{|t=0} = \sum_{i=1}^n \frac{\partial \hat{f}}{\partial x^i}(\hat{c}(0)) \frac{dx^i}{dt}(0) =$$

$$= \left(\sum_{i=1}^n \dot{x}^i(0) \left(\frac{\partial}{\partial x^i} \right)_{\varphi^{-1}(p)} \right)(\hat{f}).$$

Hence we can write

$$\dot{c}(0) = \sum_{i=1}^{n} \dot{x}^i(0) \left(\frac{\partial}{\partial x^i} \right)_p,$$

where $\left(\frac{\partial}{\partial x^i} \right)_p$ denotes the operator associated to the vector tangent to the curve c_i at p given in local coordinates by

$$\hat{c}_i(t) = \left(x^1, \ldots, x^{i-1}, x^i + t, x^{i+1}, \ldots, x^n \right),$$

with $\left(x^1, \ldots, x^n \right) = \varphi^{-1}(p)$.

Example 4.2 The map $\psi : (0, \pi) \times (-\pi, \pi) \to S^2$ given by

$$\psi(\theta, \varphi) = (\sin \theta \cos \varphi, \sin \theta \sin \varphi, \cos \theta)$$

parameterizes a neighborhood of the point $(1, 0, 0) = \psi \left(\frac{\pi}{2}, 0 \right)$. Consequently, $\left(\frac{\partial}{\partial \theta} \right)_{(1,0,0)} = \dot{c}_\theta(0)$ and $\left(\frac{\partial}{\partial \varphi} \right)_{(1,0,0)} = \dot{c}_\varphi(0)$, where

$$c_\theta(t) = \psi \left(\frac{\pi}{2} + t, 0 \right) = (\cos t, 0, -\sin t);$$
$$c_\varphi(t) = \psi \left(\frac{\pi}{2}, t \right) = (\cos t, \sin t, 0).$$

Note that, in the notation above,

$$\hat{c}_\theta(t) = \left(\frac{\pi}{2} + t, 0 \right) \quad \text{and} \quad \hat{c}_\varphi(t) = \left(\frac{\pi}{2}, t \right).$$

Moreover, since c_θ and c_φ are curves in $\mathbb{R}^3$, $\left(\frac{\partial}{\partial \theta} \right)_{(1,0,0)}$ and $\left(\frac{\partial}{\partial \varphi} \right)_{(1,0,0)}$ can be identified with the vectors $(0, 0, -1)$ and $(0, 1, 0)$.

Proposition 4.3 *The tangent space to M at p is an n-dimensional vector space.*

Proof Consider a parameterization $\varphi : U \subset \mathbb{R}^n \to M$ around p and take the vector space generated by the operators $\left(\frac{\partial}{\partial x^i} \right)_p$,

$$\mathcal{D}_p := \operatorname{span} \left\{ \left(\frac{\partial}{\partial x^1} \right)_p, \ldots, \left(\frac{\partial}{\partial x^n} \right)_p \right\}.$$

It is easy to show [cf. Exercise 4.9(1)] that these operators are linearly independent. Moreover, each tangent vector to M at p can be represented by a linear combination of these operators, so the tangent space $T_p M$ is a subset of $\mathcal{D}_p$. We will now see that $\mathcal{D}_p \subset T_p M$. Let $v \in \mathcal{D}_p$; then v can be written as

$$v = \sum_{i=1}^{n} v^i \left(\frac{\partial}{\partial x^i} \right)_p.$$

If we consider the curve $c : (-\varepsilon, \varepsilon) \to M$, defined by

$$c(t) = \varphi \left(x^1 + v^1 t, \ldots, x^n + v^n t \right)$$

(where $\left(x^1, \ldots, x^n \right) = \varphi^{-1}(p)$), then

$$\hat{c}(t) = \left(x^1 + v^1 t, \ldots, x^n + v^n t \right)$$

and so $\dot{x}^i(0) = v^i$, implying that $\dot{c}(0) = v$. Therefore $v \in T_p M$. $\square$

Remark 4.4

(1) The basis $\left\{ \left(\frac{\partial}{\partial x^i} \right)_p \right\}_{i=1}^{n}$ determined by the chosen parameterization around p is called the **associated basis** to that parameterization.
(2) Note that the definition of tangent space at p only uses functions that are differentiable on a neighborhood of p. Hence, if U is an open set of M containing p, the tangent space $T_p U$ is naturally identified with $T_p M$.

If we consider the disjoint union of all tangent spaces $T_p M$ at all points of M, we obtain the space

$$TM = \bigcup_{p \in M} T_p M = \{ v \in T_p M \mid p \in M \},$$

which admits a differentiable structure naturally determined by the one on M [cf. Exercise 4.9(8)]. With this differentiable structure, this space is called the **tangent bundle**. Note that there is a natural projection $\pi : TM \to M$ which takes $v \in T_p M$ to p (cf. Sect. 1.10.3).

Now that we have defined tangent space, we can define the **derivative at a point** p of a differentiable map $f : M \to N$ between smooth manifolds. We want this derivative to be a linear transformation

$$(df)_p : T_p M \to T_{f(p)} N$$

of the corresponding tangent spaces, to be the usual derivative (Jacobian) of f when M and N are Euclidean spaces, and to satisfy the chain rule.

Definition 4.5 Let $f : M \to N$ be a differentiable map between smooth manifolds. For $p \in M$, the derivative of f at p is the map

$$(df)_p : T_p M \to T_{f(p)} N$$
$$v \mapsto \frac{d(f \circ c)}{dt}(0),$$

where $c : (-\varepsilon, \varepsilon) \to M$ is a curve satisfying $c(0) = p$ and $\dot{c}(0) = v$.

Proposition 4.6 *The map* $(df)_p : T_pM \to T_{f(p)}N$ *defined above is a linear transformation that does not depend on the choice of the curve* c.

Proof Let (U, φ) and (V, ψ) be two parameterizations around p and $f(p)$ such that $f(\varphi(U)) \subset \psi(V)$ (cf. Fig. 1.13). Consider a vector $v \in T_pM$ and a curve $c : (-\varepsilon, \varepsilon) \to M$ such that $c(0) = p$ and $\dot{c}(0) = v$. If, in local coordinates, the curve c is given by

$$\hat{c}(t) := \left(\varphi^{-1} \circ c\right)(t) = \left(x^1(t), \ldots, x^m(t)\right),$$

and the curve $\gamma := f \circ c : (-\varepsilon, \varepsilon) \to N$ is given by

$$\hat{\gamma}(t) := \left(\psi^{-1} \circ \gamma\right)(t) = \left(\psi^{-1} \circ f \circ \varphi\right)\left(x^1(t), \ldots, x^m(t)\right)$$
$$= \left(y^1(x(t)), \ldots, y^n(x(t))\right),$$

then $\dot{\gamma}(0)$ is the tangent vector in $T_{f(p)}N$ given by

$$\dot{\gamma}(0) = \sum_{i=1}^n \frac{d}{dt}\left(y^i\left(x^1(t), \ldots, x^m(t)\right)\right)_{|t=0} \left(\frac{\partial}{\partial y^i}\right)_{f(p)}$$
$$= \sum_{i=1}^n \left\{\sum_{k=1}^m \dot{x}^k(0) \left(\frac{\partial y^i}{\partial x^k}\right)(x(0))\right\} \left(\frac{\partial}{\partial y^i}\right)_{f(p)}$$
$$= \sum_{i=1}^n \left\{\sum_{k=1}^m v^k \left(\frac{\partial y^i}{\partial x^k}\right)(x(0))\right\} \left(\frac{\partial}{\partial y^i}\right)_{f(p)},$$

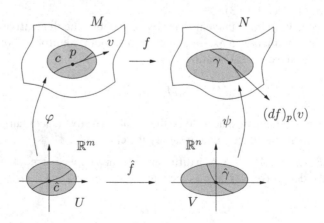

Fig. 1.13 Derivative of a differentiable map

where the v^k are the components of v in the basis associated to (U, φ). Hence $\dot{\gamma}(0)$ does not depend on the choice of c, as long as $\dot{c}(0) = v$. Moreover, the components of $w = (df)_p(v)$ in the basis associated to (V, ψ) are

$$w^i = \sum_{j=1}^{m} \frac{\partial y^i}{\partial x^j} v^j,$$

where $\left(\frac{\partial y^i}{\partial x^j}\right)$ is an $n \times m$ matrix (the Jacobian matrix of the local representation of f at $\varphi^{-1}(p)$). Therefore, $(df)_p : T_pM \to T_{f(p)}N$ is the linear transformation which, on the basis associated to the parameterizations φ and ψ, is represented by this matrix. $\square$

Remark 4.7 The derivative $(df)_p$ is sometimes called **differential** of f at p. Several other notations are often used for df, as for example f_*, Df, Tf and f'.

Example 4.8 Let $\varphi : U \subset \mathbb{R}^n \to M$ be a parameterization around a point $p \in M$. We can view φ as a differentiable map between two smooth manifolds and we can compute its derivative at $x = \varphi^{-1}(p)$

$$(d\varphi)_x : T_xU \to T_pM.$$

For $v \in T_xU \cong \mathbb{R}^n$, the ith component of $(d\varphi)_x(v)$ is

$$\sum_{j=1}^{n} \frac{\partial x^i}{\partial x^j} v^j = v^i$$

(where $\left(\frac{\partial x^i}{\partial x^j}\right)$ is the identity matrix). Hence, $(d\varphi)_x(v)$ is the vector in T_pM which, in the basis $\left\{ \left(\frac{\partial}{\partial x^i}\right)_p \right\}$ associated to the parameterization φ, is represented by v.

Given a differentiable map $f : M \to N$ we can also define a global derivative df (also called **push–forward** and denoted f_*) between the corresponding tangent bundles:

$$df : TM \to TN$$
$$T_pM \ni v \mapsto (df)_p(v) \in T_{f(p)}N.$$

Exercise 4.9

(1) Show that the operators $\left(\frac{\partial}{\partial x^i}\right)_p$ are linearly independent.

(2) Let M be a smooth manifold, p a point in M and v a vector tangent to M at p. Show that if v can be written as $v = \sum_{i=1}^{n} a^i \left(\frac{\partial}{\partial x^i}\right)_p$ and $v = \sum_{i=1}^{n} b^i \left(\frac{\partial}{\partial y^i}\right)_p$ for two basis associated to different parameterizations around p, then

$$b^j = \sum_{i=1}^{n} \frac{\partial y^j}{\partial x^i} a^i.$$

(3) Let M be an n-dimensional differentiable manifold and $p \in M$. Show that the following sets can be canonically identified with T_pM (and therefore constitute alternative definitions of the tangent space):

(a) $\mathcal{C}_p/ \sim$, where $\mathcal{C}_p$ is the set of differentiable curves $c : I \subset \mathbb{R} \to M$ such that $c(0) = p$ and $\sim$ is the equivalence relation defined by

$$c_1 \sim c_2 \Leftrightarrow \frac{d}{dt}\left(\varphi^{-1} \circ c_1\right)(0) = \frac{d}{dt}\left(\varphi^{-1} \circ c_2\right)(0)$$

for some parameterization $\varphi : U \subset \mathbb{R}^n \to M$ of a neighborhood of p.

(b) $\{(\alpha, v_\alpha) \mid p \in \varphi_\alpha(U_\alpha)$ and $v_\alpha \in \mathbb{R}^n\}/ \sim$, where $\mathcal{A} = \{(U_\alpha, \varphi_\alpha)\}$ is the differentiable structure and $\sim$ is the equivalence relation defined by

$$(\alpha, v_\alpha) \sim (\beta, v_\beta) \Leftrightarrow v_\beta = d\left(\varphi_\beta^{-1} \circ \varphi_\alpha\right)_{\varphi_\alpha^{-1}(p)} (v_\alpha).$$

(4) (Chain rule) Let $f : M \to N$ and $g : N \to P$ be two differentiable maps. Then $g \circ f : M \to P$ is also differentiable [cf. Exercise 3.2(3)]. Show that for $p \in M$,

$$(d(g \circ f))_p = (dg)_{f(p)} \circ (df)_p.$$

(5) Let $\phi : (0, +\infty) \times (0, \pi) \times (0, 2\pi) \to \mathbb{R}^3$ be the parameterization of $U = \mathbb{R}^3 \setminus \{(x, 0, z) \mid x \geq 0$ and $z \in \mathbb{R}\}$ by spherical coordinates,

$$\phi(r, \theta, \varphi) = (r \sin\theta \cos\varphi, r \sin\theta \sin\varphi, r \cos\theta).$$

Determine the Cartesian components of $\frac{\partial}{\partial r}$, $\frac{\partial}{\partial \theta}$ and $\frac{\partial}{\partial \varphi}$ at each point of U.

(6) Compute the derivative $(df)_N$ of the antipodal map $f : S^n \to S^n$ at the north pole N.

(7) Let W be a coordinate neighborhood on M, let $x : W \to \mathbb{R}^n$ be a coordinate chart and consider a smooth function $f : M \to \mathbb{R}$. Show that for $p \in W$, the derivative $(df)_p$ is given by

$$(df)_p = \frac{\partial \hat{f}}{\partial x^1}(x(p)) \left(dx^1\right)_p + \cdots + \frac{\partial \hat{f}}{\partial x^n}(x(p)) \left(dx^n\right)_p,$$

where $\hat{f} := f \circ x^{-1}$.

(8) (Tangent bundle) Let $\{(U_\alpha, \varphi_\alpha)\}$ be a differentiable structure on M and consider the maps

$$\Phi_\alpha : U_\alpha \times \mathbb{R}^n \to TM$$
$$(x, v) \mapsto (d\varphi_\alpha)_x(v) \in T_{\varphi_\alpha(x)}M.$$

Show that the family $\{(U_\alpha \times \mathbb{R}^n, \Phi_\alpha)\}$ defines a differentiable structure for TM. Conclude that, with this differentiable structure, TM is a smooth manifold of dimension $2 \times \dim M$.

(9) Let $f : M \to N$ be a differentiable map between smooth manifolds. Show that:

(a) $df : TM \to TN$ is also differentiable;
(b) if $f : M \to M$ is the identity map then $df : TM \to TM$ is also the identity;
(c) if f is a diffeomorphism then $df : TM \to TN$ is also a diffeomorphism and $(df)^{-1} = df^{-1}$.

(10) Let M_1, M_2 be two differentiable manifolds and

$$\pi_1 : M_1 \times M_2 \twoheadrightarrow M_1$$
$$\pi_2 : M_1 \times M_2 \twoheadrightarrow M_2$$

the corresponding canonical projections.

(a) Show that $d\pi_1 \times d\pi_2$ is a diffeomorphism between the tangent bundle $T(M_1 \times M_2)$ and the product manifold $TM_1 \times TM_2$.
(b) Show that if N is a smooth manifold and $f_i : N \to M_i$ $(i = 1, 2)$ are differentiable maps, then $d(f_1 \times f_2) = df_1 \times df_2$.

1.5 Immersions and Embeddings

In this section we will study the local behavior of differentiable maps $f : M \to N$ between smooth manifolds. We have already seen that f is said to be a local diffeomorphism at a point $p \in M$ if $\dim M = \dim N$ and f transforms a neighborhood of p diffeomorphically onto a neighborhood of $f(p)$. In this case, its derivative $(df)_p : T_pM \to T_{f(p)}N$ must necessarily be an isomorphism [cf. Exercise 4.9(9)]. Conversely, if $(df)_p$ is an isomorphism then the inverse function theorem implies that f is a local diffeomorphism (cf. Sect. 1.10.4). Therefore, to check whether f maps a neighborhood of p diffeomorphically onto a neighborhood of $f(p)$, one just has to check that the determinant of the local representation of $(df)_p$ is nonzero.

When $\dim M < \dim N$, the best we can hope for is that $(df)_p : T_pM \to T_{f(p)}N$ is injective. The map f is then called an **immersion at** p. If f is an immersion at every point in M, it is called an **immersion**. Locally, every immersion is (up to a diffeomorphism) the **canonical immersion** of $\mathbb{R}^m$ into $\mathbb{R}^n$ $(m < n)$ where a point $(x^1, \ldots, x^m)$ is mapped to $(x^1, \ldots, x^m, 0, \ldots, 0)$. This result is known as the **local immersion theorem**.

Theorem 5.1 *Let* $f : M \to N$ *be an immersion at* $p \in M$. *Then there exist local coordinates around* p *and* $f(p)$ *on which* f *is the canonical immersion.*

Proof Let (U, φ) and (V, ψ) be parameterizations around p and $q = f(p)$. Let us assume for simplicity that $\varphi(0) = p$ and $\psi(0) = q$. Since f is an immersion, $(d\hat{f})_0 : \mathbb{R}^m \to \mathbb{R}^n$ is injective (where $\hat{f} := \psi^{-1} \circ f \circ \varphi$ is the expression of f in local coordinates). Hence we can assume (changing basis on $\mathbb{R}^n$ if necessary) that this linear transformation is represented by the $n \times m$ matrix

$$\begin{pmatrix} I_{m \times m} \\ --- \\ 0 \end{pmatrix},$$

where $I_{m \times m}$ is the $m \times m$ identity matrix. Therefore, the map

$$F : U \times \mathbb{R}^{n-m} \to \mathbb{R}^n$$
$$\left(x^1, \ldots, x^n\right) \mapsto \hat{f}\left(x^1, \ldots, x^m\right) + \left(0, \ldots, 0, x^{m+1}, \ldots, x^n\right),$$

has derivative $(dF)_0 : \mathbb{R}^n \to \mathbb{R}^n$ given by the matrix

$$\begin{pmatrix} I_{m \times m} & | & 0 \\ --- & + & --- \\ 0 & | & I_{(n-m) \times (n-m)} \end{pmatrix} = I_{n \times n}.$$

Applying the inverse function theorem, we conclude that F is a local diffeomorphism at 0. This implies that $\psi \circ F$ is also a local diffeomorphism at 0, and so $\psi \circ F$ is another parameterization of N around q. Denoting the canonical immersion of $\mathbb{R}^m$ into $\mathbb{R}^n$ by j, we have $\hat{f} = F \circ j \Leftrightarrow f = \psi \circ F \circ j \circ \varphi^{-1}$, implying that the following diagram commutes:

$$
\begin{array}{ccc}
M \supset \varphi(\tilde{U}) & \xrightarrow{f} & (\psi \circ F)(\tilde{V}) \subset N \\
\varphi \uparrow & & \uparrow \psi \circ F \\
\mathbb{R}^m \supset \tilde{U} & \xrightarrow{j} & \tilde{V} \subset \mathbb{R}^n
\end{array}
$$

(for possibly smaller open sets $\tilde{U} \subset U$ and $\tilde{V} \subset V$). Hence, on these new coordinates, f is the canonical immersion. $\square$

Remark 5.2 As a consequence of the local immersion theorem, any immersion at a point $p \in M$ is an immersion on a neighborhood of p.

When an immersion $f : M \to N$ is also a homeomorphism onto its image $f(M) \subset N$ with its subspace topology, it is called an **embedding**. We leave as an exercise to show that the local immersion theorem implies that, locally, any immersion is an embedding.

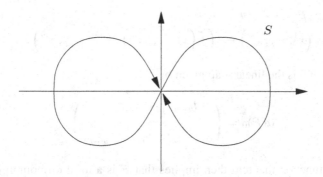

Fig. 1.14 Injective immersion which is not an embedding

Example 5.3

(1) The map $f : \mathbb{R} \to \mathbb{R}^2$ given by $f(t) = (t^2, t^3)$ is not an immersion at $t = 0$.
(2) The map $f : \mathbb{R} \to \mathbb{R}^2$ defined by $f(t) = (\cos t, \sin 2t)$ is an immersion but it is not an embedding (it is not injective).
(3) Let $g : \mathbb{R} \to \mathbb{R}$ be the function $g(t) = 2 \arctan(t) + \pi/2$. If f is the map defined in (5.3) then $h := f \circ g$ is an injective immersion which is not an embedding. Indeed, the set $S = h(\mathbb{R})$ in Fig. 1.14 is not the image of an embedding of $\mathbb{R}$ into $\mathbb{R}^2$. The arrows in the figure mean that the line approaches itself arbitrarily close at the origin but never self-intersects. If we consider the usual topologies on $\mathbb{R}$ and on $\mathbb{R}^2$, the image of a bounded open set in $\mathbb{R}$ containing 0 is not an open set in $h(\mathbb{R})$ for the subspace topology, and so h^{-1} is not continuous.
(4) The map $f : \mathbb{R} \to \mathbb{R}^2$ given by $f(t) = (e^t \cos t, e^t \sin t)$ is an embedding of $\mathbb{R}$ into $\mathbb{R}^2$.

If $M \subset N$ and the inclusion map $i : M \hookrightarrow N$ is an embedding, M is said to be a **submanifold** of N. Therefore, an embedding $f : M \to N$ maps M diffeomorphically onto a submanifold of N. Charts on $f(M)$ are just restrictions of appropriately chosen charts on N to $f(M)$ [cf. Exercise 5.9(3)].

A differentiable map $f : M \to N$ for which $(df)_p$ is surjective is called a **submersion at** p. Note that, in this case, we necessarily have $m \geq n$. If f is a submersion at every point in M it is called a **submersion**. Locally, every submersion is the standard projection of $\mathbb{R}^m$ onto the first n factors.

Theorem 5.4 *Let $f : M \to N$ be a submersion at $p \in M$. Then there exist local coordinates around p and $f(p)$ for which f is the standard projection.*

Proof Let us consider parameterizations (U, φ) and (V, ψ) around p and $f(p)$, such that $f(\varphi(U)) \subset \psi(V)$, $\varphi(0) = p$ and $\psi(0) = f(p)$. In local coordinates f is given by $\hat{f} := \psi^{-1} \circ f \circ \varphi$ and, as $(df)_p$ is surjective, $(d\hat{f})_0 : \mathbb{R}^m \to \mathbb{R}^n$ is a surjective linear transformation. By a linear change of coordinates on $\mathbb{R}^n$ we may assume that $(d\hat{f})_0 = (I_{n \times n} \mid *)$. As in the proof of the local immersion theorem, we will use an auxiliary map F that will allow us to use the inverse function theorem,

$$F : U \subset \mathbb{R}^m \to \mathbb{R}^m$$
$$\left(x^1, \dots, x^m\right) \mapsto \left(\hat{f}\left(x^1, \dots, x^m\right), x^{n+1}, \dots, x^m\right).$$

Its derivative at 0 is the linear map given by

$$(dF)_0 = \left(\begin{array}{c|c} I_{n \times n} & * \\ \hline 0 & I_{(m-n) \times (m-n)} \end{array} \right).$$

The inverse function theorem then implies that F is a local diffeomorphism at 0, meaning that it maps some open neighborhood of this point $\tilde{U} \subset U$, diffeomorphically onto an open set W of $\mathbb{R}^m$ containing 0. If $\pi_1 : \mathbb{R}^m \to \mathbb{R}^n$ is the standard projection onto the first n factors, we have $\pi_1 \circ F = \hat{f}$, and hence

$$\hat{f} \circ F^{-1} = \pi_1 : W \to \mathbb{R}^n.$$

Therefore, replacing φ by $\tilde{\varphi} := \varphi \circ F^{-1}$, we obtain coordinates for which f is the standard projection π_1 onto the first n factors:

$$\psi^{-1} \circ f \circ \tilde{\varphi} = \psi^{-1} \circ f \circ \varphi \circ F^{-1} = \hat{f} \circ F^{-1} = \pi_1.$$

$\square$

Remark 5.5 This result is often stated together with the local immersion theorem in what is known as the **rank theorem** (see for instance [Boo03]).

Let $f : M \to N$ be a differentiable map between smooth manifolds of dimensions m and n, respectively. A point $p \in M$ is called a **regular point** of f if $(df)_p$ is surjective. A point $q \in N$ is called a **regular value** of f if every point in $f^{-1}(q)$ is a regular point. A point $p \in M$ which is not regular is called a **critical point** of f. The corresponding value $f(p)$ is called a **critical value**. Note that if there exists a regular value of f then $m \geq n$. We can obtain differentiable manifolds by taking inverse images of regular values.

Theorem 5.6 *Let $q \in N$ be a regular value of $f : M \to N$ and assume that the level set $L := f^{-1}(q) = \{p \in M \mid f(p) = q\}$ is nonempty. Then L is a submanifold of M and $T_p L = \ker(df)_p \subset T_p M$ for all $p \in L$.*

Proof For each point $p \in f^{-1}(q)$, we choose parameterizations (U, φ) and (V, ψ) around p and q for which f is the standard projection π_1 onto the first n factors, $\varphi(0) = p$ and $\psi(0) = q$ (cf. Theorem 5.4). We then construct a differentiable structure for $L := f^{-1}(q)$ in the following way: take the sets U from each of these parameterizations of M; since $f \circ \varphi = \psi \circ \pi_1$, we have

$$\varphi^{-1}\left(f^{-1}(q)\right) = \pi_1^{-1}\left(\psi^{-1}(q)\right) = \pi_1^{-1}(0)$$

$$= \left\{\left(0, \ldots, 0, x^{n+1}, \ldots, x^m\right) \mid x^{n+1}, \ldots, x^m \in \mathbb{R}\right\},$$

and so

$$\widetilde{U} := \varphi^{-1}(L) = \left\{\left(x^1, \ldots, x^m\right) \in U \mid x^1 = \cdots = x^n = 0\right\};$$

hence, taking $\pi_2 : \mathbb{R}^m \to \mathbb{R}^{m-n}$, the standard projection onto the last $m - n$ factors, and $j : \mathbb{R}^{m-n} \to \mathbb{R}^m$, the immersion given by

$$j\left(x^1, \ldots, x^{m-n}\right) = \left(0, \ldots, 0, x^1, \ldots, x^{m-n}\right),$$

the family $\{(\pi_2(\widetilde{U}), \varphi \circ j)\}$ is an atlas for L.

Moreover, the inclusion map $i : L \to M$ is an embedding. In fact, if A is an open set in L contained in a coordinate neighborhood then

$$A = \varphi\left(\left(\mathbb{R}^n \times (\varphi \circ j)^{-1}(A)\right) \cap U\right) \cap L$$

is an open set for the subspace topology on L.

We will now show that $T_p L = \ker (df)_p$. For that, for each $v \in T_p L$, we consider a curve c on L such that $c(0) = p$ and $\dot{c}(0) = v$. Then $(f \circ c)(t) = q$ for every t and so

$$\frac{d}{dt}(f \circ c)(0) = 0 \Leftrightarrow (df)_p \dot{c}(0) = (df)_p v = 0,$$

implying that $v \in \ker (df)_p$. As $\dim T_p L = \dim \left(\ker (df)_p\right) = m - n$, the result follows. $\square$

Given a differentiable manifold, we can ask ourselves if it can be embedded into $\mathbb{R}^K$ for some $K \in \mathbb{N}$. The following theorem, which was proved by Whitney in [Whi44a, Whi44b] answers this question and is known as the **Whitney embedding theorem**.

Theorem 5.7 (Whitney) *Any smooth manifold M of dimension n can be embedded in $\mathbb{R}^{2n}$ (and, provided that $n > 1$, immersed in $\mathbb{R}^{2n-1}$).* $\square$

Remark 5.8 By the Whitney embedding theorem, any smooth manifold M of dimension n is diffeomorphic to a submanifold of $\mathbb{R}^{2n}$.

Exercise 5.9

(1) Show that any parameterization $\varphi : U \subset \mathbb{R}^m \to M$ is an embedding of U into M.

(2) Show that, locally, any immersion is an embedding, i.e. if $f : M \to N$ is an immersion and $p \in M$, then there is an open set $W \subset M$ containing p such that $f|_W$ is an embedding.

(3) Let N be a manifold. Show that $M \subset N$ is a submanifold of N of dimension m if and only if, for each $p \in M$, there is a coordinate system $x : W \to \mathbb{R}^n$ around p on N, for which $M \cap W$ is defined by the equations $x^{m+1} = \cdots = x^n = 0$.

(4) Consider the sphere

$$S^n = \left\{ x \in \mathbb{R}^{n+1} \mid \left(x^1\right)^2 + \cdots + \left(x^{n+1}\right)^2 = 1 \right\}.$$

Show that S^n is an n-dimensional submanifold of $\mathbb{R}^{n+1}$ and that

$$T_x S^n = \left\{ v \in \mathbb{R}^{n+1} \mid \langle x, v \rangle = 0 \right\},$$

where $\langle \cdot, \cdot \rangle$ is the usual inner product on $\mathbb{R}^n$.

(5) Let $f : M \to N$ be a differentiable map between smooth manifolds and consider submanifolds $V \subset M$ and $W \subset N$. Show that if $f(V) \subset W$ then $f : V \to W$ is also a differentiable map.

(6) Let $f : M \to N$ be an injective immersion. Show that if M is compact then $f(M)$ is a submanifold of N.

1.6 Vector Fields

A **vector field** on a smooth manifold M is a map that to each point $p \in M$ assigns a vector tangent to M at p:

$$X : M \to TM$$
$$p \mapsto X(p) := X_p \in T_p M.$$

The vector field is said to be **differentiable** if this map is differentiable. The set of all differentiable vector fields on M is denoted by $\mathfrak{X}(M)$. Locally we have:

Proposition 6.1 *Let W be a coordinate neighborhood on M (that is, $W = \varphi(U)$ for some parameterization $\varphi : U \to M$), and let $x := \varphi^{-1} : W \to \mathbb{R}^n$ be the corresponding coordinate chart. Then a map $X : W \to TW$ is a differentiable vector field on W if and only if,*

$$X_p = X^1(p) \left(\frac{\partial}{\partial x^1} \right)_p + \cdots + X^n(p) \left(\frac{\partial}{\partial x^n} \right)_p,$$

for some differentiable functions $X^i : W \to \mathbb{R}$ $(i = 1, \ldots, n)$.

Proof Let us consider the coordinate chart $x = \left(x^1, \ldots, x^n\right)$. As $X_p \in T_p M$, we have

$$X_p = X^1(p) \left(\frac{\partial}{\partial x^1} \right)_p + \cdots + X^n(p) \left(\frac{\partial}{\partial x^n} \right)_p$$

for some functions $X^i : W \to \mathbb{R}$. In the local chart associated with the parameterization $(U \times \mathbb{R}^n, d\varphi)$ of TM, the local representation of the map X is

$$\hat{X}\left(x^1, \ldots, x^n \right) = \left(x^1, \ldots, x^n, \hat{X}^1 \left(x^1, \ldots, x^n \right), \ldots, \hat{X}^n \left(x^1, \ldots, x^n \right) \right).$$

Therefore X is differentiable if and only if the functions $\hat{X}^i : U \to \mathbb{R}$ are differentiable, i.e. if and only if the functions $X^i : W \to \mathbb{R}$ are differentiable. $\square$

A vector field X is differentiable if and only if, given any differentiable function $f : M \to \mathbb{R}$, the function

$$X \cdot f : M \to \mathbb{R}$$
$$p \mapsto X_p \cdot f := X_p(f)$$

is also differentiable [cf. Exercise 6.11(1)]. This function $X \cdot f$ is called the **directional derivative** of f along X. Thus one can view $X \in \mathfrak{X}(M)$ as a linear operator $X : C^\infty(M) \to C^\infty(M)$.

Let us now take two vector fields $X, Y \in \mathfrak{X}(M)$. In general, the operators $X \circ Y$ and $Y \circ X$ will involve derivatives of order two, and will not correspond to vector fields. However, the commutator $X \circ Y - Y \circ X$ does define a vector field.

Proposition 6.2 *Given two differentiable vector fields $X, Y \in \mathfrak{X}(M)$ on a smooth manifold M, there exists a unique differentiable vector field $Z \in \mathfrak{X}(M)$ such that*

$$Z \cdot f = (X \circ Y - Y \circ X) \cdot f$$

for every differentiable function $f \in C^\infty(M)$.

Proof Considering a coordinate chart $x : W \subset M \to \mathbb{R}^n$, we have

$$X = \sum_{i=1}^n X^i \frac{\partial}{\partial x^i} \quad \text{and} \quad Y = \sum_{i=1}^n Y^i \frac{\partial}{\partial x^i}.$$

Then,

$$
\begin{aligned}
(X \circ Y - Y \circ X) \cdot f &= X \cdot \left(\sum_{i=1}^n Y^i \frac{\partial \hat{f}}{\partial x^i} \right) - Y \cdot \left(\sum_{i=1}^n X^i \frac{\partial \hat{f}}{\partial x^i} \right) \\
&= \sum_{i=1}^n \left(\left(X \cdot Y^i \right) \frac{\partial \hat{f}}{\partial x^i} - \left(Y \cdot X^i \right) \frac{\partial \hat{f}}{\partial x^i} \right)
\end{aligned}
$$

$$+ \sum_{i,j=1}^{n} \left(X^j Y^i \frac{\partial^2 \hat{f}}{\partial x^j \partial x^i} - Y^j X^i \frac{\partial^2 \hat{f}}{\partial x^j \partial x^i} \right)$$

$$= \left(\sum_{i=1}^{n} \left(X \cdot Y^i - Y \cdot X^i \right) \frac{\partial}{\partial x^i} \right) \cdot f,$$

and so, at each point $p \in W$, one has $((X \circ Y - Y \circ X) \cdot f)(p) = Z_p \cdot f$, where

$$Z_p = \sum_{i=1}^{n} \left(X \cdot Y^i - Y \cdot X^i \right)(p) \left(\frac{\partial}{\partial x^i} \right)_p.$$

Hence, the operator $X \circ Y - Y \circ X$ defines a vector field. Note that this vector field is differentiable, as $(X \circ Y - Y \circ X) \cdot f$ is smooth for every smooth function $f : M \to \mathbb{R}$. ☐

The vector field Z is called the **Lie bracket** of X and Y, and is denoted by $[X, Y]$. In local coordinates it is given by

$$[X, Y] = \sum_{i=1}^{n} \left(X \cdot Y^i - Y \cdot X^i \right) \frac{\partial}{\partial x^i}. \tag{1.1}$$

We say that two vector fields $X, Y \in \mathfrak{X}(M)$ **commute** if $[X, Y] = 0$. The Lie bracket as has the following properties.

Proposition 6.3 *Given* $X, Y, Z \in \mathfrak{X}(M)$*, we have:*

(i) **Bilinearity:** *for any* $\alpha, \beta \in \mathbb{R}$*,*

$$[\alpha X + \beta Y, Z] = \alpha[X, Z] + \beta[Y, Z]$$
$$[X, \alpha Y + \beta Z] = \alpha[X, Y] + \beta[X, Z];$$

(ii) **Antisymmetry:**
$$[X, Y] = -[Y, X];$$

(iii) **Jacobi identity:**

$$[[X, Y], Z] + [[Y, Z], X] + [[Z, X], Y] = 0;$$

(iv) **Leibniz rule:** *For any* $f, g \in C^\infty(M)$*,*

$$[f X, g Y] = fg[X, Y] + f(X \cdot g)Y - g(Y \cdot f)X.$$

Proof Exercise 6.11(2). ☐

The space $\mathfrak{X}(M)$ of vector fields on M is a particular case of a **Lie algebra**:

Definition 6.4 A vector space V equipped with an antisymmetric bilinear map $[\cdot, \cdot]$: $V \times V \to V$ (called a **Lie bracket**) satisfying the Jacobi identity is called a **Lie algebra**. A linear map $F : V \to W$ between Lie algebras is called a **Lie algebra homomorphism** if $F([v_1, v_2]) = [F(v_1), F(v_2)]$ for all $v_1, v_2 \in V$. If F is bijective then it is called a **Lie algebra isomorphism**.

Given a vector field $X \in \mathfrak{X}(M)$ and a diffeomorphism $f : M \to N$ between smooth manifolds, we can naturally define a vector field on N using the derivative of f. This vector field, the **push-forward** of X, is denoted by f_*X and is defined in the following way: given $p \in M$,

$$(f_*X)_{f(p)} := (df)_p X_p.$$

This makes the following diagram commute:

$$
\begin{array}{ccc}
TM & \overset{df}{\to} & TN \\
X \uparrow & & \uparrow f_*X \\
M & \overset{f}{\to} & N
\end{array}
$$

Let us now turn to the definition of an integral curve. If $X \in \mathfrak{X}(M)$ is a smooth vector field, an **integral curve** of X is a smooth curve $c : (-\varepsilon, \varepsilon) \to M$ such that $\dot{c}(t) = X_{c(t)}$. If this curve has initial value $c(0) = p$, we denote it by c_p and we say that c_p is an **integral curve of X at p**.

Considering a parameterization $\varphi : U \subset \mathbb{R}^n \to M$ on M, the integral curve c is locally given by $\hat{c} := \varphi^{-1} \circ c$. Applying $(d\varphi^{-1})_{c(t)}$ to both sides of the equation defining c, we obtain

$$\dot{\hat{c}}(t) = \hat{X}\left(\hat{c}(t)\right),$$

where $\hat{X} = d\varphi^{-1} \circ X \circ \varphi$ is the local representation of X with respect to the parameterizations (U, φ) and $(TU, d\varphi)$ on M and on TM (cf. Fig. 1.15). This equation is just a system of n ordinary differential equations:

$$\frac{d\hat{c}^i}{dt}(t) = \hat{X}^i\left(\hat{c}(t)\right), \quad \text{for } i = 1, \ldots, n. \tag{1.2}$$

The (local) existence and uniqueness of integral curves is then determined by the Picard–Lindelöf theorem of ordinary differential equations (see for example [Arn92]), and we have

Theorem 6.5 *Let M be a smooth manifold and let $X \in \mathfrak{X}(M)$ be a smooth vector field on M. Given $p \in M$, there exists an integral curve $c_p : I \to M$ of X at p (that is, $\dot{c}_p(t) = X_{c_p(t)}$ for $t \in I = (-\varepsilon, \varepsilon)$ and $c_p(0) = p$). Moreover, this curve is unique, meaning that any two such curves agree on the intersection of their domains.* $\square$

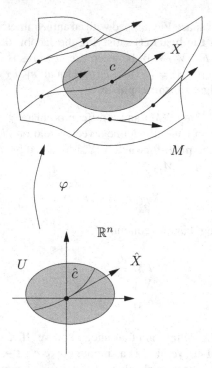

Fig. 1.15 Integral curves of a vector field

This integral curve, obtained by solving (1.2), depends smoothly on the initial point p (see [Arn92]).

Theorem 6.6 *Let $X \in \mathfrak{X}(M)$. For each $p \in M$ there exists a neighborhood W of p, an interval $I = (-\varepsilon, \varepsilon)$ and a mapping $F : W \times I \to M$ such that:*

(i) for a fixed $q \in W$ the curve $F(q, t)$, $t \in I$, is an integral curve of X at q, that is, $F(q, 0) = q$ and $\frac{\partial F}{\partial t}(q, t) = X_{F(q,t)}$;
(ii) the map F is differentiable. □

The map $F : W \times I \to M$ defined above is called the **local flow** of X at p. Let us now fix $t \in I$ and consider the map

$$\psi_t : W \to M$$
$$q \mapsto F(q, t) = c_q(t).$$

defined by the local flow. The following proposition then holds:

Proposition 6.7 *The maps $\psi_t : W \to M$ above are local diffeomorphisms and satisfy*

$$(\psi_t \circ \psi_s)(q) = \psi_{t+s}(q), \tag{1.3}$$

whenever $t, s, t + s \in I$ and $\psi_s(q) \in W$.

Proof First we note that

$$\frac{dc_q}{dt}(t) = X_{c_q(t)}$$

and so

$$\frac{d}{dt}(c_q(t + s)) = X_{c_q(t+s)}.$$

Hence, as $c_q(t+s)|_{t=0} = c_q(s)$, the curve $c_{c_q(s)}(t)$ is just $c_q(t+s)$, that is, $\psi_{t+s}(q) = \psi_t(\psi_s(q))$. We can use this formula to extend ψ_t to $\psi_s(W)$ for all $s \in I$ such that $t+s \in I$. In particular, ψ_{-t} is well defined on $\psi_t(W)$, and $(\psi_{-t} \circ \psi_t)(q) = \psi_0(q) = c_q(0) = q$ for all $q \in W$. Thus the map ψ_{-t} is the inverse of ψ_t, which consequently is a local diffeomorphism (it maps W diffeomorphically onto its image). $\qquad\square$

A collection of diffeomorphisms $\{\psi_t : M \rightarrow M\}_{t \in I}$, where $I = (-\varepsilon, \varepsilon)$, satisfying (1.3) is called a **local 1-parameter group of diffeomorphisms**. When the interval of definition I of c_q is $\mathbb{R}$, this local 1-parameter group of diffeomorphisms becomes a **group of diffeomorphisms**. A vector field X whose local flow defines a 1-parameter group of diffeomorphisms is said to be **complete**. This happens for instance when the vector field X has compact support.

Theorem 6.8 *If $X \in \mathfrak{X}(M)$ is a smooth vector field with compact support then it is complete.*

Proof For each $p \in M$ we can take a neighborhood W and an interval $I = (-\varepsilon, \varepsilon)$ such that the local flow of X at p, $F(q, t) = c_q(t)$, is defined on $W \times I$. We can therefore cover the support of X (which is compact) by a finite number of such neighborhoods W_k and consider an interval $I_0 = (-\varepsilon_0, \varepsilon_0)$ contained in the intersection of the corresponding intervals I_k. If q is not in supp(X), then $X_q = 0$ and so $c_q(t)$ is trivially defined on I_0. Hence we can extend the map F to $M \times I_0$. Moreover, condition (1.3) is true for each $-\varepsilon_0/2 < s, t < \varepsilon_0/2$, and we can again extend the map F, this time to $M \times \mathbb{R}$. In fact, for any $t \in \mathbb{R}$, we can write $t = k\varepsilon_0/2 + s$, where $k \in \mathbb{Z}$ and $0 \le s < \varepsilon_0/2$, and define $F(q, t) := F^k(F(q, s), \varepsilon_0/2)$. $\qquad\square$

Corollary 6.9 *If M is compact then all smooth vector fields on M are complete.* $\quad\square$

We finish this section with an important result.

Theorem 6.10 *Let $X_1, X_2 \in \mathfrak{X}(M)$ be two complete vector fields. Then their flows ψ_1, ψ_2 commute (i.e. $\psi_{1,t} \circ \psi_{2,s} = \psi_{2,s} \circ \psi_{1,t}$ for all $s, t \in \mathbb{R}$) if and only if $[X_1, X_2] = 0$.*

Proof Exercise 6.11(13). $\qquad\square$

Exercise 6.11

(1) Let $X : M \to TM$ be a differentiable vector field on M and, for a smooth function $f : M \to \mathbb{R}$, consider its directional derivative along X defined by

$$X \cdot f : M \to \mathbb{R}$$
$$p \mapsto X_p \cdot f.$$

Show that:

(a) $(X \cdot f)(p) = (df)_p X_p$;
(b) the vector field X is smooth if and only if $X \cdot f$ is a differentiable function for any smooth function $f : M \to \mathbb{R}$;
(c) the directional derivative satisfies the following properties: for $f, g \in C^\infty(M)$ and $\alpha \in \mathbb{R}$,
  (i) $X \cdot (f + g) = X \cdot f + X \cdot g$;
  (ii) $X \cdot (\alpha f) = \alpha(X \cdot f)$;
  (iii) $X \cdot (fg) = fX \cdot g + gX \cdot f$.

(2) Prove Proposition 6.3.
(3) Show that $(\mathbb{R}^3, \times)$ is a Lie algebra, where $\times$ is the cross product on $\mathbb{R}^3$.
(4) Compute the flows of the vector fields $X, Y, Z \in \mathfrak{X}(\mathbb{R}^2)$ defined by

$$X_{(x,y)} = \frac{\partial}{\partial x}; \quad Y_{(x,y)} = x\frac{\partial}{\partial x} + y\frac{\partial}{\partial y}; \quad Z_{(x,y)} = -y\frac{\partial}{\partial x} + x\frac{\partial}{\partial y}.$$

(5) Let $X_1, X_2, X_3 \in \mathfrak{X}(\mathbb{R}^3)$ be the vector fields defined by

$$X_1 = y\frac{\partial}{\partial z} - z\frac{\partial}{\partial y}, \quad X_2 = z\frac{\partial}{\partial x} - x\frac{\partial}{\partial z}, \quad X_3 = x\frac{\partial}{\partial y} - y\frac{\partial}{\partial x},$$

where (x, y, z) are the usual Cartesian coordinates.

(a) Compute the Lie brackets $[X_i, X_j]$ for $i, j = 1, 2, 3$.
(b) Show that $\operatorname{span}\{X_1, X_2, X_3\}$ is a Lie subalgebra of $\mathfrak{X}(\mathbb{R}^3)$, isomorphic to $(\mathbb{R}^3, \times)$.
(c) Compute the flows $\psi_{1,t}, \psi_{2,t}, \psi_{3,t}$ of X_1, X_2, X_3.
(d) Show that $\psi_{i,\frac{\pi}{2}} \circ \psi_{j,\frac{\pi}{2}} \neq \psi_{j,\frac{\pi}{2}} \circ \psi_{i,\frac{\pi}{2}}$ for $i \neq j$.

(6) Give an example of a non-complete vector field.
(7) Let N be a differentiable manifold, $M \subset N$ a submanifold and $X, Y \in \mathfrak{X}(N)$ vector fields tangent to M, i.e. such that $X_p, Y_p \in T_pM$ for all $p \in M$. Show that $[X, Y]$ is also tangent to M, and that its restriction to M coincides with the Lie bracket of the restrictions of X and Y to M.
(8) Let $f : M \to N$ be a smooth map between manifolds. Two vector fields $X \in \mathfrak{X}(M)$ and $Y \in \mathfrak{X}(N)$ are said to be f-**related** (and we write $Y = f_*X$) if, for each $q \in N$ and $p \in f^{-1}(q) \subset M$, we have $(df)_p X_p = Y_q$. Show that:

(a) given f and X it is possible that no vector field Y is f-related to X;
(b) the vector field X is f-related to Y if and only if, for any differentiable function g defined on some open subset W of N, $(Y \cdot g) \circ f = X \cdot (g \circ f)$ on the inverse image $f^{-1}(W)$ of the domain of g;
(c) for differentiable maps $f : M \to N$ and $g : N \to P$ between smooth manifolds and vector fields $X \in \mathfrak{X}(M)$, $Y \in \mathfrak{X}(N)$ and $Z \in \mathfrak{X}(P)$, if X is f-related to Y and Y is g-related to Z, then X is $(g \circ f)$-related to Z.

(9) Let $f : M \to N$ be a diffeomorphism between smooth manifolds. Show that $f_*[X, Y] = [f_*X, f_*Y]$ for every $X, Y \in \mathfrak{X}(M)$. Therefore, f_* induces a Lie algebra isomorphism between $\mathfrak{X}(M)$ and $\mathfrak{X}(N)$.

(10) Let $f : M \to N$ be a differentiable map between smooth manifolds and consider two vector fields $X \in \mathfrak{X}(M)$ and $Y \in \mathfrak{X}(N)$. Show that:

(a) if the vector field Y is f-related to X then any integral curve of X is mapped by f into an integral curve of Y;
(b) the vector field Y is f-related to X if and only if the local flows F_X and F_Y satisfy $f(F_X(p, t)) = F_Y(f(p), t)$ for all (t, p) for which both sides are defined.

(11) (*Lie derivative of a function*) Given a vector field $X \in \mathfrak{X}(M)$, we define the **Lie derivative** of a smooth function $f : M \to \mathbb{R}$ in the direction of X as

$$L_X f(p) := \frac{d}{dt} \left((f \circ \psi_t)(p) \right) \Big|_{t=0},$$

where $\psi_t = F(\cdot, t)$, for F the local flow of X at p. Show that $L_X f = X \cdot f$, meaning that the Lie derivative of f in the direction of X is just the directional derivative of f along X.

(12) (*Lie derivative of a vector field*) For two vector fields $X, Y \in \mathfrak{X}(M)$ we define the **Lie derivative** of Y in the direction of X as

$$L_X Y := \frac{d}{dt} \left((\psi_{-t})_* Y \right) \Big|_{t=0},$$

where $\{\psi_t\}_{t \in I}$ is the local flow of X. Show that:

(a) $L_X Y = [X, Y]$;
(b) $L_X[Y, Z] = [L_X Y, Z] + [Y, L_X Z]$, for $X, Y, Z \in \mathfrak{X}(M)$;
(c) $L_X \circ L_Y - L_Y \circ L_X = L_{[X,Y]}$.

(13) Let $X, Y \in \mathfrak{X}(M)$ be two complete vector fields with flows ψ, ϕ. Show that:

(a) given a diffeomorphism $f : M \to M$, we have $f_* X = X$ if and only if $f \circ \psi_t = \psi_t \circ f$ for all $t \in \mathbb{R}$;
(b) $\psi_t \circ \phi_s = \phi_s \circ \psi_t$ for all $s, t \in \mathbb{R}$ if and only if $[X, Y] = 0$.

1.7 Lie Groups

A **Lie group** G is a smooth manifold which is at the same time a group, in such a way that the group operations

$$\begin{matrix} G \times G \to G \\ (g, h) \mapsto gh \end{matrix} \quad \text{and} \quad \begin{matrix} G \to G \\ g \mapsto g^{-1} \end{matrix}$$

are differentiable maps (where we consider the standard differentiable structure of the product on $G \times G$ – cf. Exercise 2.5(6)).

Example 7.1

(1) $(\mathbb{R}^n, +)$ is trivially an abelian Lie group.
(2) The **general linear group**

$$GL(n) = \{n \times n \text{ invertible real matrices}\}$$

is the most basic example of a nontrivial Lie group. We have seen in Example 2.3(4) that it is a smooth manifold of dimension n^2. Moreover, the group multiplication is just the restriction to

$$GL(n) \times GL(n)$$

of the usual multiplication of $n \times n$ matrices, whose coordinate functions are quadratic polynomials; the inversion is just the restriction to $GL(n)$ of the usual inversion of nonsingular matrices which, by Cramer's rule, is a map with rational coordinate functions and nonzero denominators (only the determinant appears on the denominator).

(3) The **orthogonal group**

$$O(n) = \{A \in \mathcal{M}_{n \times n} \mid A^t A = I\}$$

of orthogonal transformations of $\mathbb{R}^n$ is also a Lie group. We can show this by considering the map $f : A \mapsto A^t A$ from $\mathcal{M}_{n \times n} \cong \mathbb{R}^{n^2}$ to the space $\mathcal{S}_{n \times n} \cong \mathbb{R}^{\frac{1}{2}n(n+1)}$ of symmetric $n \times n$ matrices. Its derivative at a point $A \in O(n)$, $(df)_A$, is a surjective map from $T_A \mathcal{M}_{n \times n} \cong \mathcal{M}_{n \times n}$ onto $T_{f(A)} \mathcal{S}_{n \times n} \cong \mathcal{S}_{n \times n}$. Indeed,

$$\begin{aligned} (df)_A(B) &= \lim_{h \to 0} \frac{f(A + hB) - f(A)}{h} \\ &= \lim_{h \to 0} \frac{(A + hB)^t(A + hB) - A^t A}{h} \\ &= B^t A + A^t B, \end{aligned}$$

and any symmetric matrix S can be written as $B^t A + A^t B$ with $B = \frac{1}{2}(A^{-1})^t S = \frac{1}{2} AS$. In particular, the identity I is a regular value of f and so, by Theorem 5.6, we have that $O(n) = f^{-1}(I)$ is a submanifold of $\mathcal{M}_{n\times n}$ of dimension $\frac{1}{2}n(n-1)$. Moreover, it is also a Lie group as the group multiplication and inversion are restrictions of the same operations on $GL(n)$ to $O(n)$ (a submanifold) and have values on $O(n)$ [cf. Exercise 5.9(5)].

(4) The map $f : GL(n) \to \mathbb{R}$ given by $f(A) = \det A$ is differentiable, and the level set $f^{-1}(1)$ is

$$SL(n) = \{A \in \mathcal{M}_{n\times n} \mid \det A = 1\},$$

the **special linear group**. Again, the derivative of f is surjective at a point $A \in GL(n)$, making $SL(n)$ into a Lie group. Indeed, it is easy to see that

$$(df)_I(B) = \lim_{h\to 0} \frac{\det(I + hB) - \det I}{h} = \operatorname{tr} B$$

implying that

$$
\begin{aligned}
(df)_A(B) &= \lim_{h\to 0} \frac{\det(A + hB) - \det A}{h} \\
&= \lim_{h\to 0} \frac{(\det A)\det(I + hA^{-1}B) - \det A}{h} \\
&= (\det A) \lim_{h\to 0} \frac{\det(I + hA^{-1}B) - 1}{h} \\
&= (\det A)(df)_I(A^{-1}B) = (\det A)\operatorname{tr}(A^{-1}B).
\end{aligned}
$$

Since $\det(A) = 1$, for any $k \in \mathbb{R}$, we can take the matrix $B = \frac{k}{n}A$ to obtain $(df)_A(B) = \operatorname{tr}\left(\frac{k}{n}I\right) = k$. Therefore, $(df)_A$ is surjective for every $A \in SL(n)$, and so 1 is a regular value of f. Consequently, $SL(n)$ is a submanifold of $GL(n)$. As in the preceding example, the group multiplication and inversion are differentiable, and so $SL(n)$ is a Lie group.

(5) The map $A \mapsto \det A$ is a differentiable map from $O(n)$ to $\{-1, 1\}$, and the level set $f^{-1}(1)$ is

$$SO(n) = \{A \in O(n) \mid \det A = 1\},$$

the **special orthogonal group** or the **rotation group** in $\mathbb{R}^n$, which is then an open subset of $O(n)$, and therefore a Lie group of the same dimension.

(6) We can also consider the space $\mathcal{M}_{n\times n}(\mathbb{C})$ of complex $n \times n$ matrices, and the space $GL(n, \mathbb{C})$ of complex $n \times n$ invertible matrices. This is a Lie group of real dimension $2n^2$. Moreover, similarly to what was done above for $O(n)$, we can take the group of unitary transformations on $\mathbb{C}^n$,

$$U(n) = \{A \in \mathcal{M}_{n \times n}(\mathbb{C}) \mid A^* A = I\},$$

where A^* is the adjoint of A. This group is a submanifold of $\mathcal{M}_{n \times n}(\mathbb{C}) \cong \mathbb{C}^{n^2} \cong \mathbb{R}^{2n^2}$, and a Lie group, called the **unitary group**. This can be seen from the fact that I is a regular value of the map $f : A \mapsto A^* A$ from $\mathcal{M}_{n \times n}(\mathbb{C})$ to the space of self-adjoint matrices. As any element of $\mathcal{M}_{n \times n}(\mathbb{C})$ can be uniquely written as a sum of a self-adjoint with an anti-self-adjoint matrix, and the map $A \to iA$ is an isomorphism from the space of self-adjoint matrices to the space of anti-self-adjoint matrices, we conclude that these two spaces have real dimension $\frac{1}{2} \dim_{\mathbb{R}} \mathcal{M}_{n \times n}(\mathbb{C}) = n^2$. Hence, $\dim U(n) = n^2$.

(7) The **special unitary group**

$$SU(n) = \{A \in U(n) \mid \det A = 1\},$$

is also a Lie group now of dimension $n^2 - 1$ (note that $A \mapsto \det(A)$ is now a differentiable map from $U(n)$ to S^1).

As a Lie group G is, by definition, a manifold, we can consider the tangent space at one of its points. In particular, the tangent space at the identity e is usually denoted by

$$\mathfrak{g} := T_e G.$$

For $g \in G$, we have the maps

$$\begin{matrix} L_g : G \to G \\ h \mapsto g \cdot h \end{matrix} \quad \text{and} \quad \begin{matrix} R_g : G \to G \\ h \mapsto h \cdot g \end{matrix}$$

which correspond to **left multiplication** and **right multiplication** by g.

A vector field on G is called **left-invariant** if $(L_g)_* X = X$ for every $g \in G$, that is,

$$((L_g)_* X)_{gh} = X_{gh} \quad \text{or} \quad (dL_g)_h X_h = X_{gh},$$

for every $g, h \in G$. There is, of course, a vector space isomorphism between $\mathfrak{g}$ and the space of left-invariant vector fields on G that, to each $V \in \mathfrak{g}$, assigns the vector field X^V defined by

$$X_g^V := (dL_g)_e V,$$

for any $g \in G$. This vector field is left-invariant as

$$(dL_g)_h X_h^V = (dL_g)_h (dL_h)_e V = (d(L_g \circ L_h))_e V = (dL_{gh})_e V = X_{gh}^V.$$

Note that, given a left-invariant vector field X, the corresponding element of $\mathfrak{g}$ is X_e. As the space $\mathfrak{X}_L(G)$ of left-invariant vector fields is closed under the Lie bracket of vector fields (because, from Exercise 6.11(9) $(L_g)_*[X, Y] = [(L_g)_* X, (L_g)_* Y]$), it is a Lie subalgebra of the Lie algebra of vector fields (see Definition 6.4).

The isomorphism $\mathfrak{X}_L(G) \cong \mathfrak{g}$ then determines a Lie algebra structure on $\mathfrak{g}$. We call $\mathfrak{g}$ the **Lie algebra** of the Lie group G.

Example 7.2

(1) If $G = GL(n)$, then $\mathfrak{gl}(n) = T_I GL(n) = \mathcal{M}_{n \times n}$ is the space of $n \times n$ matrices with real coefficients, and the Lie bracket on $\mathfrak{gl}(n)$ is the commutator of matrices

$$[A, B] = AB - BA.$$

In fact, if $A, B \in \mathfrak{gl}(n)$ are two $n \times n$ matrices, the corresponding left-invariant vector fields are given by

$$X_g^A = (dL_g)_I(A) = \sum_{i,k,j} x^{ik} a^{kj} \frac{\partial}{\partial x^{ij}}$$

$$X_g^B = (dL_g)_I(B) = \sum_{i,k,j} x^{ik} b^{kj} \frac{\partial}{\partial x^{ij}},$$

where $g \in GL(n)$ is a matrix with components x^{ij}. The ij-component of $[X^A, X^B]_g$ is given by $X_g^A \cdot (X^B)^{ij} - X_g^B \cdot (X^A)^{ij}$, i.e.

$$[X^A, X^B]^{ij}(g) = \left(\sum_{l,m,p} x^{lp} a^{pm} \frac{\partial}{\partial x^{lm}} \right) \left(\sum_k x^{ik} b^{kj} \right)$$

$$- \left(\sum_{l,m,p} x^{lp} b^{pm} \frac{\partial}{\partial x^{lm}} \right) \left(\sum_k x^{ik} a^{kj} \right)$$

$$= \sum_{k,l,m,p} x^{lp} a^{pm} \delta_{il} \delta_{km} b^{kj} - \sum_{k,l,m,p} x^{lp} b^{pm} \delta_{il} \delta_{km} a^{kj}$$

$$= \sum_{m,p} x^{ip} (a^{pm} b^{mj} - b^{pm} a^{mj})$$

$$= \sum_p x^{ip} (AB - BA)^{pj}$$

(where $\delta_{ij} = 1$ if $i = j$ and $\delta_{ij} = 0$ if $i \neq j$ is the **Kronecker symbol**). Making $g = I$, we obtain

$$[A, B] = [X^A, X^B]_I = AB - BA.$$

From Exercise 6.11(7) we see that this will always be the case when G is a matrix group, that is, when G is a subgroup of $GL(n)$ for some n.

(2) If $G = O(n)$ then its Lie algebra is

$$\mathfrak{o}(n) = \{A \in \mathcal{M}_{n \times n} \mid A^t + A = 0\}.$$

In fact, we have seen in Example 7.1(3) that $O(n) = f^{-1}(I)$, where the identity I is a regular value of the map

$$f : \mathcal{M}_{n \times n} \to \mathcal{S}_{n \times n}$$
$$A \mapsto A^t A.$$

Hence, $\mathfrak{o}(n) = T_I G = \ker(df)_I = \{A \in \mathcal{M}_{n \times n} \mid A^t + A = 0\}$ is the space of skew-symmetric matrices.

(3) If $G = SL(n)$ then its Lie algebra is

$$\mathfrak{sl}(n) = \{A \in \mathcal{M}_{n \times n} \mid \operatorname{tr} A = 0\}.$$

In fact, we have seen in Example 7.1(4) that $SL(n) = f^{-1}(1)$, where 1 is a regular value of the map

$$f : \mathcal{M}_{n \times n} \to \mathbb{R}$$
$$A \mapsto \det A.$$

Hence, $\mathfrak{sl}(n) = T_I G = \ker(df)_I = \{A \in \mathcal{M}_{n \times n} \mid \operatorname{tr} A = 0\}$ is the space of traceless matrices.

(4) If $G = SO(n) = \{A \in O(n) \mid \det A = 1\}$, then its Lie algebra is

$$\mathfrak{so}(n) = T_I SO(n) = T_I O(n) = \mathfrak{o}(n).$$

(5) Similarly to Example 7.2(2), the Lie algebra of $U(n)$ is

$$\mathfrak{u}(n) = \{A \in \mathcal{M}_{n \times n}(\mathbb{C}) \mid A^* + A = 0\},$$

the space of skew-hermitian matrices.

(6) To determine the Lie algebra of $SU(n)$, we see that $SU(n)$ is the level set $f^{-1}(1)$, where $f(A) = \det A$, and so

$$\mathfrak{su}(n) = \ker(df)_I = \{A \in \mathfrak{u}(n) \mid \operatorname{tr}(A) = 0\}.$$

We now study the flow of a left-invariant vector field.

Proposition 7.3 *Let F be the local flow of a left-invariant vector field X at a point $h \in G$. Then the map ψ_t defined by F (that is, $\psi_t(g) = F(g, t)$) satisfies $\psi_t = R_{\psi_t(e)}$. Moreover, the flow of X is globally defined for all $t \in \mathbb{R}$.*

Proof For $g \in G$, $R_{\psi_t(e)}(g) = g \cdot \psi_t(e) = L_g(\psi_t(e))$. Hence,

$$R_{\psi_0(e)}(g) = g \cdot e = g$$

and

$$\frac{d}{dt}\left(R_{\psi_t(e)}(g)\right) = \frac{d}{dt}\left(L_g(\psi_t(e))\right) = (dL_g)_{\psi_t(e)}\left(\frac{d}{dt}(\psi_t(e))\right)$$
$$= (dL_g)_{\psi_t(e)}\left(X_{\psi_t(e)}\right) = X_{g\cdot\psi_t(e)}$$
$$= X_{R_{\psi_t(e)}(g)},$$

implying that $R_{\psi_t(e)}(g) = c_g(t) = \psi_t(g)$ is the integral curve of X at g. Consequently, if $\psi_t(e)$ is defined for $t \in (-\varepsilon, \varepsilon)$, then $\psi_t(g)$ is defined for $t \in (-\varepsilon, \varepsilon)$ and $g \in G$. Moreover, condition (1.3) in Sect. 1.6 is true for each $-\varepsilon/2 < s, t < \varepsilon/2$ and we can extend the map F to $G \times \mathbb{R}$ as before: for any $t \in \mathbb{R}$, we write $t = k\varepsilon/2 + s$ where $k \in \mathbb{Z}$ and $0 \le s < \varepsilon/2$, and define $F(g, t) := F^k(F(g, s), \varepsilon/2) = gF(e, s)F^k(e, \varepsilon/2)$. $\square$

Remark 7.4 A homomorphism $F : G_1 \to G_2$ between Lie groups is called a **Lie group homomorphism** if, besides being a group homomorphism, it is also a differentiable map. Since

$$\psi_{t+s}(e) = \psi_s(\psi_t(e)) = R_{\psi_s(e)}\psi_t(e) = \psi_t(e) \cdot \psi_s(e),$$

the integral curve $t \mapsto \psi_t(e)$ defines a group homomorphism between $(\mathbb{R}, +)$ and $(G, \cdot)$.

Definition 7.5 The **exponential map** $\exp : \mathfrak{g} \to G$ is the map that, to each $V \in \mathfrak{g}$, assigns the value $\psi_1(e)$, where ψ_t is the flow of the left-invariant vector field X^V.

Remark 7.6 If $c_g(t)$ is the integral curve of X at g and $s \in \mathbb{R}$, it is easy to check that $c_g(st)$ is the integral curve of sX at g. On the other hand, for $V \in \mathfrak{g}$ one has $X^{sV} = sX^V$. Consequently,

$$\psi_t(e) = c_e(t) = c_e(t \cdot 1) = F(e, 1) = \exp(tV),$$

where F is the flow of $tX^V = X^{tV}$.

Example 7.7 If G is a group of matrices, then for $A \in \mathfrak{g}$,

$$\exp A = e^A = \sum_{k=0}^{\infty} \frac{A^k}{k!}.$$

In fact, this series converges for any matrix A and the map $h(t) = e^{At}$ satisfies

$$h(0) = e^0 = I$$
$$\frac{dh}{dt}(t) = e^{At}A = h(t)A.$$

Hence, h is the flow of X^A at the identity (that is, $h(t) = \psi_t(e)$), and so $\exp A = \psi_1(e) = e^A$.

Let now G be any group and M be any set. We say that G **acts on** M if there is a homomorphism ϕ from G to the group of bijective mappings from M to M, or, equivalently, writing

$$\phi(g)(p) = A(g, p),$$

if there is a mapping $A : G \times M \to M$ satisfying the following conditions:

  (i) if e is the identity in G, then $A(e, p) = p$, $\forall p \in M$;
 (ii) if $g, h \in G$, then $A(g, A(h, p)) = A(gh, p)$, $\forall p \in M$.

Usually we denote $A(g, p)$ by $g \cdot p$.

Example 7.8

 (1) Let G be a group and $H \subset G$ a subgroup. Then H acts on G by left multiplication: $A(h, g) = h \cdot g$ for $h \in H, g \in G$.
 (2) $GL(n)$ acts on $\mathbb{R}^n$ through $A \cdot x = Ax$ for $A \in GL(n)$ and $x \in \mathbb{R}^n$. The same is true for any subgroup $G \subset GL(n)$.

For each $p \in M$ we can define the **orbit of p** as the set $G \cdot p := \{g \cdot p \mid g \in G\}$. If $G \cdot p = \{p\}$ then p is called a **fixed point** of G. If there is a point $p \in M$ whose orbit is all of M (i.e. $G \cdot p = M$), then the action is said to be **transitive**. Note that when this happens, there is only one orbit and, for every $p, q \in M$ with $p \neq q$, there is always an element of the group $g \in G$ such that $q = g \cdot p$. The manifold M is then called a **homogeneous space** of G. The **stabilizer** (or **isotropy subgroup**) of a point $p \in M$ is the group

$$G_p = \{g \in G \mid g \cdot p = p\}.$$

The action is called **free** if all the stabilizers are trivial.

If G is a Lie group and M is a smooth manifold, we say that the action is **smooth** if the map $A : G \times M \to M$ is differentiable. In this case, the map $p \mapsto g \cdot p$ is a diffeomorphism. We will always assume the action of a Lie group on a differentiable manifold to be smooth. A smooth action is said to be **proper** if the map

$$G \times M \to M \times M$$
$$(g, p) \mapsto (g \cdot p, p)$$

is proper (recall that a map is called **proper** if the preimage of any compact set is compact—cf. Sect. 1.10.5).

Remark 7.9 Note that a smooth action is proper if and only if, given two convergent sequences $\{p_n\}$ and $\{g_n \cdot p_n\}$ in M, there exists a convergent subsequence $\{g_{n_k}\}$ in G. If G is compact this condition is always satisfied.

The orbits of the action of G on M are equivalence classes of the equivalence relation $\sim$ given by $p \sim q \Leftrightarrow q \in G \cdot p$ (cf. Sect. 1.10.1). For that reason, the quotient (topological) space $M/\sim$ is usually called the **orbit space** of the action, and denoted by M/G.

Proposition 7.10 *If the action of a Lie group G on a differentiable manifold M is proper, then the orbit space M/G is a Hausdorff space.*

Proof The relation $p \sim q \Leftrightarrow q \in G \cdot p$ is an open equivalence relation (cf. Sect. 1.10.1). Indeed, since $p \mapsto g \cdot p$ is a homeomorphism, the set $\pi^{-1}(\pi(U)) = \{g \cdot p \mid p \in U \text{ and } g \in G\} = \bigcup_{g \in G} g \cdot U$ is an open subset of M for any open set U in M, meaning that $\pi(U)$ is open (here $\pi : M \to M/G$ is the quotient map). Therefore we just have to show that the set

$$R = \{(p, q) \in M \times M \mid p \sim q\}$$

is closed (cf. Proposition 10.2). This follows from the fact that R is the image of the map

$$G \times M \to M \times M$$
$$(g, p) \mapsto (g \cdot p, p)$$

which is continuous and proper, hence closed (cf. Sect. 1.10.5).   □

Under certain conditions the orbit space M/G is naturally a differentiable manifold.

Theorem 7.11 *Let M be a differentiable manifold equipped with a free proper action of a Lie group G. Then the orbit space M/G is naturally a differentiable manifold of dimension $\dim M - \dim G$, and the quotient map $\pi : M \to M/G$ is a submersion.*

Proof By the previous proposition, the quotient M/G is Hausdorff. Moreover, this quotient satisfies the second countability axiom because M does so and the equivalence relation defined by G is open. It remains to be shown that M/G has a natural differentiable structure for which the quotient map is a submersion. We do this only in the case of a discrete (i.e. zero-dimensional) Lie group (cf. Remark 1.2); the proof for general Lie groups can be found in [DK99].

In our case, we just have to prove that for each point $p \in M$ there exists a neighborhood $U \ni p$ such that $g \cdot U \cap h \cdot U = \varnothing$ for $g \neq h$. This guarantees that each point $[p] \in M/G$ has a neighborhood $[U]$ homeomorphic to U, which we can assume to be a coordinate neighborhood. Since G acts by diffeomorphisms, the differentiable structure defined in this way does not depend on the choice of $p \in [p]$. Since the charts of M/G are obtained from charts of M, the overlap maps are smooth. Therefore M/G has a natural differentiable structure for which $\pi : M \to M/G$ is a local diffeomorphism (as the coordinate expression of $\pi|_U : U \to [U]$ is the identity map).

Showing that $g \cdot U \cap h \cdot U = \varnothing$ for $g \neq h$ is equivalent to showing that $g \cdot U \cap U = \varnothing$ for $g \neq e$. Assume that this did not happen for any neighborhood $U \ni p$. Then there would exist a sequence of open sets $U_n \ni p$ with $U_{n+1} \subset U_n$, $\bigcap_{n=1}^{+\infty} U_n = \{p\}$ and a sequence $g_n \in G \setminus \{e\}$ such that $g_n \cdot U_n \cap U_n \neq \varnothing$. Choose $p_n \in g_n \cdot U_n \cap U_n$. Then $p_n = g_n \cdot q_n$ for some $q_n \in U_n$. We have $p_n \to p$ and $q_n \to p$. Since the action is proper, g_n admits a convergent subsequence g_{n_k}. Let g be its limit. Making $k \to +\infty$ in $q_{n_k} = g_{n_k} \cdot p_{n_k}$ yields $g \cdot p = p$, implying that $g = e$ (the action is free). Because G is discrete, we would then have $g_{n_k} = e$ for sufficiently large k, which is a contradiction.   □

Example 7.12

(1) Let $S^n = \left\{ x \in \mathbb{R}^{n+1} \mid \sum_{i=1}^{n} \left(x^i \right)^2 = 1 \right\}$ be equipped with the action of $G = \mathbb{Z}_2 = \{-I, I\}$ given by $-I \cdot x = -x$ (antipodal map). This action is proper and free, and so the orbit space S^n / G is an n-dimensional manifold. This space is the real projective space $\mathbb{R}P^n$ [cf. Exercise 2.5(8)].

(2) The group $G = \mathbb{R} \setminus \{0\}$ acts on $M = \mathbb{R}^{n+1} \setminus \{0\}$ by multiplication: $t \cdot x = tx$. This action is proper and free, and so M/G is a differentiable manifold of dimension n (which is again $\mathbb{R}P^n$).

(3) Consider $M = \mathbb{R}^n$ equipped with an action of $G = \mathbb{Z}^n$ defined by:

$$\left(k^1, \ldots, k^n \right) \cdot \left(x^1, \ldots, x^n \right) = \left(x^1 + k^1, \ldots, x^n + k^n \right).$$

This action is proper and free, and so the quotient M/G is a manifold of dimension n. This space with the quotient differentiable structure defined in Theorem 7.11 is called the n-**torus** and is denoted by $\mathbb{T}^n$. It is diffeomorphic to the product manifold $S^1 \times \cdots \times S^1$ and, when $n = 2$, is diffeomorphic to the torus of revolution in $\mathbb{R}^3$.

Quotients by discrete group actions determine **coverings** of manifolds.

Definition 7.13 A smooth **covering** of a differentiable manifold B is a pair (M, π), where M is a connected differentiable manifold, $\pi : M \to B$ is a surjective local diffeomorphism, and, for each $p \in B$, there exists a connected neighborhood U of p in B such that $\pi^{-1}(U)$ is the union of disjoint open sets $U_\alpha \subset M$ (called **slices**), and the restrictions π_α of π to U_α are diffeomorphisms onto U. The map π is called a **covering map** and M is called a **covering manifold**.

Remark 7.14

(1) It is clear that we must have $\dim M = \dim B$.
(2) Note that the collection of mutually disjoint open sets $\{U_\alpha\}$ must be countable (M has a countable basis).
(3) The **fibers** $\pi^{-1}(p) \subset M$ have the discrete topology. Indeed, as each slice U_α is open and intersects $\pi^{-1}(p)$ in exactly one point, this point is open in the subspace topology.

Example 7.15

(1) The map $\pi : \mathbb{R} \to S^1$ given by

$$\pi(t) = (\cos(2\pi t), \sin(2\pi t))$$

is a smooth covering of S^1. However, the restriction of this map to $(0, +\infty)$ is a surjective local diffeomorphism which is not a covering map.

(2) The product of covering maps is clearly a covering map. Thus we can generalize the above example and obtain a covering of $\mathbb{T}^n \cong S^1 \times \cdots \times S^1$ by $\mathbb{R}^n$.

(3) In Example 7.12(1) we have a covering of $\mathbb{R}P^n$ by S^n.

A diffeomorphism $h : M \to M$, where M is a covering manifold, is called a **deck transformation** (or **covering transformation**) if $\pi \circ h = \pi$, or, equivalently, if each set $\pi^{-1}(p)$ is carried to itself by h. It can be shown that the group G of all covering transformations is a discrete Lie group whose action on M is free and proper.

If the covering manifold M is **simply connected** (cf. Sect. 1.10.5), the covering is said to be a **universal covering**. In this case, B is diffeomorphic to M/G. Moreover, G is isomorphic to the **fundamental group** $\pi_1(B)$ of B (cf. Sect. 1.10.5).

The **Lie theorem** (see for instance [DK99]) states that for a given Lie algebra $\mathfrak{g}$ there exists a unique simply connected Lie group $\widetilde{G}$ whose Lie algebra is $\mathfrak{g}$. If a Lie group G also has $\mathfrak{g}$ as its Lie algebra, then there exists a unique Lie group homomorphism $\pi : \widetilde{G} \to G$ which is a covering map. The group of deck transformations is, in this case, simply $\ker(\pi)$, and hence G is diffeomorphic to $\widetilde{G}/\ker(\pi)$. In fact, G is also isomorphic to $\widetilde{G}/\ker(\pi)$, which has a natural group structure ($\ker(\pi)$ is a normal subgroup).

Example 7.16

(1) In the universal covering of S^1 of Example 7.15(1) the deck transformations are translations $h_k : t \mapsto t + k$ by an integer k, and so the fundamental group of S^1 is $\mathbb{Z}$.

(2) Similarly, the deck transformations of the universal covering of $\mathbb{T}^n$ are translations by integer vectors [cf. Example 7.15(2)], and so the fundamental group of $\mathbb{T}^n$ is $\mathbb{Z}^n$.

(3) In the universal covering of $\mathbb{R}P^n$ from Example 7.15(3), the only deck transformations are the identity and the antipodal map, and so the fundamental group of $\mathbb{R}P^n$ is $\mathbb{Z}_2$.

Exercise 7.17

(1) (a) Given two Lie groups G_1, G_2, show that $G_1 \times G_2$ (the direct product of the two groups) is a Lie group with the standard differentiable structure on the product.

(b) The circle S^1 can be identified with the set of complex numbers of absolute value 1. Show that S^1 is a Lie group and conclude that the n-torus $T^n \cong S^1 \times \ldots \times S^1$ is also a Lie group.

(2) (a) Show that $(\mathbb{R}^n, +)$ is a Lie group, determine its Lie algebra and write an expression for the exponential map.

(b) Prove that, if G is an abelian Lie group, then $[V, W] = 0$ for all $V, W \in \mathfrak{g}$.

(3) We can identify each point in

$$H = \left\{ (x, y) \in \mathbb{R}^2 \mid y > 0 \right\}$$

with the invertible affine map $h : \mathbb{R} \to \mathbb{R}$ given by $h(t) = yt + x$. The set of all such maps is a group under composition; consequently, our identification induces a group structure on H.

(a) Show that the induced group operation is given by

$$(x, y) \cdot (z, w) = (yz + x, yw),$$

and that H, with this group operation, is a Lie group.

(b) Show that the derivative of the left translation map $L_{(x,y)} : H \to H$ at a point $(z, w) \in H$ is represented in the above coordinates by the matrix

$$\left(dL_{(x,y)} \right)_{(z,w)} = \begin{pmatrix} y & 0 \\ 0 & y \end{pmatrix}.$$

Conclude that the left-invariant vector field $X^V \in \mathfrak{X}(H)$ determined by the vector

$$V = \xi \frac{\partial}{\partial x} + \eta \frac{\partial}{\partial y} \in \mathfrak{h} \equiv T_{(0,1)}H \qquad (\xi, \eta \in \mathbb{R})$$

is given by

$$X^V_{(x,y)} = \xi y \frac{\partial}{\partial x} + \eta y \frac{\partial}{\partial y}.$$

(c) Given $V, W \in \mathfrak{h}$, compute $[V, W]$.

(d) Determine the flow of the vector field X^V, and give an expression for the exponential map $\exp : \mathfrak{h} \to H$.

(e) Confirm your results by first showing that H is the subgroup of $GL(2)$ formed by the matrices

$$\begin{pmatrix} y & x \\ 0 & 1 \end{pmatrix}$$

with $y > 0$.

(4) Consider the group

$$SL(2) = \left\{ \begin{pmatrix} a & b \\ c & d \end{pmatrix} \mid ad - bc = 1 \right\},$$

which we already know to be a 3-manifold. Making

$$a = p + q, d = p - q, b = r + s, c = r - s,$$

show that $SL(2)$ is diffeomorphic to $S^1 \times \mathbb{R}^2$.

(5) Give examples of matrices $A, B \in \mathfrak{gl}(2)$ such that $e^{A+B} \neq e^A e^B$.
(6) For $A \in \mathfrak{gl}(n)$, consider the differentiable map

$$h : \mathbb{R} \to \mathbb{R} \backslash \{0\}$$
$$t \mapsto \det e^{At}$$

and show that:

(a) this map is a group homomorphism between $(\mathbb{R}, +)$ and
     $(\mathbb{R} \backslash \{0\}, \cdot)$;
(b) $h'(0) = \operatorname{tr} A$;
(c) $\det(e^A) = e^{\operatorname{tr} A}$.

(7) (a) If $A \in \mathfrak{sl}(2)$, show that there is a $\lambda \in \mathbb{R} \cup i\mathbb{R}$ such that

$$e^A = \cosh \lambda \, I + \frac{\sinh \lambda}{\lambda} A.$$

(b) Show that $\exp : \mathfrak{sl}(2) \to SL(2)$ is not surjective.
(8) Consider the vector field $X \in \mathfrak{X}(\mathbb{R}^2)$ defined by

$$X = \sqrt{x^2 + y^2} \frac{\partial}{\partial x}.$$

(a) Show that the flow of X defines a free action of $\mathbb{R}$ on $M = \mathbb{R}^2 \backslash \{0\}$.
(b) Describe the topological quotient space $M/\mathbb{R}$. Is the action above proper?

(9) Let $M = S^2 \times S^2$ and consider the diagonal S^1-action on M given by

$$e^{i\theta} \cdot (u, v) = \left(e^{i\theta} \cdot u, e^{2i\theta} \cdot v \right),$$

where, for $u \in S^2 \subset \mathbb{R}^3$ and $e^{i\beta} \in S^1$, $e^{i\beta} \cdot u$ denotes the rotation of u by an angle β around the z-axis.

(a) Determine the fixed points for this action.
(b) What are the possible nontrivial stabilizers?

(10) Let G be a Lie group and H a closed Lie subgroup, i.e. a subgroup of G which is also a closed submanifold of G. Show that the action of H in G defined by $A(h, g) = h \cdot g$ is free and proper.
(11) (*Grassmannian*) Consider the set $H \subset GL(n)$ of invertible matrices of the form

$$\begin{pmatrix} A & 0 \\ C & B \end{pmatrix},$$

where $A \in GL(k)$, $B \in GL(n - k)$ and $C \in M_{(n-k) \times k}$.

(a) Show that H is a closed Lie subgroup of $GL(n)$. Therefore H acts freely and properly on $GL(n)$ [cf. Exercise 7.17(10)].

(b) Show that the quotient manifold

$$Gr(n, k) := GL(n)/H$$

can be identified with the set of k-dimensional subspaces of $\mathbb{R}^n$ (in particular $Gr(n, 1)$ is just the projective space $\mathbb{R}P^{n-1}$).

(c) The manifold $Gr(n, k)$ is called the **Grassmannian** of k-planes in $\mathbb{R}^n$. What is its dimension?

(12) Let G and H be Lie groups and $F : G \to H$ a Lie group homomorphism. Show that:

(a) $(dF)_e : \mathfrak{g} \to \mathfrak{h}$ is a Lie algebra homomorphism;

(b) if $(dF)_e$ is an isomorphism then F is a local diffeomorphism;

(c) if F is a surjective local diffeomorphism then F is a covering map.

(13) (a) Show that $\mathbb{R} \cdot SU(2)$ is a four-dimensional real linear subspace of $M_{2 \times 2}(\mathbb{C})$, closed under matrix multiplication, with basis

$$1 = \begin{pmatrix} 1 & 0 \\ 0 & 1 \end{pmatrix}, \quad i = \begin{pmatrix} i & 0 \\ 0 & -i \end{pmatrix},$$

$$j = \begin{pmatrix} 0 & 1 \\ -1 & 0 \end{pmatrix}, \quad k = \begin{pmatrix} 0 & i \\ i & 0 \end{pmatrix},$$

satisfying $i^2 = j^2 = k^2 = ijk = -1$. Therefore this space can be identified with the **quaternions** (cf. Sect. 1.10.1). Show that $SU(2)$ can be identified with the quaternions of Euclidean norm equal to 1, and is therefore diffeomorphic to S^3.

(b) Show that if $n \in \mathbb{R}^3$ is a unit vector, which we identify with a quaternion with zero real part, then

$$\exp\left(\frac{n\theta}{2}\right) = 1\cos\left(\frac{\theta}{2}\right) + n\sin\left(\frac{\theta}{2}\right)$$

is also a unit quaternion.

(c) Again identifying $\mathbb{R}^3$ with quaternions with zero real part, show that the map

$$\mathbb{R}^3 \to \mathbb{R}^3$$
$$v \mapsto \exp\left(\frac{n\theta}{2}\right) \cdot v \cdot \exp\left(-\frac{n\theta}{2}\right)$$

is a rotation by an angle θ about the axis defined by n.

(d) Show that there exists a surjective homomorphism $F : SU(2) \to SO(3)$, and use this to conclude that $SU(2)$ is the universal covering of $SO(3)$.

(e) What is the fundamental group of $SO(3)$?

1.8 Orientability

Let V be a finite dimensional vector space and consider two ordered bases $\beta = \{b_1, \ldots, b_n\}$ and $\beta' = \{b'_1, \ldots, b'_n\}$. There is a unique linear transformation $S : V \to V$ such that $b'_i = S b_i$ for every $i = 1, \ldots, n$. We say that the two bases are **equivalent** if $\det S > 0$. This defines an equivalence relation that divides the set of all ordered bases of V into two equivalence classes. An **orientation** for V is an assignment of a positive sign to the elements of one equivalence class and a negative sign to the elements of the other. The sign assigned to a basis is called its **orientation** and the basis is said to be **positively oriented** or **negatively oriented** according to its sign. It is clear that there are exactly two possible orientations for V.

Remark 8.1

(1) The ordering of the basis is very important. If we interchange the positions of two basis vectors we obtain a different ordered basis with the opposite orientation.

(2) An orientation for a zero-dimensional vector space is just an assignment of a sign $+1$ or -1.

(3) We call the **standard orientation** of $\mathbb{R}^n$ to the orientation that assigns a positive sign to the standard ordered basis.

An isomorphism $A : V \to W$ between two oriented vector spaces carries equivalent ordered bases of V to equivalent ordered bases of W. Hence, for any ordered basis β, the sign of the image $A\beta$ is either always the same as the sign of β or always the opposite. In the first case, the isomorphism A is said to be **orientation-preserving**, and in the latter it is called **orientation reversing**.

An **orientation** of a smooth manifold consists of a choice of orientations for all tangent spaces $T_p M$. If $\dim M = n \geq 1$, these orientations have to fit together smoothly, meaning that for each point $p \in M$ there exists a parameterization (U, φ) around p such that

$$(d\varphi)_x : \mathbb{R}^n \to T_{\varphi(x)} M$$

preserves the standard orientation of $\mathbb{R}^n$ at each point $x \in U$.

Remark 8.2 If the dimension of M is zero, an orientation is just an assignment of a sign ($+1$ or -1), called **orientation number**, to each point $p \in M$.

Definition 8.3 A smooth manifold M is said to be **orientable** if it admits an orientation.

Proposition 8.4 *If a smooth manifold M is connected and orientable then it admits precisely two orientations.*

Proof We will show that the set of points where two orientations agree and the set of points where they disagree are both open. Hence, one of them has to be M and the other the empty set. Let p be a point in M and let $(U_\alpha, \varphi_\alpha)$, (U_β, φ_β) be two parameterizations centered at p such that $d\varphi_\alpha$ is orientation-preserving for the first orientation and $d\varphi_\beta$ is orientation-preserving for the second. The map $\left(d(\varphi_\beta^{-1} \circ \varphi_\alpha)\right)_0 : \mathbb{R}^n \to \mathbb{R}^n$ is either orientation-preserving (if the two orientations agree at p) or reversing. In the first case, it has positive determinant at 0, and so, by continuity, $\left(d(\varphi_\beta^{-1} \circ \varphi_\alpha)\right)_x$ has positive determinant for x in a neighborhood of 0, implying that the two orientations agree in a neighborhood of p. Similarly, if $\left(d(\varphi_\beta^{-1} \circ \varphi_\alpha)\right)_0$ is orientation reversing, the determinant of $\left(d(\varphi_\beta^{-1} \circ \varphi_\alpha)\right)_x$ is negative in a neighborhood of 0, and so the two orientations disagree in a neighborhood of p.

Let O be an orientation for M (i.e. a smooth choice of an orientation O_p of T_pM for each $p \in M$), and $-O$ the opposite orientation (corresponding to taking the opposite orientation $-O_p$ at each tangent space T_pM). If O' is another orientation for M, then, for a given point $p \in M$, we know that O'_p agrees either with O_p or with $-O_p$ (because a vector space has just two possible orientations). Consequently, O' agrees with either O or $-O$ on M.   □

An alternative characterization of orientability is given by the following proposition.

Proposition 8.5 *A smooth manifold M is orientable if and only if there exists an atlas $\mathcal{A} = \{(U_\alpha, \varphi_\alpha)\}$ for which all the overlap maps $\varphi_\beta^{-1} \circ \varphi_\alpha$ are orientation-preserving.*

Proof Exercise 8.6(2)   □

An **oriented** manifold is an orientable manifold together with a choice of an orientation. A map $f : M \to N$ between two oriented manifolds with the same dimension is said to be **orientation-preserving** if $(df)_p$ is orientation-preserving at all points $p \in M$, and **orientation reversing** if $(df)_p$ is orientation reversing at all points $p \in M$.

Exercise 8.6

(1) Prove that the relation of "being equivalent" between ordered bases of a finite dimensional vector space described above is an equivalence relation.
(2) Show that a differentiable manifold M is orientable *iff* there exists an atlas $\mathcal{A} = \{(U_\alpha, \varphi_\alpha)\}$ for which all the overlap maps $\varphi_\beta^{-1} \circ \varphi_\alpha$ are orientation-preserving.

(3) (a) Show that if a manifold M is covered by two coordinate neighborhoods V_1 and V_2 such that $V_1 \cap V_2$ is connected, then M is orientable.

(b) Show that S^n is orientable.

(4) Let M be an oriented n-dimensional manifold and $c : I \to M$ a differentiable curve. A **smooth vector field along** c is a differentiable map $V : I \to TM$ such that $V(t) \in T_{c(t)}M$ for all $t \in I$ (cf. Sect. 3.2 in Chap. 3). Show that if $V_1, \ldots, V_n : I \to M$ are smooth vector fields along c such that $\{V_1(t), \ldots, V_n(t)\}$ is a basis of $T_{c(t)}M$ for all $t \in I$ then all these bases have the same orientation.

(5) We can see the Möbius band as the 2-dimensional submanifold of $\mathbb{R}^3$ given by the image of the immersion $g : (-1, 1) \times \mathbb{R} \to \mathbb{R}^3$ defined by

$$g(t, \varphi) = \left(\left(1 + t \cos\left(\frac{\varphi}{2}\right)\right) \cos\varphi, \left(1 + t \cos\left(\frac{\varphi}{2}\right)\right) \sin\varphi, t \sin\left(\frac{\varphi}{2}\right)\right).$$

Show that the Möbius band is not orientable.

(6) Let $f : M \to N$ be a diffeomorphism between two smooth manifolds. Show that M is orientable if and only if N is orientable. If, in addition, both manifolds are connected and oriented, and $(df)_p : T_pM \to T_{f(p)}N$ preserves orientation at one point $p \in M$, show that f is orientation-preserving.

(7) Let M and N be two oriented manifolds. We define an orientation on the product manifold $M \times N$ (called **product orientation**) in the following way: If $\alpha = \{a_1, \ldots, a_m\}$ and $\beta = \{b_1, \ldots, b_n\}$ are ordered bases of T_pM and T_qN, we consider the ordered basis $\{(a_1, 0), \ldots, (a_m, 0), (0, b_1), \ldots, (0, b_n)\}$ of $T_{(p,q)}(M \times N) \cong T_pM \times T_qN$. We then define an orientation on this space by setting the sign of this basis equal to the product of the signs of α and β. Show that this orientation does not depend on the choice of α and β.

(8) Show that the tangent bundle TM is always orientable, even if M is not.

(9) (*Orientable double covering*) Let M be a non-orientable n-dimensional manifold. For each point $p \in M$ we consider the set $\mathcal{O}_p$ of the (two) equivalence classes of bases of T_pM. Let $\overline{M}$ be the set

$$\overline{M} = \{(p, O_p) \mid p \in M, O_p \in \mathcal{O}_p\}.$$

Given a parameterization (U, φ) of M consider the maps $\overline{\varphi} : U \to \overline{M}$ defined by

$$\overline{\varphi}\left(x^1, \ldots, x^n\right) = \left(\varphi\left(x^1, \ldots, x^n\right), \left[\left(\frac{\partial}{\partial x^1}\right)_{\varphi(x)}, \ldots, \left(\frac{\partial}{\partial x^n}\right)_{\varphi(x)}\right]\right),$$

where $x = (x^1, \ldots, x^n) \in U$ and $\left[\left(\frac{\partial}{\partial x^1}\right)_{\varphi(x)}, \ldots, \left(\frac{\partial}{\partial x^n}\right)_{\varphi(x)}\right]$ represents the equivalence class of the basis $\left\{\left(\frac{\partial}{\partial x^1}\right)_{\varphi(x)}, \ldots, \left(\frac{\partial}{\partial x^n}\right)_{\varphi(x)}\right\}$ of $T_{\varphi(x)}M$.

(a) Show that these maps determine the structure of an orientable differentiable manifold of dimension n on $\overline{M}$.

(b) Consider the map $\pi : \overline{M} \to M$ defined by $\pi(p, O_p) = p$. Show that π is differentiable and surjective. Moreover, show that, for each $p \in M$, there exists a neighborhood V of p with $\pi^{-1}(V) = W_1 \cup W_2$, where W_1 and W_2 are two disjoint open subsets of $\overline{M}$, such that π restricted to W_i ($i = 1, 2$) is a diffeomorphism onto V.

(c) Show that $\overline{M}$ is connected ($\overline{M}$ is therefore called the **orientable double covering** of M).

(d) Let $\sigma : \overline{M} \to \overline{M}$ be the map defined by $\sigma(p, O_p) = (p, -O_p)$, where $-O_p$ represents the orientation of T_pM opposite to O_p. Show that σ is a diffeomorphism which reverses orientations satisfying $\pi \circ \sigma = \pi$ and $\sigma \circ \sigma = \mathrm{id}$.

(e) Show that any simply connected manifold is orientable.

1.9 Manifolds with Boundary

Let us consider again the **closed half space**

$$\mathbb{H}^n = \left\{ \left(x^1, \ldots, x^n \right) \in \mathbb{R}^n \mid x^n \geq 0 \right\}$$

with the topology induced by the usual topology of $\mathbb{R}^n$. Recall that a map $f : U \to \mathbb{R}^m$ defined on an open set $U \subset \mathbb{H}^n$ is said to be **differentiable** if it is the restriction to U of a differentiable map $\tilde{f}$ defined on an open subset of $\mathbb{R}^n$ containing U (cf. Sect. 1.10.2). In this case, the derivative $(df)_p$ is defined to be $(d\tilde{f})_p$. Note that this derivative is independent of the extension used since any two extensions have to agree on U.

Definition 9.1 A smooth n-**manifold with boundary** is a topological manifold with boundary of dimension n and a family of parameterizations $\varphi_\alpha : U_\alpha \subset \mathbb{H}^n \to M$ (that is, homeomorphisms of open sets U_α of $\mathbb{H}^n$ onto open sets of M), such that:

(i) the coordinate neighborhoods cover M, meaning that $\bigcup_\alpha \varphi_\alpha(U_\alpha) = M$;

(ii) for each pair of indices α, β such that

$$W := \varphi_\alpha(U_\alpha) \cap \varphi_\beta(U_\beta) \neq \varnothing,$$

the overlap maps

$$\varphi_\beta^{-1} \circ \varphi_\alpha : \varphi_\alpha^{-1}(W) \to \varphi_\beta^{-1}(W)$$
$$\varphi_\alpha^{-1} \circ \varphi_\beta : \varphi_\beta^{-1}(W) \to \varphi_\alpha^{-1}(W)$$

are smooth;

(iii) the family $\mathcal{A} = \{(U_\alpha, \varphi_\alpha)\}$ is maximal with respect to (i) and (ii), meaning that, if $\varphi_0 : U_0 \to M$ is a parameterization such that $\varphi_0 \circ \varphi^{-1}$ and $\varphi^{-1} \circ \varphi_0$ are C^∞ for all φ in $\mathcal{A}$, then φ_0 is in $\mathcal{A}$.

Recall that a point in M is said to be a **boundary point** if it is on the image of $\partial \mathbb{H}^n$ under some parameterization (that is, if there is a parameterization $\varphi : U \subset \mathbb{H}^n \to M$ such that $\varphi(x^1, \ldots, x^{n-1}, 0) = p$ for some $(x^1, \ldots, x^{n-1}) \in \mathbb{R}^{n-1}$), and that the set ∂M of all such points is called the **boundary** of M. Notice that differentiable manifolds are particular cases of differentiable manifolds with boundary, for which $\partial M = \varnothing$.

Proposition 9.2 *The boundary of a smooth n-manifold with boundary is a differentiable manifold of dimension $n - 1$.*

Proof Suppose that p is a boundary point of M (an n-manifold with boundary) and choose a parameterization $\varphi_\alpha : U_\alpha \subset \mathbb{H}^n \to M$ around p. Letting $V_\alpha := \varphi_\alpha(U_\alpha)$, we claim that $\varphi_\alpha(\partial U_\alpha) = \partial V_\alpha$, where $\partial U_\alpha = U_\alpha \cap \partial \mathbb{H}^n$ and $\partial V_\alpha = V_\alpha \cap \partial M$. By definition of boundary, we already know that $\varphi_\alpha(\partial U_\alpha) \subset \partial V_\alpha$, so we just have to show that $\partial V_\alpha \subset \varphi_\alpha(\partial U_\alpha)$. Let $q \in \partial V_\alpha$ and consider a parameterization $\varphi_\beta : U_\beta \to V_\alpha$ around q, mapping an open subset of $\mathbb{H}^n$ to an open subset of M and such that $q \in \varphi_\beta(\partial U_\beta)$. If we show that $\varphi_\beta(\partial U_\beta) \subset \varphi_\alpha(\partial U_\alpha)$ we are done. For that, we prove that $\left(\varphi_\alpha^{-1} \circ \varphi_\beta\right) (\partial U_\beta) \subset \partial U_\alpha$. Indeed, suppose that this map $\varphi_\alpha^{-1} \circ \varphi_\beta$ takes a point $x \in \partial U_\beta$ to an interior point (in $\mathbb{R}^n$) of U_α. As this map is a diffeomorphism, x would be an interior point (in $\mathbb{R}^n$) of U_β. This, of course, contradicts the assumption that $x \in \partial U_\beta$. Hence, $\left(\varphi_\alpha^{-1} \circ \varphi_\beta\right) (\partial U_\beta) \subset \partial U_\alpha$ and so $\varphi_\beta(\partial U_\beta) \subset \varphi_\alpha(\partial U_\alpha)$.

The map φ_α then restricts to a diffeomorphism from ∂U_α onto ∂V_α, where we identify ∂U_α with an open subset of $\mathbb{R}^{n-1}$. We obtain in this way a parameterization around p in ∂M, and it is easily seen that these parameterizations define a differentiable structure on ∂M. $\qquad\square$

Remark 9.3 In the above proof we saw that the definition of a boundary point does not depend on the parameterization chosen, meaning that, if there exists a parameterization around p such that p is an image of a point in $\partial \mathbb{H}^n$, then any parameterization around p maps a boundary point of $\mathbb{H}^n$ to p.

The definition of orientability can easily be extended to manifolds with boundary. We then have the following result.

Proposition 9.4 *Let M be an orientable manifold with boundary. Then ∂M is also orientable.*

Proof If M is orientable we can choose an atlas $\{(U_\alpha, \varphi_\alpha)\}$ on M for which the determinants of the derivatives of all overlap maps are positive. With this atlas we can obtain an atlas $\{(\partial U_\alpha, \widetilde{\varphi}_\alpha)\}$ for ∂M in the way described in the proof of Proposition 9.2. For any overlap map

$$\left(\varphi_\beta^{-1} \circ \varphi_\alpha\right) \left(x^1, \ldots, x^n\right) = \left(y^1\left(x^1, \ldots, x^n\right), \ldots, y^n\left(x^1, \ldots, x^n\right)\right)$$

we have

$$\left(\varphi_\beta^{-1} \circ \varphi_\alpha\right)\left(x^1, \ldots, x^{n-1}, 0\right) =$$
$$= \left(y^1\left(x^1, \ldots, x^{n-1}, 0\right), \ldots, y^{n-1}\left(x^1, \ldots, x^{n-1}, 0\right), 0\right)$$

and

$$\left(\widetilde{\varphi}_\beta^{-1} \circ \widetilde{\varphi}_\alpha\right)\left(x^1, \ldots, x^{n-1}\right) = \left(y^1\left(x^1, \ldots, x^{n-1}, 0\right), \ldots, y^{n-1}\left(x^1, \ldots, x^{n-1}, 0\right)\right).$$

Consequently, denoting $\left(x^1, \ldots, x^{n-1}, 0\right)$ by $(\widetilde{x}, 0)$,

$$\left(d\left(\varphi_\beta^{-1} \circ \varphi_\alpha\right)\right)_{(\widetilde{x},0)} = \left(\begin{array}{c|c} \left(d\left(\widetilde{\varphi}_\beta^{-1} \circ \widetilde{\varphi}_\alpha\right)\right)_{\widetilde{x}} & * \\ \hline 0 & \frac{\partial y^n}{\partial x^n}(\widetilde{x}, 0) \end{array} \right)$$

and so

$$\det\left(d\left(\varphi_\beta^{-1} \circ \varphi_\alpha\right)\right)_{(\widetilde{x},0)} = \frac{\partial y^n}{\partial x^n}(\widetilde{x}, 0) \det\left(d\left(\widetilde{\varphi}_\beta^{-1} \circ \widetilde{\varphi}_\alpha\right)\right)_{\widetilde{x}}.$$

However, fixing $x^1, \cdots, x^{n-1}$, we have that y^n is positive for positive values of x^n and is zero for $x^n = 0$. Consequently, $\frac{\partial y^n}{\partial x^n}(\widetilde{x}, 0) > 0$, and so

$$\det\left(d\left(\widetilde{\varphi}_\beta^{-1} \circ \widetilde{\varphi}_\alpha\right)\right)_{\widetilde{x}} > 0.$$

$\square$

Hence, choosing an orientation on a manifold with boundary M induces an orientation on the boundary ∂M. The convenient choice, called the **induced orientation**, can be obtained in the following way. For $p \in \partial M$ the tangent space $T_p(\partial M)$ is a subspace of $T_p M$ of codimension 1. As we have seen above, considering a coordinate system $x : W \to \mathbb{R}^n$ around p, we have $x^n(p) = 0$ and $\left(x^1, \ldots, x^{n-1}\right)$ is a coordinate system around p in ∂M. Setting $n_p := -\left(\frac{\partial}{\partial x^n}\right)_p$ (called an **outward pointing vector** at p), the induced orientation on ∂M is defined by assigning a positive sign to an ordered basis β of $T_p(\partial M)$ whenever the ordered basis $\{n_p, \beta\}$ of $T_p M$ is positive, and negative otherwise. Note that, since $\frac{\partial y^n}{\partial x^n}\left(\varphi^{-1}(p)\right) > 0$ (in the above notation), the sign of the last component of n_p does not depend on the choice of coordinate system. In general, the induced orientation is **not** the one obtained from the charts of M by simply dropping the last coordinate (in fact, it is $(-1)^n$ times this orientation).

Exercise 9.5

(1) Show with an example that the product of two manifolds with boundary is not always a manifold with boundary.
(2) Let M be a manifold without boundary and N a manifold with boundary. Show that the product $M \times N$ is a manifold with boundary. What is $\partial(M \times N)$?
(3) Show that a diffeomorphism between two manifolds with boundary M and N maps the boundary ∂M diffeomorphically onto ∂N.

1.10   Notes

1.10.1   Section 1.1

We begin by briefly reviewing the main concepts and results from general topology that we will need (see [Mun00] for a detailed exposition).

(1) A **topology** on a set M is a collection $\mathcal{T}$ of subsets of M having the following properties:

  (i)   the sets $\varnothing$ and M are in $\mathcal{T}$;
  (ii) the union of the elements of any sub-collection of $\mathcal{T}$ is in $\mathcal{T}$;
  (iii) the intersection of the elements of any finite sub-collection of $\mathcal{T}$ is in $\mathcal{T}$.

    A set M equipped with a topology $\mathcal{T}$ is called a **topological space**. We say that a subset $U \subset M$ is an **open set** of M if it belongs to the topology $\mathcal{T}$. A **neighborhood** of a point $p \in M$ is simply an open set $U \in \mathcal{T}$ containing p. A **closed set** $F \subset M$ is a set whose complement $M \setminus F$ is open. The **interior** int A of a subset $A \subset M$ is the largest open set contained in A, and its **closure** $\overline{A}$ is the smallest closed set containing A. Finally, the **subspace topology** on $A \subset M$ is $\mathcal{T}_A := \{U \cap A\}_{U \in \mathcal{T}}$.

(2) A topological space $(M, \mathcal{T})$ is said to be **Hausdorff** if for each pair of distinct points $p_1, p_2 \in M$ there exist neighborhoods U_1, U_2 of p_1 and p_2 such that $U_1 \cap U_2 = \varnothing$.

(3) A **basis** for a topology $\mathcal{T}$ on M is a collection $\mathcal{B} \subset \mathcal{T}$ such that, for each point $p \in M$ and each open set U containing p, there exists a basis element $B \in \mathcal{B}$ for which $p \in B \subset U$. If $\mathcal{B}$ is a basis for a topology $\mathcal{T}$ then any element of $\mathcal{T}$ is a union of elements of $\mathcal{B}$. A topological space $(M, \mathcal{T})$ is said to satisfy the **second countability axiom** if $\mathcal{T}$ has a countable basis.

(4) A map $f : M \to N$ between two topological spaces is said to be **continuous** if for each open set $U \subset N$ the preimage $f^{-1}(U)$ is an open subset of M. A bijection f is called a **homeomorphism** if both f and its inverse f^{-1} are continuous.

(5) An **open cover** for a topological space $(M, \mathcal{T})$ is a collection $\{U_\alpha\} \subset \mathcal{T}$ such that $\bigcup_\alpha U_\alpha = M$. A **subcover** is a sub-collection $\{V_\beta\} \subset \{U_\alpha\}$ which is still

an open cover. A topological space is said to be **compact** if every open cover
admits a finite subcover. A subset $A \subset M$ is said to be a **compact subset** if it
is a compact topological space for the subspace topology. It is easily seen that
continuous maps carry compact sets to compact sets.

(6) A topological space is said to be **connected** if the only subsets of M which
   are simultaneously open and closed are $\varnothing$ and M. A subset $A \subset M$ is said to
   be a **connected subset** if it is a connected topological space for the subspace
   topology. It is easily seen that continuous maps carry connected sets to connected
   sets.

(7) Let $(M, \mathcal{T})$ be a topological space. A sequence $\{p_n\}$ in M is said to **converge**
   to $p \in M$ if for each neighborhood V of p there exists an $N \in \mathbb{N}$ for which
   $p_n \in V$ for $n > N$. If $(M, \mathcal{T})$ is Hausdorff, then a convergent sequence has a
   unique limit. If in addition $(M, \mathcal{T})$ is second countable, then $F \subset M$ is closed
   if and only if every convergent sequence in F has limit in F, and $K \subset M$ is
   compact if and only if every sequence in K has a sublimit in K.

(8) If M and N are topological spaces, the set of all Cartesian products of open
   subsets of M by open subsets of N is a basis for a topology on $M \times N$, called
   the **product topology**. Note that with this topology the canonical projections are
   continuous maps.

(9) An **equivalence relation** $\sim$ on a set M is a relation with the following properties:

   (i) *reflexivity*: $p \sim p$ for every $p \in M$;
   (ii) *symmetry*: if $p \sim q$ then $q \sim p$;
   (iii) *transitivity*: if $p \sim q$ and $q \sim r$ then $p \sim r$.

   Given a point $p \in M$, we define the **equivalence class** of p as the set

   $$[p] = \{q \in M \mid q \sim p\}.$$

   Note that $p \in [p]$ by reflexivity. Whenever we have an equivalence relation $\sim$ on
   a set M, the corresponding set of equivalence classes is called the **quotient space**,
   and is denoted by $M/\!\!\sim$. There is a canonical projection $\pi : M \to M/\!\!\sim$, which
   maps each element of M to its equivalence class. If M is a topological space,
   we can define a topology on the quotient space (called the **quotient topology**)
   by letting a subset $V \subset M/\!\!\sim$ be open if and only if the set $\pi^{-1}(V)$ is open
   in M. The map π is then continuous for this topology. We will be interested in
   knowing whether some quotient spaces are Hausdorff. For that, the following
   definition will be helpful.

   Definition 10.1 An equivalence relation $\sim$ on a topological space M is called
   open if the map $\pi : M \to M/\!\!\sim$ is open, i.e. if for every open set $U \subset M$, the
   set $[U] := \pi(U)$ is open.

   We then have

   Proposition 10.2 *Let $\sim$ be an open equivalence relation on M and let $R = \{(p, q) \in M \times M \mid p \sim q\}$. Then the quotient space is Hausdorff if and only if R is closed in $M \times M$.*

Proof Assume that R is closed. Let $[p], [q] \in M/\sim$ with $[p] \neq [q]$. Then $p \nsim q$, and $(p, q) \notin R$. As R is closed, there are open sets U, V containing p, q, respectively, such that $(U \times V) \cap R = \varnothing$. This implies that $[U] \cap [V] = \varnothing$. In fact, if there were a point $[r] \in [U] \cap [V]$, then r would be equivalent to points $p' \in U$ and $q' \in V$ (that is $p' \sim r$ and $r \sim q'$). Therefore we would have $p' \sim q'$ (implying that $(p', q') \in R$), and so $(U \times V) \cap R$ would not be empty. Since $[U]$ and $[V]$ are open (as $\sim$ is an open equivalence relation), we conclude that $M/\sim$ is Hausdorff.

Conversely, let us assume that $M/\sim$ is Hausdorff. If $(p, q) \notin R$, then $p \nsim q$ and $[p] \neq [q]$, implying the existence of open sets $\widetilde{U}, \widetilde{V} \subset M/\sim$ containing $[p]$ and $[q]$, such that $\widetilde{U} \cap \widetilde{V} = \varnothing$. The sets $U := \pi^{-1}(\widetilde{U})$ and $V := \pi^{-1}(\widetilde{V})$ are open in M and $(U \times V) \cap R = \varnothing$. In fact, if that was not so, there would exist points $p' \in U$ and $q' \in V$ such that $p' \sim q'$. Then we would have $[p'] = [q']$, contradicting the fact that $\widetilde{U} \cap \widetilde{V} = \varnothing$ (as $[p'] \in \pi(U) = \widetilde{U}$ and $[q'] \in \pi(V) = \widetilde{V}$). Since $(p, q) \in U \times V \subset (M \times M) \setminus R$ and $U \times V$ is open, we conclude that $(M \times M) \setminus R$ is open, and hence R is closed. $\qquad\square$

1.10.2 Section 1.2

(1) Let us begin by reviewing some facts about differentiability of maps on $\mathbb{R}^n$. A function $f : U \to \mathbb{R}$ defined on an open subset U of $\mathbb{R}^n$ is said to be **continuously differentiable** on U if all partial derivatives $\frac{\partial f}{\partial x^1}, \ldots, \frac{\partial f}{\partial x^n}$ exist and are continuous on U. In this book, the words *differentiable* and *smooth* will be used to mean **infinitely differentiable**, that is, all partial derivatives $\frac{\partial^k f}{\partial x^{i_1} \ldots \partial x^{i_k}}$ exist and are continuous on U. Similarly, a map $f : U \to \mathbb{R}^m$, defined on an open subset of $\mathbb{R}^n$, is said to be **differentiable** or **smooth** if all coordinate functions f^i have the same property, that is, if they all possess continuous partial derivatives of all orders. If the map f is differentiable on U, its **derivative** at each point of U is the linear map $Df : \mathbb{R}^n \to \mathbb{R}^m$ represented in the canonical bases of $\mathbb{R}^n$ and $\mathbb{R}^m$ by the **Jacobian matrix**

$$Df = \begin{bmatrix} \frac{\partial f^1}{\partial x^1} & \cdots & \frac{\partial f^1}{\partial x^n} \\ \vdots & & \vdots \\ \frac{\partial f^m}{\partial x^1} & \cdots & \frac{\partial f^m}{\partial x^n} \end{bmatrix}.$$

A map $f : A \to \mathbb{R}^m$ defined on an arbitrary set $A \subset \mathbb{R}^n$ (not necessarily open) is said to be **differentiable** on A is there exists an open set $U \supset A$ and a differentiable map $\widetilde{f} : U \to \mathbb{R}^m$ such that $f = \widetilde{f}|_A$.

1.10.3 Section 1.4

(1) Let E, B and F be smooth manifolds and $\pi : E \to B$ a differentiable map. Then $\pi : E \to B$ is called a **fiber bundle** with **basis** B, **total space** E and **fiber** F if

   (i) the map π is surjective;
   (ii) there is a covering of B by open sets $\{U_\alpha\}$ and diffeomorphisms $\psi_\alpha : \pi^{-1}(U_\alpha) \to U_\alpha \times F$ such that for every $b \in U_\alpha$ we have $\psi_\alpha(\pi^{-1}(b)) = \{b\} \times F$.

1.10.4 Section 1.5

(1) (*The Inverse function theorem*) Let $f : U \subset \mathbb{R}^n \to \mathbb{R}^n$ be a smooth function and $p \in U$ such that $(df)_p$ is a linear isomorphism. Then there exists an open subset $V \subset U$ containing p such that $f|_V : V \to f(V)$ is a diffeomorphism. Moreover,

$$(d(f|_V)^{-1})_{f(q)} = ((d(f|_V))_q)^{-1}$$

for all $q \in V$.

1.10.5 Section 1.7

(1) A **group** is a set G equipped with a binary operation $\cdot : G \times G \to G$ satisfying:

   (i) **Associativity:** $g_1 \cdot (g_2 \cdot g_3) = (g_1 \cdot g_2) \cdot g_3$ for all $g_1, g_2, g_3 \in G$;
   (ii) **Existence of identity:** There exists an element $e \in G$ such that $e \cdot g = g \cdot e = g$ for all $g \in G$;
   (iii) **Existence of inverses:** For all $g \in G$ there exists $g^{-1} \in G$ such that $g \cdot g^{-1} = g^{-1} \cdot g = e$.

If the group operation is commutative, meaning that $g_1 \cdot g_2 = g_2 \cdot g_1$ for all $g_1, g_2 \in G$, the group is said to be **abelian**. A subset $H \subset G$ is said to be a **subgroup** of G if the restriction of $\cdot$ to $H \times H$ is a binary operation on H, and H with this operation is a group. A subgroup $H \subset G$ is said to be **normal** if $ghg^{-1} \in H$ for all $g \in G$, $h \in H$. A map $f : G \to H$ between two groups G and H is said to be a **group homomorphism** if $f(g_1 \cdot g_2) = f(g_1) \cdot f(g_2)$ for all $g_1, g_2 \in G$. An **isomorphism** is a bijective homomorphism. The **kernel** of a group homomorphism $f : G \to H$ is the subset $\ker(f) = \{g \in G \mid f(g) = e\}$, and is easily seen to be a normal subgroup of G.

(2) Let M and N be topological manifolds. A map $f : M \to N$ is called **proper** if the preimage $f^{-1}(K)$ of any compact set $K \subset N$ is compact. If f is also continuous then f is **closed**, i.e. f maps closed sets to closed sets. To see this,

let $F \subset M$ be a closed set, and consider a convergent sequence $\{q_n\}$ in $f(F)$ with $q_n \to q$. It is easily seen that the closure K of the set $\{q_n \mid n \in \mathbb{N}\}$ is compact, and since f is proper, then so is $f^{-1}(K)$. For each $n \in \mathbb{N}$ choose $p_n \in F$ such that $f(p_n) = q_n$. Then $p_n \in f^{-1}(K)$, and so $\{p_n\}$ must have a sublimit $p \in F$ (since F is closed). If $\{p_{n_k}\}$ is a subsequence which converges to p we have $q_{n_k} = f(p_{n_k}) \to f(p)$ (because f is continuous). Therefore $q = f(p) \in f(F)$, and $f(F)$ is closed.

(3) Let $f, g : X \to Y$ be two continuous maps between topological spaces and let $I = [0, 1]$. We say that f is **homotopic** to g if there exists a continuous map $H : I \times X \to Y$ such that $H(0, x) = f(x)$ and $H(1, x) = g(x)$ for every $x \in X$. This map is called a **homotopy**.

Homotopy of maps forms an equivalence relation in the set of continuous maps between X and Y. As an application, let us fix a **base point** p on a manifold M and consider the homotopy classes of continuous maps $f : I \to M$ such that $f(0) = f(1) = p$ (these maps are called **loops** based at p), with the additional restriction that $H(t, 0) = H(t, 1) = p$ for all $t \in I$. This set of homotopy classes is called the **fundamental group** of M relative to the base point p, and is denoted by $\pi_1(M, p)$. Among its elements there is the class of the **constant loop based at** p, given by $f(t) = p$ for every $t \in I$. Note that the set $\pi_1(M, p)$ is indeed a group with operation $*$ (**composition** of loops) defined by $[f] * [g] := [h]$, where $h : I \to M$ is given by

$$h(t) = \begin{cases} f(2t) & \text{if } t \in [0, \frac{1}{2}] \\ g(2t - 1) & \text{if } t \in [\frac{1}{2}, 1] \end{cases}.$$

The identity element of this group is the equivalence class of the constant loop based at p.

If M is connected and this is the only class in $\pi_1(M, p)$, M is said to be **simply connected**. This means that every loop through p can be continuously deformed to the constant loop. This property does not depend on the choice of point p, and is equivalent to the condition that any closed path may be continuously deformed to a constant loop in M.

(4) **Quaternions** are a generalization of the complex numbers introduced by Hamilton in 1843, when he considered numbers of the form $a + bi + cj + dk$ with $a, b, c, d \in \mathbb{R}$ and
$$i^2 = j^2 = k^2 = ijk = -1.$$

Formally, the set $\mathbb{H}$ of quaternions is simply $\mathbb{R}^4$ with

$$1 = (1, 0, 0, 0)$$
$$i = (0, 1, 0, 0)$$
$$j = (0, 0, 1, 0)$$

$$k = (0, 0, 0, 1)$$

and the bilinear associative product defined by the Hamilton formulas (and the
assumption that 1 is the identity). With these definitions, $\mathbb{H}$ is a division ring, that
is, $(\mathbb{H} \setminus \{0\}, \cdot)$ is a (non-commutative) group and multiplication is distributive
with respect to addition.

The **real part** of a quaternion $a + bi + cj + ik$ is a, whereas its **vector part** is
$bi + cj + dk$. Quaternions with zero vector part are identified with real numbers,
while quaternions with zero real part are identified with vectors in $\mathbb{R}^3$. The **norm**
of a quaternion is the usual Euclidean norm.

1.10.6 Bibliographical Notes

The material in this chapter is completely standard, and can be found in almost
any book on differential geometry (e.g. [Boo03, dC93, GHL04]). Immersions and
embeddings are the starting point of **differential topology**, which is studied in [GP73,
Mil97]. Lie groups and Lie algebras are a huge field of mathematics, to which we
could not do justice. See for instance [BtD03, DK99, War83]. More details on the
fundamental group and covering spaces can be found for instance in [Mun00].

References

[Arn92]   Arnold, V.I.: Ordinary Differential Equations. Springer, Berlin (1992)
[Blo96]   Bloch, E.: A First Course in Geometric Topology and Differential Geometry. Birkäuser,
          Boston (1996)
[Boo03]   Boothby, W.: An Introduction to Differentiable Manifolds and Riemannian Geometry.
          Academic Press, Orlando (2003)
[BtD03]   Bröcker, T., tom Diek, T.: Representations of Compact Lie Groups. Springer, Berlin
          (2003)
[dC93]   do Carmo, M.: Riemannian Geometry. Birkhäuser, Boston (1993)
[DK99]   Duistermaat, J., Kolk, J.: Lie Groups. Springer, Berlin (1999)
[Fre82]   Freedman, M.: The topology of four-dimensional manifolds. J. Differ. Geom. **17**, 357–
          453 (1982)
[GHL04]   Gallot, S., Hulin, D., Lafontaine, J.: Riemannian Geometry. Springer, Berlin (2004)
[GP73]   Guillemin, V., Pollack, A.: Differential Topology. Prentice-Hall, Englewood Cliffs
          (1973)
[Gom83]   Gompf, R.: Three exotic $\mathbb{R}^4$'s and other anomalies. J. Differ. Geom. **18**, 317–328 (1983)
[Ker60]   Kervaire, M.: A manifold wich does not admit any differentiable structure. Comment.
          Math. Helv. **34**, 257–270 (1960)
[KM63]   Kervaire, M., Milnor, J.: Groups of homotopy spheres. I. Ann. Math. **77**(2), 504–537
          (1963)
[Mil56]   Milnor, J.: On manifolds homeomorphic to the 7-sphere. Ann. Math. **64**, 399–405 (1956)
[Mil59]   Milnor, J.: Differentiable structures on spheres. Amer. J. Math. **81**, 962–972 (1959)

[Mil97]   Milnor, J.: 1413 Topology from the Differentiable Viewpoint. Princeton University Press, Princeton (1997)

[Mil07]   Milnor, J.: On the relationship between differentiable manifolds and combinatorial manifolds, Collected papers. III. Differential topology. pp. 19–28, American Mathematical Society, Providence (2007)

[Mun00]   Munkres, J.: Topology. Prentice-Hall, Upper Saddle River (2000)

[Nov65]   Novikov, S.P.: Topological invariance of rational Pontrjagin classes. Soviet Math. Dokl. **6**, 921–923 (1965)

[Sma60]   Smale, S.: The generalized Poincaré conjecture in higher dimensions. Bull. AMS **66**, 373–375 (1960)

[War83]   Warner, F.: Foundations of Differentiable Manifolds and Lie Groups. Springer, Berlin (1983)

[Whi44a]   Whitney, H.: The selfintersections of a smooth n-manifold in $(2n-1)$-space. Ann. Math. **45**, 247–293 (1944)

[Whi44b]   Whitney, H.: The selfintersections of a smooth n-manifold in $2n$-space. Ann. Math. **45**, 220–246 (1944)

Chapter 2
Differential Forms

This chapter discusses integration on differentiable manifolds. Because there is no canonical choice of local coordinates, there is no natural notion of volume, and so only objects with appropriate transformation properties under coordinate changes can be integrated. These objects, called **differential forms**, were introduced by Élie Cartan in 1899; they come equipped with natural algebraic and differential operations, making them a fundamental tool of differential geometry.

Besides their role in integration, differential forms occur in many other places in differential geometry and physics: for instance, they can be used as a very efficient device for computing the curvature of Riemannian (Chap. 4) or Lorentzian (Chap. 6) manifolds; to formulate Hamiltonian mechanics (Chap. 5); or to write Maxwell's equations of electromagnetism in a compact and elegant form.

The algebraic structure of differential forms is set up in Sect. 2.1, which reviews the notions of **tensors** and **tensor product**, and introduces **alternating tensors** and their **exterior product**.

Tensor fields, which are natural generalizations of vector fields, are discussed in Sect. 2.2, where a new operation, the **pull-back** of a covariant tensor field by a smooth map, is defined. Differential forms are introduced in Sect. 2.3 as fields of alternating tensors, along with their **exterior derivative**. Important ideas which will not be central to the remainder of this book, such as the **Poincaré lemma**, **de Rham cohomology** or the **Lie derivative**, are discussed in the exercises.

The **integral** of a differential form on a smooth manifold in defined in Sect. 2.4. This makes use of another basic tool of differential geometry, namely the existence of **partitions of unity**.

The celebrated **Stokes theorem**, generalizing the fundamental theorems of vector calculus (Green's theorem, the divergence theorem and the classical Stokes theorem for vector fields) is proved in Sect. 2.5. Some of its consequences, such as **invariance by homotopy** of the integral of closed forms, or **Brouwer's fixed point theorem**, are explored in the exercises.

Finally, Sect. 2.6 studies the relation between orientability and the existence of special differential forms, called **volume forms**, which can be used to define a notion of volume on orientable manifolds.

© Springer International Publishing Switzerland 2014
L. Godinho and J. Natário, *An Introduction to Riemannian Geometry*, Universitext,
DOI 10.1007/978-3-319-08666-8_2

2.1 Tensors

Let V be an n-dimensional vector space. A k-**tensor** on V is a real multilinear function (meaning linear in each variable) defined on the product $V \times \cdots \times V$ of k copies of V. The set of all k-tensors is itself a vector space and is usually denoted by $\mathcal{T}^k(V^*)$.

Example 1.1

(1) The space of 1-tensors $\mathcal{T}^1(V^*)$ is equal to V^*, the **dual space** of V, that is, the space of real-valued linear functions on V.
(2) The usual inner product on $\mathbb{R}^n$ is an example of a 2-tensor.
(3) The determinant is an n-tensor on $\mathbb{R}^n$.

Given a k-tensor T and an m-tensor S, we define their **tensor product** as the $(k+m)$-tensor $T \otimes S$ given by

$$T \otimes S(v_1, \ldots, v_k, v_{k+1}, \ldots, v_{k+m}) := T(v_1, \ldots, v_k) \cdot S(v_{k+1}, \ldots, v_{k+m}).$$

This operation is bilinear and associative, but not commutative [cf. Exercise 1.15(1)].

Proposition 1.2 *If* $\{T_1, \ldots, T_n\}$ *is a basis for* $\mathcal{T}^1(V^*) = V^*$ *(the dual space of V),* *then the set* $\{T_{i_1} \otimes \cdots \otimes T_{i_k} \mid 1 \leq i_1, \ldots, i_k \leq n\}$ *is a basis of* $\mathcal{T}^k(V^*)$, *and therefore* $\dim \mathcal{T}^k(V^*) = n^k$.

Proof We will first show that the elements of this set are linearly independent. If

$$T := \sum_{i_1, \ldots, i_k} a_{i_1 \ldots i_k} T_{i_1} \otimes \cdots \otimes T_{i_k} = 0,$$

then, taking the basis $\{v_1, \ldots, v_n\}$ of V dual to $\{T_1, \ldots, T_n\}$, meaning that $T_i(v_j) = \delta_{ij}$ (cf. Sect. 2.7.1), we have $T(v_{j_1}, \ldots, v_{j_k}) = a_{j_1 \ldots j_k} = 0$ for every $1 \leq j_1, \ldots, j_k \leq n$.

To show that $\{T_{i_1} \otimes \cdots \otimes T_{i_k} \mid 1 \leq i_1, \ldots, i_k \leq n\}$ spans $\mathcal{T}^k(V^*)$, we take any element $T \in \mathcal{T}^k(V^*)$ and consider the k-tensor S defined by

$$S := \sum_{i_1, \ldots, i_k} T(v_{i_1}, \ldots, v_{i_k}) T_{i_1} \otimes \cdots \otimes T_{i_k}.$$

Clearly, $S(v_{i_1}, \ldots, v_{i_k}) = T(v_{i_1}, \ldots, v_{i_k})$ for every $1 \leq i_1, \ldots, i_k \leq n$, and so, by linearity, $S = T$. $\qquad\square$

If we consider k-tensors on V^*, instead of V, we obtain the space $\mathcal{T}^k(V)$ (note that $(V^*)^* = V$, as shown in Sect. 2.7.1). These tensors are called **contravariant tensors** on V, while the elements of $\mathcal{T}^k(V^*)$ are called **covariant tensors** on V. Note that the contravariant tensors on V are the covariant tensors on V^*. The words covariant and contravariant are related to the transformation behavior of the tensor components under a change of basis in V, as explained in Sect. 2.7.1.

We can also consider **mixed** (k, m)-tensors on V, that is, multilinear functions defined on the product $V \times \cdots \times V \times V^* \times \cdots \times V^*$ of k copies of V and m copies of V^*. A (k, m)-tensor is then k times covariant and m times contravariant on V. The space of all (k, m)-tensors on V is denoted by $T^{k,m}(V^*, V)$.

Remark 1.3

(1) We can identify the space $T^{1,1}(V^*, V)$ with the space of linear maps from V to V. Indeed, for each element $T \in T^{1,1}(V^*, V)$, we define the linear map from V to V, given by $v \mapsto T(v, \cdot)$. Note that $T(v, \cdot) : V^* \to \mathbb{R}$ is a linear function on V^*, that is, an element of $(V^*)^* = V$.

(2) Generalizing the above definition of tensor product to tensors defined on different vector spaces, we can define the spaces $T^k(V^*) \otimes T^m(W^*)$ generated by the tensor products of elements of $T^k(V^*)$ by elements of $T^m(W^*)$. Note that $T^{k,m}(V^*, V) = T^k(V^*) \otimes T^m(V)$. We leave it as an exercise to find a basis for this space.

A tensor is called **alternating** if, like the determinant, it changes sign every time two of its variables are interchanged, that is, if

$$T(v_1, \ldots, v_i, \ldots, v_j, \ldots, v_k) = -T(v_1, \ldots, v_j, \ldots, v_i, \ldots, v_k).$$

The space of all alternating k-tensors is a vector subspace $\Lambda^k(V^*)$ of $T^k(V^*)$. Note that, for any alternating k-tensor T, we have $T(v_1, \ldots, v_k) = 0$ if $v_i = v_j$ for some $i \neq j$.

Example 1.4

(1) All 1-tensors are trivially alternating, that is, $\Lambda^1(V^*) = T^1(V^*) = V^*$.
(2) The determinant is an alternating n-tensor on $\mathbb{R}^n$.

Consider now S_k, the group of all possible permutations of $\{1, \ldots, k\}$. If $\sigma \in S_k$, we set $\sigma(v_1, \ldots, v_k) = (v_{\sigma(1)}, \ldots, v_{\sigma(k)})$. Given a k-tensor $T \in T^k(V^*)$ we can define a new alternating k-tensor, called $\mathrm{Alt}(T)$, in the following way:

$$\mathrm{Alt}(T) := \frac{1}{k!} \sum_{\sigma \in S_k} (\mathrm{sgn}\, \sigma)\, (T \circ \sigma),$$

where $\mathrm{sgn}\, \sigma$ is $+1$ or -1 according to whether σ is an even or an odd permutation. We leave it as an exercise to show that $\mathrm{Alt}(T)$ is in fact alternating.

Example 1.5 If $T \in T^3(V^*)$,

$$\mathrm{Alt}(T)(v_1, v_2, v_3) = \tfrac{1}{6} \left(T(v_1, v_2, v_3) + T(v_3, v_1, v_2) + T(v_2, v_3, v_1) \right.$$
$$\left. - T(v_1, v_3, v_2) - T(v_2, v_1, v_3) - T(v_3, v_2, v_1) \right).$$

We will now define the **wedge product** between alternating tensors: if $T \in \Lambda^k(V^*)$ and $S \in \Lambda^m(V^*)$, then $T \wedge S \in \Lambda^{k+m}(V^*)$ is given by

$$T \wedge S := \frac{(k+m)!}{k!\,m!}\,\mathrm{Alt}(T \otimes S).$$

Example 1.6 If $T, S \in \Lambda^1(V^*) = V^*$, then

$$T \wedge S = 2\,\mathrm{Alt}(T \otimes S) = T \otimes S - S \otimes T,$$

implying that $T \wedge S = -S \wedge T$ and $T \wedge T = 0$.

It is easy to verify that this product is bilinear. To prove associativity we need the following proposition.

Proposition 1.7

(i) Let $T \in T^k(V^*)$ and $S \in T^m(V^*)$. If $\mathrm{Alt}(T) = 0$ then

$$\mathrm{Alt}(T \otimes S) = \mathrm{Alt}(S \otimes T) = 0;$$

(ii) $\mathrm{Alt}(\mathrm{Alt}(T \otimes S) \otimes R) = \mathrm{Alt}(T \otimes S \otimes R) = \mathrm{Alt}(T \otimes \mathrm{Alt}(S \otimes R))$.

Proof

(i) Let us consider

$$(k+m)!\ \mathrm{Alt}(T \otimes S)(v_1, \ldots, v_{k+m}) =$$
$$= \sum_{\sigma \in S_{k+m}} (\mathrm{sgn}\,\sigma)\,T(v_{\sigma(1)}, \ldots, v_{\sigma(k)})S(v_{\sigma(k+1)}, \ldots, v_{\sigma(k+m)}).$$

Taking the subgroup G of S_{k+m} formed by the permutations of $\{1, \ldots, k+m\}$ that leave $k+1, \ldots, k+m$ fixed, we have

$$\sum_{\sigma \in G} (\mathrm{sgn}\,\sigma)T(v_{\sigma(1)}, \ldots, v_{\sigma(k)})S(v_{\sigma(k+1)}, \ldots, v_{\sigma(k+m)}) =$$
$$= \left(\sum_{\sigma \in G} (\mathrm{sgn}\,\sigma)T(v_{\sigma(1)}, \ldots, v_{\sigma(k)}) \right) S(v_{k+1}, \ldots, v_{k+m})$$
$$= k!\ (\mathrm{Alt}(T) \otimes S)(v_1, \ldots, v_{k+m}) = 0.$$

Then, since G decomposes S_{k+m} into disjoint right cosets $G \cdot \tilde{\sigma} := \{\sigma\tilde{\sigma} \mid \sigma \in G\}$, and for each coset

$$\sum_{\sigma \in G \cdot \widetilde{\sigma}} (\operatorname{sgn} \sigma)(T \otimes S)(v_{\sigma(1)}, \ldots, v_{\sigma(k+m)}) =$$

$$= (\operatorname{sgn} \widetilde{\sigma}) \sum_{\sigma \in G} (\operatorname{sgn} \sigma) \, (T \otimes S)(v_{\sigma(\widetilde{\sigma}(1))}, \ldots, v_{\sigma(\widetilde{\sigma}(k+m))})$$

$$= (\operatorname{sgn} \widetilde{\sigma}) k! \, (\operatorname{Alt}(T) \otimes S)(v_{\widetilde{\sigma}(1)}, \ldots, v_{\widetilde{\sigma}(k+m)}) = 0,$$

we have that $\operatorname{Alt}(T \otimes S) = 0$. Similarly, we prove that $\operatorname{Alt}(S \otimes T) = 0$.

(ii) By linearity of the operator Alt and the fact that $\operatorname{Alt} \circ \operatorname{Alt} = \operatorname{Alt}$ [cf. Exercise 1.15(3)], we have

$$\operatorname{Alt}(\operatorname{Alt}(S \otimes R) - S \otimes R) = 0.$$

Hence, by (i),

$$0 = \operatorname{Alt}(T \otimes (\operatorname{Alt}(S \otimes R) - S \otimes R))$$
$$= \operatorname{Alt}(T \otimes \operatorname{Alt}(S \otimes R)) - \operatorname{Alt}(T \otimes S \otimes R),$$

and the result follows.   □

Using these properties we can show the following.

Proposition 1.8 $(T \wedge S) \wedge R = T \wedge (S \wedge R)$.

Proof By Proposition 1.7, for $T \in \Lambda^k(V^*)$, $S \in \Lambda^m(V^*)$ and $R \in \Lambda^l(V^*)$, we have

$$(T \wedge S) \wedge R = \frac{(k+m+l)!}{(k+m)! \, l!} \operatorname{Alt}((T \wedge S) \otimes R)$$
$$= \frac{(k+m+l)!}{k! \, m! \, l!} \operatorname{Alt}(T \otimes S \otimes R)$$

and

$$T \wedge (S \wedge R) = \frac{(k+m+l)!}{k! \, (m+l)!} \operatorname{Alt}(T \otimes (S \wedge R))$$
$$= \frac{(k+m+l)!}{k! \, m! \, l!} \operatorname{Alt}(T \otimes S \otimes R).$$ □

We can now prove the following theorem.

Theorem 1.9 *If* $\{T_1, \ldots, T_n\}$ *is a basis for* V^*, *then the set*

$$\{T_{i_1} \wedge \cdots \wedge T_{i_k} \mid 1 \le i_1 < \ldots < i_k \le n\}$$

is a basis for $\Lambda^k(V^*)$, and

$$\dim \Lambda^k(V^*) = \binom{n}{k} = \frac{n!}{k!(n-k)!}.$$

Proof Let $T \in \Lambda^k(V^*) \subset \mathcal{T}^k(V^*)$. By Proposition 1.2,

$$T = \sum_{i_1,\ldots,i_k} a_{i_1\ldots i_k} T_{i_1} \otimes \cdots \otimes T_{i_k}$$

and, since T is alternating,

$$T = \text{Alt}(T) = \sum_{i_1,\ldots,i_k} a_{i_1\cdots i_k} \text{Alt}(T_{i_1} \otimes \cdots \otimes T_{i_k}).$$

We can show by induction that $\text{Alt}(T_{i_1} \otimes \cdots \otimes T_{i_k}) = \frac{1}{k!} T_{i_1} \wedge T_{i_2} \wedge \cdots \wedge T_{i_k}$. Indeed, for $k = 1$, the result is trivially true, and, assuming it is true for k basis tensors, we have, by Proposition 1.7, that

$$\begin{aligned}
\text{Alt}(T_{i_1} \otimes \cdots \otimes T_{i_{k+1}}) &= \text{Alt}(\text{Alt}(T_{i_1} \otimes \cdots \otimes T_{i_k}) \otimes T_{i_{k+1}}) \\
&= \frac{k!}{(k+1)!} \text{Alt}(T_{i_1} \otimes \cdots \otimes T_{i_k}) \wedge T_{i_{k+1}} \\
&= \frac{1}{(k+1)!} T_{i_1} \wedge T_{i_2} \wedge \cdots \wedge T_{i_{k+1}}.
\end{aligned}$$

Hence,

$$T = \frac{1}{k!} \sum_{i_1,\ldots,i_k} a_{i_1\ldots i_k} T_{i_1} \wedge T_{i_2} \wedge \cdots \wedge T_{i_k}.$$

However, the tensors $T_{i_1} \wedge \cdots \wedge T_{i_k}$ are not linearly independent. Indeed, due to anticommutativity, if two sequences $(i_1, \ldots i_k)$ and $(j_1, \ldots j_k)$ differ only in their orderings, then $T_{i_1} \wedge \cdots \wedge T_{i_k} = \pm T_{j_1} \wedge \cdots \wedge T_{j_k}$. In addition, if any two of the indices are equal, then $T_{i_1} \wedge \cdots \wedge T_{i_k} = 0$. Hence, we can avoid repeating terms by considering only increasing index sequences:

$$T = \sum_{i_1 < \cdots < i_k} b_{i_1\ldots i_k} T_{i_1} \wedge \cdots \wedge T_{i_k}$$

and so the set $\{T_{i_1} \wedge \cdots \wedge T_{i_k} \mid 1 \leq i_1 < \ldots < i_k \leq n\}$ spans $\Lambda^k(V^*)$. Moreover, the elements of this set are linearly independent. Indeed, if

$$0 = T = \sum_{i_1 < \cdots < i_k} b_{i_1\ldots i_k} T_{i_1} \wedge \cdots \wedge T_{i_k},$$

then, taking a basis $\{v_1, \ldots, v_n\}$ of V dual to $\{T_1, \ldots, T_n\}$ and an increasing index sequence $(j_1, \ldots, j_k)$, we have

$$0 = T(v_{j_1}, \ldots, v_{j_k}) = k! \sum_{i_1 < \cdots < i_k} b_{i_1 \ldots i_k} \, \mathrm{Alt}(T_{i_1} \otimes \cdots \otimes T_{i_k})(v_{j_1}, \ldots, v_{j_k})$$

$$= \sum_{i_1 < \cdots < i_k} b_{i_1 \ldots i_k} \sum_{\sigma \in S_k} (\mathrm{sgn}\,\sigma) \, T_{i_1}(v_{j_{\sigma(1)}}) \cdots T_{i_k}(v_{j_{\sigma(k)}}).$$

Since $(i_1, \ldots, i_k)$ and $(j_1, \ldots, j_k)$ are both increasing, the only term of the second sum that may be different from zero is the one for which $\sigma = \mathrm{id}$. Consequently,

$$0 = T(v_{j_1}, \ldots, v_{j_k}) = b_{j_1 \ldots j_k}.$$

$\square$

The following result is clear from the anticommutativity shown in Example 1.6.

Proposition 1.10 *If $T \in \Lambda^k(V^*)$ and $S \in \Lambda^m(V^*)$, then*

$$T \wedge S = (-1)^{km} S \wedge T.$$

Proof Exercise 1.15(4) $\square$

Remark 1.11

(1) Another consequence of Theorem 1.9 is that $\dim(\Lambda^n(V^*)) = 1$. Hence, if $V = \mathbb{R}^n$, any alternating n-tensor in $\mathbb{R}^n$ is a multiple of the determinant.
(2) It is also clear that $\Lambda^k(V^*) = 0$ if $k > n$. Moreover, the set $\Lambda^0(V^*)$ is defined to be equal to $\mathbb{R}$ (identified with the set of constant functions on V).

A linear transformation $F : V \to W$ induces a linear transformation $F^* : T^k(W^*) \to T^k(V^*)$ defined by

$$(F^*T)(v_1, \ldots, v_k) = T(F(v_1), \ldots, F(v_k)).$$

This map has the following properties.

Proposition 1.12 *Let V, W, Z be vector spaces, let $F : V \to W$ and $H : W \to Z$ be linear maps, and let $T \in T^k(W^*)$ and $S \in T^m(W^*)$. We have:*

(1) $F^*(T \otimes S) = (F^*T) \otimes (F^*S)$;
(2) If T is alternating then so is F^*T;
(3) $F^*(T \wedge S) = (F^*T) \wedge (F^*S)$;
(4) $(F \circ H)^* = H^* \circ F^*$.

Proof Exercise 1.15(5)   □

Another important fact about alternating tensors is the following.

Theorem 1.13 *Let* $F : V \to V$ *be a linear map and let* $T \in \Lambda^n(V^*)$. *Then* $F^*T = (\det A)T$, *where* A *is any matrix representing* F.

Proof As $\Lambda^n(V^*)$ is 1-dimensional and F^* is a linear map, F^* is just multiplication by some constant C. Let us consider an isomorphism H between V and $\mathbb{R}^n$. Then, H^* det is an alternating n-tensor in V, and so F^*H^* det $= CH^*$ det. Hence

$$(H^{-1})^* F^* H^* \det = C \det \Leftrightarrow (H \circ F \circ H^{-1})^* \det = C \det \Leftrightarrow A^* \det = C \det,$$

where A is the matrix representation of F induced by H. Taking the standard basis in $\mathbb{R}^n$, $\{e_1, \ldots, e_n\}$, we have

$$A^* \det (e_1, \ldots, e_n) = C \det(e_1, \ldots, e_n) = C,$$

and so

$$\det (Ae_1, \ldots, Ae_n) = C,$$

implying that $C = \det A$.   □

Remark 1.14 By the above theorem, if $T \in \Lambda^n(V^*)$ and $T \neq 0$, then two ordered basis $\{v_1, \ldots, v_n\}$ and $\{w_1, \ldots, w_n\}$ are equivalently oriented if and only if $T(v_1, \ldots, v_n)$ and $T(w_1, \ldots, w_n)$ have the same sign.

Exercise 1.15

(1) Show that the tensor product is bilinear and associative but not commutative.
(2) Find a basis for the space $T^{k,m}(V^*, V)$ of mixed (k, m)-tensors.
(3) If $T \in T^k(V^*)$, show that

   (a) $\text{Alt}(T)$ is an alternating tensor;
   (b) if T is alternating then $\text{Alt}(T) = T$;
   (c) $\text{Alt}(\text{Alt}(T)) = \text{Alt}(T)$.

(4) Prove Proposition 1.10.
(5) Prove Proposition 1.12.
(6) Let $T_1, \ldots, T_k \in V^*$. Show that

$$(T_1 \wedge \cdots \wedge T_k)(v_1, \ldots, v_k) = \det [T_i(v_j)].$$

(7) Show that Let $T_1, \ldots, T_k \in \Lambda^1(V^*) = V^*$ are linearly independent if and only if $T_1 \wedge \cdots \wedge T_k \neq 0$.

(8) Let $T \in \Lambda^k(V^*)$ and let $v \in V$. We define **contraction** of T by v, $\iota(v)T$, as the $(k-1)$-tensor given by

$$(\iota(v)T)(v_1, \ldots, v_{k-1}) = T(v, v_1, \ldots, v_{k-1}).$$

Show that:

(a) $\iota(v_1)(\iota(v_2)T) = -\iota(v_2)(\iota(v_1)T)$;
(b) if $T \in \Lambda^k(V^*)$ and $S \in \Lambda^m(V^*)$ then

$$\iota(v)(T \wedge S) = (\iota(v)T) \wedge S + (-1)^k T \wedge (\iota(v)S).$$

2.2 Tensor Fields

The definition of a vector field can be generalized to tensor fields of general type. For that, we denote by $T_p^* M$ the dual of the tangent space $T_p M$ at a point p in M (usually called the **cotangent space** to M at p).

Definition 2.1 A (k, m)-**tensor field** is a map that to each point $p \in M$ assigns a tensor $T \in T^{k,m}(T_p^* M, T_p M)$.

Example 2.2 A vector field is a $(0, 1)$-tensor field (or a 1-contravariant tensor field), that is, a map that to each point $p \in M$ assigns the 1-contravariant tensor $X_p \in T_p M$.

Example 2.3 Let $f : M \to \mathbb{R}$ be a differentiable function. We can define a $(1, 0)$-tensor field df which carries each point $p \in M$ to $(df)_p$, where

$$(df)_p : T_p M \to \mathbb{R}$$

is the derivative of f at p. This tensor field is called the **differential** of f. For any $v \in T_p M$ we have $(df)_p(v) = v \cdot f$ (the directional derivative of f at p along the vector v). Considering a coordinate system $x : W \to \mathbb{R}^n$, we can write $v = \sum_{i=1}^n v^i \left(\frac{\partial}{\partial x^i} \right)_p$, and so

$$(df)_p(v) = \sum_i v^i \frac{\partial \hat{f}}{\partial x^i}(x(p)),$$

where $\hat{f} = f \circ x^{-1}$. Taking the coordinate functions $x^i : W \to \mathbb{R}$, we can obtain 1-forms dx^i defined on W. These satisfy

$$(dx^i)_p \left(\left(\frac{\partial}{\partial x^j} \right)_p \right) = \delta_{ij}$$

and so they form a basis of each cotangent space $T_p^* M$, dual to the coordinate basis $\left\{ \left(\frac{\partial}{\partial x^1} \right)_p, \ldots, \left(\frac{\partial}{\partial x^n} \right)_p \right\}$ of $T_p M$. Hence, any $(1, 0)$-tensor field on W can be written as $\omega = \sum_i \omega_i dx^i$, where $\omega_i : W \to \mathbb{R}$ is such that $\omega_i(p) = \omega_p \left(\left(\frac{\partial}{\partial x^i} \right)_p \right)$. In particular, df can be written in the usual way

$$(df)_p = \sum_{i=1}^{n} \frac{\partial \hat{f}}{\partial x^i}(x(p))(dx^i)_p.$$

Remark 2.4 Similarly to what was done for the tangent bundle, we can consider the disjoint union of all cotangent spaces and obtain the manifold

$$T^* M = \bigcup_{p \in M} T_p^* M$$

called the **cotangent bundle** of M. Note that a $(1, 0)$-tensor field is just a map from M to $T^* M$ defined by

$$p \mapsto \omega_p \in T_p^* M.$$

This construction can be easily generalized for arbitrary tensor fields.

The space of (k, m)-tensor fields is clearly a vector space, since linear combinations of (k, m)-tensors are still (k, m)-tensors. If W is a coordinate neighborhood of M, we know that $\{(dx^i)_p\}$ is a basis for $T_p^* M$ and that $\left\{ \left(\frac{\partial}{\partial x^i} \right)_p \right\}$ is a basis for $T_p M$. Hence, the value of a (k, m)-tensor field T at a point $p \in W$ can be written as the tensor

$$T_p = \sum a_{i_1 \ldots i_k}^{j_1 \ldots j_m}(p)(dx^{i_1})_p \otimes \cdots \otimes (dx^{i_k})_p \otimes \left(\frac{\partial}{\partial x^{j_1}} \right)_p \otimes \cdots \otimes \left(\frac{\partial}{\partial x^{j_m}} \right)_p$$

where the $a_{i_1 \ldots i_k}^{j_1 \ldots j_m} : W \to \mathbb{R}$ are functions which at each $p \in W$ give us the components of T_p relative to these bases of $T_p^* M$ and $T_p M$. Just as we did with vector fields, we say that a tensor field is **differentiable** if all these functions are differentiable for all coordinate systems of the maximal atlas. Again, we only need to consider the coordinate systems of an atlas, since all overlap maps are differentiable [cf. Exercise 2.8(1)].

Example 2.5 The differential of a smooth function $f : M \to \mathbb{R}$ is clearly a differentiable $(1, 0)$-tensor field, since its components $\frac{\partial \hat{f}}{\partial x^i} \circ x$ on a given coordinate system $x : W \to \mathbb{R}^n$ are smooth.

An important operation on covariant tensors is the **pull-back** by a smooth map.

Definition 2.6 Let $f : M \to N$ be a differentiable map between smooth manifolds. Then, each differentiable k-covariant tensor field T on N defines a k-covariant tensor field f^*T on M in the following way:

$$(f^*T)_p(v_1, \ldots, v_k) = T_{f(p)}((df)_p v_1, \ldots, (df)_p v_k),$$

for $v_1, \ldots, v_k \in T_p M$.

Remark 2.7 Notice that $(f^*T)_p$ is just the image of $T_{f(p)}$ by the linear map $(df)_p^* :$ $T^k(T_{f(p)}^* N) \to T^k(T_p^* M)$ induced by $(df)_p : T_p M \to T_{f(p)} N$ (cf. Sect. 2.1). Therefore the properties $f^*(\alpha T + \beta S) = \alpha(f^*T) + \beta(f^*S)$ and $f^*(T \otimes S) = (f^*T) \otimes (f^*S)$ hold for all $\alpha, \beta \in \mathbb{R}$ and all appropriate covariant tensor fields T, S. We will see in Exercise 2.8(2) that the pull-back of a differentiable covariant tensor field is still a differentiable covariant tensor field.

Exercise 2.8

(1) Find the relation between coordinate functions of a tensor field in two overlapping coordinate systems.
(2) Show that the pull-back of a differentiable covariant tensor field is still a differentiable covariant tensor field.
(3) (*Lie derivative of a tensor field*) Given a vector field $X \in \mathfrak{X}(M)$, we define the **Lie derivative of a k-covariant tensor field** T along X as

$$L_X T := \frac{d}{dt}(\psi_t^* T)\Big|_{t=0},$$

where $\psi_t = F(\cdot, t)$ with F the local flow of X at p.

(a) Show that

$$L_X (T(Y_1, \ldots, Y_k)) = (L_X T)(Y_1, \ldots, Y_k)$$
$$+ T(L_X Y_1, \ldots, Y_k) + \ldots + T(Y_1, \ldots, L_X Y_k),$$

i.e. show that

$$X \cdot (T(Y_1, \ldots, Y_k)) = (L_X T)(Y_1, \ldots, Y_k)$$
$$+ T([X, Y_1], \ldots, Y_k) + \ldots + T(Y_1, \ldots, [X, Y_k]),$$

for all vector fields $Y_1, \ldots, Y_k$ [cf. Exercises 6.11(11) and 6.11(12) in Chap. 1].
(b) How would you define the Lie derivative of a (k, m)-tensor field?

2.3 Differential Forms

Fields of alternating tensors are very important objects called **forms**.

Definition 3.1 Let M be a smooth manifold. A **form of degree** k (or k-**form**) on M is a field of alternating k-tensors defined on M, that is, a map ω that, to each point $p \in M$, assigns an element $\omega_p \in \Lambda^k(T_p^*M)$.

The space of k-forms on M is clearly a vector space. By Theorem 1.9, given a coordinate system $x : W \to \mathbb{R}^n$, any k-form on W can be written as

$$\omega = \sum_I \omega_I dx^I$$

where $I = (i_1, \ldots, i_k)$ denotes any increasing index sequence of integers in $\{1, \ldots, n\}$, dx^I is the form $dx^{i_1} \wedge \cdots \wedge dx^{i_k}$, and the ω_I are functions defined on W. It is easy to check that the components of ω in the basis $\{dx^{i_1} \otimes \cdots \otimes dx^{i_k}\}$ are $\pm\omega_I$. Therefore ω is a differentiable $(k, 0)$-tensor (in which case it is called a **differential form**) if the functions ω_I are smooth for all coordinate systems of the maximal atlas. The set of differential k-forms on M is represented by $\Omega^k(M)$. From now on we will use the word "form" to mean a differential form.

Given a smooth map $f : M \to N$ between differentiable manifolds, we can induce forms on M from forms on N using the pull-back operation (cf. Definition 2.6), since the pull-back of a field of alternating tensors is still a field of alternating tensors.

Remark 3.2 If $g : N \to \mathbb{R}$ is a 0-form, that is, a function, the pull-back is defined as $f^*g = g \circ f$.

It is easy to verify that the pull-back of forms satisfies the following properties.

Proposition 3.3 *Let* $f : M \to N$ *be a differentiable map and* α, β *forms on* N. *Then,*

(i) $f^*(\alpha + \beta) = f^*\alpha + f^*\beta$;
(ii) $f^*(g\alpha) = (g \circ f)f^*\alpha = (f^*g)(f^*\alpha)$ *for any function* $g \in C^\infty(N)$;
(iii) $f^*(\alpha \wedge \beta) = (f^*\alpha) \wedge (f^*\beta)$;
(iv) $g^*(f^*\alpha) = (f \circ g)^*\alpha$ *for any map* $g \in C^\infty(L, M)$, *where* L *is a differentiable manifold.*

Proof Exercise 3.8(1)   □

Example 3.4 If $f : M \to N$ is differentiable and we consider coordinate systems $x : V \to \mathbb{R}^m$, $y : W \to \mathbb{R}^n$ respectively on M and N, we have $y^i = \hat{f}^i(x^1, \ldots, x^m)$ for $i = 1, \ldots, n$ and $\hat{f} = y \circ f \circ x^{-1}$ the local representation of f. If $\omega = \sum_I \omega_I dy^I$ is a k-form on W, then by Proposition 3.3,

$$f^*\omega = f^*\left(\sum_I \omega_I dy^I\right) = \sum_I (f^*\omega_I)(f^*dy^I) = \sum_I (\omega_I \circ f)(f^*dy^{i_1}) \wedge \cdots \wedge (f^*dy^{i_k}).$$

Moreover, for $v \in T_p M$,

$$(f^*(dy^i))_p(v) = (dy^i)_{f(p)}((df)_p v) = \left(d(y^i \circ f)\right)_p (v),$$

that is, $f^*(dy^i) = d(y^i \circ f)$. Hence,

$$f^*\omega = \sum_I (\omega_I \circ f)\, d(y^{i_1} \circ f) \wedge \cdots \wedge d(y^{i_k} \circ f)$$

$$= \sum_I (\omega_I \circ f)\, d(\hat{f}^{i_1} \circ x) \wedge \cdots \wedge d(\hat{f}^{i_k} \circ x).$$

If $k = \dim M = \dim N = n$, then the pull-back $f^*\omega$ can easily be computed from Theorem 1.13, according to which

$$(f^*(dy^1 \wedge \cdots \wedge dy^n))_p = \det (d\hat{f})_{x(p)}(dx^1 \wedge \cdots \wedge dx^n)_p. \qquad (2.1)$$

Given any form ω on M and a parameterization $\varphi : U \to M$, we can consider the pull-back of ω by φ and obtain a form defined on the open set U, called the **local representation** of ω on that parameterization.

Example 3.5 Let $x : W \to \mathbb{R}^n$ be a coordinate system on a smooth manifold M and consider the 1-form dx^i defined on W. The pull-back $\varphi^* dx^i$ by the corresponding parameterization $\varphi := x^{-1}$ is a 1-form on an open subset U of $\mathbb{R}^n$ satisfying

$$(\varphi^* dx^i)_x(v) = (\varphi^* dx^i)_x \left(\sum_{j=1}^n v^j \left(\frac{\partial}{\partial x^j}\right)_x\right) = (dx^i)_p \left(\sum_{j=1}^n v^j (d\varphi)_x \left(\frac{\partial}{\partial x^j}\right)_x\right)$$

$$= (dx^i)_p \left(\sum_{j=1}^n v^j \left(\frac{\partial}{\partial x^j}\right)_p\right) = v^i = (dx^i)_x(v),$$

for $x \in U$, $p = \varphi(x)$ and $v = \sum_{j=1}^n v^j \left(\frac{\partial}{\partial x^j}\right)_x \in T_x U$. Hence, just as we had $\left(\frac{\partial}{\partial x^i}\right)_p = (d\varphi)_x \left(\frac{\partial}{\partial x^i}\right)_x$, we now have $(dx^i)_x = \varphi^*(dx^i)_p$, and so $(dx^i)_p$ is the 1-form in W whose local representation on U is $(dx^i)_x$.

If $\omega = \sum_I \omega_I dx^I$ is a k-form defined on an open subset of $\mathbb{R}^n$, we define a $(k+1)$-form called **exterior derivative** of ω as

$$d\omega := \sum_I d\omega_I \wedge dx^I.$$

Example 3.6 Consider the form $\omega = -\frac{y}{x^2+y^2}\,dx + \frac{x}{x^2+y^2}\,dy$ defined on $\mathbb{R}^2\backslash\{0\}$. Then,

$$dw = d\left(-\frac{y}{x^2+y^2}\right) \wedge dx + d\left(\frac{x}{x^2+y^2}\right) \wedge dy$$

$$= \frac{y^2-x^2}{(x^2+y^2)^2}\,dy \wedge dx + \frac{y^2-x^2}{(x^2+y^2)^2}\,dx \wedge dy = 0.$$

The exterior derivative satisfies the following properties:

Proposition 3.7 *If $\alpha, \omega, \omega_1, \omega_2$ are forms on $\mathbb{R}^n$, then*

(i) $d(\omega_1 + \omega_2) = d\omega_1 + d\omega_2$;
(ii) *if ω is k-form*, $d(\omega \wedge \alpha) = d\omega \wedge \alpha + (-1)^k \omega \wedge d\alpha$;
(iii) $d(d\omega) = 0$;
(iv) *if $f : \mathbb{R}^m \to \mathbb{R}^n$ is smooth*, $d(f^*\omega) = f^*(d\omega)$.

Proof Property (i) is obvious. Using (i), it is enough to prove (ii) for $\omega = a_I dx^I$ and $\alpha = b_J dx^J$:

$$d(\omega \wedge \alpha) = d(a_I b_J \, dx^I \wedge dx^J) = d(a_I b_J) \wedge dx^I \wedge dx^J$$
$$= (b_J \, da_I + a_I \, db_J) \wedge dx^I \wedge dx^J$$
$$= b_J \, da_I \wedge dx^I \wedge dx^J + a_I \, db_J \wedge dx^I \wedge dx^J$$
$$= d\omega \wedge \alpha + (-1)^k a_I dx^I \wedge db_J \wedge dx^J$$
$$= d\omega \wedge \alpha + (-1)^k \omega \wedge d\alpha.$$

Again, to prove (iii), it is enough to consider forms $\omega = a_I dx^I$. Since

$$d\omega = da_I \wedge dx^I = \sum_{i=1}^{n} \frac{\partial a_I}{\partial x^i}\, dx^i \wedge dx^I,$$

we have

$$d(d\omega) = \sum_{j=1}^{n}\sum_{i=1}^{n} \frac{\partial^2 a_I}{\partial x^j \partial x^i} dx^j \wedge dx^i \wedge dx^I$$

$$= \sum_{i=1}^{n}\sum_{j<i} \left(\frac{\partial^2 a_I}{\partial x^j \partial x^i} - \frac{\partial^2 a_I}{\partial x^i \partial x^j}\right) dx^j \wedge dx^i \wedge dx^I = 0.$$

To prove (iv), we first consider a 0-form g:

$$f^*(dg) = f^*\left(\sum_{i=1}^{n} \frac{\partial g}{\partial x^i} dx^i\right) = \sum_{i=1}^{n} \left(\frac{\partial g}{\partial x^i} \circ f\right) df^i = \sum_{i,j=1}^{n} \left(\left(\frac{\partial g}{\partial x^i} \circ f\right) \frac{\partial f^i}{\partial x^j}\right) dx^j$$

$$= \sum_{j=1}^{n} \frac{\partial(g \circ f)}{\partial x^j} dx^j = d(g \circ f) = d(f^*g).$$

Then, if $\omega = a_I dx^I$, we have

$$d(f^*\omega) = d((f^*a_I)df^I) = d(f^*a_I) \wedge df^I + (f^*a_I)d(df^I) = d(f^*a_I) \wedge df^I$$
$$= (f^*da_I) \wedge (f^*dx^I) = f^*(da_I \wedge dx^I) = f^*(d\omega)$$

(where df^I denotes the form $df^{i_1} \wedge \cdots \wedge df^{i_k}$), and the result follows. $\square$

Suppose now that ω is a differential k-form on a smooth manifold M. We define the $(k+1)$-form $d\omega$ as the smooth form that is locally represented by $d\omega_\alpha$ for each parameterization $\varphi_\alpha : U_\alpha \to M$, where $\omega_\alpha := \varphi_\alpha^* \omega$ is the local representation of ω, that is, $d\omega = (\varphi_\alpha^{-1})^*(d\omega_\alpha)$ on $\varphi_\alpha(U)$. Given another parameterization $\varphi_\beta : U_\beta \to M$ such that $W := \varphi_\alpha(U_\alpha) \cap \varphi_\beta(U_\beta) \neq \varnothing$, it is easy to verify that

$$(\varphi_\alpha^{-1} \circ \varphi_\beta)^* \omega_\alpha = \omega_\beta.$$

Setting f equal to $\varphi_\alpha^{-1} \circ \varphi_\beta$, we have

$$f^*(d\omega_\alpha) = d(f^*\omega_\alpha) = d\omega_\beta.$$

Consequently,

$$(\varphi_\beta^{-1})^* d\omega_\beta = (\varphi_\beta^{-1})^* f^*(d\omega_\alpha)$$
$$= (f \circ \varphi_\beta^{-1})^*(d\omega_\alpha)$$
$$= (\varphi_\alpha^{-1})^*(d\omega_\alpha),$$

and so the two definitions agree on the overlapping set W. Therefore $d\omega$ is well defined. We leave it as an exercise to show that the exterior derivative defined for forms on smooth manifolds also satisfies the properties of Proposition 3.7.

Exercise 3.8

(1) Prove Proposition 3.3.
(2) (*Exterior derivative*) Let M be a smooth manifold. Given a k-form ω in M we can define its exterior derivative $d\omega$ without using local coordinates: given $k+1$ vector fields $X_1, \ldots, X_{k+1} \in \mathfrak{X}(M)$,

$$d\omega(X_1, \ldots, X_{k+1}) := \sum_{i=1}^{k+1} (-1)^{i-1} X_i \cdot \omega(X_1, \ldots, \hat{X}_i, \ldots, X_{k+1})$$
$$+ \sum_{i<j} (-1)^{i+j} \omega([X_i, X_j], X_1, \ldots, \hat{X}_i, \ldots, \hat{X}_j, \ldots, X_{k+1}),$$

where the hat indicates an omitted variable.

(a) Show that $d\omega$ defined above is in fact a $(k+1)$-form in M, that is,
  (i) $d\omega(X_1, \ldots, X_i + Y_i, \ldots, X_{k+1}) =$
      $d\omega(X_1, \ldots, X_i, \ldots, X_{k+1}) + d\omega(X_1, \ldots, Y_i, \ldots, X_{k+1})$;
  (ii) $d\omega(X_1, \ldots, fX_j, \ldots, X_{k+1}) = f d\omega(X_1, \ldots, X_{k+1})$ for any differentiable function f;
  (iii) $d\omega$ is alternating;
  (iv) $d\omega(X_1, \ldots, X_{k+1})(p)$ depends only on $(X_1)_p, \ldots, (X_{k+1})_p$.

(b) Let $x : W \to \mathbb{R}^n$ be a coordinate system of M and let $\omega = \sum_I a_I dx^{i_1} \wedge \cdots \wedge dx^{i_k}$ be the expression of ω in these coordinates (where the a_I are smooth functions). Show that the local expression of $d\omega$ is the same as the one used in the local definition of exterior derivative, that is,

$$dw = \sum_I da_I \wedge dx^{i_1} \wedge \cdots \wedge dx^{i_k}.$$

(3) Show that the exterior derivative defined for forms on smooth manifolds satisfies the properties of Proposition 3.7.

(4) Show that:

(a) if $\omega = f^1 dx + f^2 dy + f^3 dz$ is a 1-form on $\mathbb{R}^3$ then

$$d\omega = g^1 dy \wedge dz + g^2 dz \wedge dx + g^3 dx \wedge dy,$$

where $(g^1, g^2, g^3) = \mathrm{curl}(f^1, f^2, f^3)$;

(b) if $\omega = f^1 dy \wedge dz + f^2 dz \wedge dx + f^3 dx \wedge dy$ is a 2-form on $\mathbb{R}^3$, then

$$d\omega = \mathrm{div}(f^1, f^2, f^3) \, dx \wedge dy \wedge dz.$$

(5) (*De Rham cohomology*) A k-form ω is called **closed** if $d\omega = 0$. If it exists a $(k-1)$-form β such that $\omega = d\beta$ then ω is called **exact**. Note that every exact form is closed. Let Z^k be the set of all closed k-forms on M and define a relation between forms on Z^k as follows: $\alpha \sim \beta$ if and only if they differ by an exact form, that is, if $\beta - \alpha = d\theta$ for some $(k-1)$-form θ.

(a) Show that this relation is an equivalence relation.

(b) Let $H^k(M)$ be the corresponding set of equivalence classes (called the k-dimensional **de Rham cohomology space** of M). Show that addition and scalar multiplication of forms define indeed a vector space structure on $H^k(M)$.

(c) Let $f : M \to N$ be a smooth map. Show that:
  (i) the pull-back f^* carries closed forms to closed forms and exact forms to exact forms;
  (ii) if $\alpha \sim \beta$ on N then $f^*\alpha \sim f^*\beta$ on M;

(iii) f^* induces a linear map on cohomology $f^\sharp : H^k(N) \to H^k(M)$
naturally defined by $f^\sharp[\omega] = [f^*\omega]$;

(iv) if $g : L \to M$ is another smooth map, then $(f \circ g)^\sharp = g^\sharp \circ f^\sharp$.

(d) Show that the dimension of $H^0(M)$ is equal to the number of connected components of M.

(e) Show that $H^k(M) = 0$ for every $k > \dim M$.

(6) Let M be a manifold of dimension n, let U be an open subset of $\mathbb{R}^n$ and let ω be a k-form on $\mathbb{R} \times U$. Writing ω as

$$\omega = dt \wedge \sum_I a_I dx^I + \sum_J b_J dx^J,$$

where $I = (i_1, \ldots, i_{k-1})$ and $J = (j_1, \ldots, j_k)$ are increasing index sequences, $(x^1, \ldots, x^n)$ are coordinates in U and t is the coordinate in $\mathbb{R}$, consider the operator $\mathcal{Q}$ defined by

$$\mathcal{Q}(\omega)_{(t,x)} = \sum_I \left(\int_{t_0}^t a_I ds \right) dx^I,$$

which transforms k-forms ω in $\mathbb{R} \times U$ into $(k-1)$-forms.

(a) Let $f : V \to U$ be a diffeomorphism between open subsets of $\mathbb{R}^n$. Show that the induced diffeomorphism $\widetilde{f} := \mathrm{id} \times f : \mathbb{R} \times V \to \mathbb{R} \times U$ satisfies

$$\widetilde{f}^* \circ \mathcal{Q} = \mathcal{Q} \circ \widetilde{f}^*.$$

(b) Using (a), construct an operator $\mathcal{Q}$ which carries k-forms on $\mathbb{R} \times M$ into $(k-1)$-forms and, for any diffeomorphism $f : M \to N$, the induced diffeomorphism $\widetilde{f} := \mathrm{id} \times f : \mathbb{R} \times M \to \mathbb{R} \times N$ satisfies $\widetilde{f}^* \circ \mathcal{Q} = \mathcal{Q} \circ \widetilde{f}^*$. Show that this operator is linear.

(c) Considering the operator $\mathcal{Q}$ defined in (b) and the inclusion $i_{t_0} : M \to \mathbb{R} \times M$ of M at the "level" t_0, defined by $i_{t_0}(p) = (t_0, p)$, show that $\omega - \pi^* i_{t_0}^* \omega = d\mathcal{Q}\omega + \mathcal{Q}d\omega$, where $\pi : \mathbb{R} \times M \to M$ is the projection on M.

(d) Show that the maps $\pi^\sharp : H^k(M) \to H^k(\mathbb{R} \times M)$ and $i_{t_0}^\sharp : H^k(\mathbb{R} \times M) \to H(M)$ are inverses of each other (and so $H^k(M)$ is isomorphic to $H^k(\mathbb{R} \times M)$).

(e) Use (d) to show that, for $k > 0$ and $n > 0$, every closed k-form in $\mathbb{R}^n$ is exact, that is, $H^k(\mathbb{R}^n) = 0$ if $k > 0$.

(f) Use (d) to show that, if $f, g : M \to N$ are two **smoothly homotopic maps** between smooth manifolds (meaning that there exists a smooth map $H : \mathbb{R} \times M \to N$ such that $H(t_0, p) = f(p)$ and $H(t_1, p) = g(p)$ for some fixed $t_0, t_1 \in \mathbb{R}$), then $f^\sharp = g^\sharp$.

(g) We say that M is **contractible** if the identity map $\mathrm{id} : M \to M$ is smoothly homotopic to a constant map. Show that $\mathbb{R}^n$ is contractible.

(h) (*Poincaré lemma*) Let M be a contractible smooth manifold. Show that every closed form on M is exact, that is, $H^k(M) = 0$ for all $k > 0$.

(**Remark:** This exercise is based on an exercise in [GP73]).

(7) (*Lie derivative of a differential form*) Given a vector field $X \in \mathfrak{X}(M)$, we define the **Lie derivative of a form** ω along X as

$$L_X\omega := \frac{d}{dt}(\psi_t{}^*\omega)\Big|_{t=0} ,$$

where $\psi_t = F(\cdot, t)$ with F the local flow of X at p [cf. Exercise 2.8(3)]. Show that the Lie derivative satisfies the following properties:

(a) $L_X(\omega_1 \wedge \omega_2) = (L_X\omega_1) \wedge \omega_2 + \omega_1 \wedge (L_X\omega_2)$;
(b) $d(L_X\omega) = L_X(d\omega)$;
(c) **Cartan formula:** $L_X\omega = \iota(X)d\omega + d(\iota(X)\omega)$;
(d) $L_X(\iota(Y)\omega) = \iota(L_XY)\omega + \iota(Y)L_X\omega$
    [cf. Exercise 6.11(12) on Chap. 1 and Exercise 1.15(8)].

2.4 Integration on Manifolds

Before we see how to integrate differential forms on manifolds, we will start by studying the $\mathbb{R}^n$ case. For that let us consider an n-form ω defined on an open subset U of $\mathbb{R}^n$. We already know that ω can be written as

$$\omega_x = a(x)\,dx^1 \wedge \cdots \wedge dx^n,$$

where $a : U \to \mathbb{R}$ is a smooth function. The **support** of ω is, by definition, the closure of the set where $\omega \neq 0$ that is,

$$\operatorname{supp} \omega = \overline{\{x \in \mathbb{R}^n \mid \omega_x \neq 0\}}.$$

We will assume that this set is compact (in which case ω is said to be **compactly supported**). We define

$$\int_U \omega = \int_U a(x)\,dx^1 \wedge \cdots \wedge dx^n := \int_U a(x)\,dx^1 \cdots dx^n,$$

where the integral on the right is a multiple integral on a subset of $\mathbb{R}^n$. This definition is almost well-behaved with respect to changes of variables in $\mathbb{R}^n$. Indeed, if $f : V \to U$ is a diffeomorphism of open sets of $\mathbb{R}^n$, we have from (2.1) that

$$f^*\omega = (a \circ f)(\det df)dy^1 \wedge \cdots \wedge dy^n,$$

and so

$$\int_V f^*\omega = \int_V (a \circ f)(\det df)dy^1 \cdots dy^n.$$

If f is orientation-preserving, then det $(df) > 0$, and the integral on the right is, by the change of variables theorem for multiple integrals in $\mathbb{R}^n$ (cf. Sect. 2.7.2), equal to $\int_U \omega$. For this reason, we will only consider orientable manifolds when integrating forms on manifolds. Moreover, we will also assume that supp ω is always compact to avoid convergence problems.

Let M be an oriented manifold, and let $\mathcal{A} = \{(U_\alpha, \varphi_\alpha)\}$ be an atlas whose parameterizations are orientation-preserving. Suppose that supp ω is contained in some coordinate neighborhood $W_\alpha = \varphi_\alpha(U_\alpha)$. Then we define

$$\int_M \omega := \int_{U_\alpha} \varphi_\alpha^*\omega = \int_{U_\alpha} \omega_\alpha.$$

Note that this does not depend on the choice of coordinate neighborhood: if supp ω is contained in some other coordinate neighborhood $W_\beta = \varphi_\beta(U_\beta)$, then $\omega_\beta = f^*\omega_\alpha$, where $f := \varphi_\alpha^{-1} \circ \varphi_\beta$ is orientation-preserving, and hence

$$\int_{U_\beta} \omega_\beta = \int_{U_\beta} f^*\omega_\alpha = \int_{U_\alpha} \omega_\alpha.$$

To define the integral in the general case we use a **partition of unity** (cf. Sect. 2.7.2) subordinate to the cover $\{W_\alpha\}$ of M, i.e. a family of differentiable functions on M, $\{\rho_i\}_{i \in I}$, such that:

(i) for every point $p \in M$, there exists a neighborhood V of p such that $V \cap$ supp $\rho_i = \varnothing$ except for a finite number of ρ_i;
(ii) for every point $p \in M$, $\sum_{i \in I} \rho_i(p) = 1$;
(iii) $0 \leq \rho_i \leq 1$ and supp $\rho_i \subset W_{\alpha_i}$ for some element W_{α_i} of the cover.

Because of property (i), supp ω (being compact) intersects the supports of only finitely many ρ_i. Hence we can assume that I is finite, and then

$$\omega = \left(\sum_{i \in I} \rho_i\right)\omega = \sum_{i \in I} \rho_i\omega = \sum_{i \in I} \omega_i$$

with $\omega_i := \rho_i\omega$ and supp $\omega_i \subset W_{\alpha_i}$. Consequently we define:

$$\int_M \omega := \sum_{i \in I} \int_M \omega_i = \sum_{i \in I} \int_{U_{\alpha_i}} \varphi_{\alpha_i}^*\omega_i.$$

Remark 4.1

(1) When supp ω is contained in one coordinate neighborhood W, the two definitions above agree. Indeed,

$$\int_M \omega = \int_W \omega = \int_W \sum_{i \in I} \omega_i = \int_U \varphi^* \left(\sum_{i \in I} \omega_i \right)$$
$$= \int_U \sum_{i \in I} \varphi^* \omega_i = \sum_{i \in I} \int_U \varphi^* \omega_i = \sum_{i \in I} \int_M \omega_i,$$

where we used the linearity of the pull-back and of integration on $\mathbb{R}^n$.

(2) The definition of integral is independent of the choice of partition of unity and the choice of cover. Indeed, if $\{\tilde{\rho}_j\}_{j \in J}$ is another partition of unity subordinate to another cover $\{\tilde{W}_\beta\}$ compatible with the same orientation, we have by (1)

$$\sum_{i \in I} \int_M \rho_i \omega = \sum_{i \in I} \sum_{j \in J} \int_M \tilde{\rho}_j \rho_i \omega$$

and

$$\sum_{j \in J} \int_M \tilde{\rho}_j \omega = \sum_{j \in J} \sum_{i \in I} \int_M \rho_i \tilde{\rho}_j \omega.$$

(3) It is also easy to verify the linearity of the integral, that is,

$$\int_M a\omega_1 + b\omega_2 = a \int_M \omega_1 + b \int_M \omega_2.$$

for $a, b \in \mathbb{R}$ and ω_1, ω_2 two n-forms on M.

(4) The definition of integral can easily be extended to oriented manifolds with boundary.

Exercise 4.2

(1) Let M be an n-dimensional differentiable manifold. A subset $N \subset M$ is said to have **zero measure** if the sets $\varphi_\alpha^{-1}(N) \subset U_\alpha$ have zero measure for every parameterization $\varphi_\alpha : U_\alpha \to M$ in the maximal atlas.

   (a) Prove that in order to show that $N \subset M$ has zero measure it suffices to check that the sets $\varphi_\alpha^{-1}(N) \subset U_\alpha$ have zero measure for the parameterizations in an arbitrary atlas.

   (b) Suppose that M is oriented. Let $\omega \in \Omega^n(M)$ be compactly supported and let $W = \varphi(U)$ be a coordinate neighborhood such that $M \backslash W$ has zero measure. Show that

$$\int_M \omega = \int_U \varphi^* \omega,$$

where the integral on the right-hand side is defined as above and always exists.

(2) Let x, y, z be the restrictions of the Cartesian coordinate functions in $\mathbb{R}^3$ to S^2, oriented so that $\{(1, 0, 0); (0, 1, 0)\}$ is a positively oriented basis of $T_{(0,0,1)}S^2$, and consider the 2-form

$$\omega = x\,dy \wedge dz + y\,dz \wedge dx + z\,dx \wedge dy \in \Omega^2(S^2).$$

Compute the integral

$$\int_{S^2} \omega$$

using the parameterizations corresponding to

(a) spherical coordinates;
(b) stereographic projection.

(3) Consider the manifolds

$$S^3 = \left\{(x, y, z, w) \in \mathbb{R}^4 \mid x^2 + y^2 + z^2 + w^2 = 2\right\};$$
$$T^2 = \left\{(x, y, z, w) \in \mathbb{R}^4 \mid x^2 + y^2 = z^2 + w^2 = 1\right\}.$$

The submanifold $T^2 \subset S^3$ splits S^3 into two connected components. Let M be one of these components and let ω be the 3-form

$$\omega = z\,dx \wedge dy \wedge dw - x\,dy \wedge dz \wedge dw.$$

Compute the two possible values of $\int_M \omega$.

(4) Let M and N be n-dimensional manifolds, $f : M \to N$ an orientation-preserving diffeomorphism and $\omega \in \Omega^n(N)$ a compactly supported form. Prove that

$$\int_N \omega = \int_M f^*\omega.$$

2.5 Stokes Theorem

In this section we will prove a very important theorem.

Theorem 5.1 (Stokes) *Let M be an n-dimensional oriented smooth manifold with boundary, let ω be a $(n-1)$-differential form on M with compact support, and let $i : \partial M \to M$ be the inclusion of the boundary ∂M in M. Then*

$$\int_{\partial M} i^*\omega = \int_M d\omega,$$

where we consider ∂M with the induced orientation (cf. Sect. 9 in Chap. 1).

Proof Let us take a partition of unity $\{\rho_i\}_{i \in I}$ subordinate to an open cover of M by coordinate neighborhoods compatible with the orientation. Then $\omega = \sum_{i \in I} \rho_i \omega$, where we can assume I to be finite (ω is compactly supported), and hence

$$d\omega = d \sum_{i \in I} \rho_i \omega = \sum_{i \in I} d(\rho_i \omega).$$

By linearity of the integral we then have,

$$\int_M d\omega = \sum_{i \in I} \int_M d(\rho_i \omega) \text{ and } \int_{\partial M} i^*\omega = \sum_{i \in I} \int_{\partial M} i^*(\rho_i \omega).$$

Hence, to prove this theorem, it is enough to consider the case where $\mathrm{supp}\,\omega$ is contained inside one coordinate neighborhood of the cover. Let us then consider an $(n-1)$-form ω with compact support contained in a coordinate neighborhood W. Let $\varphi : U \to W$ be the corresponding parameterization, where we can assume U to be bounded ($\mathrm{supp}(\varphi^*\omega)$ is compact). Then, the representation of ω on U can be written as

$$\varphi^*\omega = \sum_{j=1}^{n} a_j \, dx^1 \wedge \cdots \wedge dx^{j-1} \wedge dx^{j+1} \wedge \cdots \wedge dx^n,$$

(where each $a_j : U \to \mathbb{R}$ is a C^∞-function), and

$$\varphi^* d\omega = d\varphi^*\omega = \sum_{j=1}^{n} (-1)^{j-1} \frac{\partial a_j}{\partial x^j} \, dx^1 \wedge \cdots \wedge dx^n.$$

The functions a_j can be extended to C^∞-functions on $\mathbb{H}^n$ by letting

$$a_j(x^1, \cdots, x^n) = \begin{cases} a_j(x^1, \cdots, x^n) \text{ if } (x^1, \ldots, x^n) \in U \\ \qquad 0 \qquad \text{ if } (x^1, \ldots, x^n) \in \mathbb{H}^n \backslash U. \end{cases}$$

If $W \cap \partial M = \varnothing$, then $i^*\omega = 0$. Moreover, if we consider a rectangle I in $\mathbb{H}$ containing U defined by equations $b_j \leq x^j \leq c_j$ ($j = 1, \ldots, n$), we have

$$\int_M d\omega = \int_U \left(\sum_{j=1}^{n} (-1)^{j-1} \frac{\partial a_j}{\partial x^j} \right) dx^1 \cdots dx^n = \sum_{j=1}^{n} (-1)^{j-1} \int_I \frac{\partial a_j}{\partial x^j} \, dx^1 \cdots dx^n$$

$$= \sum_{j=1}^{n}(-1)^{j-1}\int_{\mathbb{R}^{n-1}}\left(\int_{b_j}^{c_j}\frac{\partial a_j}{\partial x^j}dx^j\right)dx^1\cdots dx^{j-1}dx^{j+1}\cdots dx^n$$

$$= \sum_{j=1}^{n}(-1)^{j-1}\int_{\mathbb{R}^{n-1}}\left(a_j(x^1,\ldots,x^{j-1},c_j,x^{j+1},\ldots,x^n)-\right.$$

$$\left.-a_j(x^1,\ldots,x^{j-1},b_j,x^{j+1},\ldots,x^n)\right)dx^1\cdots dx^{j-1}dx^{j+1}\cdots dx^n = 0,$$

where we used the Fubini theorem (cf. Sect. 2.7.3), the fundamental theorem of Calculus and the fact that the a_j are zero outside U. We conclude that, in this case, $\int_{\partial M} i^*\omega = \int_M d\omega = 0$.

If, on the other hand, $W \cap \partial M \neq \varnothing$ we take a rectangle I containing U now defined by the equations $b_j \leq x^j \leq c_j$ for $j = 1,\ldots,n-1$, and $0 \leq x^n \leq c_n$. Then, as in the preceding case, we have

$$\int_M d\omega = \int_U\left(\sum_{j=1}^{n}(-1)^{j-1}\frac{\partial a_j}{\partial x^j}\right)dx^1\cdots dx^n = \sum_{j=1}^{n}(-1)^{j-1}\int_I\frac{\partial a_j}{\partial x^j}dx^1\cdots dx^n$$

$$= 0 + (-1)^{n-1}\int_{\mathbb{R}^{n-1}}\left(\int_0^{c_n}\frac{\partial a_n}{\partial x^n}dx^n\right)dx^1\cdots dx^{n-1}$$

$$= (-1)^{n-1}\int_{\mathbb{R}^{n-1}}\left(a_n(x^1,\ldots,x^{n-1},c_n)-a_n(x^1,\ldots,x^{n-1},0)\right)dx^1\cdots dx^{n-1}$$

$$= (-1)^{n}\int_{\mathbb{R}^{n-1}}a_n(x^1,\ldots,x^{n-1},0)\,dx^1\ldots dx^{n-1}.$$

To compute $\int_{\partial M} i^*\omega$ we need to consider a parameterization $\tilde{\varphi}$ of ∂M defined on an open subset of $\mathbb{R}^{n-1}$ which preserves the standard orientation on $\mathbb{R}^{n-1}$ when we consider the induced orientation on ∂M. For that, we can for instance consider the set

$$\tilde{U} = \{(x^1,\ldots,x^{n-1}) \in \mathbb{R}^{n-1} \mid ((-1)^n x^1, x^2,\ldots,x^{n-1},0) \in U\}$$

and the parameterization $\tilde{\varphi}: \tilde{U} :\to \partial M$ given by

$$\tilde{\varphi}(x^1,\ldots,x^{n-1}) := \varphi\left((-1)^n x^1, x^2,\ldots,x^{n-1},0\right).$$

Recall that the orientation on ∂M obtained from φ by just dropping the last coordinate is $(-1)^n$ times the induced orientation on ∂M (cf. Sect. 9 in Chap. 1). Therefore $\tilde{\varphi}$ gives the correct orientation. The local expression of $i : \partial M \to M$ on these coordinates $(\hat{i}: \tilde{U} \to U$ such that $\hat{i} = \varphi^{-1} \circ i \circ \tilde{\varphi})$ is given by

$$\hat{i}(x^1,\ldots,x^{n-1}) = \left((-1)^n x^1, x^2,\ldots,x^{n-1},0\right).$$

Hence,

$$\int_{\partial M} i^*\omega = \int_{\tilde U} \tilde\varphi^* i^*\omega = \int_{\tilde U} (i \circ \tilde\varphi)^*\omega = \int_{\tilde U} (\varphi \circ \hat\imath)^*\omega = \int_{\tilde U} \hat\imath^* \varphi^*\omega.$$

Moreover,

$$\hat\imath^* \varphi^*\omega = \hat\imath^* \sum_{j=1}^{n} a_j \, dx^1 \wedge \cdots \wedge dx^{j-1} \wedge dx^{j+1} \wedge \cdots \wedge dx^n$$

$$= \sum_{j=1}^{n} (a_j \circ \hat\imath) \, d\hat\imath^1 \wedge \cdots \wedge d\hat\imath^{j-1} \wedge d\hat\imath^{j+1} \wedge \cdots \wedge d\hat\imath^n$$

$$= (-1)^n (a_n \circ \hat\imath) \, dx^1 \wedge \cdots \wedge dx^{n-1},$$

since $d\hat\imath^1 = (-1)^n dx^1, d\hat\imath^n = 0$ and $d\hat\imath^j = dx^j$, for $j \neq 1$ and $j \neq n$. Consequently,

$$\int_{\partial M} i^*\omega = (-1)^n \int_{\tilde U} (a_n \circ \hat\imath) \, dx^1 \cdots dx^{n-1}$$

$$= (-1)^n \int_{\tilde U} a_n \left((-1)^n x^1, x^2, \ldots, x^{n-1}, 0 \right) dx^1 \cdots dx^{n-1}$$

$$= (-1)^n \int_{\mathbb{R}^{n-1}} a_n(x^1, x^2, \ldots, x^{n-1}, 0) \, dx^1 \cdots dx^{n-1} = \int_M d\omega$$

(where we have used the change of variables theorem). $\square$

Remark 5.2 If M is an oriented n-dimensional differentiable manifold (that is, a manifold with boundary $\partial M = \varnothing$), it is clear from the proof of the Stokes theorem that

$$\int_M d\omega = 0$$

for any $(n-1)$-differential form ω on M with compact support. This can be viewed as a particular case of the Stokes theorem if we define the integral over the empty set to be zero.

Exercise 5.3

(1) Use the Stokes theorem to confirm the result of Exercise 4.2(3).
(2) (*Homotopy invariance of the integral*) Recall that two maps $f_0, f_1 : M \to N$ are said to be smoothly homotopic if there exists a differentiable map $H : \mathbb{R} \times M \to N$ such that $H(0, p) = f_0(p)$ and $H(1, p) = f_1(p)$ [cf. Exercise 3.8(6)]. If M is a compact oriented manifold of dimension n and ω is a closed n-form on N, show that

$$\int_M f_0^*\omega = \int_M f_1^*\omega.$$

(3) (a) Let $X \in \mathfrak{X}(S^n)$ be a vector field with no zeros. Show that

$$H(t, p) = \cos(\pi t)p + \sin(\pi t)\frac{X_p}{\|X_p\|}$$

is a smooth homotopy between the identity map and the antipodal map, where we make use of the identification

$$X_p \in T_p S^n \subset T_p \mathbb{R}^{n+1} \cong \mathbb{R}^{n+1}.$$

(b) Using the Stokes theorem, show that

$$\int_{S^n} \omega > 0,$$

where

$$\omega = \sum_{i=1}^{n+1} (-1)^{i+1} x^i dx^1 \wedge \cdots \wedge dx^{i-1} \wedge dx^{i+1} \wedge \cdots \wedge dx^{n+1}$$

and $S^n = \partial\{x \in \mathbb{R}^{n+1} \mid \|x\| \leq 1\}$ has the orientation induced by the standard orientation of $\mathbb{R}^{n+1}$.

(c) Show that if n is even then X cannot exist. What about when n is odd?

(4) (*Degree of a map*) Let M, N be compact, connected oriented manifolds of dimension n, and let $f : M \to N$ be a smooth map. It can be shown that there exists a real number $\deg(f)$ (called the **degree** of f) such that, for any n-form $\omega \in \Omega^n(N)$,

$$\int_M f^*\omega = \deg(f) \int_N \omega.$$

(a) Show that if f is not surjective then $\deg(f) = 0$.

(b) Show that if f is an orientation-preserving diffeomorphism then $\deg(f) = 1$, and that if f is an orientation-reversing diffeomorphism then $\deg(f) = -1$.

(c) Let $f : M \to N$ be surjective and let $q \in N$ be a regular value of f. Show that $f^{-1}(q)$ is a finite set and that there exists a neighborhood W of q in N such that $f^{-1}(W)$ is a disjoint union of opens sets V_i of M with $f|_{V_i} : V_i \to W$ a diffeomorphism.

(d) Admitting the existence of a regular value of f, show that $\deg(f)$ is an integer. (Remark: The **Sard theorem** guarantees that the set of critical values of a differentiable map f between manifolds with the same dimension has zero measure, which in turn guarantees the existence of a regular value of f).

(e) Given $n \in \mathbb{N}$, indicate a smooth map $f : S^1 \to S^1$ of degree n.

(f) Show that homotopic maps have the same degree.

(g) Let $f : S^n \to S^n$ be an orientation-preserving diffeomorphism if n is even, or an orientation-reversing diffeomorphism if n is odd. Prove that f has a fixed point, that is, a point $p \in S^n$ such that $f(p) = p$. (**Hint:** Show that if f had no fixed points then it would be possible to construct an homotopy between f and the antipodal map).

2.6 Orientation and Volume Forms

In this section we will study the relation between orientation and differential forms.

Definition 6.1 A **volume form** (or **volume element**) on a manifold M of dimension n is an n-form ω such that $\omega_p \neq 0$ for all $p \in M$.

The existence of a volume form is equivalent to M being orientable.

Proposition 6.2 *A manifold M of dimension n is orientable if and only if there exists a volume form on M.*

Proof Let ω be a volume form on M, and consider an atlas $\{(U_\alpha, \varphi_\alpha)\}$. We can assume without loss of generality that the open sets U_α are connected. We will construct a new atlas from this one whose overlap maps have derivatives with positive determinant. Indeed, considering the representation of ω on one of these open sets $U_\alpha \subset \mathbb{R}^n$, we have

$$\varphi_\alpha^* \omega = a_\alpha dx_\alpha^1 \wedge \cdots \wedge dx_\alpha^n,$$

where the function a_α cannot vanish, and hence must have a fixed sign. If a_α is positive, we keep the corresponding parameterization. If not, we construct a new parameterization by composing φ_α with the map

$$(x^1, \ldots, x^n) \mapsto (-x^1, x^2, \ldots, x^n).$$

Clearly, in these new coordinates, the new function a_α is positive. Repeating this for all coordinate neighborhoods we obtain a new atlas for which all the functions a_α are positive, which we will also denote by $\{(U_\alpha, \varphi_\alpha)\}$. Moreover, whenever $W := \varphi_\alpha(U_\alpha) \cap \varphi_\beta(U_\beta) \neq \varnothing$, we have $\omega_\alpha = (\varphi_\beta^{-1} \circ \varphi_\alpha)^* \omega_\beta$. Hence,

$$a_\alpha dx_\alpha^1 \wedge \cdots \wedge dx_\alpha^n = (\varphi_\beta^{-1} \circ \varphi_\alpha)^* (a_\beta \, dx_\beta^1 \wedge \cdots \wedge dx_\beta^n)$$

$$= (a_\beta \circ \varphi_\beta^{-1} \circ \varphi_\alpha)(\det(d(\varphi_\beta^{-1} \circ \varphi_\alpha))) \, dx_\alpha^1 \wedge \cdots \wedge dx_\alpha^n$$

and so $\det(d(\varphi_\beta^{-1} \circ \varphi_\alpha)) > 0$. We conclude that M is orientable.

Conversely, if M is orientable, we consider an atlas $\{(U_\alpha, \varphi_\alpha)\}$ for which the overlap maps $\varphi_\beta^{-1} \circ \varphi_\alpha$ are such that $\det d(\varphi_\beta^{-1} \circ \varphi_\alpha) > 0$. Taking a partition of unity $\{\rho_i\}_{i \in I}$ subordinate to the cover of M by the corresponding coordinate neighborhoods, we may define the forms

$$\omega_i := \rho_i dx_i^1 \wedge \cdots \wedge dx_i^n$$

with $\operatorname{supp} \omega_i = \operatorname{supp} \rho_i \subset \varphi_{\alpha_i}(U_{\alpha_i})$. Extending these forms to M by making them zero outside $\operatorname{supp} \rho_i$, we may define the form $\omega := \sum_{i \in I} \omega_i$. Clearly ω is a well-defined n-form on M so we just need to show that $\omega_p \neq 0$ for all $p \in M$. Let p be a point in M. There is an $i \in I$ such that $\rho_i(p) > 0$, and so there exist linearly independent vectors $v_1, \ldots, v_n \in T_p M$ such that $(\omega_i)_p(v_1, \ldots, v_n) > 0$. Moreover, for all other $j \in I \backslash \{i\}$ we have $(\omega_j)_p(v_1, \ldots, v_n) \geq 0$. Indeed, if $p \notin \varphi_{\alpha_j}(U_{\alpha_j})$, then $(\omega_j)_p(v_1, \ldots, v_n) = 0$. On the other hand, if $p \in \varphi_{\alpha_j}(U_{\alpha_j})$, then by (2.1)

$$dx_j^1 \wedge \cdots \wedge dx_j^n = \det(d(\varphi_{\alpha_j}^{-1} \circ \varphi_{\alpha_i})) dx_i^1 \wedge \cdots \wedge dx_i^n$$

and hence

$$(\omega_j)_p(v_1, \ldots, v_n) = \frac{\rho_j(p)}{\rho_i(p)} (\det(d(\varphi_{\alpha_j}^{-1} \circ \varphi_{\alpha_i}))) (\omega_i)_p(v_1, \ldots, v_n) \geq 0.$$

Consequently, $\omega_p(v_1, \ldots, v_n) > 0$, and so ω is a volume form. $\qquad \square$

Remark 6.3 Sometimes we call a volume form an orientation. In this case the orientation on M is the one for which a basis $\{v_1, \ldots, v_n\}$ of $T_p M$ is positive if and only if $\omega_p(v_1, \ldots, v_n) > 0$.

If we fix a volume form $\omega \in \Omega^n(M)$ on an orientable manifold M, we can define the **integral** of any compactly supported function $f \in C^\infty(M, \mathbb{R})$ as

$$\int_M f := \int_M f\omega$$

(where the orientation of M is determined by ω). If M is compact, we define its **volume** to be

$$\operatorname{vol}(M) := \int_M 1 = \int_M \omega.$$

Exercise 6.4

(1) Show that $M \times N$ is orientable if and only if both M and N are orientable.
(2) Let M be a compact oriented manifold with volume element $\omega \in \Omega^n(M)$. Prove that if $f > 0$ then $\int_M f\omega > 0$. (**Remark:** In particular, the volume of a compact manifold is always positive).
(3) Let M be a compact orientable manifold of dimension n, and let ω be an $(n-1)$-form in M.

  (a) Show that there exists a point $p \in M$ for which $(d\omega)_p = 0$.

(b) Prove that there exists no immersion $f : S^1 \to \mathbb{R}$ of the unit circle into $\mathbb{R}$.

(4) Let $f : S^n \to S^n$ be the antipodal map. Recall that the n-dimensional projective space is the differential manifold $\mathbb{R}P^n = S^n/\mathbb{Z}_2$, where the group $\mathbb{Z}_2 = \{1, -1\}$ acts on S^n through $1 \cdot x = x$ and $(-1) \cdot x = f(x)$. Let $\pi : S^n \to \mathbb{R}P^n$ be the natural projection.

(a) Prove that $\omega \in \Omega^k(S^n)$ is of the form $\omega = \pi^*\theta$ for some $\theta \in \Omega^k(\mathbb{R}P^n)$ if and only if $f^*\omega = \omega$.

(b) Show that $\mathbb{R}P^n$ is orientable if and only if n is odd, and that in this case,

$$\int_{S^n} \pi^*\theta = 2 \int_{\mathbb{R}P^n} \theta.$$

(c) Show that for n even the sphere S^n is the orientable double covering of $\mathbb{R}P^n$ [cf. Exercise 8.6(9) in Chap. 1].

(5) Let M be a compact oriented manifold with boundary and $\omega \in \Omega^n(M)$ a volume element. The **divergence** of a vector field $X \in \mathfrak{X}(M)$ is the function $\operatorname{div}(X)$ such that

$$L_X\omega = (\operatorname{div}(X))\omega$$

[cf. Exercise 3.8(7)]. Show that

$$\int_M \operatorname{div}(X) = \int_{\partial M} \iota(X)\omega.$$

(6) (*Brouwer fixed point theorem*)

(a) Let M be an n-dimensional compact orientable manifold with boundary $\partial M \neq \varnothing$. Show that there exists no smooth map $f : M \to \partial M$ satisfying $f|_{\partial M} = \operatorname{id}$.

(b) Prove the **Brouwer fixed point theorem**: Any smooth map $g : B \to B$ of the closed ball $B := \{x \in \mathbb{R}^n \mid \|x\| \leq 1\}$ to itself has a fixed point, that is, a point $p \in B$ such that $g(p) = p$. (**Hint:** For each point $x \in B$, consider the ray r_x starting at $g(x)$ and passing through x. There is only one point $f(x)$ different from $g(x)$ on $r_x \cap \partial B$. Consider the map $f : B \to \partial B$).

2.7 Notes

2.7.1 Section 2.1

(1) Given a finite dimensional vector space V we define its **dual space** as the space of linear functionals on V.

Proposition 7.1 If $\{v_1, \ldots, v_n\}$ is a basis for V then there is a unique basis $\{T_1, \ldots, T_n\}$ of V^* dual to $\{v_1, \ldots, v_n\}$, that is, such that $T_i(v_j) = \delta_{ij}$.

Proof By linearity, the equations $T_i(v_j) = \delta_{ij}$ define a unique set of functionals $T_i \in V^*$. Indeed, for any $v \in V$, we have $v = \sum_{j=1}^{n} a_j v_j$ and so

$$T_i(v) = \sum_{j=1}^{n} a_j T_i(v_j) = \sum_{j=1}^{n} a_j \delta_{ij} = a_i.$$

Moreover, these uniquely defined functionals are linearly independent. In fact, if

$$T := \sum_{i=1}^{n} b_i T_i = 0,$$

then, for each $j = 1, \ldots, n$, we have

$$0 = T(v_j) = \sum_{i=1}^{n} b_i T_i(v_j) = b_j.$$

To show that $\{T_1, \ldots, T_n\}$ generates V^*, we take any $S \in V^*$ and set $b_i := S(v_i)$. Then, defining $T := \sum_{i=1}^{n} b_i T_i$, we see that $S(v_j) = T(v_j)$ for all $j = 1, \ldots, n$. Since $\{v_1, \ldots, v_n\}$ is a basis for V, we have $S = T$.   □

Moreover, if $\{v_1, \ldots, v_n\}$ is a basis for V and $\{T_1, \ldots, T_n\}$ is its dual basis, then, for any $v = \sum a_j v_j \in V$ and $T = \sum b_i T_i \in V^*$, we have

$$T(v) = \sum_{j=i}^{n} b_i T_i(v) = \sum_{i,j=1}^{n} a_j b_i T_i(v_j) = \sum_{i,j=1}^{n} a_j b_i \delta_{ij} = \sum_{i=1}^{n} a_i b_i.$$

If we now consider a linear functional F on V^*, that is, an element of $(V^*)^*$, we have $F(T) = T(v_0)$ for some fixed vector $v_0 \in V$. Indeed, let $\{v_1, \ldots, v_n\}$ be a basis for V and let $\{T_1, \ldots, T_n\}$ be its dual basis. Then if $T = \sum_{i=1}^{n} b_i T_i$, we have $F(T) = \sum_{i=1}^{n} b_i F(T_i)$. Denoting the values $F(T_i)$ by a_i, we get $F(T) = \sum_{i=1}^{n} a_i b_i = T(v_0)$ for $v_0 = \sum_{i=1}^{n} a_i v_i$. This establishes a one-to-one correspondence between $(V^*)^*$ and V, and allows us to view V as the space of linear functionals on V^*. For $v \in V$ and $T \in V^*$, we write $v(T) = T(v)$.

(2) Changing from a basis $\{v_1, \ldots, v_n\}$ to a new basis $\{v_1', \ldots, v_n'\}$ in V, we obtain a **change of basis matrix** S, whose jth column is the vector of coordinates of the new basis vector v_j' in the old basis. We can then write the symbolic matrix equation

$$(v_1', \ldots, v_n') = (v_1, \ldots, v_n)S.$$

The coordinate (column) vectors a and b of a vector $v \in V$ (a contravariant 1-tensor on V) with respect to the old basis and to the new basis are related by

$$b = \begin{pmatrix} b_1 \\ \vdots \\ b_n \end{pmatrix} = S^{-1} \begin{pmatrix} a_1 \\ \vdots \\ a_n \end{pmatrix} = S^{-1} a,$$

since we must have $(v'_1, \ldots, v'_n)b = (v_1, \ldots, v_n)a = (v'_1, \ldots, v'_n)S^{-1}a$. On the other hand, if $\{T_1, \ldots, T_n\}$ and $\{T'_1, \ldots, T'_n\}$ are the dual bases of $\{v_1, \ldots, v_n\}$ and $\{v'_1, \ldots, v'_n\}$, we have

$$\begin{pmatrix} T_1 \\ \vdots \\ T_n \end{pmatrix} (v_1, \ldots, v_n) = \begin{pmatrix} T'_1 \\ \vdots \\ T'_n \end{pmatrix} (v'_1, \ldots, v'_n) = I$$

(where, in the symbolic matrix multiplication above, each coordinate is obtained by applying the covectors to the vectors). Hence,

$$\begin{pmatrix} T_1 \\ \vdots \\ T_n \end{pmatrix} (v'_1, \ldots, v'_n) S^{-1} = I \Leftrightarrow S^{-1} \begin{pmatrix} T_1 \\ \vdots \\ T_n \end{pmatrix} (v'_1, \ldots, v'_n) = I,$$

implying that

$$\begin{pmatrix} T'_1 \\ \vdots \\ T'_n \end{pmatrix} = S^{-1} \begin{pmatrix} T_1 \\ \vdots \\ T_n \end{pmatrix}.$$

The coordinate (row) vectors $a = (a_1, \ldots, a_n)$ and $b = (b_1, \ldots, b_n)$ of a 1-tensor $T \in V^*$ (a covariant 1-tensor on V) with respect to the old basis $\{T_1, \ldots, T_n\}$ and to the new basis $\{T'_1, \ldots, T'_n\}$ are related by

$$a \begin{pmatrix} T_1 \\ \vdots \\ T_n \end{pmatrix} = b \begin{pmatrix} T'_1 \\ \vdots \\ T'_n \end{pmatrix} \Leftrightarrow aS \begin{pmatrix} T'_1 \\ \vdots \\ T'_n \end{pmatrix} = b \begin{pmatrix} T'_1 \\ \vdots \\ T'_n \end{pmatrix}$$

and so $b = aS$. Note that the coordinate vectors of the covariant 1-tensors on V transform like the basis vectors of V (that is, by means of the matrix S) whereas the coordinate vectors of the contravariant 1-tensors on V transform by means of the inverse of this matrix. This is the origin of the terms "covariant" and "contravariant".

2.7.2 Section 2.4

(1) *(Change of variables theorem)* Let $U, V \subset \mathbb{R}^n$ be open sets, let $g : U \to V$ be a diffeomorphism and let $f : V \to \mathbb{R}$ be an integrable function. Then

$$\int_V f = \int_U (f \circ g)|\det dg|.$$

(2) To define smooth objects on manifolds it is often useful to define them first on coordinate neighborhoods and then glue the pieces together by means of a **partition of unity**.

Theorem 7.1 *Let M be a smooth manifold and $\mathcal{V}$ an open cover of M. Then there is a family of differentiable functions on M, $\{\rho_i\}_{i \in I}$, such that:*

(i) *for every point $p \in M$, there exists a neighborhood U of p such that $U \cap \operatorname{supp} \rho_i = \varnothing$ except for a finite number of ρ_i;*
(ii) *for every point $p \in M$, $\sum_{i \in I} \rho_i(p) = 1$;*
(iii) $0 \le \rho_i \le 1$ *and* $\operatorname{supp} \rho_i \subset V$ *for some element* $V \in \mathcal{V}$.

Remark 7.2 This collection ρ_i of smooth functions is called partition of unity subordinate to the cover $\mathcal{V}$.

Proof Let us first assume that M is compact. For every point $p \in M$ we consider a coordinate neighborhood $W_p = \varphi_p(U_p)$ around p contained in an element V_p of $\mathcal{V}$, such that $\varphi_p(0) = p$ and $B_3(0) \subset U_p$ (where $B_3(0)$ denotes the ball of radius 3 around 0). Then we consider the C^∞-functions (cf. Fig. 2.1)

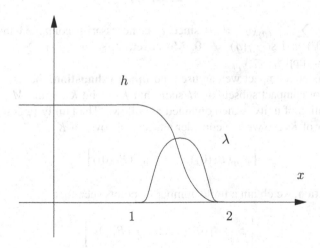

Fig. 2.1 Graphs of the functions λ and h

$$\lambda : \mathbb{R} \to \mathbb{R}$$

$$x \mapsto \begin{cases} e^{\frac{1}{(x-1)(x-2)}} & \text{if } 1 < x < 2 \\ 0 & \text{otherwise} \end{cases},$$

$$h : \mathbb{R} \to \mathbb{R}$$

$$x \mapsto \frac{\int_x^2 \lambda(t)\,dt}{\int_1^2 \lambda(t)\,dt},$$

$$\beta : \mathbb{R}^n \to \mathbb{R}$$

$$x \mapsto h(\|x\|).$$

Notice that h is a decreasing function with values $0 \le h(x) \le 1$, equal to zero for $x \ge 2$ and equal to 1 for $x \le 1$. Hence, we can consider **bump functions** $\gamma_p : M \to [0, 1]$ defined by

$$\gamma_p(q) = \begin{cases} \beta(\varphi_p^{-1}(q)) & \text{if } q \in \varphi_p(U_p) \\ 0 & \text{otherwise.} \end{cases}$$

Then $\operatorname{supp} \gamma_p = \overline{\varphi_p(B_2(0))} \subset \varphi_p(B_3(0)) \subset W_p$ is contained inside an element V_p of the cover. Moreover, $\{\varphi_p(B_1(0))\}_{p \in M}$ is an open cover of M and so we can consider a finite subcover $\{\varphi_{p_i}(B_1(0))\}_{i=1}^k$ such that $M = \cup_{i=1}^k \varphi_{p_i}(B_1(0))$. Finally we take the functions

$$\rho_i = \frac{\gamma_{p_i}}{\sum_{j=1}^k \gamma_{p_j}}.$$

Note that $\sum_{j=1}^k \gamma_{p_j}(q) \ne 0$ since q is necessarily contained inside some $\varphi_{p_i}(B_1(0))$ and so $\gamma_i(q) \ne 0$. Moreover, $0 \le \rho_i \le 1$, $\sum \rho_i = 1$ and $\operatorname{supp} \rho_i = \operatorname{supp} \gamma_{p_i} \subset V_{p_i}$.

If M is not compact we can use a **compact exhaustion**, that is, a sequence $\{K_i\}_{i \in \mathbb{N}}$ of compact subsets of M such that $K_i \subset \operatorname{int} K_{i+1}$ and $M = \cup_{i=1}^\infty K_i$. The partition of unity is then obtained as follows. The family $\{\varphi_p(B_1(0))\}_{p \in M}$ is a cover of K_1, so we can consider a finite subcover of K_1,

$$\left\{ \varphi_{p_1}(B_1(0)), \ldots, \varphi_{p_{k_1}}(B_1(0)) \right\}.$$

By induction, we obtain a finite number of points such that

$$\left\{ \varphi_{p_1^i}(B_1(0)), \ldots, \varphi_{p_{k_i}^i}(B_1(0)) \right\}$$

covers $K_i \setminus \operatorname{int} K_{i-1}$ (a compact set). Then, for each i, we consider the corresponding bump functions

$$\gamma_{p_1^i}, \dots, \gamma_{p_{k_i}^i} : M \to [0, 1].$$

Note that $\gamma_{p_1^i} + \cdots + \gamma_{p_{k_i}^i} > 0$ for every $q \in K_i \setminus \operatorname{int} K_{i-1}$ (as there is always one of these functions which is different from zero). As in the compact case, we can choose these bump functions so that $\operatorname{supp} \gamma_{p_j^i}$ is contained in some element of $\mathcal{V}$. We will also choose them so that $\operatorname{supp} \gamma_{p_j^i} \subset \int K_{i+1} \setminus K_{i-2}$ (an open set). Hence, $\{\gamma_{p_j^i}\}_{i \in \mathbb{N}, 1 \leq j \leq k_i}$ is locally finite, meaning that, given a point $p \in M$, there exists an open neighborhood V of p such that only a finite number of these functions is different from zero in V. Consequently, the sum $\sum_{i=1}^\infty \sum_{j=1}^{k_i} \gamma_{p_j^i}$ is a positive, differentiable function on M. Finally, making

$$\rho_j^i = \frac{\gamma_{p_j^i}}{\sum_{i=1}^\infty \sum_{j=1}^{k_i} \gamma_{p_j^i}},$$

we obtain the desired partition of unity (subordinate to $\mathcal{V}$). $\qquad\square$

Remark 7.3 Compact exhaustions always exist on manifolds. In fact, if U is a bounded open set of $\mathbb{R}^n$, one can easily construct a compact exhaustion $\{K_i\}_{i \in \mathbb{N}}$ for U by setting

$$K_i = \left\{ x \in U \mid \operatorname{dist}(x, \partial U) \geq \frac{1}{n} \right\}.$$

If M is a differentiable manifold, one can always take a countable atlas $\mathcal{A} = \{(U_j, \varphi_j)\}_{j \in \mathbb{N}}$ such that each U_j is a bounded open set, thus admitting a compact exhaustion $\{K_i^j\}_{i \in \mathbb{N}}$. Therefore

$$\left\{ \bigcup_{i+j=l} \varphi_j \left(K_i^j \right) \right\}_{l \in \mathbb{N}}$$

is a compact exhaustion of M.

2.7.3 Section 2.5

(Fubini theorem) Let $A \subset \mathbb{R}^n$ and $B \subset \mathbb{R}^m$ be compact intervals and let $f : A \times B \to \mathbb{R}$ be a continuous function. Then

$$\int_{A \times B} f = \int_A \left(\int_B f(x,y) dy^1 \cdots dy^m \right) dx^1 \cdots dx^n$$
$$= \int_B \left(\int_A f(x,y) dx^1 \cdots dx^n \right) dy^1 \cdots dy^m.$$

2.7.4 Bibliographical Notes

The material in this chapter can be found in most books on differential geometry
(e.g. [Boo03, GHL04]). A text entirely dedicated to differential forms and their
applications is [dC94]. The study of de Rham cohomology leads to a beautiful and
powerful theory, whose details can be found for instance in [BT82].

References

[Boo03]   Boothby, W.: An Introduction to Differentiable Manifolds and Riemannian Geometry.
          Academic Press, Boston (2003)
[BT82]   Bott, R., Tu, L.: Differential Forms in Algebraic Topology. Springer, New York (1982)
[dC94]   do Carmo, M.: Differential Forms and Applications. Springer, New York (1994)
[GHL04]   Gallot, S., Hulin, D., Lafontaine, J.: Riemannian Geometry. Springer, Berlin (2004)
[GP73]   Guillemin, V., Pollack, A.: Differential Topology. Prentice-Hall, Englewood Cliffs (1973)

Chapter 3
Riemannian Manifolds

The metric properties of $\mathbb{R}^n$ (distances and angles) are determined by the canonical Cartesian coordinates. In a general differentiable manifold, however, there are no such preferred coordinates; to define distances and angles one must add more structure by choosing a special 2-tensor field, called a **Riemannian metric** (much in the same way as a volume form must be selected to determine a notion of volume). This idea was introduced by Riemann in his 1854 habilitation lecture "On the hypotheses which underlie geometry", following the discovery (around 1830) of non-Euclidean geometry by Gauss, Bolyai and Lobachevsky (in fact, it was Gauss who suggested the subject of Riemann's lecture). It proved to be an extremely fruitful concept, having led, among other things, to the development of Einstein's general theory of relativity.

This chapter initiates the study of Riemannian geometry. Section 3.1 introduces Riemannian metrics as tensor fields determining an inner product at each tangent space. This naturally leads to a number of concepts, such as the **length** of a vector (or a curve), the **angle** between two vectors, the **Riemannian volume form** (which assigns unit volume to any orthonormal basis) and the **gradient** of a function.

Section 3.2 discusses differentiation of vector fields. This concept also requires introducing some additional structure, called an **affine connection**, since vector fields on a differentiable manifold do not have preferred Cartesian components to be differentiated. It provides a notion of **parallelism** of vectors along curves, and consequently of **geodesics**, that is, curves whose tangent vector is parallel. Riemannian manifolds come equipped with a special affine connection, called the **Levi-Civita connection** (Sect. 3.3), whose geodesics have distance-minimizing properties (Sect. 3.4). This is in line with the intuitive idea that the shortest distance route between two points is one that does not turn.

Finally, the **Hopf-Rinow theorem**, relating the properties of a Riemannian manifold as a metric space to the properties of its geodesics, is proved in Sect. 3.5. This theorem completely characterizes the important class of **complete** Riemannian manifolds.

© Springer International Publishing Switzerland 2014
L. Godinho and J. Natário, *An Introduction to Riemannian Geometry*, Universitext,
DOI 10.1007/978-3-319-08666-8_3

3.1 Riemannian Manifolds

To define Riemannian manifolds we must first take a closer look at 2-tensors.

Definition 1.1 A tensor $g \in T^2(T_p^* M)$ is said to be

(i) **symmetric** if $g(v, w) = g(w, v)$ for all $v, w \in T_p M$;
(ii) **nondegenerate** if $g(v, w) = 0$ for all $w \in T_p M$ implies $v = 0$;
(iii) **positive definite** if $g(v, v) > 0$ for all $v \in T_p M \setminus \{0\}$.

A covariant 2-tensor field g is said to be symmetric, nondegenerate or positive definite if g_p is symmetric, nondegenerate or positive definite for all $p \in M$. If $x : V \to \mathbb{R}^n$ is a local chart, we have

$$g = \sum_{i,j=1}^{n} g_{ij} dx^i \otimes dx^j$$

in V, where

$$g_{ij} = g\left(\frac{\partial}{\partial x^i}, \frac{\partial}{\partial x^j}\right).$$

It is easy to see that g is symmetric, nondegenerate or positive definite if and only if the matrix (g_{ij}) has these properties [see Exercise 1.10(1)].

Definition 1.2 A **Riemannian metric** on a smooth manifold M is a symmetric positive definite smooth covariant 2-tensor field g. A smooth manifold M equipped with a Riemannian metric g is called a **Riemannian manifold**, and is denoted by (M, g).

A Riemannian metric is therefore a smooth assignment of an inner product to each tangent space. It is usual to write

$$g_p(v, w) = \langle v, w \rangle_p.$$

Example 1.3 (Euclidean n-space) It should be clear that $M = \mathbb{R}^n$ and

$$g = \sum_{i=1}^{n} dx^i \otimes dx^i$$

define a Riemannian manifold.

Proposition 1.4 *Let (N, g) be a Riemannian manifold and $f : M \to N$ an immersion. Then $f^* g$ is a Riemannian metric in M (called the* **induced metric***).*

Proof We just have to prove that $f^* g$ is symmetric and positive definite. Let $p \in M$ and $v, w \in T_p M$. Since g is symmetric,

$$(f^*g)_p(v, w) = g_{f(p)}((df)_p v, (df)_p w) = g_{f(p)}((df)_p w, (df)_p v) = (f^*g)_p(w, v).$$

On the other hand, it is clear that $(f^*g)_p(v, v) \geq 0$, and

$$(f^*g)_p(v, v) = 0 \Rightarrow g_{f(p)}((df)_p v, (df)_p v) = 0 \Rightarrow (df)_p v = 0 \Rightarrow v = 0$$

(as $(df)_p$ is injective). $\square$

In particular, any submanifold M of a Riemannian manifold (N, g) is itself a Riemannian manifold. Notice that, in this case, the induced metric at each point $p \in M$ is just the restriction of g_p to $T_p M \subset T_p N$. Since $\mathbb{R}^n$ is a Riemannian manifold (cf. Example 1.3), we see that any submanifold of $\mathbb{R}^n$ is a Riemannian manifold. The Whitney theorem then implies that any manifold admits a Riemannian metric.

It was proved in 1954 by John Nash [Nas56] that any compact n-dimensional Riemannian manifold can be isometrically embedded in $\mathbb{R}^N$ for $N = \frac{n(3n+11)}{2}$ (that is, embedded in such a way that its metric is induced by the Euclidean metric of $\mathbb{R}^N$). Gromov [GR70] later proved that one can take $N = \frac{(n+2)(n+3)}{2}$. Notice that, for $n = 2$, Nash's result gives an isometric embedding of any compact surface in $\mathbb{R}^{17}$, and Gromov's in $\mathbb{R}^{10}$. In fact, Gromov has further showed that any surface isometrically embeds in $\mathbb{R}^5$. This result cannot be improved, as the real projective plane with the standard metric [see Exercise 1.10(3)] cannot be isometrically embedded into $\mathbb{R}^4$.

Example 1.5 The **standard metric** on

$$S^n = \{x \in \mathbb{R}^{n+1} \mid \|x\| = 1\}$$

is the metric induced on S^n by the Euclidean metric on $\mathbb{R}^{n+1}$. A parameterization of the open set

$$U = \{x \in S^n \mid x^{n+1} > 0\}$$

is for instance

$$\varphi(x^1, \ldots, x^n) = \left(x^1, \ldots, x^n, \sqrt{1 - (x^1)^2 - \cdots - (x^n)^2}\right),$$

and the corresponding coefficients of the metric tensor are

$$g_{ij} = \left\langle \frac{\partial \varphi}{\partial x^i}, \frac{\partial \varphi}{\partial x^j} \right\rangle = \delta_{ij} + \frac{x^i x^j}{1 - (x^1)^2 - \cdots - (x^n)^2}.$$

Two Riemannian manifolds will be regarded as the same if they are **isometric**.

Definition 1.6 Let (M, g) and (N, h) be Riemannian manifolds. A diffeomorphism $f : M \to N$ is said to be an **isometry** if $f^*h = g$. Similarly, a local diffeomorphism $f : M \to N$ is said to be a **local isometry** if $f^*h = g$.

A Riemannian metric allows us to compute the **length** $\|v\| = \langle v, v \rangle^{\frac{1}{2}}$ of any vector $v \in TM$ (as well as the **angle** between two vectors with the same base point). Therefore we can measure the length of curves.

Definition 1.7 If $(M, \langle \cdot, \cdot \rangle)$ is a Riemannian manifold and $c : [a, b] \to M$ is a differentiable curve, the **length** of c is

$$l(c) = \int_a^b \|\dot{c}(t)\| dt.$$

The length of a curve segment does not depend on the parameterization [see Exercise 1.10(5)].

Recall that if M is an orientable n-dimensional manifold then it possesses volume elements, that is, differential forms $\omega \in \Omega^n(M)$ such that $\omega_p \neq 0$ for all $p \in M$. Clearly, there are as many volume elements as differentiable functions $f \in C^\infty(M)$ without zeros.

Definition 1.8 If (M, g) is an orientable Riemannian manifold, $\omega \in \Omega^n(M)$ is said to be a **Riemannian volume element** if

$$\omega_p(v_1, \ldots, v_n) = \pm 1$$

for any orthonormal basis $\{v_1, \ldots, v_n\}$ of $T_p M$ and all $p \in M$.

Notice that if M is connected there exist exactly two Riemannian volume elements (one for each choice of orientation). Moreover, if ω is a Riemannian volume element and $x : V \to \mathbb{R}$ is a chart compatible with the orientation induced by ω, one has

$$\omega = f dx^1 \wedge \ldots \wedge dx^n$$

for some positive function

$$f = \omega \left(\frac{\partial}{\partial x^1}, \ldots, \frac{\partial}{\partial x^n} \right).$$

If S is the matrix whose columns are the components of $\frac{\partial}{\partial x^1}, \ldots, \frac{\partial}{\partial x^n}$ on some orthonormal basis with the same orientation, we have

$$f = \det S = \left(\det \left(S^2 \right) \right)^{\frac{1}{2}} = \left(\det \left(S^t S \right) \right)^{\frac{1}{2}} = \left(\det(g_{ij}) \right)^{\frac{1}{2}}$$

since clearly $S^t S$ is the matrix whose (i, j)th entry is the inner product $g \left(\frac{\partial}{\partial x^i}, \frac{\partial}{\partial x^j} \right) = g_{ij}$.

A Riemannian metric $\langle \cdot, \cdot \rangle$ on M determines a linear isomorphism $\Phi_g : T_p M \to T_p^* M$ for all $p \in M$ defined by $\Phi_g(v)(w) = \langle v, w \rangle$ for all $v, w \in T_p M$. This extends to an isomorphism between $\mathfrak{X}(M)$ and $\Omega^1(M)$. In particular, we have

Definition 1.9 Let (M, g) be a Riemannian manifold and $f : M \to \mathbb{R}$ a smooth function. The **gradient** of f is the vector field grad f associated to the 1-form df through the isomorphism determined by g.

Exercise 1.10

(1) Let $g = \sum_{i,j=1}^n g_{ij} \, dx^i \otimes dx^j \in T^2(T_p^* M)$. Show that:

   (a) g is symmetric if and only if $g_{ij} = g_{ji}$ $(i, j = 1, \ldots, n)$;
   (b) g is nondegenerate if and only if $\det(g_{ij}) \neq 0$;
   (c) g is positive definite if and only if (g_{ij}) is a positive definite matrix;
   (d) if g is nondegenerate, the map $\Phi_g : T_p M \to T_p^* M$ given by $\Phi_g(v)(w) = g(v, w)$ for all $v, w \in T_p M$ is a linear isomorphism;
   (e) if g is positive definite then g is nondegenerate.

(2) Prove that any differentiable manifold admits a Riemannian structure without invoking the Whitney theorem. (**Hint:** Use partitions of unity).

(3) (a) Let (M, g) be a Riemannian manifold and let G be a discrete Lie group acting freely and properly on M by isometries. Show that M/G has a natural Riemannian structure (called the **quotient** structure).
   (b) How would you define the **flat square metric** on the n-torus $T^n = \mathbb{R}^n / \mathbb{Z}^n$?
   (c) How would you define the **standard metric** on the real projective n-space $\mathbb{R}P^n = S^n / \mathbb{Z}_2$?

(4) Recall that given a Lie group G and $x \in G$, the left translation by x is the diffeomorphism $L_x : G \to G$ given by $L_x(y) = xy$ for all $y \in G$. A Riemannian metric g on G is said to be **left-invariant** if L_x is an isometry for all $x \in G$. Show that:

   (a) $g(\cdot, \cdot) \equiv \langle \cdot, \cdot \rangle$ is left-invariant if and only if

$$\langle v, w \rangle_x = \left\langle \left(dL_{x^{-1}} \right)_x v, \left(dL_{x^{-1}} \right)_x w \right\rangle_e$$

   for all $x \in G$ and $v, w \in T_x G$, where $e \in G$ is the identity and $\langle \cdot, \cdot \rangle_e$ is an inner product on the Lie algebra $\mathfrak{g} = T_e G$;
   (b) the standard metric on $S^3 \cong SU(2)$ is left-invariant;
   (c) the metric induced on $O(n)$ by the Euclidean metric of $\mathcal{M}_{n \times n} \cong \mathbb{R}^{n^2}$ is left-invariant.

(5) We say that a differentiable curve $\gamma : [\alpha, \beta] \to M$ is obtained from the curve $c : [a, b] \to M$ by **reparameterization** if there exists a smooth bijection $f : [\alpha, \beta] \to [a, b]$ (the reparameterization) such that $\gamma = c \circ f$. Show that if γ is obtained from c by reparameterization then $l(\gamma) = l(c)$.

(6) Let (M, g) be a Riemannian manifold and $f \in C^\infty(M)$. Show that if $a \in \mathbb{R}$ is a regular value of f then grad(f) is orthogonal to the submanifold $f^{-1}(a)$.

3.2 Affine Connections

If X and Y are vector fields in Euclidean space, we can define the **directional derivative** $\nabla_X Y$ of Y along X. This definition, however, uses the existence of Cartesian
coordinates, which no longer holds in a general manifold. To overcome this difficulty
we must introduce more structure:

Definition 2.1 Let M be a differentiable manifold. An **affine connection** on M is a
map $\nabla : \mathfrak{X}(M) \times \mathfrak{X}(M) \to \mathfrak{X}(M)$ such that

   (i) $\nabla_{fX+gY} Z = f \nabla_X Z + g \nabla_Y Z$;
  (ii) $\nabla_X (Y + Z) = \nabla_X Y + \nabla_X Z$;
 (iii) $\nabla_X (fY) = (X \cdot f) Y + f \nabla_X Y$

for all $X, Y, Z \in \mathfrak{X}(M)$ and $f, g \in C^\infty(M, \mathbb{R})$ (we write $\nabla_X Y := \nabla(X, Y)$).

The vector field $\nabla_X Y$ is sometimes known as the **covariant derivative** of Y
along X.

Proposition 2.2 *Let ∇ be an affine connection on M, let $X, Y \in \mathfrak{X}(M)$ and $p \in M$.
Then $(\nabla_X Y)_p \in T_p M$ depends only on X_p and on the values of Y along a curve
tangent to X at p. Moreover, if $x : W \to \mathbb{R}^n$ are local coordinates on some open set
$W \subset M$ and*

$$X = \sum_{i=1}^n X^i \frac{\partial}{\partial x^i}, \quad Y = \sum_{i=1}^n Y^i \frac{\partial}{\partial x^i}$$

on this set, we have

$$\nabla_X Y = \sum_{i=1}^n \left(X \cdot Y^i + \sum_{j,k=1}^n \Gamma^i_{jk} X^j Y^k \right) \frac{\partial}{\partial x^i} \tag{3.1}$$

*where the n^3 differentiable functions $\Gamma^i_{jk} : W \to \mathbb{R}$, called the **Christoffel symbols**,
are defined by*

$$\nabla_{\frac{\partial}{\partial x^j}} \frac{\partial}{\partial x^k} = \sum_{i=1}^n \Gamma^i_{jk} \frac{\partial}{\partial x^i}. \tag{3.2}$$

Proof It is easy to show that an affine connection is **local**, that is, if $X, Y \in \mathfrak{X}(M)$
coincide with $\widetilde{X}, \widetilde{Y} \in \mathfrak{X}(M)$ in some open set $W \subset M$ then $\nabla_X Y = \nabla_{\widetilde{X}} \widetilde{Y}$ on W
[see Exercise 2.6(1)]. Consequently, we can compute $\nabla_X Y$ for vector fields X, Y
defined on W only. Let W be a coordinate neighborhood for the local coordinates $x :
W \to \mathbb{R}^n$, and define the Christoffel symbols associated with these local coordinates
through (3.2). Writing out

$$\nabla_X Y = \nabla_{\left(\sum_{i=1}^n X^i \frac{\partial}{\partial x^i} \right)} \left(\sum_{j=1}^n Y^j \frac{\partial}{\partial x^j} \right)$$

and using the properties listed in definition (2.1) yields (3.1). This formula shows that $(\nabla_X Y)_p$ depends only on $X^i(p)$, $Y^i(p)$ and $(X \cdot Y^i)(p)$. Moreover, $X^i(p)$ and $Y^i(p)$ depend only on X_p and Y_p, and $(X \cdot Y^i)(p) = \frac{d}{dt} Y^i(c(t))|_{t=0}$ depends only on the values of Y^i (or Y) along a curve c whose tangent vector at $p = c(0)$ is X_p.

$\square$

Remark 2.3 Locally, an affine connection is uniquely determined by specifying its Christoffel symbols on a coordinate neighborhood. However, the choices of Christoffel symbols on different charts are not independent, as the covariant derivative must agree on the overlap.

A **vector field defined along a differentiable curve** $c : I \to M$ is a differentiable map $V : I \to TM$ such that $V(t) \in T_{c(t)}M$ for all $t \in I$. An obvious example is the tangent vector $\dot{c}(t)$. If V is a vector field defined along the differentiable curve $c : I \to M$ with $\dot{c} \neq 0$, its **covariant derivative** along c is the vector field defined along c given by

$$\frac{DV}{dt}(t) := \nabla_{\dot{c}(t)} V = (\nabla_X Y)_{c(t)}$$

for any vector fields $X, Y \in \mathfrak{X}(M)$ such that $X_{c(t)} = \dot{c}(t)$ and $Y_{c(s)} = V(s)$ with $s \in (t - \varepsilon, t + \varepsilon)$ for some $\varepsilon > 0$. Note that if $\dot{c}(t) \neq 0$ such extensions always exist. Proposition 2.2 guarantees that $(\nabla_X Y)_{c(t)}$ does not depend on the choice of X, Y. In fact, if in local coordinates $x : W \to \mathbb{R}^n$ we have $x^i(t) := x^i(c(t))$ and

$$V(t) = \sum_{i=1}^{n} V^i(t) \left(\frac{\partial}{\partial x^i}\right)_{c(t)},$$

then

$$\frac{DV}{dt}(t) = \sum_{i=1}^{n} \left(\dot{V}^i(t) + \sum_{j,k=1}^{n} \Gamma^i_{jk}(c(t))\dot{x}^j(t)V^k(t)\right) \left(\frac{\partial}{\partial x^i}\right)_{c(t)}.$$

Definition 2.4 A vector field V defined along a curve $c : I \to M$ is said to be **parallel along** c if

$$\frac{DV}{dt}(t) = 0$$

for all $t \in I$. The curve c is called a **geodesic** of the connection ∇ if $\dot{c}$ is parallel along c, i.e. if

$$\frac{D\dot{c}}{dt}(t) = 0$$

for all $t \in I$.

In local coordinates $x : W \to \mathbb{R}^n$, the condition for V to be parallel along c is written as

$$\dot{V}^i + \sum_{j,k=1}^n \Gamma^i_{jk} \dot{x}^j V^k = 0 \quad (i = 1, \ldots, n). \tag{3.3}$$

This is a system of first-order linear ODEs for the components of V. By the Picard–Lindelöf theorem, together with the global existence theorem for linear ODEs [Arn92], given a curve $c : I \to M$, a point $p \in c(I)$ and a vector $v \in T_pM$, there exists a unique vector field $V : I \to TM$ parallel along c such that $V(0) = v$, which is called the **parallel transport** of v along c.

Moreover, the geodesic equations are

$$\ddot{x}^i + \sum_{j,k=1}^n \Gamma^i_{jk} \dot{x}^j \dot{x}^k = 0 \quad (i = 1, \ldots, n). \tag{3.4}$$

This is a system of second-order (nonlinear) ODEs for the coordinates of $c(t)$. Therefore the Picard–Lindelöf theorem implies that, given a point $p \in M$ and a vector $v \in T_pM$, there exists a unique geodesic $c : I \to M$, defined on a maximal open interval I such that $0 \in I$, satisfying $c(0) = p$ and $\dot{c}(0) = v$.

We will now define the torsion of an affine connection ∇. For that, we note that, in local coordinates $x : W \to \mathbb{R}^n$, we have

$$\nabla_X Y - \nabla_Y X = \sum_{i=1}^n \left(X \cdot Y^i - Y \cdot X^i + \sum_{j,k=1}^n \Gamma^i_{jk} \left(X^j Y^k - Y^j X^k \right) \right) \frac{\partial}{\partial x^i}$$

$$= [X, Y] + \sum_{i,j,k=1}^n \left(\Gamma^i_{jk} - \Gamma^i_{kj} \right) X^j Y^k \frac{\partial}{\partial x^i}.$$

Definition 2.5 The **torsion operator** of a connection ∇ on M is the operator $T : \mathfrak{X}(M) \times \mathfrak{X}(M) \to \mathfrak{X}(M)$ given by

$$T(X, Y) = \nabla_X Y - \nabla_Y X - [X, Y],$$

for all $X, Y \in \mathfrak{X}(M)$. The connection is said to be **symmetric** if $T = 0$.

The local expression of $T(X, Y)$ makes it clear that $T(X, Y)_p$ depends linearly on X_p and Y_p. In other words, T is the $(2, 1)$-tensor field on M given in local coordinates by

$$T = \sum_{i,j,k=1}^n \left(\Gamma^i_{jk} - \Gamma^i_{kj} \right) dx^j \otimes dx^k \otimes \frac{\partial}{\partial x^i}$$

(recall from Remark 1.3 in Chap. 2 that any $(2, 1)$-tensor $T \in \mathcal{T}^{2,1}(V^*, V)$ is naturally identified with a bilinear map $\Phi_T : V^* \times V^* \to V \cong V^{**}$ through $\Phi_T(v, w)(\alpha) := T(v, w, \alpha)$ for all $v, w \in V$, $\alpha \in V^*$).

Notice that the connection is symmetric if and only if $\nabla_X Y - \nabla_Y X = [X, Y]$ for all $X, Y \in \mathfrak{X}(M)$. In local coordinates, the condition for the connection to be symmetric is

$$\Gamma^i_{jk} = \Gamma^i_{kj} \quad (i, j, k = 1, \ldots, n)$$

(hence the name).

Exercise 2.6

(1) (a) Show that if $X, Y \in \mathfrak{X}(M)$ coincide with $\widetilde{X}, \widetilde{Y} \in \mathfrak{X}(M)$ in some open set $W \subset M$ then $\nabla_X Y = \nabla_{\widetilde{X}} \widetilde{Y}$ on W. (**Hint:** Use bump functions with support contained on W and the properties listed in definition 2.1).
   (b) Obtain the local coordinate expression (3.1) for $\nabla_X Y$.
   (c) Obtain the local coordinate Eq. (3.3) for the parallel transport law.
   (d) Obtain the local coordinate Eq. (3.4) for the geodesics of the connection ∇.
(2) Determine all affine connections on $\mathbb{R}^n$. Of these, determine the connections whose geodesics are straight lines $c(t) = at + b$ (with $a, b \in \mathbb{R}^n$).
(3) Let ∇ be an affine connection on M. If $\omega \in \Omega^1(M)$ and $X \in \mathfrak{X}(M)$, we define the **covariant derivative** of ω along X, $\nabla_X \omega \in \Omega^1(M)$, by

$$\nabla_X \omega(Y) = X \cdot (\omega(Y)) - \omega(\nabla_X Y)$$

for all $Y \in \mathfrak{X}(M)$.

   (a) Show that this formula defines indeed a 1-form, i.e. show that $(\nabla_X \omega(Y))(p)$ is a linear function of Y_p.
   (b) Show that
      (i) $\nabla_{fX+gY} \omega = f \nabla_X \omega + g \nabla_Y \omega$;
      (ii) $\nabla_X (\omega + \eta) = \nabla_X \omega + \nabla_X \eta$;
      (iii) $\nabla_X (f\omega) = (X \cdot f)\omega + f \nabla_X \omega$
      for all $X, Y \in \mathfrak{X}(M)$, $f, g \in C^\infty(M)$ and $\omega, \eta \in \Omega^1(M)$.
   (c) Let $x : W \to \mathbb{R}^n$ be local coordinates on an open set $W \subset M$, and take

$$\omega = \sum_{i=1}^n \omega_i dx^i.$$

   Show that

$$\nabla_X \omega = \sum_{i=1}^n \left(X \cdot \omega_i - \sum_{j,k=1}^n \Gamma^k_{ji} X^j \omega_k \right) dx^i.$$

   (d) Define the covariant derivative $\nabla_X T$ for an arbitrary tensor field T in M, and write its expression in local coordinates.

3.3 Levi-Civita Connection

In the case of a Riemannian manifold, there is a particular choice of connection, called the **Levi-Civita connection**, with special geometric properties.

Definition 3.1 A connection ∇ in a Riemannian manifold $(M, \langle \cdot, \cdot \rangle)$ is said to be **compatible** with the metric if

$$X \cdot \langle Y, Z \rangle = \langle \nabla_X Y, Z \rangle + \langle Y, \nabla_X Z \rangle$$

for all $X, Y, Z \in \mathfrak{X}(M)$.

If ∇ is compatible with the metric, then the inner product of two vector fields V_1 and V_2, parallel along a curve, is constant along the curve:

$$\frac{d}{dt} \langle V_1(t), V_2(t) \rangle = \langle \nabla_{\dot{c}(t)} V_1(t), V_2(t) \rangle + \langle V_1(t), \nabla_{\dot{c}(t)} V_2(t) \rangle = 0.$$

In particular, parallel transport preserves lengths of vectors and angles between vectors. Therefore, if $c : I \to M$ is a geodesic, then $\|\dot{c}(t)\| = k$ is constant. If $a \in I$, the length s of the geodesic between a and t is

$$s = \int_a^t \|\dot{c}(v)\| \, dv = \int_a^t k \, dv = k(t - a).$$

In other words, t is an affine function of the arclength s (and is therefore called an **affine parameter**). In particular, this shows that the parameters of two geodesics with the same image are affine functions of each other).

Theorem 3.2 (Levi–Civita) *If $(M, \langle \cdot, \cdot \rangle)$ is a Riemannian manifold then there exists a unique connection ∇ on M which is symmetric and compatible with $\langle \cdot, \cdot \rangle$. In local coordinates $(x^1, \ldots, x^n)$, the Christoffel symbols for this connection are*

$$\Gamma^i_{jk} = \frac{1}{2} \sum_{l=1}^n g^{il} \left(\frac{\partial g_{kl}}{\partial x^j} + \frac{\partial g_{jl}}{\partial x^k} - \frac{\partial g_{jk}}{\partial x^l} \right) \tag{3.5}$$

where $\left(g^{ij} \right) = \left(g_{ij} \right)^{-1}$.

Proof Let $X, Y, Z \in \mathfrak{X}(M)$. If the Levi-Civita connection exists then we must have

$$X \cdot \langle Y, Z \rangle = \langle \nabla_X Y, Z \rangle + \langle Y, \nabla_X Z \rangle;$$
$$Y \cdot \langle X, Z \rangle = \langle \nabla_Y X, Z \rangle + \langle X, \nabla_Y Z \rangle;$$
$$-Z \cdot \langle X, Y \rangle = -\langle \nabla_Z X, Y \rangle - \langle X, \nabla_Z Y \rangle,$$

as ∇ is compatible with the metric. Moreover, since ∇ is symmetric, we must also have

$$-\langle [X, Z], Y \rangle = -\langle \nabla_X Z, Y \rangle + \langle \nabla_Z X, Y \rangle;$$
$$-\langle [Y, Z], X \rangle = -\langle \nabla_Y Z, X \rangle + \langle \nabla_Z Y, X \rangle;$$
$$\langle [X, Y], Z \rangle = \langle \nabla_X Y, Z \rangle - \langle \nabla_Y X, Z \rangle.$$

Adding these six equalities, we obtain the **Koszul formula**

$$2 \langle \nabla_X Y, Z \rangle = X \cdot \langle Y, Z \rangle + Y \cdot \langle X, Z \rangle - Z \cdot \langle X, Y \rangle$$
$$- \langle [X, Z], Y \rangle - \langle [Y, Z], X \rangle + \langle [X, Y], Z \rangle.$$

Since $\langle \cdot, \cdot \rangle$ is nondegenerate and Z is arbitrary, this formula determines $\nabla_X Y$. Thus, if the Levi-Civita connection exists, it must be unique.

To prove existence, we **define** $\nabla_X Y$ through the Koszul formula. It is not difficult to show that this indeed defines a connection [cf. Exercise 3.3(1)]. Also, using this formula, we obtain

$$2 \langle \nabla_X Y - \nabla_Y X, Z \rangle = 2 \langle \nabla_X Y, Z \rangle - 2 \langle \nabla_Y X, Z \rangle = 2 \langle [X, Y], Z \rangle$$

for all $X, Y, Z \in \mathfrak{X}(M)$, and hence ∇ is symmetric. Finally, again using the Koszul formula, we have

$$2 \langle \nabla_X Y, Z \rangle + 2 \langle Y, \nabla_X Z \rangle = 2 X \cdot \langle Y, Z \rangle$$

and therefore the connection defined by this formula is compatible with the metric.

Choosing local coordinates $(x^1, \ldots, x^n)$, we have

$$\left[\frac{\partial}{\partial x^i}, \frac{\partial}{\partial x^j} \right] = 0 \quad \text{and} \quad \left\langle \frac{\partial}{\partial x^i}, \frac{\partial}{\partial x^j} \right\rangle = g_{ij}.$$

Therefore the Koszul formula yields

$$2 \left\langle \nabla_{\frac{\partial}{\partial x^j}} \frac{\partial}{\partial x^k}, \frac{\partial}{\partial x^l} \right\rangle = \frac{\partial}{\partial x^j} \cdot g_{kl} + \frac{\partial}{\partial x^k} \cdot g_{jl} - \frac{\partial}{\partial x^l} \cdot g_{jk}$$

$$\Leftrightarrow \left\langle \sum_{i=1}^{n} \Gamma_{jk}^i \frac{\partial}{\partial x^i}, \frac{\partial}{\partial x^l} \right\rangle = \frac{1}{2} \left(\frac{\partial g_{kl}}{\partial x^j} + \frac{\partial g_{jl}}{\partial x^k} - \frac{\partial g_{jk}}{\partial x^l} \right)$$

$$\Leftrightarrow \sum_{i=1}^{n} g_{il} \Gamma_{jk}^i = \frac{1}{2} \left(\frac{\partial g_{kl}}{\partial x^j} + \frac{\partial g_{jl}}{\partial x^k} - \frac{\partial g_{jk}}{\partial x^l} \right).$$

This linear system is easily solved to give (3.5). $\square$

Exercise 3.3

(1) Show that the Koszul formula defines a connection.

(2) We introduce in $\mathbb{R}^3$, with the usual Euclidean metric $\langle \cdot, \cdot \rangle$, the connection ∇ defined in Cartesian coordinates (x^1, x^2, x^3) by

$$\Gamma^i_{jk} = \omega \varepsilon_{ijk},$$

where $\omega : \mathbb{R}^3 \to \mathbb{R}$ is a smooth function and

$$\varepsilon_{ijk} = \begin{cases} +1 & \text{if } (i, j, k) \text{ is an even permutation of } (1, 2, 3) \\ -1 & \text{if } (i, j, k) \text{ is an odd permutation of } (1, 2, 3) \\ 0 & \text{otherwise.} \end{cases}$$

Show that:

(a) ∇ is compatible with $\langle \cdot, \cdot \rangle$;
(b) the geodesics of ∇ are straight lines;
(c) the torsion of ∇ is not zero in all points where $\omega \neq 0$ (therefore ∇ is not the Levi-Civita connection unless $\omega \equiv 0$);
(d) the parallel transport equation is

$$\dot{V}^i + \sum_{j,k=1}^{3} \omega \varepsilon_{ijk} \dot{x}^j V^k = 0 \Leftrightarrow \dot{V} + \omega(\dot{x} \times V) = 0$$

(where $\times$ is the cross product in $\mathbb{R}^3$); therefore, a vector parallel along a straight line rotates about it with angular velocity $-\omega \dot{x}$.

(3) Let (M, g) and $(N, \tilde{g})$ be isometric Riemannian manifolds with Levi-Civita connections ∇ and $\tilde{\nabla}$, and let $f : M \to N$ be an isometry. Show that:

(a) $f_* \nabla_X Y = \tilde{\nabla}_{f_* X} f_* Y$ for all $X, Y \in \mathfrak{X}(M)$;
(b) if $c : I \to M$ is a geodesic then $f \circ c : I \to N$ is also a geodesic.

(4) Consider the usual local coordinates (θ, φ) in $S^2 \subset \mathbb{R}^3$ defined by the parameterization $\phi : (0, \pi) \times (0, 2\pi) \to \mathbb{R}^3$ given by

$$\phi(\theta, \varphi) = (\sin \theta \cos \varphi, \sin \theta \sin \varphi, \cos \theta).$$

(a) Using these coordinates, determine the expression of the Riemannian metric induced on S^2 by the Euclidean metric of $\mathbb{R}^3$.
(b) Compute the Christoffel symbols for the Levi-Civita connection in these coordinates.
(c) Show that the equator is the image of a geodesic.
(d) Show that any rotation about an axis through the origin in $\mathbb{R}^3$ induces an isometry of S^2.

(e) Show that the images of geodesics of S^2 are great circles.

(f) Find a **geodesic triangle** (i.e. a triangle whose sides are images of geodesics) whose internal angles add up to $\frac{3\pi}{2}$.

(g) Let $c : \mathbb{R} \to S^2$ be given by $c(t) = (\sin\theta_0 \cos t, \sin\theta_0 \sin t, \cos\theta_0)$, where $\theta_0 \in (0, \frac{\pi}{2})$ (therefore c is not a geodesic). Let V be a vector field parallel along c such that $V(0) = \frac{\partial}{\partial\theta}$ ($\frac{\partial}{\partial\theta}$ is well defined at $(\sin\theta_0, 0, \cos\theta_0)$ by continuity). Compute the angle by which V is rotated when it returns to the initial point. (**Remark:** The angle you have computed is exactly the angle by which the oscillation plane of the **Foucault pendulum** rotates during a day in a place at latitude $\frac{\pi}{2} - \theta_0$, as it tries to remain fixed with respect to the stars on a rotating Earth).

(h) Use this result to prove that no open set $U \subset S^2$ is isometric to an open set $W \subset \mathbb{R}^2$ with the Euclidean metric.

(i) Given a geodesic $c : \mathbb{R} \to \mathbb{R}^2$ of $\mathbb{R}^2$ with the Euclidean metric and a point $p \notin c(\mathbb{R})$, there exists a unique geodesic $\tilde{c} : \mathbb{R} \to \mathbb{R}^2$ (up to reparameterization) such that $p \in \tilde{c}(\mathbb{R})$ and $c(\mathbb{R}) \cap \tilde{c}(\mathbb{R}) = \emptyset$ (**parallel postulate**). Is this true in S^2?

(5) Recall that identifying each point in

$$H = \{(x, y) \in \mathbb{R}^2 \mid y > 0\}$$

with the invertible affine map $h : \mathbb{R} \to \mathbb{R}$ given by $h(t) = yt + x$ induces a Lie group structure on H [cf. Exercise 7.17(3) in Chap. 1].

(a) Show that the left-invariant metric induced by the Euclidean inner product $dx \otimes dx + dy \otimes dy$ in $\mathfrak{h} = T_{(0,1)}H$ is

$$g = \frac{1}{y^2} (dx \otimes dx + dy \otimes dy)$$

[cf. Exercise 1.10(4)]. (**Remark:** H endowed with this metric is called the **hyperbolic plane**).

(b) Compute the Christoffel symbols of the Levi-Civita connection in the coordinates (x, y).

(c) Show that the curves $\alpha, \beta : \mathbb{R} \to H$ given in these coordinates by

$$\alpha(t) = (0, e^t)$$

$$\beta(t) = \left(\tanh t, \frac{1}{\cosh t}\right)$$

are geodesics. What are the sets $\alpha(\mathbb{R})$ and $\beta(\mathbb{R})$?

(d) Determine all images of geodesics.

(e) Show that, given two points $p, q \in H$, there exists a unique geodesic through them (up to reparameterization).

(f) Give examples of connected Riemannian manifolds containing two points through which there are (i) infinitely many geodesics (up to reparameterization); (ii) no geodesics.

(g) Show that no open set $U \subset H$ is isometric to an open set $V \subset \mathbb{R}^2$ with the Euclidean metric. (**Hint:** Show that in any neighborhood of any point $p \in H$ there is always a geodesic quadrilateral whose internal angles add up to less than 2π).

(h) Does the parallel postulate hold in the hyperbolic plane?

(6) Let $(M, \langle \cdot, \cdot \rangle)$ be a Riemannian manifold with Levi-Civita connection $\widetilde{\nabla}$, and let $(N, \langle\langle \cdot, \cdot \rangle\rangle)$ be a submanifold with the induced metric and Levi-Civita connection ∇.

(a) Show that

$$\nabla_X Y = \left(\widetilde{\nabla}_{\widetilde{X}} \widetilde{Y} \right)^\top$$

for all $X, Y \in \mathfrak{X}(N)$, where $\widetilde{X}, \widetilde{Y}$ are any extensions of X, Y to $\mathfrak{X}(M)$ and $^\top : TM|_N \to TN$ is the orthogonal projection.

(b) Use this result to indicate curves that are, and curves that are not, geodesics of the following surfaces in $\mathbb{R}^3$:
   (i) the sphere S^2;
   (ii) the torus of revolution;
   (iii) the surface of a cone;
   (iv) a general surface of revolution.

(c) Show that if two surfaces in $\mathbb{R}^3$ are tangent along a curve, then the parallel transport of vectors along this curve in both surfaces coincides.

(d) Use this result to compute the angle $\Delta\theta$ by which a vector V is rotated when it is parallel transported along a circle on the sphere. (**Hint:** Consider the cone which is tangent to the sphere along the circle (cf. Fig. 3.1); notice that the cone minus a ray through the vertex is isometric to an open set of the Euclidean plane).

(7) Let (M, g) be a Riemannian manifold with Levi-Civita connection ∇. Show that g is **parallel** along any curve, i.e. show that

$$\nabla_X g = 0$$

for all $X \in \mathfrak{X}(M)$ [cf. Exercise 2.6(3)].

(8) Let (M, g) be a Riemannian manifold with Levi-Civita connection ∇, and let $\psi_t : M \to M$ be a 1-parameter group of isometries. The vector field $X \in \mathfrak{X}(M)$ defined by

$$X_p := \frac{d}{dt}\bigg|_{t=0} \psi_t(p)$$

is called the **Killing vector field** associated to ψ_t. Show that:

(a) $L_X g = 0$ [cf. Exercise 2.8(3)];
(b) X satisfies $\langle \nabla_Y X, Z \rangle + \langle \nabla_Z X, Y \rangle = 0$ for all vector fields $Y, Z \in \mathfrak{X}(M)$;

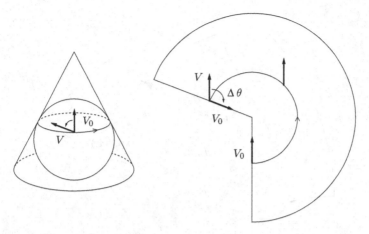

Fig. 3.1 Parallel transport along a circle on the sphere

(c) if $c : I \to M$ is a geodesic then $\langle \dot{c}(t), X_{c(t)} \rangle$ is constant.

(9) Recall that if M is an oriented differential manifold with volume element $\omega \in \Omega^n(M)$, the **divergence** of X is the function $\text{div}(X)$ such that

$$L_X \omega = (\text{div}(X))\omega$$

[cf. Exercise 6.4(5) in Chap. 2]. Suppose that M has a Riemannian metric and that ω is a Riemannian volume element. Show that at each point $p \in M$,

$$\text{div}(X) = \sum_{i=1}^{n} \langle \nabla_{Y_i} X, Y_i \rangle,$$

where $\{Y_1, \ldots, Y_n\}$ is an orthonormal basis of $T_p M$ and ∇ is the Levi-Civita connection.

3.4 Minimizing Properties of Geodesics

Let M be a differentiable manifold with an affine connection ∇. As we saw in Sect. 3.2, given a point $p \in M$ and a tangent vector $v \in T_p M$, there exists a unique geodesic $c_v : I \to M$, defined on a maximal open interval $I \subset \mathbb{R}$, such that $0 \in I$, $c_v(0) = p$ and $\dot{c}_v(0) = v$. Consider now the curve $\gamma : J \to M$ defined by $\gamma(t) = c_v(at)$, where $a \in \mathbb{R}$ and J is the inverse image of I by the map $t \mapsto at$. We have

$$\dot{\gamma}(t) = a\dot{c}_v(at),$$

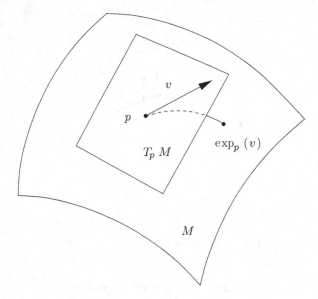

Fig. 3.2 The exponential map

and consequently

$$\nabla_{\dot\gamma}\dot\gamma = \nabla_{a\dot c_v}(a\dot c_v) = a^2\nabla_{\dot c_v}\dot c_v = 0.$$

Thus γ is also a geodesic. Since $\gamma(0) = c_v(0) = p$ and $\dot\gamma(0) = a\dot c_v(0) = av$, we see that γ is the unique geodesic with initial velocity $av \in T_pM$ (that is, $\gamma = c_{av}$). Therefore, we have $c_{av}(t) = c_v(at)$ for all $t \in I$. This property is sometimes referred to as the **homogeneity** of geodesics. Notice that we can make the interval J arbitrarily large by making a sufficiently small. If $1 \in I$, we define $\exp_p(v) = c_v(1)$. By homogeneity of geodesics, we can define $\exp_p(v)$ for v in some open neighborhood U of the origin in T_pM. The map $\exp_p : U \subset T_pM \to M$ thus obtained is called the **exponential map** at p (Fig. 3.2).

Proposition 4.1 *There exists an open set $U \subset T_pM$ containing the origin such that $\exp_p : U \to M$ is a diffeomorphism onto some open set $V \subset M$ containing p (called a **normal neighborhood**).*

Proof The exponential map is clearly differentiable as a consequence of the smooth dependence of the solution of an ODE on its initial data (cf. [Arn92]). If $v \in T_pM$ is such that $\exp_p(v)$ is defined, we have, by homogeneity, that $\exp_p(tv) = c_{tv}(1) = c_v(t)$. Consequently,

$$\left(d\exp_p\right)_0 v = \frac{d}{dt}\exp_p(tv)_{|t=0} = \frac{d}{dt}c_v(t)_{|t=0} = v.$$

We conclude that $(d\exp_p)_0 : T_0(T_pM) \cong T_pM \to T_pM$ is the identity map. By the inverse function theorem, $\exp_p$ is then a diffeomorphism of some open neighborhood U of $0 \in T_pM$ onto some open set $V \subset M$ containing $p = \exp_p(0)$. $\square$

Example 4.2 Consider the Levi-Civita connection in S^2 with the standard metric, and let $p \in S^2$. Then $\exp_p(v)$ is well defined for all $v \in T_pS^2$, but it is not a diffeomorphism, as it is clearly not injective. However, its restriction to the open ball $B_\pi(0) \subset T_pS^2$ is a diffeomorphism onto $S^2 \setminus \{-p\}$.

Now let $(M, \langle \cdot, \cdot \rangle)$ be a Riemannian manifold and ∇ its Levi-Civita connection. Since $\langle \cdot, \cdot \rangle$ defines an inner product in T_pM, we can think of T_pM as the Euclidean n-space $\mathbb{R}^n$. Let E be the vector field defined on $T_pM \setminus \{0\}$ by

$$E_v = \frac{v}{\|v\|},$$

and define $X := (\exp_p)_* E$ on $V \setminus \{p\}$, where $V \subset M$ is a normal neighborhood. We have

$$X_{\exp_p(v)} = \left(d\exp_p\right)_v E_v = \frac{d}{dt}\exp_p\left(v + t\frac{v}{\|v\|}\right)_{|t=0}$$

$$= \frac{d}{dt}c_v\left(1 + \frac{t}{\|v\|}\right)_{|t=0} = \frac{1}{\|v\|}\dot{c}_v(1).$$

Since $\|\dot{c}_v(1)\| = \|\dot{c}_v(0)\| = \|v\|$, we see that $X_{\exp_p(v)}$ is the unit tangent vector to the geodesic c_v. In particular, X must satisfy

$$\nabla_X X = 0.$$

For $\varepsilon > 0$ such that $\overline{B_\varepsilon(0)} \subset U := \exp_p^{-1}(V)$, we define the **normal ball** with center p and radius ε as the open set $B_\varepsilon(p) := \exp_p(B_\varepsilon(0))$, and the **normal sphere** of radius ε centered at p as the compact submanifold $S_\varepsilon(p) := \exp_p(\partial B_\varepsilon(0))$. We will now prove that X is (and hence the geodesics through p are) orthogonal to normal spheres. For that, we choose a local parameterization $\varphi : W \subset \mathbb{R}^{n-1} \to S^{n-1} \subset T_pM$, and use it to define a parameterization $\widetilde{\varphi} : (0, +\infty) \times W \to T_pM$ through

$$\widetilde{\varphi}(r, \theta^1, \dots, \theta^{n-1}) = r\varphi(\theta^1, \dots, \theta^{n-1})$$

(hence $(r, \theta^1, \dots, \theta^{n-1})$ are spherical coordinates on T_pM). Notice that

$$\frac{\partial}{\partial r} = E,$$

since

$$E_{\widetilde{\varphi}(r,\theta)} = E_{r\varphi(\theta)} = \varphi(\theta) = \frac{\partial\widetilde{\varphi}}{\partial r}(r, \theta),$$

and so

$$X = (\exp_p)_* \frac{\partial}{\partial r}. \tag{3.6}$$

Since $\frac{\partial}{\partial \theta^i}$ is tangent to $\{r = \varepsilon\}$, the vector fields

$$Y_i := (\exp_p)_* \frac{\partial}{\partial \theta^i} \tag{3.7}$$

are tangent to $S_\varepsilon(p)$. Notice also that $\left\| \frac{\partial}{\partial \theta^i} \right\| = \left\| \frac{\partial \tilde{\varphi}}{\partial \theta^i} \right\| = r \left\| \frac{\partial \varphi}{\partial \theta^i} \right\|$ is proportional to r, and consequently $\frac{\partial}{\partial \theta^i} \to 0$ as $r \to 0$, implying that $(Y_i)_q \to 0_p$ as $q \to p$. Since $\exp_p$ is a local diffeomorphism, the vector fields X and Y_i are linearly independent at each point. Also,

$$[X, Y_i] = \left[(\exp_p)_* \frac{\partial}{\partial r}, (\exp_p)_* \frac{\partial}{\partial \theta^i} \right] = (\exp_p)_* \left[\frac{\partial}{\partial r}, \frac{\partial}{\partial \theta^i} \right] = 0$$

[cf. Exercise 6.11(9) in Chap. 1], or, since the Levi-Civita connection is symmetric,

$$\nabla_X Y_i = \nabla_{Y_i} X.$$

To prove that X is orthogonal to the normal spheres $S_\varepsilon(p)$, we show that X is orthogonal to each of the vector fields Y_i. In fact, since $\nabla_X X = 0$ and $\|X\| = 1$, we have

$$X \cdot \langle X, Y_i \rangle = \langle \nabla_X X, Y_i \rangle + \langle X, \nabla_X Y_i \rangle = \langle X, \nabla_{Y_i} X \rangle = \frac{1}{2} Y_i \cdot \langle X, X \rangle = 0,$$

and hence $\langle X, Y_i \rangle$ is constant along each geodesic through p. Consequently,

$$\langle X, Y_i \rangle(\exp_p v) = \left\langle X_{\exp_p(v)}, (Y_i)_{\exp_p(v)} \right\rangle = \lim_{t \to 0} \left\langle X_{\exp_p(tv)}, (Y_i)_{\exp_p(tv)} \right\rangle = 0$$

(as $\|X\| = 1$ and $(Y_i)_q \to 0_p$ as $q \to p$), and so every geodesic through p is orthogonal to all normal spheres centered at p. Using this we obtain the following result.

Proposition 4.3 *Let $\gamma : [0, 1] \to M$ be a differentiable curve such that $\gamma(0) = p$ and $\gamma(1) \in S_\varepsilon(p)$, where $S_\varepsilon(p)$ is a normal sphere. Then $l(\gamma) \geq \varepsilon$, and $l(\gamma) = \varepsilon$ if and only if γ is a reparameterized geodesic.*

Proof We can assume that $\gamma(t) \neq p$ for all $t \in (0, 1)$, since otherwise we could easily construct a curve $\tilde{\gamma} : [0, 1] \to M$ with $\tilde{\gamma}(0) = p$, $\tilde{\gamma}(1) = \gamma(1) \in S_\varepsilon(p)$ and $l(\tilde{\gamma}) < l(\gamma)$. For the same reason, we can assume that $\gamma([0, 1)) \subset B_\varepsilon(p)$. We can then write

$$\gamma(t) := \exp_p(r(t)n(t)),$$

where $r(t) \in (0, \varepsilon]$ and $n(t) \in S^{n-1}$ are well defined for $t \in (0, 1]$. Note that $r(t)$ can be extended to $[0, 1]$ as a smooth function. Then

$$\dot{\gamma}(t) = (\exp_p)_* (\dot{r}(t)n(t) + r(t)\dot{n}(t)).$$

Since $\langle n(t), n(t) \rangle = 1$, we have $\langle \dot{n}(t), n(t) \rangle = 0$, and consequently $\dot{n}(t)$ is tangent to $\partial B_{r(t)}(0)$. Noticing that $n(t) = \left(\frac{\partial}{\partial r}\right)_{r(t)n(t)}$, we conclude that

$$\dot{\gamma}(t) = \dot{r}(t)X_{\gamma(t)} + Y(t),$$

where $X = (\exp_p)_* \frac{\partial}{\partial r}$ and $Y(t) = r(t)(\exp_p)_* \dot{n}(t)$ is tangent to $S_{r(t)}(p)$, and hence orthogonal to $X_{\gamma(t)}$. Consequently,

$$l(\gamma) = \int_0^1 \langle \dot{r}(t)X_{\gamma(t)} + Y(t), \dot{r}(t)X_{\gamma(t)} + Y(t) \rangle^{\frac{1}{2}} dt$$

$$= \int_0^1 \left(\dot{r}(t)^2 + \|Y(t)\|^2\right)^{\frac{1}{2}} dt$$

$$\geq \int_0^1 \dot{r}(t)dt = r(1) - r(0) = \varepsilon.$$

It should be clear that $l(\gamma) = \varepsilon$ if and only if $\|Y(t)\| \equiv 0$ and $\dot{r}(t) \geq 0$ for all $t \in [0, 1]$. In this case, $\dot{n}(t) = 0$ (implying that n is constant), and $\gamma(t) = \exp_p(r(t)n) = c_{r(t)n}(1) = c_n(r(t))$ is, up to reparameterization, the geodesic through p with initial condition $n \in T_pM$. $\qquad \square$

Definition 4.4 A **piecewise differentiable curve** is a continuous map $c : [a, b] \to M$ such that the restriction of c to $[t_{i-1}, t_i]$ is differentiable for $i = 1, \ldots, n$, where $a = t_0 < t_1 < \cdots < t_{n-1} < t_n = b$. We say that c **connects** $p \in M$ to $q \in M$ if $c(a) = p$ and $c(b) = q$.

The definition of **length** of a piecewise differentiable curve offers no difficulties. It should also be clear that Proposition 4.3 easily extends to piecewise differentiable curves, if we now allow for piecewise differentiable reparameterizations. Using this extended version of Proposition 4.3 as well as the properties of the exponential map and the invariance of length under reparameterization, one easily shows the following result.

Theorem 4.5 *Let $(M, \langle \cdot, \cdot \rangle)$ be a Riemannian manifold, $p \in M$ and $B_\varepsilon(p)$ a normal ball centered at p. Then, for each point $q \in B_\varepsilon(p)$, there exists a geodesic $c : I \to M$ connecting p to q. Moreover, if $\gamma : J \to M$ is any other piecewise differentiable curve connecting p to q, then $l(\gamma) \geq l(c)$, and $l(\gamma) = l(c)$ if and only if γ is a reparameterization of c.* $\qquad \square$

Conversely, we have

Theorem 4.6 *Let $(M, \langle \cdot, \cdot \rangle)$ be a Riemannian manifold and $p, q \in M$. If $c : I \to M$ is a piecewise differentiable curve connecting p to q and $l(c) \leq l(\gamma)$ for any piecewise differentiable curve $\gamma : J \to M$ connecting p to q then c is a reparameterized geodesic.*

Proof To prove this theorem, we need the following definition.

Definition 4.7 A normal neighborhood $V \subset M$ is called a **totally normal neighborhood** if there exists $\varepsilon > 0$ such that $V \subset B_\varepsilon(p)$ for all $p \in V$.

We will now prove that totally normal neighborhoods always exist. To do so, we recall that local coordinates $(x^1, \ldots, x^n)$ on M yield local coordinates $(x^1, \ldots, x^n, v^1, \ldots, v^n)$ on TM labeling the vector

$$v^1 \frac{\partial}{\partial x^1} + \cdots + v^n \frac{\partial}{\partial x^n}.$$

The geodesic equations,

$$\ddot{x}^i + \sum_{j,k=1}^n \Gamma^i_{jk} \dot{x}^j \dot{x}^k = 0 \quad (i = 1, \ldots, n),$$

correspond to the system of first-order ODEs

$$\begin{cases} \dot{x}^i = v^i \\ \dot{v}^i = -\sum_{j,k=1}^n \Gamma^i_{jk} v^j v^k \end{cases} \quad (i = 1, \ldots, n).$$

These equations define the local flow of the vector field $X \in \mathfrak{X}(TM)$ given in local coordinates by

$$X = \sum_{i=1}^n v^i \frac{\partial}{\partial x^i} - \sum_{i,j,k=1}^n \Gamma^i_{jk} v^j v^k \frac{\partial}{\partial v^i},$$

called the **geodesic flow**. As it was seen in Chap. 1, for each point $v \in TM$ there exists an open neighborhood $W \subset TM$ and an open interval $I \subset \mathbb{R}$ containing 0 such that the local flow $F : W \times I \to TM$ of X is well defined. In particular, for each point $p \in M$ we can choose an open neighborhood U containing p and $\varepsilon > 0$ such that the geodesic flow is well defined in $W \times I$ with

$$W = \{v \in T_q M \mid q \in U, \|v\| < \varepsilon\}.$$

Using homogeneity of geodesics, we can make the interval I as large as we want by making ε sufficiently small. Therefore, for ε small enough we can define a map $G : W \to M \times M$ by $G(v) := (q, \exp_q(v))$. Since $\exp_q(0) = q$, the matrix representation of $(dG)_0$ in the above local coordinates is $\begin{pmatrix} I & 0 \\ I & I \end{pmatrix}$, and hence G is a local diffeomorphism. Reducing U and ε if necessary, we can therefore assume that G is a diffeomorphism onto its image $G(W)$, which contains the point $(p, p) = G(0_p)$. Choosing an open neighborhood V of p such that $V \times V \subset G(W)$, it is clear that V is a totally normal neighborhood: for each point $q \in V$ we have $\{q\} \times \exp_q(B_\varepsilon(0)) = G(W) \cap (\{q\} \times M) \supset \{q\} \times V$, that is, $\exp_q(B_\varepsilon(0)) \supset V$.

Notice that given any two points p, q in a totally normal neighborhood V, there exists a geodesic $c : I \to M$ connecting p to q such that any other piecewise differentiable curve $\gamma : J \to M$ connecting p to q satisfies $l(\gamma) \geq l(c)$ (and $l(\gamma) = l(c)$ if and only if γ is a reparameterization of c). The proof of Theorem 4.6 is now an immediate consequence of the following observation: if $c : I \to M$ is a piecewise differentiable curve connecting p to q such that $l(c) \leq l(\gamma)$ for any curve $\gamma : J \to M$ connecting p to q, then c must be a reparameterized geodesic in each totally normal neighborhood it intersects. $\qquad\square$

Exercise 4.8

(1) Let (M, g) be a Riemannian manifold and $f : M \to \mathbb{R}$ a smooth function. Show that if $\| \operatorname{grad}(f)\| \equiv 1$ then the integral curves of $\operatorname{grad}(f)$ are geodesics, using:
  (a) the definition of geodesic;
  (b) the minimizing properties of geodesics.

(2) Let M be a Riemannian manifold and ∇ the Levi–Civita connection on M. Given $p \in M$ and a basis $\{v_1, \ldots, v_n\}$ for T_pM, we consider the parameterization $\varphi : U \subset \mathbb{R}^n \to M$ of a normal neighborhood given by

$$\varphi(x^1, \ldots, x^n) = \exp_p(x^1 v_1 + \cdots + x^n v_n)$$

(the local coordinates $(x^1, \ldots, x^n)$ are called **normal coordinates**). Show that:

  (a) in these coordinates, $\Gamma^i_{jk}(p) = 0$ (Hint: Consider the geodesic equation);
  (b) if $\{v_1, \ldots, v_n\}$ is an orthonormal basis then $g_{ij}(p) = \delta_{ij}$.

(3) Let G be a Lie group endowed with a **bi-invariant Riemannian metric** (i.e. such that L_g and R_g are isometries for all $g \in G$), and let $i : G \to G$ be the diffeomorphism defined by $i(g) = g^{-1}$.

  (a) Compute $(di)_e$ and show that

$$(di)_g = \left(dR_{g^{-1}}\right)_e (di)_e \left(dL_{g^{-1}}\right)_g$$

  for all $g \in G$. Conclude that i is an isometry.
  (b) Let $v \in \mathfrak{g} = T_eG$ and c_v be the geodesic satisfying $c_v(0) = e$ and $\dot{c}_v(0) = v$. Show that if t is sufficiently small then $c_v(-t) = (c_v(t))^{-1}$. Conclude that c_v is defined in $\mathbb{R}$ and satisfies $c_v(t + s) = c_v(t)c_v(s)$ for all $t, s \in \mathbb{R}$. (Hint: Recall that any two points in a totally normal neighborhood are connected by a unique geodesic in that neighborhood).

(c) Show that the geodesics of G are the integral curves of left-invariant vector fields, and that the maps exp (the Lie group exponential) and $\exp_e$ (the geodesic exponential at the identity) coincide.

(d) Let ∇ be the Levi-Civita connection of the bi-invariant metric and X, Y two left-invariant vector fields. Show that

$$\nabla_X Y = \frac{1}{2}[X, Y].$$

(e) Check that the left-invariant metrics Exercise 1.10(4) are actually bi-invariant.

(f) Show that any compact Lie group admits a bi-invariant metric. (**Hint:** Take the average of a left-invariant metric over all right translations).

(4) Use Theorem 4.6 to prove that if $f : M \to N$ is an isometry and $c : I \to M$ is a geodesic then $f \circ c : I \to N$ is also a geodesic.

(5) Let $f : M \to M$ be an isometry whose set of fixed points is a connected 1-dimensional submanifold $N \subset M$. Show that N is the image of a geodesic.

(6) Let $(M, \langle \cdot, \cdot \rangle)$ be a Riemannian manifold whose geodesics can be extended for all values of their parameters, and let $p \in M$.

(a) Let X and Y_i be the vector fields defined on a normal ball centered at p as in (3.6) and (3.7). Show that Y_i satisfies the **Jacobi equation**

$$\nabla_X \nabla_X Y_i = R(X, Y_i)X,$$

where $R : \mathfrak{X}(M) \times \mathfrak{X}(M) \times \mathfrak{X}(M) \to \mathfrak{X}(M)$, defined by

$$R(X, Y)Z = \nabla_X \nabla_Y Z - \nabla_Y \nabla_X Z - \nabla_{[X,Y]}Z,$$

is called the **curvature operator** (cf. Chap. 4). (**Remark:** It can be shown that $(R(X, Y)Z)_p$ depends only on X_p, Y_p, Z_p).

(b) Consider a geodesic $c : \mathbb{R} \to M$ parameterized by the arclength such that $c(0) = p$. A vector field Y along c is called a **Jacobi field** if it satisfies the Jacobi equation along c,

$$\frac{D^2 Y}{dt^2} = R(\dot{c}, Y)\dot{c}.$$

Show that Y is a Jacobi field with $Y(0) = 0$ if and only if

$$Y(t) = \frac{\partial}{\partial s} \exp_p(tv(s)) \Big|_{s=0}$$

with $v : (-\varepsilon, \varepsilon) \to T_p M$ satisfying $v(0) = \dot{c}(0)$.

(c) A point $q \in M$ is said to be **conjugate** to p if it is a critical value of $\exp_p$. Show that q is conjugate to p if and only if there exists a nonzero Jacobi field Y along a geodesic c connecting $p = c(0)$ to $q = c(b)$ such that $Y(0) = Y(b) = 0$. Conclude that if q is conjugate to p then p is conjugate to q.

(d) The manifold M is said to have **nonpositive curvature** if $\langle R(X, Y)X, Y \rangle \geq 0$ for all $X, Y \in \mathfrak{X}(M)$. Show that for such a manifold no two points are conjugate.

(e) Given a geodesic $c : I \to M$ parameterized by the arclength such that $c(0) = p$, let t_c be the supremum of the set of values of t such that c is the minimizing curve connecting p to $c(t)$ (hence $t_c > 0$). The **cut locus** of p is defined to be the set of all points of the form $c(t_c)$ for $t_c < +\infty$. Determine the cut locus of a given point $p \in M$ when M is:

   (i) the torus T^n with the flat square metric;
   (ii) the sphere S^n with the standard metric;
   (iii) the projective space $\mathbb{R}P^n$ with the standard metric.

   Check in these examples that any point in the cut locus is either conjugate to p or joined to p by two geodesics with the same length but different images.

   (**Remark:** This is a general property of the cut locus—see [dC93] or [GHL04] for a proof).

3.5 Hopf-Rinow Theorem

Let (M, g) be a Riemannian manifold. The existence of totally normal neighborhoods implies that it is always possible to connect two sufficiently close points $p, q \in M$ by a minimizing geodesic. We now address the same question globally.

Example 5.1

(1) Given two distinct points $p, q \in \mathbb{R}^n$ there exists a unique (up to reparameterization) geodesic for the Euclidean metric connecting them.

(2) Given two distinct points $p, q \in S^n$ there exist at least two geodesics for the standard metric connecting them which are not reparameterizations of each other.

(3) If $p \neq 0$ then there exists no geodesic for the Euclidean metric in $\mathbb{R}^n \setminus \{0\}$ connecting p to $-p$.

   In many cases (for example in $\mathbb{R}^n \setminus \{0\}$) there exist geodesics which cannot be extended for all values of its parameter. In other words, $\exp_p(v)$ is not defined for all $v \in T_pM$.

Definition 5.2 A Riemannian manifold $(M, \langle \cdot, \cdot \rangle)$ is said to be **geodesically complete** if, for every point $p \in M$, the map $\exp_p$ is defined in T_pM.

   There exists another notion of completeness of a connected Riemannian manifold, coming from the fact that any such manifold is naturally a metric space.

Definition 5.3 Let $(M, \langle \cdot, \cdot \rangle)$ be a connected Riemannian manifold and $p, q \in M$. The **distance** between p and q is defined as

$$d(p, q) = \inf\{l(\gamma) \mid \gamma \text{ is a piecewise differentiable curve connecting } p \text{ to } q\}.$$

Notice that if there exists a minimizing geodesic c connecting p to q then $d(p, q) = l(c)$. The function $d : M \times M \to [0, +\infty)$ is indeed a distance, as stated in the following proposition.

Proposition 5.4 (M, d) *is a metric space, that is, d satisfies:*

(i) **Positivity:** $d(p, q) \geq 0$ *and* $d(p, q) = 0$ *if and only if* $p = q$;
(ii) **Symmetry:** $d(p, q) = d(q, p)$;
(iii) **Triangle inequality:** $d(p, r) \leq d(p, q) + d(q, r)$,

for all $p, q, r \in M$. *The metric space topology induced on M coincides with the topology of M as a differentiable manifold.*

Proof Exercise 5.8(1). $\qquad\qquad\qquad\qquad\qquad\qquad\qquad\qquad\qquad\qquad\qquad$ □

Therefore we can discuss the completeness of M as a metric space (that is, whether Cauchy sequences converge). The fact that completeness and geodesic completeness are equivalent is the content of the following theorem.

Theorem 5.5 (Hopf-Rinow) *Let* $(M, \langle \cdot, \cdot \rangle)$ *be a connected Riemannian manifold and* $p \in M$. *The following assertions are equivalent:*

(i) *M is geodesically complete;*
(ii) *(M, d) is a complete metric space;*
(iii) *$\exp_p$ is defined in $T_p M$.*

Moreover, if $(M, \langle \cdot, \cdot \rangle)$ *is geodesically complete then for all* $q \in M$ *there exists a geodesic c connecting p to q with* $l(c) = d(p, q)$.

Proof It is clear that $(i) \Rightarrow (iii)$.

We begin by showing that if (iii) holds then for all $q \in M$ there exists a geodesic c connecting p to q with $l(c) = d(p, q)$. Let $d(p, q) = \rho$. If $\rho = 0$ then $q = p$ and there is nothing to prove. If $\rho > 0$, let $\varepsilon \in (0, \rho)$ be such that $S_\varepsilon(p)$ is a normal sphere (which is a compact submanifold of M). The continuous function $x \mapsto d(x, q)$ will then have a minimum point $x_0 \in S_\varepsilon(p)$. Moreover, $x_0 = \exp_p(\varepsilon v)$, where $\|v\| = 1$. Let us consider the geodesic $c_v(t) = \exp_p(tv)$. We will show that $q = c_v(\rho)$. For that, we consider the set

$$A = \{t \in [0, \rho] \mid d(c_v(t), q) = \rho - t\}.$$

Since the map $t \mapsto d(c_v(t), q)$ is continuous, A is a closed set. Moreover, $A \neq \varnothing$, as clearly $0 \in A$. We will now show that no point $t_0 \in [0, \rho)$ can be the maximum of A, which implies that the maximum of A must be ρ (hence $d(c_v(\rho), q) = 0$, that is,

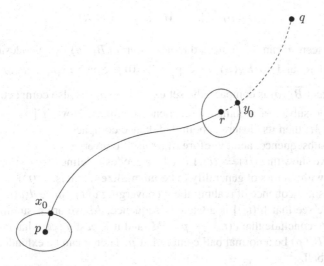

Fig. 3.3 Proof of the Hopf-Rinow theorem

$c_v(\rho) = q$, and so c_v connects p to q and $l(c_v) = \rho$). Let $t_0 \in A \cap [0, \rho), r = c_v(t_0)$
and $\delta \in (0, \rho - t_0)$ such that $S_\delta(r)$ is a normal sphere. Let y_0 be a minimum point
of the continuous function $y \mapsto d(y, q)$ on the compact set $S_\delta(r)$ (Fig 3.3). We will
show that $y_0 = c_v(t_0 + \delta)$. In fact, we have

$$\rho - t_0 = d(r, q) = \delta + \min_{y \in S_\delta(r)} d(y, q) = \delta + d(y_0, q),$$

and so

$$d(y_0, q) = \rho - t_0 - \delta. \tag{3.8}$$

The triangle inequality then implies that

$$d(p, y_0) \geq d(p, q) - d(y_0, q) = \rho - (\rho - t_0 - \delta) = t_0 + \delta,$$

and, since the piecewise differentiable curve which connects p to r through c_v and
r to y_0 through a geodesic has length $t_0 + \delta$, we conclude that this is a minimizing
curve, hence a (reparameterized) geodesic. Thus, as promised, $y_0 = c_v(t_0 + \delta)$.
Consequently, Eq. (3.8) can be written as

$$d(c_v(t_0 + \delta), q) = \rho - (t_0 + \delta),$$

implying that $t_0 + \delta \in A$, and so t_0 cannot be the maximum of A.

We can now prove that $(iii) \Rightarrow (ii)$. To do so, we begin by showing that any
bounded closed subset $K \subset M$ is compact. Indeed, if K is bounded then $K \subset B_R(p)$
for some $R > 0$, where

$$B_R(p) = \{q \in M \mid d(p,q) < R\}.$$

As we have seen, p can be connected to any point in $B_R(p)$ by a geodesic of length smaller than R, and so $B_R(p) \subset \exp_p\left(\overline{B_R(0)}\right)$. Since $\exp_p : T_pM \to M$ is continuous and $\overline{B_R(0)}$ is compact, the set $\exp_p\left(\overline{B_R(0)}\right)$ is also compact. Therefore K is a closed subset of a compact set, hence compact. Now, if $\{p_n\}$ is a Cauchy sequence in M, then its closure is bounded, hence compact. Thus $\{p_n\}$ must have a convergent subsequence, and therefore must itself converge.

Finally, we show that $(ii) \Rightarrow (i)$. Let c be a geodesic defined for $t < t_0$, which we can assume without loss of generality to be **normalized**, that is, $\|\dot{c}(t)\| = 1$. Let $\{t_n\}$ be an increasing sequence of real numbers converging to t_0. Since $d(c(t_m), c(t_n)) \leq |t_m - t_n|$, we see that $\{c(t_n)\}$ is a Cauchy sequence. As we are assuming M to be complete, we conclude that $c(t_n) \to p \in M$, and it is easily seen that $c(t) \to p$ as $t \to t_0$. Let $B_\varepsilon(p)$ be a normal ball centered at p. Then c can be extended past t_0 in this normal ball. $\qquad\square$

Corollary 5.6 *If M is compact then M is geodesically complete.*

Proof Any compact metric space is complete. $\qquad\square$

Corollary 5.7 *If M is a closed connected submanifold of a complete connected Riemannian manifold with the induced metric then M is complete.*

Proof Let M be a closed connected submanifold of a complete connected Riemannian manifold N. Let d be the distance determined by the metric on N, and let d^* be the distance determined by the induced metric on M. Then $d \leq d^*$, as any curve on M is also a curve on N. Let $\{p_n\}$ be a Cauchy sequence on (M, d^*). Then $\{p_n\}$ is a Cauchy sequence on (N, d), and consequently converges in N to a point $p \in M$ (as N is complete and M is closed). Since the topology of M is induced by the topology of N, we conclude that $p_n \to p$ on M. $\qquad\square$

Exercise 5.8

(1) Prove Proposition 5.4.
(2) Consider $\mathbb{R}^2 \setminus \{(x, 0) \mid -3 \leq x \leq 3\}$ with the Euclidean metric. Determine $B_7(0, 4)$.
(3) (a) Prove that a connected Riemannian manifold is complete if and only if the compact sets are the closed bounded sets.
    (b) Give an example of a connected Riemannian manifold containing a non-compact closed bounded set.
(4) A Riemannian manifold $(M, \langle \cdot, \cdot \rangle)$ is said to be **homogeneous** if, given any two points $p, q \in M$, there exists an isometry $f : M \to M$ such that $f(p) = q$. Show that:
    (a) any homogeneous Riemannian manifold is complete;
    (b) if G is a Lie group admitting a bi-invariant metric [cf. Exercise 4.8(3)] then the exponential map $\exp : \mathfrak{g} \to G$ is surjective;

(c) $SL(2, \mathbb{R})$ does not admit a bi-invariant metric.

(5) Let (M, g) be a complete Riemannian manifold. Show that:

   (a) (*Ambrose theorem*) if (N, h) is a Riemannian manifold and $f : M \to N$ is a local isometry then f is a covering map;
   (b) there exist surjective local isometries which are not covering maps;
   (c) (*Cartan–Hadamard theorem*) if (M, g) has nonpositive curvature [cf. Exercise 4.8(6)] then for each point $p \in M$ the exponential map $\exp_p : T_pM \to M$ is a covering map. (**Remark:** In particular, if M is simply connected then M must be diffeomorphic to $\mathbb{R}^n$).

3.6 Notes

3.6.1 Section 3.5

In this section we use several definitions and results about metric spaces, which we now discuss. A **metric space** is a pair (M, d), where M is a set and $d : M \times M \to [0, +\infty)$ is a map satisfying the properties enumerated in Proposition 5.4. The set

$$B_\varepsilon(p) = \{q \in M \mid d(p, q) < \varepsilon\}$$

is called the **open ball** with center p and radius ε. The family of all such balls is a basis for a Hausdorff topology on M, called the **metric topology**. Notice that in this topology $p_n \to p$ if and only if $d(p_n, p) \to 0$. Although a metric space (M, d) is not necessarily second countable, it is still true that $F \subset M$ is closed if and only if every convergent sequence in F has limit in F, and $K \subset M$ is compact if and only if every sequence in K has a sublimit in K.

A sequence $\{p_n\}$ in M is said to be a **Cauchy sequence** if for all $\varepsilon > 0$ there exists $N \in \mathbb{N}$ such that $d(p_n, p_m) < \varepsilon$ for all $m, n > N$. It is easily seen that all convergent sequences are Cauchy sequences. The converse, however, is not necessarily true (but if a Cauchy sequence has a convergent subsequence then it must converge). A metric space is said to be **complete** if all its Cauchy sequences converge. A closed subset of a complete metric space is itself complete.

A set is said to be bounded if it is a subset of some ball. For instance, the set of all terms of a Cauchy sequence is bounded. It is easily shown that if $K \subset M$ is compact then K must be bounded and closed (but the converse is not necessarily true). A compact metric space is necessarily complete.

3.6.2 Bibliographical Notes

The material in this chapter can be found in most books on Riemannian geometry
(e.g. [Boo03, dC93, GHL04]). For more details on general affine connections, see
[KN96]. Bi-invariant metrics on a Lie group are examples of symmetric spaces,
whose beautiful theory is studied for instance in [Hel01].

References

[Arn92]   Arnold, V.I.: Ordinary Differential Equations. Springer, Berlin (1992)
[Boo03]   Boothby, W.: An Introduction to Differentiable Manifolds and Riemannian Geometry.
          Academic Press, New York (2003)
[dC93]   do Carmo, M.: Riemannian Geometry. Birkhäuser, Boston (1993)
[GHL04]   Gallot, S., Hulin, D., Lafontaine, J.: Riemannian Geometry. Springer, Berlin (2004)
[GR70]   Gromov, M., Rohlin, V.: Imbeddings and Immersions in Riemannian Geometry. Usp.
          Mat. Nauk 25, 3–62 (1970)
[Hel01]   Helgasson, S.: Differential Geometry, Lie Groups and Symmetric Spaces. American
          Mathematical Society, Providence (2001)
[KN96]   Kobayashi, S., Nomizu, K.: Foundations of Differential Geometry, vol. I and II. Wiley,
          New York (1996)
[Nas56]   Nash, J.: The Imbedding Problem for Riemannian Manifolds. Ann. Math. 63, 20–63
          (1956)

Chapter 4
Curvature

The local geometry of a general Riemannian manifold differs from the flat geometry of the Euclidean space $\mathbb{R}^n$: for example, the internal angles of a geodesic triangle in the 2-sphere S^2 (with the standard metric) always add up to **more** than π. A measure of this difference is provided by the notion of **curvature**, introduced by Gauss in his 1827 paper "General investigations of curved surfaces", and generalized to arbitrary Riemannian manifolds by Riemann himself (in 1854). It can appear under many guises: the rate of deviation of geodesics, the degree of non-commutativity of covariant derivatives along different vector fields, the difference between the sum of the internal angles of a geodesic triangle and π, or the angle by which a vector is rotated when parallel-transported along a closed curve.

This chapter addresses the various characterizations and properties of curvature. Section 4.1 introduces the **curvature operator** of a general affine connection, and, for Riemannian manifolds, the equivalent (more geometric) notion of **sectional curvature**. The **Ricci curvature tensor** and the **scalar curvature**, obtained from the curvature tensor by **contraction**, are also defined. These quantities are fundamental in general relativity to formulate Einstein's equation (Chap. 6).

Section 4.2 establishes the **Cartan structure equations**, a powerful computational method which employs differential forms to calculate the curvature. These equations are used in Sect. 4.3 to prove the **Gauss–Bonnet theorem**, relating the curvature of a compact surface to its topology. This theorem provides a simple example of how the curvature of a complete Riemannian manifold can constrain its topology.

Complete Riemannian manifolds with **constant curvature** are discussed in Sect. 4.4. These provide important examples of curved geometries, including the negatively curved non-Euclidean geometry of Gauss, Bolyai and Lobachevsky.

Finally, the relation between the curvature of a Riemannian manifold and the curvature of a submanifold (with the induced metric) is studied in Sect. 4.5. This generalizes Gauss's investigations of curved surfaces, including his celebrated **Theorema Egregium**.

© Springer International Publishing Switzerland 2014
L. Godinho and J. Natário, *An Introduction to Riemannian Geometry*, Universitext,
DOI 10.1007/978-3-319-08666-8_4

4.1 Curvature

As we saw in Exercise 3.3(4) of Chap. 3, no open set of the 2-sphere S^2 with the standard metric is isometric to an open set of the Euclidean plane. The geometric object that locally distinguishes these two Riemannian manifolds is the so-called **curvature operator**, which appears in many other situations [cf. Exercise 4.8(6) in Chap. 3].

Definition 1.1 The **curvature** R of a connection ∇ is a correspondence that to each pair of vector fields $X, Y \in \mathfrak{X}(M)$ associates the map $R(X, Y) : \mathfrak{X}(M) \to \mathfrak{X}(M)$ defined by

$$R(X, Y)Z = \nabla_X \nabla_Y Z - \nabla_Y \nabla_X Z - \nabla_{[X,Y]} Z.$$

Hence, R is a way of measuring the non-commutativity of the connection. We leave it as an exercise to show that this defines a $(3, 1)$-tensor (called the **Riemann tensor**), since

(i) $R(fX_1 + gX_2, Y)Z = fR(X_1, Y)Z + gR(X_2, Y)Z,$
(ii) $R(X, fY_1 + gY_2)Z = fR(X, Y_1)Z + gR(X, Y_2)Z,$
(iii) $R(X, Y)(fZ_1 + gZ_2) = fR(X, Y)Z_1 + gR(X, Y)Z_2,$

for all vector fields $X, X_1, X_2, Y, Y_1, Y_2, Z, Z_1, Z_2 \in \mathfrak{X}(M)$ and all smooth functions $f, g \in C^\infty(M)$ [cf. Exercise 1.12(1)]. Choosing a coordinate system $x : V \to \mathbb{R}^n$ on M, this tensor can be locally written as

$$R = \sum_{i,j,k,l=1}^{n} R_{ijk}{}^l \, dx^i \otimes dx^j \otimes dx^k \otimes \frac{\partial}{\partial x^l},$$

where each coefficient $R_{ijk}{}^l$ is the l-coordinate of the vector field $R\left(\frac{\partial}{\partial x^i}, \frac{\partial}{\partial x^j}\right)\frac{\partial}{\partial x^k}$, that is,

$$R\left(\frac{\partial}{\partial x^i}, \frac{\partial}{\partial x^j}\right)\frac{\partial}{\partial x^k} = \sum_{l=1}^{n} R_{ijk}{}^l \frac{\partial}{\partial x^l}.$$

Using the fact that $\left[\frac{\partial}{\partial x^i}, \frac{\partial}{\partial x^j}\right] = 0$, we have

$$R\left(\frac{\partial}{\partial x^i}, \frac{\partial}{\partial x^j}\right)\frac{\partial}{\partial x^k} = \nabla_{\frac{\partial}{\partial x^i}} \nabla_{\frac{\partial}{\partial x^j}} \frac{\partial}{\partial x^k} - \nabla_{\frac{\partial}{\partial x^j}} \nabla_{\frac{\partial}{\partial x^i}} \frac{\partial}{\partial x^k}$$

$$= \nabla_{\frac{\partial}{\partial x^i}} \left(\sum_{m=1}^{n} \Gamma_{jk}^m \frac{\partial}{\partial x^m}\right) - \nabla_{\frac{\partial}{\partial x^j}} \left(\sum_{m=1}^{n} \Gamma_{ik}^m \frac{\partial}{\partial x^m}\right)$$

$$= \sum_{m=1}^{n} \left(\frac{\partial}{\partial x^i} \cdot \Gamma_{jk}^m - \frac{\partial}{\partial x^j} \cdot \Gamma_{ik}^m\right)\frac{\partial}{\partial x^m} + \sum_{l,m=1}^{n} \left(\Gamma_{jk}^m \Gamma_{im}^l - \Gamma_{ik}^m \Gamma_{jm}^l\right)\frac{\partial}{\partial x^l}$$

$$= \sum_{l=1}^{n} \left(\frac{\partial \Gamma_{jk}^l}{\partial x^i} - \frac{\partial \Gamma_{ik}^l}{\partial x^j} + \sum_{m=1}^{n} \Gamma_{jk}^m \Gamma_{im}^l - \sum_{m=1}^{n} \Gamma_{ik}^m \Gamma_{jm}^l \right) \frac{\partial}{\partial x^l},$$

and so

$$R_{ijk}{}^l = \frac{\partial \Gamma_{jk}^l}{\partial x^i} - \frac{\partial \Gamma_{ik}^l}{\partial x^j} + \sum_{m=1}^{n} \Gamma_{jk}^m \Gamma_{im}^l - \sum_{m=1}^{n} \Gamma_{ik}^m \Gamma_{jm}^l.$$

Example 1.2 Consider $M = \mathbb{R}^n$ with the Euclidean metric and the corresponding Levi–Civita connection (that is, with Christoffel symbols $\Gamma_{ij}^k \equiv 0$). Then $R_{ijk}{}^l = 0$, and the curvature R is zero. Thus, we can also interpret the curvature as a measure of how much a connection on a given manifold differs from the Levi–Civita connection of the Euclidean space.

When the connection is symmetric (as in the case of the Levi–Civita connection), the tensor R satisfies the so-called **Bianchi identity**.

Proposition 1.3 (Bianchi identity) *If M is a manifold with a symmetric connection then the associated curvature satisfies*

$$R(X, Y)Z + R(Y, Z)X + R(Z, X)Y = 0.$$

Proof This property is a direct consequence of the Jacobi identity of vector fields. Indeed,

$$
\begin{aligned}
R(X, &Y)Z + R(Y, Z)X + R(Z, X)Y \\
&= \nabla_X \nabla_Y Z - \nabla_Y \nabla_X Z - \nabla_{[X,Y]} Z + \nabla_Y \nabla_Z X - \nabla_Z \nabla_Y X - \nabla_{[Y,Z]} X \\
&\quad + \nabla_Z \nabla_X Y - \nabla_X \nabla_Z Y - \nabla_{[Z,X]} Y \\
&= \nabla_X (\nabla_Y Z - \nabla_Z Y) + \nabla_Y (\nabla_Z X - \nabla_X Z) + \nabla_Z (\nabla_X Y - \nabla_Y X) \\
&\quad - \nabla_{[X,Y]} Z - \nabla_{[Y,Z]} X - \nabla_{[Z,X]} Y,
\end{aligned}
$$

and so, since the connection is symmetric, we have

$$
\begin{aligned}
R(X, &Y)Z + R(Y, Z)X + R(Z, X)Y \\
&= \nabla_X [Y, Z] + \nabla_Y [Z, X] + \nabla_Z [X, Y] - \nabla_{[Y,Z]} X - \nabla_{[Z,X]} Y - \nabla_{[X,Y]} Z \\
&= [X, [Y, Z]] + [Y, [Z, X]] + [Z, [X, Y]] = 0. \qquad \square
\end{aligned}
$$

We will assume from this point on that (M, g) is a Riemannian manifold and ∇ its Levi–Civita connection. We can define a new covariant 4-tensor, known as the **curvature tensor**:

$$R(X, Y, Z, W) := g(R(X, Y)Z, W).$$

Notice that because the metric is nondegenerate the curvature tensor contains the same information as the Riemann tensor. Again, choosing a coordinate system $x : V \to \mathbb{R}^n$ on M, we can write this tensor as

$$R(X, Y, Z, W) = \left(\sum_{i,j,k,l=1}^{n} R_{ijkl} \, dx^i \otimes dx^j \otimes dx^k \otimes dx^l \right) (X, Y, Z, W)$$

where

$$R_{ijkl} = g\left(R\left(\frac{\partial}{\partial x^i}, \frac{\partial}{\partial x^j} \right) \frac{\partial}{\partial x^k}, \frac{\partial}{\partial x^l} \right) = g\left(\sum_{m=1}^{n} R_{ijk}{}^m \frac{\partial}{\partial x^m}, \frac{\partial}{\partial x^l} \right) = \sum_{m=1}^{n} R_{ijk}{}^m g_{ml}.$$

This tensor satisfies the following symmetry properties.

Proposition 1.4 *If X, Y, Z, W are vector fields in M and ∇ is the Levi–Civita connection, then*

(i) $R(X, Y, Z, W) + R(Y, Z, X, W) + R(Z, X, Y, W) = 0$;
(ii) $R(X, Y, Z, W) = -R(Y, X, Z, W)$;
(iii) $R(X, Y, Z, W) = -R(X, Y, W, Z)$;
(iv) $R(X, Y, Z, W) = R(Z, W, X, Y)$.

Proof Property (i) is an immediate consequence of the Bianchi identity, and property (ii) holds trivially.

Property (iii) is equivalent to showing that $R(X, Y, Z, Z) = 0$. Indeed, if (iii) holds then clearly $R(X, Y, Z, Z) = 0$. Conversely, if this is true, we have

$$R(X, Y, Z + W, Z + W) = 0 \Leftrightarrow R(X, Y, Z, W) + R(X, Y, W, Z) = 0.$$

Now, using the fact that the Levi–Civita connection is compatible with the metric, we have

$$X \cdot \langle \nabla_Y Z, Z \rangle = \langle \nabla_X \nabla_Y Z, Z \rangle + \langle \nabla_Y Z, \nabla_X Z \rangle$$

and

$$[X, Y] \cdot \langle Z, Z \rangle = 2 \langle \nabla_{[X,Y]} Z, Z \rangle.$$

Hence,

$$\begin{aligned} R(X, Y, Z, Z) &= \langle \nabla_X \nabla_Y Z, Z \rangle - \langle \nabla_Y \nabla_X Z, Z \rangle - \langle \nabla_{[X,Y]} Z, Z \rangle \\ &= X \cdot \langle \nabla_Y Z, Z \rangle - \langle \nabla_Y Z, \nabla_X Z \rangle - Y \cdot \langle \nabla_X Z, Z \rangle \\ &\quad + \langle \nabla_X Z, \nabla_Y Z \rangle - \frac{1}{2}[X, Y] \cdot \langle Z, Z \rangle \end{aligned}$$

$$= \frac{1}{2}X \cdot (Y \cdot \langle Z, Z \rangle) - \frac{1}{2}Y \cdot (X \cdot \langle Z, Z \rangle) - \frac{1}{2}[X, Y] \cdot \langle Z, Z \rangle$$

$$= \frac{1}{2}[X, Y] \cdot \langle Z, Z \rangle - \frac{1}{2}[X, Y] \cdot \langle Z, Z \rangle = 0.$$

To show (iv), we use (i) to get

$$R(X, Y, Z, W) + R(Y, Z, X, W) + R(Z, X, Y, W) = 0$$
$$R(Y, Z, W, X) + R(Z, W, Y, X) + R(W, Y, Z, X) = 0$$
$$R(Z, W, X, Y) + R(W, X, Z, Y) + R(X, Z, W, Y) = 0$$
$$R(W, X, Y, Z) + R(X, Y, W, Z) + R(Y, W, X, Z) = 0$$

and so, adding these and using (iii), we have

$$R(Z, X, Y, W) + R(W, Y, Z, X) + R(X, Z, W, Y) + R(Y, W, X, Z) = 0.$$

Using (ii) and (iii), we obtain

$$2R(Z, X, Y, W) - 2R(Y, W, Z, X) = 0. \qquad \square$$

An equivalent way of encoding the information about the curvature of a Riemannian manifold is by considering the following definition.

Definition 1.5 Let Π be a 2-dimensional subspace of T_pM and let X_p, Y_p be two linearly independent elements of Π. Then, the **sectional curvature** of Π is defined as

$$K(\Pi) := -\frac{R(X_p, Y_p, X_p, Y_p)}{\|X_p\|^2 \|Y_p\|^2 - \langle X_p, Y_p \rangle^2}.$$

Note that $\|X_p\|^2 \|Y_p\|^2 - \langle X_p, Y_p \rangle^2$ is the square of the area of the parallelogram in T_pM spanned by X_p, Y_p, and so the above definition of sectional curvature does not depend on the choice of the linearly independent vectors X_p, Y_p. Indeed, when we change the basis on Π, both $R(X_p, Y_p, X_p, Y_p)$ and $\|X_p\|^2 \|Y_p\|^2 - \langle X_p, Y_p \rangle^2$ change by the square of the determinant of the change of basis matrix [cf. Exercise 1.12(4)]. We will now see that knowing the sectional curvature of every section of T_pM completely determines the curvature tensor on this space.

Proposition 1.6 *The Riemannian curvature tensor at p is uniquely determined by the values of the sectional curvatures of sections (that is, 2-dimensional subspaces) of T_pM.*

Proof Let us consider two covariant 4-tensors R_1, R_2 on T_pM satisfying the symmetry properties of Proposition 1.4. Then the tensor $T := R_1 - R_2$ also satisfies these symmetry properties. We will see that, if the values $R_1(X_p, Y_p, X_p, Y_p)$ and $R_2(X_p, Y_p, X_p, Y_p)$ agree for every $X_p, Y_p \in T_pM$ (that is, if $T(X_p, Y_p, X_p, Y_p) = 0$ for every $X_p, Y_p \in T_pM$), then $R_1 = R_2$ (that is, $T \equiv 0$). Indeed, for all vectors $X_p, Y_p, Z_p \in T_pM$, we have

$$0 = T(X_p + Z_p, Y_p, X_p + Z_p, Y_p) = T(X_p, Y_p, Z_p, Y_p) + T(Z_p, Y_p, X_p, Y_p)$$
$$= 2T(X_p, Y_p, Z_p, Y_p),$$

and so

$$0 = T(X_p, Y_p + W_p, Z_p, Y_p + W_p) = T(X_p, Y_p, Z_p, W_p) + T(X_p, W_p, Z_p, Y_p)$$
$$= T(Z_p, W_p, X_p, Y_p) - T(W_p, X_p, Z_p, Y_p),$$

that is, $T(Z_p, W_p, X_p, Y_p) = T(W_p, X_p, Z_p, Y_p)$. Hence T is invariant by cyclic permutations of the first three elements and so, by the Bianchi identity, we have $3T(X_p, Y_p, Z_p, W_p) = 0$. □

A Riemannian manifold is called **isotropic at a point** $p \in M$ if its sectional curvature is a constant K_p for every section $\Pi \subset T_pM$. Moreover, it is called **isotropic** if it is isotropic at all points. Note that every 2-dimensional manifold is trivially isotropic. Its sectional curvature $K(p) := K_p$ is called the **Gauss curvature**.

Remark 1.7 As we will see later, the Gauss curvature measures how much the local geometry of the surface differs from the geometry of the Euclidean plane. For instance, its integral over a disk D on the surface gives the angle by which a vector is rotated when parallel-transported around the boundary of D [cf. Exercise 2.8(7)]. Alternatively, its integral over the interior of a geodesic triangle Δ is equal to the difference between the sum of the inner angles of Δ and π [cf. Exercise 3.6(6)]. We will also see that the sectional curvature of an n-dimensional Riemannian manifold is actually the Gauss curvature of special 2-dimensional submanifolds, formed by the geodesics tangent to the sections [cf. Exercise 5.7(5)].

Proposition 1.8 *If M is isotropic at p and $x : V \to \mathbb{R}^n$ is a coordinate system around p, then the coefficients of the Riemannian curvature tensor at p are given by*

$$R_{ijkl}(p) = -K_p(g_{ik}\, g_{jl} - g_{il}\, g_{jk}).$$

Proof We first define a covariant 4-tensor A on T_pM as

$$A := \sum_{i,j,k,l=1}^{n} -K_p(g_{ik}\, g_{jl} - g_{il}\, g_{jk})\, dx^i \otimes dx^j \otimes dx^k \otimes dx^l.$$

We leave it as an exercise to check that A satisfies the symmetry properties of Proposition 1.4. Moreover,

$$A(X_p, Y_p, X_p, Y_p) = \sum_{i,j,k,l=1}^{n} -K_p(g_{ik}\, g_{jl} - g_{il}\, g_{jk})\, X_p^i\, Y_p^j\, X_p^k\, Y_p^l$$
$$= -K_p\left(\langle X_p, X_p\rangle\langle Y_p, Y_p\rangle - \langle X_p, Y_p\rangle^2\right)$$

$$= R(X_p, Y_p, X_p, Y_p),$$

and so we conclude from Proposition 1.6 that $A = R$. $\qquad\square$

Definition 1.9 A Riemannian manifold is called a manifold of **constant curvature** if it is isotropic and K_p is the same at all points of M.

Example 1.10 The Euclidean space is a manifold of constant curvature $K_p \equiv 0$. We will see the complete classification of (complete, connected) manifolds of constant curvature in Sect. 4.4.

Another geometric object, very important in general relativity, is the so-called **Ricci tensor**.

Definition 1.11 The **Ricci curvature tensor** is the covariant 2-tensor locally defined as

$$Ric(X, Y) := \sum_{k=1}^{n} dx^k \left(R\left(\frac{\partial}{\partial x^k}, X \right) Y \right).$$

The above definition is independent of the choice of coordinates. Indeed, we can see $Ric_p(X_p, Y_p)$ as the trace of the linear map from T_pM to T_pM given by $Z_p \mapsto R(Z_p, X_p)Y_p$, hence independent of the choice of basis. Moreover, this tensor is symmetric. In fact, choosing an orthonormal basis $\{E_1, \ldots, E_n\}$ of T_pM we have

$$Ric_p(X_p, Y_p) = \sum_{k=1}^{n} \langle R(E_k, X_p)Y_p, E_k \rangle = \sum_{k=1}^{n} R(E_k, X_p, Y_p, E_k)$$

$$= \sum_{k=1}^{n} R(Y_p, E_k, E_k, X_p) = \sum_{k=1}^{n} R(E_k, Y_p, X_p, E_k)$$

$$= Ric_p(Y_p, X_p).$$

Locally, we can write

$$Ric = \sum_{i,j=1}^{n} R_{ij} dx^i \otimes dx^j$$

where the coefficients R_{ij} are given by

$$R_{ij} := Ric\left(\frac{\partial}{\partial x^i}, \frac{\partial}{\partial x^j} \right) = \sum_{k=1}^{n} dx^k \left(R\left(\frac{\partial}{\partial x^k}, \frac{\partial}{\partial x^i} \right) \frac{\partial}{\partial x^j} \right) = \sum_{k=1}^{n} R_{kij}{}^k,$$

that is, $R_{ij} = \sum_{k=1}^{n} R_{kij}{}^k$.

Incidentally, note that we obtained a (2, 0)-tensor from a (3, 1)-tensor. This is an example of a general procedure called **contraction**, where we obtain a $(k-1, m-1)$-tensor from a (k, m)-tensor. To do so, we choose two indices on the components of the (k, m)-tensor, one covariant and other contravariant, set them equal and then sum over them, thus obtaining the components of a $(k-1, m-1)$-tensor. On the example of the Ricci tensor, we took the (3, 1)-tensor $\widetilde{R}$ defined by the curvature,

$$\widetilde{R}(X, Y, Z, \omega) = \omega(R(X, Y)Z),$$

chose the first covariant index and the first contravariant index, set them equal and summed over them:

$$Ric(X, Y) = \sum_{k=1}^{n} \widetilde{R}\left(\frac{\partial}{\partial x^k}, X, Y, dx^k\right).$$

Similarly, we can use contraction to obtain a function (0-tensor) from the Ricci tensor (a covariant 2-tensor). For that, we first need to define a new (1, 1)-tensor field T using the metric,

$$T(X, \omega) := Ric(X, Y),$$

where Y is such that $\omega(Z) = \langle Y, Z \rangle$ for every vector field Z. Then, we set the covariant index equal to the contravariant one and add, obtaining a function $S : M \to \mathbb{R}$ called the **scalar curvature**. Locally, choosing a coordinate system $x : V \to \mathbb{R}^n$, we have

$$S(p) := \sum_{k=1}^{n} T\left(\frac{\partial}{\partial x^k}, dx^k\right) = \sum_{k=1}^{n} Ric\left(\frac{\partial}{\partial x^k}, Y_k\right),$$

where, for every vector field Z on V,

$$Z^k = dx^k(Z) = \langle Z, Y_k \rangle = \sum_{i,j=1}^{n} g_{ij} Z^i Y_k^j.$$

Therefore, we must have $Y_k^j = g^{jk}$ (where $(g^{ij}) = (g_{ij})^{-1}$), and hence $Y_k = \sum_{i=1}^{n} g^{ik} \frac{\partial}{\partial x^i}$. We conclude that the scalar curvature is locally given by

$$S(p) = \sum_{k=1}^{n} Ric\left(\frac{\partial}{\partial x^k}, \sum_{i=1}^{n} g^{ik} \frac{\partial}{\partial x^i}\right) = \sum_{i,k=1}^{n} R_{ki} g^{ik} = \sum_{i,k=1}^{n} g^{ik} R_{ik}.$$

(since Ric is symmetric).

Exercise 1.12

(1) (a) Show that the curvature operator satisfies
   (i) $R(fX_1 + gX_2, Y)Z = fR(X_1, Y)Z + gR(X_2, Y)Z$;
   (ii) $R(X, fY_1 + gY_2)Z = fR(X, Y_1)Z + gR(X, Y_2)Z$;
   (iii) $R(X, Y)(fZ_1 + gZ_2) = fR(X, Y)Z_1 + gR(X, Y)Z_2$,
   for all vector fields $X, X_1, X_2, Y, Y_1, Y_2, Z, Z_1, Z_2 \in \mathfrak{X}(M)$ and smooth functions $f, g \in C^\infty(M)$.

   (b) Show that $(R(X, Y)Z)_p \in T_pM$ depends only on X_p, Y_p, Z_p. Conclude that R defines a $(3, 1)$-tensor. (Hint: Choose local coordinates around $p \in M$).

(2) Let (M, g) be an n-dimensional Riemannian manifold and $p \in M$. Show that if $(x^1, \dots, x^n)$ are normal coordinates centered at p [cf. Exercise 4.8(2) in Chap. 3] then

$$R_{ijkl}(p) = \frac{1}{2}\left(\frac{\partial^2 g_{jl}}{\partial x^i \partial x^k} - \frac{\partial^2 g_{il}}{\partial x^j \partial x^k} - \frac{\partial^2 g_{jk}}{\partial x^i \partial x^l} + \frac{\partial^2 g_{ik}}{\partial x^j \partial x^l} \right)(p).$$

(3) Recall that if G is a Lie group endowed with a bi-invariant Riemannian metric, ∇ is the Levi–Civita connection and X, Y are two left-invariant vector fields then

$$\nabla_X Y = \frac{1}{2}[X, Y]$$

[cf. Exercise 4.8(3) in Chap. 3]. Show that if Z is also left-invariant, then

$$R(X, Y)Z = \frac{1}{4}[Z, [X, Y]].$$

(4) Show that $\|X_p\|^2 \|Y_p\|^2 - \langle X_p, Y_p \rangle^2$ gives us the square of the area of the parallelogram in T_pM spanned by X_p, Y_p. Conclude that the sectional curvature does not depend on the choice of the linearly independent vectors X_p, Y_p, that is, when we change the basis on Π, both $R(X_p, Y_p, X_p, Y_p)$ and $\|X_p\|^2 \|Y_p\|^2 - \langle X_p, Y_p \rangle^2$ change by the square of the determinant of the change of basis matrix.

(5) Show that Ric is the only independent contraction of the curvature tensor: choosing any other two indices and contracting, one either gets $\pm Ric$ or 0.

(6) Let M be a 3-dimensional Riemannian manifold. Show that the curvature tensor is entirely determined by the Ricci tensor.

(7) Let (M, g) be an n-dimensional isotropic Riemannian manifold with sectional curvature K. Show that $Ric = (n - 1)Kg$ and $S = n(n - 1)K$.

(8) Let g_1, g_2 be two Riemannian metrics on a manifold M such that $g_1 = \rho g_2$, for some constant $\rho > 0$. Show that:

   (a) the corresponding sectional curvatures K_1 and K_2 satisfy $K_1(\Pi) = \rho^{-1}K_2(\Pi)$ for any 2-dimensional section of a tangent space of M;

(b) the corresponding Ricci curvature tensors satisfy $Ric_1 = Ric_2$;
(c) the corresponding scalar curvatures satisfy $S_1 = \rho^{-1} S_2$.

(9) If ∇ is not the Levi–Civita connection can we still define the Ricci curvature tensor Ric? Is it necessarily symmetric?

4.2 Cartan Structure Equations

In this section we will reformulate the properties of the Levi–Civita connection and of the Riemannian curvature tensor in terms of differential forms. For that we will take an open subset V of M where we have defined a **field of frames** $\{X_1, \ldots, X_n\}$, that is, a set of n vector fields that, at each point p of V, form a basis for $T_p M$ (for example, we can take a coordinate neighborhood V and the vector fields $X_i = \frac{\partial}{\partial x^i}$; however, in general, the X_i are **not** associated to a coordinate system). Then we consider a **field of dual coframes**, that is, 1-forms $\{\omega^1, \ldots, \omega^n\}$ on V such that $\omega^i(X_j) = \delta_{ij}$. Note that, at each point $p \in V$, $\{\omega_p^1, \ldots, \omega_p^n\}$ is a basis for $T_p^* M$. From the properties of a connection, in order to define $\nabla_X Y$ we just have to establish the values of

$$\nabla_{X_i} X_j = \sum_{k=1}^{n} \Gamma_{ij}^k X_k,$$

where Γ_{ij}^k is defined as the kth component of the vector field $\nabla_{X_i} X_j$ on the basis $\{X_i\}_{i=1}^n$. Note that if the X_i are not associated to a coordinate system then the Γ_{ij}^k cannot be computed using formula (9), and, in general, they are not even symmetric in the indices i, j [cf. Exercise 2.8(1)]. Given the values of the Γ_{ij}^k on V, we can define 1-forms ω_j^k ($j, k = 1, \ldots, n$) in the following way:

$$\omega_j^k := \sum_{i=1}^{n} \Gamma_{ij}^k \omega^i. \tag{4.1}$$

Conversely, given these forms, we can obtain the values of Γ_{ij}^k through

$$\Gamma_{ij}^k = \omega_j^k(X_i).$$

The connection is then completely determined from these forms: given two vector fields $X = \sum_{i=1}^n a^i X_i$ and $Y = \sum_{i=1}^n b^i X_i$, we have

$$\nabla_X X_j = \nabla_{\sum_{i=1}^n a^i X_i} X_j = \sum_{i=1}^{n} a^i \nabla_{X_i} X_j = \sum_{i,k=1}^{n} a^i \Gamma_{ij}^k X_k \tag{4.2}$$

$$= \sum_{i,k=1}^{n} a^i \omega_j^k(X_i) X_k = \sum_{k=1}^{n} \omega_j^k(X) X_k$$

and hence

$$\nabla_X Y = \nabla_X \left(\sum_{i=1}^n b^i X_i \right) = \sum_{i=1}^n \left((X \cdot b^i) X_i + b^i \nabla_X X_i \right) \tag{4.3}$$

$$= \sum_{j=1}^n \left(X \cdot b^j + \sum_{i=1}^n b^i \omega_i^j(X) \right) X_j.$$

Note that the values of the forms ω_j^k at X are the components of $\nabla_X X_j$ relative to the field of frames, that is,

$$\omega_j^i(X) = \omega^i \left(\nabla_X X_j \right). \tag{4.4}$$

The ω_j^k are called the **connection forms**. For the Levi–Civita connection, these forms cannot be arbitrary. Indeed, they have to satisfy certain equations corresponding to the properties of symmetry and compatibility with the metric.

Theorem 2.1 (Cartan) *Let V be an open subset of a Riemannian manifold M on which we have defined a field of frames $\{X_1, \ldots, X_n\}$. Let $\{\omega^1, \ldots, \omega^n\}$ be the corresponding field of coframes. Then the connection forms of the Levi–Civita connection are the unique solution of the equations*

(i) $d\omega^i = \sum_{j=1}^n \omega^j \wedge \omega_j^i,$
(ii) $dg_{ij} = \sum_{k=1}^n (g_{kj} \omega_i^k + g_{ki} \omega_j^k),$
where $g_{ij} = \langle X_i, X_j \rangle.$

Proof We begin by showing that the Levi–Civita connection forms, defined by (4.1), satisfy (*i*) and (*ii*). For this, we will use the following property of 1-forms [cf. Exercise 3.8(2) of Chap. 2]:

$$d\omega(X, Y) = X \cdot (\omega(Y)) - Y \cdot (\omega(X)) - \omega([X, Y]).$$

We have

$$\nabla_Y X = \nabla_Y \left(\sum_{j=1}^n \omega^j(X) X_j \right) = \sum_{j=1}^n \left(Y \cdot \omega^j(X) X_j + \omega^j(X) \nabla_Y X_j \right),$$

which implies

$$\omega^i(\nabla_Y X) = Y \cdot \omega^i(X) + \sum_{j=1}^n \omega^j(X) \omega^i(\nabla_Y X_j). \tag{4.5}$$

Using (4.4) and (4.5), we have

$$\left(\sum_{j=1}^{n}\omega^j \wedge \omega^i_j\right)(X, Y) = \sum_{j=1}^{n}\left(\omega^j(X)\,\omega^i_j(Y) - \omega^j(Y)\,\omega^i_j(X)\right)$$

$$= \sum_{j=1}^{n}\left(\omega^j(X)\,\omega^i(\nabla_Y X_j) - \omega^j(Y)\,\omega^i(\nabla_X X_j)\right)$$

$$= \omega^i(\nabla_Y X) - Y \cdot (\omega^i(X)) - \omega^i(\nabla_X Y) + X \cdot (\omega^i(Y)),$$

and so

$$\left(d\omega^i - \sum_{j=1}^{n}\omega^j \wedge \omega^i_j\right)(X, Y)$$

$$= X \cdot (\omega^i(Y)) - Y \cdot (\omega^i(X)) - \omega^i([X, Y]) - \left(\sum_{j=1}^{n}\omega^j \wedge \omega^i_j\right)(X, Y)$$

$$= \omega^i(\nabla_X Y - \nabla_Y X - [X, Y]) = 0.$$

Note that equation (i) is equivalent to symmetry of the connection. To show that (ii) holds, we notice that

$$dg_{ij}(Y) = Y \cdot \langle X_i, X_j \rangle,$$

and that, on the other hand,

$$\left(\sum_{k=1}^{n} g_{kj}\,\omega^k_i + g_{ki}\,\omega^k_j\right)(Y) = \sum_{k=1}^{n} g_{kj}\,\omega^k_i(Y) + g_{ki}\,\omega^k_j(Y)$$

$$= \left\langle \sum_{k=1}^{n}\omega^k_i(Y)\,X_k, X_j \right\rangle + \left\langle \sum_{k=1}^{n}\omega^k_j(Y)\,X_k, X_i \right\rangle$$

$$= \langle \nabla_Y X_i, X_j \rangle + \langle \nabla_Y X_j, X_i \rangle.$$

Hence, equation (ii) is equivalent to

$$Y \cdot \langle X_i, X_j \rangle = \langle \nabla_Y X_i, X_j \rangle + \langle X_i, \nabla_Y X_j \rangle,$$

for every i, j, that is, it is equivalent to compatibility with the metric [cf. Exercise 2.8(2)]. We conclude that the Levi–Civita connection forms satisfy (i) and (ii).

To prove unicity, we take 1-forms ω^j_i $(i, j = 1, \ldots, n)$ satisfying (i) and (ii). Using (4.2) and (4.3), we can define a connection, which is necessarily symmetric and compatible with the metric. By uniqueness of the Levi–Civita connection, we have uniqueness of the set of forms ω_i^j satisfying (i) and (ii) (note that each connection determines a unique set of n^2 connection forms and vice versa). $\qquad\qquad\qquad\square$

Remark 2.2 Given a field of frames on some open set, we can perform Gram–Schmidt orthogonalization to obtain a smooth field of orthonormal frames $\{E_1, \ldots, E_n\}$. Then, as $g_{ij} = \langle E_i, E_j \rangle = \delta_{ij}$, equations (i) and (ii) above become

(i) $d\omega^i = \sum_{j=1}^{n} \omega^j \wedge \omega_j^i$,

(ii) $\omega_i^j + \omega_j^i = 0$.

In addition to connection forms, we can also define **curvature forms**. Again we consider an open subset V of M where we have a field of frames $\{X_1, \ldots, X_n\}$ (hence a corresponding field of dual coframes $\{\omega^1, \ldots, \omega^n\}$). We then define 2-forms Ω_k^l $(k, l = 1, \ldots, n)$ by

$$\Omega_k^l(X, Y) := \omega^l(R(X, Y)X_k),$$

for all vector fields X, Y in V (i.e. $R(X, Y)X_k = \sum_{l=1}^{n} \Omega_k^l(X, Y)X_l$). Using the basis $\{\omega^i \wedge \omega^j\}_{i<j}$ for 2-forms, we have

$$\Omega_k^l = \sum_{i<j} \Omega_k^l(X_i, X_j)\,\omega^i \wedge \omega^j = \sum_{i<j} \omega^l(R(X_i, X_j)X_k)\,\omega^i \wedge \omega^j$$

$$= \sum_{i<j} R_{ijk}{}^l\,\omega^i \wedge \omega^j = \frac{1}{2}\sum_{i,j=1}^{n} R_{ijk}{}^l\,\omega^i \wedge \omega^j,$$

where the $R_{ijk}{}^l$ are the coefficients of the curvature relative to these frames:

$$R(X_i, X_j)X_k = \sum_{l=1}^{n} R_{ijk}{}^l X_l.$$

The curvature forms satisfy the following equation.

Proposition 2.3 *In the above notation,*

(iii) $\Omega_i^j = d\omega_i^j - \sum_{k=1}^{n} \omega_i^k \wedge \omega_k^j$, *for every* $i, j = 1, \ldots, n$.

Proof We will show that

$$R(X, Y)X_i = \sum_{j=1}^{n} \Omega_i^j(X, Y)X_j = \sum_{j=1}^{n}\left(\left(d\omega_i^j - \sum_{k=1}^{n}\omega_i^k \wedge \omega_k^j\right)(X, Y)\right)X_j.$$

Indeed,

$$R(X, Y)X_i = \nabla_X \nabla_Y X_i - \nabla_Y \nabla_X X_i - \nabla_{[X,Y]} X_i$$

$$= \nabla_X \left(\sum_{k=1}^n \omega_i^k(Y) X_k \right) - \nabla_Y \left(\sum_{k=1}^n \omega_i^k(X) X_k \right) - \sum_{k=1}^n \omega_i^k([X, Y]) X_k$$

$$= \sum_{k=1}^n \left(X \cdot (\omega_i^k(Y)) - Y \cdot (\omega_i^k(X)) - \omega_i^k([X, Y]) \right) X_k +$$

$$+ \sum_{k=1}^n \omega_i^k(Y) \nabla_X X_k - \sum_{k=1}^n \omega_i^k(X) \nabla_Y X_k$$

$$= \sum_{k=1}^n d\omega_i^k(X, Y) X_k + \sum_{k,j=1}^n \left(\omega_i^k(Y) \omega_k^j(X) X_j - \omega_i^k(X) \omega_k^j(Y) X_j \right)$$

$$= \sum_{j=1}^n \left(d\omega_i^j(X, Y) - \sum_{k=1}^n (\omega_i^k \wedge \omega_k^j)(X, Y) \right) X_j. \qquad \square$$

Equations (i), (ii) and (iii) are known as the **Cartan structure equations**. We list these equations below, as well as the main definitions.

(i) $d\omega^i = \sum_{j=1}^n \omega^j \wedge \omega_j^i$,

(ii) $dg_{ij} = \sum_{k=1}^n (g_{kj} \omega_i^k + g_{ki} \omega_j^k)$,

(iii) $d\omega_i^j = \Omega_i^j + \sum_{k=1}^n \omega_i^k \wedge \omega_k^j$,

where $\omega^i(X_j) = \delta_{ij}$, $\omega_j^k = \sum_{i=1}^n \Gamma_{ij}^k \omega^i$ and $\Omega_i^j = \sum_{k<l} R_{kli}{}^j \omega^k \wedge \omega^l$.

Remark 2.4 If we consider a field of orthonormal frames $\{E_1, \ldots, E_n\}$, the above equations become:

(i) $d\omega^i = \sum_{j=1}^n \omega^j \wedge \omega_j^i$,

(ii) $\omega_i^j + \omega_j^i = 0$,

(iii) $d\omega_i^j = \Omega_i^j + \sum_{k=1}^n \omega_i^k \wedge \omega_k^j$ (and so $\Omega_i^j + \Omega_j^i = 0$).

Example 2.5 For a field of orthonormal frames in $\mathbb{R}^n$ with the Euclidean metric, the curvature forms must vanish (as $R = 0$), and we obtain the following structure equations:

(i) $d\omega^i = \sum_{j=1}^n \omega^j \wedge \omega_j^i$,

(ii) $\omega_i^j + \omega_j^i = 0$,

(iii) $d\omega_i^j = \sum_{k=1}^n \omega_i^k \wedge \omega_k^j$.

To finish this section, we will consider in detail the special case of a 2-dimensional Riemannian manifold. In this case, the structure equations for a field of orthonormal frames are particularly simple: equation (ii) implies that there is only one independent connection form ($\omega_1^1 = \omega_2^2 = 0$ and $\omega_2^1 = -\omega_1^2$), which can be computed from equation (i):

$$dw^1 = -\omega^2 \wedge \omega_1^2;$$
$$dw^2 = \omega^1 \wedge \omega_1^2.$$

Equation (iii) then yields that there is only one independent curvature form $\Omega_1^2 = d\omega_1^2$. This form is closely related to the Gauss curvature of the manifold.

Proposition 2.6 *If M is a 2-dimensional manifold, then for an orthonormal frame we have $\Omega_1^2 = -K\omega^1 \wedge \omega^2$, where the function K is the Gauss curvature of M (that is, its sectional curvature).*

Proof Let p be a point in M and let us choose an open set containing p where we have defined a field of orthonormal frames $\{E_1, E_2\}$. Then

$$K = -R(E_1, E_2, E_1, E_2) = -R_{1212},$$

and consequently

$$\Omega_1^2 = \Omega_1^2(E_1, E_2)\,\omega^1 \wedge \omega^2 = \omega^2(R(E_1, E_2)E_1)\,\omega^1 \wedge \omega^2$$
$$= \langle R(E_1, E_2)E_1, E_2 \rangle\,\omega^1 \wedge \omega^2 = R_{1212}\,\omega^1 \wedge \omega^2 = -K\,\omega^1 \wedge \omega^2. \qquad \square$$

Note that K does not depend on the choice of the field of frames, since it is a sectional curvature (cf. Definition 1.5), and, since $\omega^1 \wedge \omega^2$ is a Riemannian volume form, neither does the curvature form (up to a sign). However, the connection forms do. Let $\{E_1, E_2\}, \{F_1, F_2\}$ be two fields of orthonormal frames on an open subset V of M. Then

$$\begin{pmatrix} F_1 & F_2 \end{pmatrix} = \begin{pmatrix} E_1 & E_2 \end{pmatrix} S$$

where $S : V \to O(2)$ has values in the orthogonal group of 2×2 matrices. Note that S has one of the following two forms

$$S = \begin{pmatrix} a & -b \\ b & a \end{pmatrix} \text{ or } S = \begin{pmatrix} a & b \\ b & -a \end{pmatrix},$$

where $a, b : V \to \mathbb{R}$ are such that $a^2 + b^2 = 1$. The determinant of S is then ± 1 depending on whether the two frames have the same orientation. We have the following proposition.

Proposition 2.7 *If $\{E_1, E_2\}$ and $\{F_1, F_2\}$ have the same orientation then, denoting by ω_1^2 and $\overline{\omega}_1^2$ the corresponding connection forms, we have $\overline{\omega}_1^2 - \omega_1^2 = \sigma$, where $\sigma := a\,db - b\,da$.*

Proof Denoting by $\{\omega^1, \omega^2\}$ and $\{\overline{\omega}^1, \overline{\omega}^2\}$ the fields of dual coframes corresponding to $\{E_1, E_2\}$ and $\{F_1, F_2\}$, we define the column vectors of 1-forms

$$\omega := \begin{pmatrix} \omega^1 \\ \omega^2 \end{pmatrix} \quad \text{and} \quad \overline{\omega} := \begin{pmatrix} \overline{\omega}^1 \\ \overline{\omega}^2 \end{pmatrix}$$

and the matrices of 1-forms

$$A := \begin{pmatrix} 0 & -\omega_1^2 \\ \omega_1^2 & 0 \end{pmatrix} \quad \text{and} \quad \overline{A} := \begin{pmatrix} 0 & -\overline{\omega}_1^2 \\ \overline{\omega}_1^2 & 0 \end{pmatrix}.$$

The relation between the frames can be written as

$$\overline{\omega} = S^{-1}\omega \Leftrightarrow \omega = S\overline{\omega}$$

(cf. Sect. 2.7.1 in Chap. 2), and the Cartan structure equations as

$$d\omega = -A \wedge \omega \quad \text{and} \quad d\overline{\omega} = -\overline{A} \wedge \overline{\omega}.$$

Therefore

$$d\omega = S\,d\overline{\omega} + dS \wedge \overline{\omega} = -S\overline{A} \wedge \overline{\omega} + dS \wedge S^{-1}\omega$$
$$= -S\overline{A} \wedge S^{-1}\omega + dS \wedge S^{-1}\omega = -\left(S\overline{A}S^{-1} - dS\,S^{-1}\right) \wedge \omega,$$

and unicity of solutions of the Cartan structure equations implies

$$A = S\overline{A}S^{-1} - dS\,S^{-1}.$$

Writing this out in full one obtains

$$\begin{pmatrix} 0 & -\omega_1^2 \\ \omega_1^2 & 0 \end{pmatrix} = \begin{pmatrix} 0 & -\overline{\omega}_1^2 \\ \overline{\omega}_1^2 & 0 \end{pmatrix} - \begin{pmatrix} a\,da + b\,db & b\,da - a\,db \\ a\,db - b\,da & a\,da + b\,db \end{pmatrix},$$

and the result follows (we also obtain $a\,da + b\,db = 0$, which is clear from $\det S = a^2 + b^2 = 1$). $\qquad\square$

Let us now give a geometric interpretation of $\sigma := a\,db - b\,da$. Locally, we can define at each point $p \in M$ the angle $\theta(p)$ between $(E_1)_p$ and $(F_1)_p$. Then the change of basis matrix S has the form

$$\begin{pmatrix} a & -b \\ b & a \end{pmatrix} = \begin{pmatrix} \cos\theta & -\sin\theta \\ \sin\theta & \cos\theta \end{pmatrix}.$$

Hence,

$$\sigma = a\,db - b\,da = \cos\theta\,d\,(\sin\theta) - \sin\theta\,d\,(\cos\theta)$$
$$= \cos^2\theta\,d\theta + \sin^2\theta\,d\theta = d\theta.$$

Therefore, integrating σ along a curve yields the angle by which F_1 rotates with respect to E_1 along the curve.

In particular, notice that σ is closed. This is also clear from

$$d\sigma = d\overline{\omega}_1^2 - d\omega_1^2 = -K\overline{\omega}^1 \wedge \overline{\omega}^2 + K\omega^1 \wedge \omega^2 = 0$$

($\overline{\omega}^1 \wedge \overline{\omega}^2 = \omega^1 \wedge \omega^2$ since the two fields of frames have the same orientation).

We can use the connection form ω_1^2 to define the **geodesic curvature** of a curve on an oriented Riemannian 2-manifold M. Let $c : I \to M$ be a smooth curve in M parameterized by its arclength s (hence $\|\dot{c}(s)\| = 1$). Let V be a neighborhood of a point $c(s)$ in this curve where we have a field of orthonormal frames $\{E_1, E_2\}$ satisfying $(E_1)_{c(s)} = \dot{c}(s)$. Note that it is always possible to consider such a field of frames: we start by extending the vector field $\dot{c}(s)$ to a unit vector field E_1 defined on a neighborhood of $c(s)$, and then consider a unit vector field E_2 orthogonal to the first, such that $\{E_1, E_2\}$ is positively oriented. Since

$$\nabla_{E_1} E_1 = \omega_1^1(E_1)E_1 + \omega_1^2(E_1)E_2 = \omega_1^2(E_1)E_2,$$

the **covariant acceleration** of c is

$$\nabla_{\dot{c}(s)}\dot{c}(s) = \nabla_{E_1(s)} E_1(s) = \omega_1^2(E_1(s))E_2(s).$$

We define the **geodesic curvature** of the curve c to be

$$k_g(s) := \omega_1^2(E_1(s))$$

(thus $|k_g(s)| = \|\nabla_{\dot{c}(s)}\dot{c}(s)\|$). It is a measure of how much the curve fails to be a geodesic at $c(s)$. In particular, c is a geodesic if and only if its geodesic curvature vanishes.

Exercise 2.8

(1) Let $\{X_1, \ldots, X_n\}$ be a field of frames on an open set V of a Riemannian manifold $(M, \langle \cdot, \cdot \rangle)$ with Levi–Civita connection ∇. The associated **structure functions** C_{ij}^k are defined by

$$[X_i, X_j] = \sum_{k=1}^n C_{ij}^k X_k.$$

Show that:

(a) $C_{jk}^i = \Gamma_{jk}^i - \Gamma_{kj}^i$;

(b) $\Gamma_{jk}^i = \frac{1}{2} \sum_{l=1}^n g^{il} \left(X_j \cdot g_{kl} + X_k \cdot g_{jl} - X_l \cdot g_{jk} \right)$
$\qquad + \frac{1}{2} C_{jk}^i - \frac{1}{2} \sum_{l,m=1}^n g^{il} \left(g_{jm} C_{kl}^m + g_{km} C_{jl}^m \right)$;

(c) $d\omega^i + \frac{1}{2} \sum_{j,k=1}^n C_{jk}^i \omega^j \wedge \omega^k = 0$, where $\{\omega^1, \ldots, \omega^n\}$ is the field of dual coframes.

(2) Let $\{X_1, \ldots, X_n\}$ be a field of frames on an open set V of a Riemannian manifold $(M, \langle \cdot, \cdot \rangle)$. Show that a connection ∇ on M is compatible with the metric on V if and only if

$$X_k \cdot \langle X_i, X_j \rangle = \langle \nabla_{X_k} X_i, X_j \rangle + \langle X_i, \nabla_{X_k} X_j \rangle$$

for all i, j, k.

(3) Compute the Gauss curvature of:

(a) the sphere S^2 with the standard metric;
(b) the hyperbolic plane, i.e. the upper half-plane

$$H = \{(x, y) \in \mathbb{R}^2 \mid y > 0\}$$

with the metric

$$g = \frac{1}{y^2}(dx \otimes dx + dy \otimes dy)$$

[cf. Exercise 3.3(5) of Chap. 3].

(4) Determine all surfaces of revolution with constant Gauss curvature.

(5) Let M be the image of the parameterization $\varphi : (0, +\infty) \times \mathbb{R} \to \mathbb{R}^3$ given by

$$\varphi(u, v) = (u \cos v, u \sin v, v),$$

and let N be the image of the parameterization $\psi : (0, +\infty) \times \mathbb{R} \to \mathbb{R}^3$ given by

$$\psi(u, v) = (u \cos v, u \sin v, \log u).$$

Consider in both M and N the Riemannian metric induced by the Euclidean metric of $\mathbb{R}^3$. Show that the map $f : M \to N$ defined by

$$f(\varphi(u, v)) = \psi(u, v)$$

preserves the Gauss curvature but is not a local isometry.

(6) Consider the metric

$$g = A^2(r)dr \otimes dr + r^2 d\theta \otimes d\theta + r^2 \sin^2 \theta \, d\varphi \otimes d\varphi$$

on $M = I \times S^2$, where r is a local coordinate on $I \subset \mathbb{R}$ and (θ, φ) are spherical local coordinates on S^2.

(a) Compute the Ricci tensor and the scalar curvature of this metric.
(b) What happens when $A(r) = (1 - r^2)^{-\frac{1}{2}}$ (that is, when M is locally isometric to S^3)?

(c) And when $A(r) = (1 + r^2)^{-\frac{1}{2}}$ (that is, when M is locally isometric to the **hyperbolic** 3-**space**)?

(d) For which functions $A(r)$ is the scalar curvature constant?

(7) Let M be an oriented Riemannian 2-manifold and let p be a point in M. Let D be a neighborhood of p in M homeomorphic to a disc, with a smooth boundary ∂D. Consider a point $q \in \partial D$ and a unit vector $X_q \in T_q M$. Let X be the parallel transport of X_q along ∂D in the positive direction. When X returns to q it makes an angle $\Delta\theta$ with the initial vector X_q. Using fields of positively oriented orthonormal frames $\{E_1, E_2\}$ and $\{F_1, F_2\}$ such that $F_1 = X$, show that

$$\Delta\theta = \int_D K.$$

Conclude that the Gauss curvature of M at p satisfies

$$K(p) = \lim_{D \to p} \frac{\Delta\theta}{\mathrm{vol}(D)}.$$

(8) Compute the geodesic curvature of a positively oriented circle on:

(a) $\mathbb{R}^2$ with the Euclidean metric and the usual orientation;

(b) S^2 with the usual metric and orientation.

(9) Let c be a smooth curve on an oriented 2-manifold M as in the definition of geodesic curvature. Let X be a vector field parallel along c and let θ be the angle between X and $\dot{c}(s)$ along c in the given orientation. Show that the geodesic curvature of c, k_g, is equal to $\frac{d\theta}{ds}$. (**Hint:** Consider two fields of orthonormal frames $\{E_1, E_2\}$ and $\{F_1, F_2\}$ positively oriented such that $E_1 = \frac{X}{\|X\|}$ and $F_1 = \dot{c}$).

4.3 Gauss–Bonnet Theorem

We will now use the Cartan structure equations to prove the **Gauss–Bonnet theorem**, relating the curvature of a compact surface to its topology. Let M be a compact, oriented, 2-dimensional manifold and X a vector field on M.

Definition 3.1 A point $p \in M$ is said to be a **singular point** of X if $X_p = 0$. A singular point is said to be an **isolated singularity** if there exists a neighborhood $V \subset M$ of p such that p is the only singular point of X in V.

Since M is compact, if all the singularities of X are isolated then they are in finite number (as otherwise they would accumulate on a non-isolated singularity).

To each isolated singularity $p \in V$ of $X \in \mathfrak{X}(M)$ one can associate an integer number, called the **index** of X at p, as follows:

(i) fix a Riemannian metric in M;

(ii) choose a positively oriented orthonormal frame $\{F_1, F_2\}$, defined on $V \backslash \{p\}$, such that

$$F_1 = \frac{X}{\|X\|},$$

let $\{\overline{\omega}^1, \overline{\omega}^2\}$ be the dual coframe and let $\overline{\omega}_1^2$ be the corresponding connection form;

(iii) possibly shrinking V, choose a positively oriented orthonormal frame $\{E_1, E_2\}$, defined on V, with dual coframe $\{\omega^1, \omega^2\}$ and connection form ω_1^2;

(iv) take a neighborhood D of p in V, homeomorphic to a disc, with a smooth boundary ∂D, endowed with the induced orientation, and define the index I_p of X at p as

$$2\pi I_p = \int_{\partial D} \sigma,$$

where $\sigma := \overline{\omega}_1^2 - \omega_1^2$ is the form in Proposition 2.7.

Recall that σ satisfies $\sigma = d\theta$, where θ is the angle between E_1 and F_1. Therefore I_p must be an integer. Intuitively, the index of a vector field X measures the number of times that X rotates as one goes around the singularity anticlockwise, counted positively if X itself rotates anticlockwise, and negatively otherwise.

Example 3.2 In $M = \mathbb{R}^2$ the following vector fields have isolated singularities at the origin with the indicated indices (cf. Fig. 4.1):

(1) $X_{(x,y)} = (x, y)$ has index 1;
(2) $Y_{(x,y)} = (-y, x)$ has index 1;
(3) $Z_{(x,y)} = (y, x)$ has index -1;
(4) $W_{(x,y)} = (x, -y)$ has index -1.

We will now check that the index is well defined. We begin by observing that, since σ is closed, I_p does not depend on the choice of D. Indeed, the boundaries of any two such discs are necessarily homotopic [cf. Exercise 5.3(2) in Chap. 2]. Next we prove that I_p does not depend on the choice of the frame $\{E_1, E_2\}$. More precisely, we will show that

$$I_p = \lim_{r \to 0} \frac{1}{2\pi} \int_{S_r(p)} \overline{\omega}_1^2,$$

where $S_r(p)$ is the normal sphere of radius r centered at p. Indeed, if $r_1 > r_2 > 0$ are radii of normal spheres, one has

$$\int_{S_{r_1}(p)} \overline{\omega}_1^2 - \int_{S_{r_2}(p)} \overline{\omega}_1^2 = \int_{\Delta_{12}} d\overline{\omega}_1^2 = -\int_{\Delta_{12}} K\overline{\omega}^1 \wedge \overline{\omega}^2 = -\int_{\Delta_{12}} K, \quad (4.6)$$

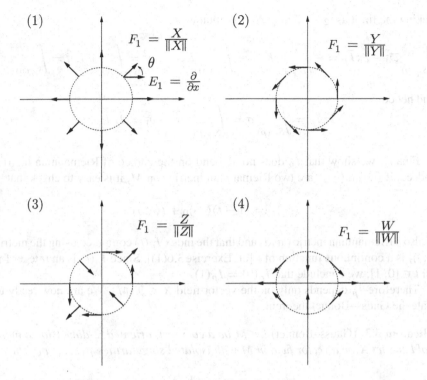

Fig. 4.1 Computing the indices of the vector fields X, Y, Z and W

where $\Delta_{12} = B_{r_1}(p) \setminus B_{r_2}(p)$. Since K is continuous, we see that

$$\left(\int_{S_{r_1}(p)} \overline{\omega}_1^2 - \int_{S_{r_2}(p)} \overline{\omega}_1^2 \right) \longrightarrow 0$$

as $r_1 \to 0$. Therefore, if $\{r_n\}$ is a decreasing sequence of positive numbers converging to zero, the sequence

$$\left\{ \int_{S_{r_n}(p)} \overline{\omega}_1^2 \right\}$$

is a Cauchy sequence, and therefore converges. Let

$$\overline{I}_p := \lim_{r \to 0} \frac{1}{2\pi} \int_{S_r(p)} \overline{\omega}_1^2.$$

Taking the limit as $r_2 \to 0$ in (4.6) one obtains

$$\int_{S_{r_1}(p)} \overline{\omega}_1^2 - 2\pi \overline{I}_p = -\int_{B_{r_1}(p)} K = -\int_{B_{r_1}(p)} K\omega^1 \wedge \omega^2 = \int_{B_{r_1}(p)} d\omega_1^2 = \int_{S_{r_1}(p)} \omega_1^2,$$

and hence

$$2\pi I_p = \int_{S_{r_1}(p)} \sigma = \int_{S_{r_1}(p)} \overline{\omega}_1^2 - \omega_1^2 = 2\pi \overline{I}_p.$$

Finally, we show that I_p does not depend on the choice of Riemannian metric. Indeed, if $\langle \cdot, \cdot \rangle_0$, $\langle \cdot, \cdot \rangle_1$ are two Riemannian metrics on M, it is easy to check that

$$\langle \cdot, \cdot \rangle_t := (1 - t)\langle \cdot, \cdot \rangle_0 + t\langle \cdot, \cdot \rangle_1$$

is also a Riemannian metric on M, and that the index $I_p(t)$ computed using the metric $\langle \cdot, \cdot \rangle_t$ is a continuous function of t [cf. Exercise 3.6(1)]. Since $I_p(t)$ is an integer for all $t \in [0, 1]$, we conclude that $I_p(0) = I_p(1)$.

Therefore I_p depends only on the vector field $X \in \mathfrak{X}(M)$. We are now ready to state the Gauss–Bonnet theorem:

Theorem 3.3 (Gauss–Bonnet) *Let M be a compact, oriented, 2-dimensional manifold and let X be a vector field in M with isolated singularities $p_1, \ldots, p_k$. Then*

$$\int_M K = 2\pi \sum_{i=1}^k I_{p_i} \tag{4.7}$$

for any Riemannian metric on M, where K is the Gauss curvature.

Proof We consider the positively oriented orthonormal frame $\{F_1, F_2\}$, with

$$F_1 = \frac{X}{\|X\|},$$

defined on $M \setminus \bigcup_{i=1}^k \{p_i\}$, with dual coframe $\{\overline{\omega}^1, \overline{\omega}^2\}$ and connection form $\overline{\omega}_1^2$. For $r > 0$ sufficiently small, we take $B_i := B_r(p_i)$ such that $\overline{B}_i \cap \overline{B}_j = \varnothing$ for $i \neq j$ and note that

$$\int_{M \setminus \bigcup_{i=1}^k B_i} K = \int_{M \setminus \bigcup_{i=1}^k B_i} K \overline{\omega}^1 \wedge \overline{\omega}^2 = -\int_{M \setminus \bigcup_{i=1}^k B_i} d\overline{\omega}_1^2$$

$$= \int_{\bigcup_{i=1}^k \partial B_i} \overline{\omega}_1^2 = \sum_{i=1}^k \int_{\partial B_i} \overline{\omega}_1^2,$$

where the ∂B_i have the orientation induced by the orientation of B_i. Taking the limit as $r \to 0$ one obtains

$$\int_M K = 2\pi \sum_{i=1}^{k} I_{p_i}.$$ □

Remark 3.4

(1) Since the right-hand side of (4.7) does not depend on the metric, we conclude that $\int_M K$ is the same for **all** Riemannian metrics on M.
(2) Since the left-hand side of (4.7) does not depend on the vector field X, we conclude that $\chi(M) := \sum_{i=1}^{k} I_{p_i}$ is the same for all vector fields on M with isolated singularities. This is the so-called **Euler characteristic** of M.
(3) Recall that a **triangulation** of M is a decomposition of M in a finite number of triangles (i.e. images of Euclidean triangles by parameterizations) such that the intersection of any two triangles is either a common edge, a common vertex or empty (it is possible to prove that such a triangulation always exists). Given a triangulation, one can construct a vector field X with the following properties (cf. Fig. 4.2):

(a) each vertex is a singularity which is a sink, that is,

$$X = -x\frac{\partial}{\partial x} - y\frac{\partial}{\partial y}$$

for certain local coordinates (x, y) centered at the singularity;

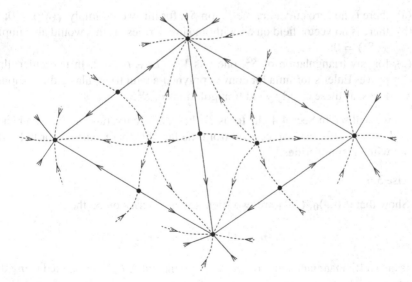

Fig. 4.2 Vector field associated to a triangulation

(b) the interior of each 2-dimensional face contains exactly one singularity which is a source, that is

$$X = x\frac{\partial}{\partial x} + y\frac{\partial}{\partial y}$$

for certain local coordinates (x, y) centered at the singularity;
(c) each edge is formed by integral curves of the vector field and contains exactly one singularity which is not a vertex.

It is easy to see that all singularities are isolated, that the singularities at the vertices and 2-dimensional faces have index 1 and that the singularities at the edges have index -1. Therefore,

$$\chi(M) = V - E + F,$$

where V is the number of vertices, E is the number of edges and F is the number of 2-dimensional faces on any triangulation. This is the definition we used in Exercise 1.8(5) of Chap. 1.

Example 3.5

(1) Choosing the standard metric in S^2, we have

$$\chi(S^2) = \frac{1}{2\pi}\int_{S^2} 1 = \frac{1}{2\pi}\text{vol}(S^2) = 2.$$

From this we can derive a number of conclusions:

(a) there is no zero curvature metric on S^2, for this would imply $\chi(S^2) = 0$;
(b) there is no vector field on S^2 without singularities, as this would also imply $\chi(S^2) = 0$;
(c) for any triangulation of S^2, one has $V - E + F = 2$. In particular, this proves Euler's formula for convex polyhedra with triangular 2-dimensional faces, as these clearly yield triangulations of S^2.

(2) As we will see in Sect. 4.4, the torus T^2 has a zero curvature metric, and hence $\chi(T^2) = 0$. This can also be seen from the fact that there exist vector fields on T^2 without singularities.

Exercise 3.6

(1) Show that if $\langle \cdot, \cdot \rangle_0, \langle \cdot, \cdot \rangle_1$ are two Riemannian metrics on M then

$$\langle \cdot, \cdot \rangle_t := (1 - t)\langle \cdot, \cdot \rangle_0 + t\langle \cdot, \cdot \rangle_1$$

is also a Riemannian metric on M, and that the index $I_p(t)$ computed using the metric $\langle \cdot, \cdot \rangle_t$ is a continuous function of t.

(2) (*Gauss–Bonnet theorem for non-orientable manifolds*) Let (M, g) be a compact, non-orientable, 2-dimensional Riemannian manifold and let $\pi : \overline{M} \to M$ be its orientable double covering [cf. Exercise 8.6(9) in Chap. 1]. Show that:

(a) $\chi(\overline{M}) = 2\chi(M)$;
(b) $\overline{K} = \pi^*K$, where $\overline{K}$ is the Gauss curvature of the Riemannian metric $\overline{g} := \pi^*g$ on $\overline{M}$;
(c) $2\pi\chi(M) = \frac{1}{2}\displaystyle\int_{\overline{M}} \overline{K}$.

(**Remark:** Even though M is not orientable, we can still define the integral of a function f on M through $\int_M f = \frac{1}{2}\int_{\overline{M}} \pi^*f$; with this definition, the Gauss–Bonnet theorem holds for non-orientable Riemannian 2-manifolds).

(3) (*Gauss–Bonnet theorem for manifolds with boundary*) Let M be a compact, oriented, 2-dimensional manifold with boundary and let X be a vector field in M **transverse** to ∂M (i.e. such that $X_p \notin T_p\partial M$ for all $p \in \partial M$), with isolated singularities $p_1, \ldots, p_k \in M\backslash\partial M$. Prove that

$$\int_M K + \int_{\partial M} k_g = 2\pi \sum_{i=1}^{k} I_{p_i}$$

for any Riemannian metric on M, where K is the Gauss curvature of M and k_g is the geodesic curvature of ∂M.

(4) Let (M, g) be a compact orientable 2-dimensional Riemannian manifold, with positive Gauss curvature. Show that any two non-self-intersecting closed geodesics must intersect each other.

(5) Let M be a differentiable manifold and $f : M \to \mathbb{R}$ a smooth function.

(a) (*Hessian*) Let $p \in M$ be a critical point of f (i.e. $(df)_p = 0$). The **Hessian** of f at p is the map $(Hf)_p : T_pM \times T_pM \to \mathbb{R}$ given by

$$(Hf)_p(v, w) = \left.\frac{\partial^2}{\partial t \partial s}\right|_{s=t=0} (f \circ \gamma)(s, t),$$

where $\gamma : U \subset \mathbb{R}^2 \to M$ is such that $\gamma(0, 0) = p$, $\frac{\partial \gamma}{\partial s}(0, 0) = v$ and $\frac{\partial \gamma}{\partial t}(0, 0) = w$. Show that $(Hf)_p$ is a well-defined symmetric 2-tensor.
(b) (*Morse theorem*) If $(Hf)_p$ is nondegenerate then p is called a **nondegenerate critical point**. Assume that M is compact and f is a **Morse function**, i.e. all its critical points are nondegenerate. Prove that there are only a finite number of critical points. Moreover, show that if M is 2-dimensional then

$$\chi(M) = m - s + n,$$

where m, n and s are the numbers of maxima, minima and saddle points respectively. (**Hint:** Choose a Riemannian metric on M and consider the vector field $X := \text{grad } f$).

(6) Let (M, g) be a 2-dimensional Riemannian manifold and $\Delta \subset M$ a **geodesic triangle**, i.e. an open set homeomorphic to an Euclidean triangle whose sides are images of geodesic arcs. Let α, β, γ be the inner angles of Δ, i.e. the angles between the geodesics at the intersection points contained in $\partial \Delta$. Prove that for small enough Δ one has

$$\alpha + \beta + \gamma = \pi + \int_\Delta K,$$

where K is the Gauss curvature of M, using:

(a) the fact that $\int_\Delta K$ is the angle by which a vector parallel-transported once around $\partial \Delta$ rotates;

(b) the Gauss–Bonnet theorem for manifolds with boundary.

(**Remark:** We can use this result to give another geometric interpretation of the Gauss curvature: $K(p) = \lim_{\Delta \to p} \frac{\alpha + \beta + \gamma - \pi}{\text{vol}(\Delta)}$.)

(7) Let (M, g) be a simply connected 2-dimensional Riemannian manifold with nonpositive Gauss curvature. Show that any two geodesics intersect at most in one point. (**Hint:** Note that if two geodesics intersected in more than one point then there would exist a **geodesic biangle**, i.e. an open set homeomorphic to a disc whose boundary is formed by the images of two geodesic arcs).

4.4 Manifolds of Constant Curvature

Recall that a manifold is said to have constant curvature if all sectional curvatures at all points have the same constant value K. There is an easy way to identify these manifolds using their curvature forms.

Lemma 4.1 *If M is a manifold of constant curvature K, then, around each point $p \in M$, all curvature forms Ω_i^j satisfy*

$$\Omega_i^j = -K\omega^i \wedge \omega^j, \tag{4.8}$$

where $\{\omega^1, \ldots, \omega^n\}$ is any field of orthonormal coframes defined on a neighborhood of p. Conversely, if on a neighborhood of each point of M there is a field of orthonormal frames $\{E_1, \ldots, E_n\}$ such that the corresponding field of coframes $\{\omega^1, \ldots, \omega^n\}$ satisfies (4.8) for some constant K, then M has constant curvature K.

Proof If M has constant curvature K then

$$\Omega_i^j = \sum_{k<l} \Omega_i^j(E_k, E_l)\, \omega^k \wedge \omega^l = \sum_{k<l} \omega^j(R(E_k, E_l)E_i)\, \omega^k \wedge \omega^l$$
$$= \sum_{k<l} \langle R(E_k, E_l)E_i, E_j \rangle \omega^k \wedge \omega^l = \sum_{k<l} R_{klij}\, \omega^k \wedge \omega^l$$

$$= -\sum_{k<l} K(\delta_{ki}\delta_{lj} - \delta_{kj}\delta_{li})\,\omega^k \wedge \omega^l = -K\omega^i \wedge \omega^j.$$

Conversely, let us assume that there is a constant K such that on a neighborhood of each point $p \in M$ we have $\Omega_i^j = -K\omega^i \wedge \omega^j$. Then, for every section Π of the tangent space $T_p M$, the corresponding sectional curvature is given by

$$K(\Pi) = -R(X, Y, X, Y)$$

where X, Y are two linearly independent vectors spanning Π (which we assume to span a parallelogram of unit area). Using the field of orthonormal frames around p, we have $X = \sum_{i=1}^n X^i E_i$ and $Y = \sum_{i=1}^n Y^i E_i$ and so,

$$K(\Pi) = -\sum_{i,j,k,l=1}^n X^i Y^j X^k Y^l R(E_i, E_j, E_k, E_l)$$

$$= -\sum_{i,j,k,l=1}^n X^i Y^j X^k Y^l \,\Omega_k^l(E_i, E_j)$$

$$= K \sum_{i,j,k,l=1}^n X^i Y^j X^k Y^l \,\omega^k \wedge \omega^l(E_i, E_j)$$

$$= K \sum_{i,j,k,l=1}^n X^i Y^j X^k Y^l \left(\omega^k(E_i)\omega^l(E_j) - \omega^k(E_j)\omega^l(E_i) \right)$$

$$= K \sum_{i,j,k,l=1}^n X^i Y^j X^k Y^l (\delta_{ik}\delta_{jl} - \delta_{jk}\delta_{il})$$

$$= K \left(\|X\|^2 \|Y\|^2 - \langle X, Y \rangle^2 \right) = K. \qquad \square$$

Let us now see an example of how we can use this lemma.

Example 4.2 The n-dimensional **hyperbolic space** of radius $a > 0$, $H^n(a)$, is the open half-space

$$\{(x^1, \dots x^n) \in \mathbb{R}^n \mid x^n > 0\}$$

equipped with the Riemannian metric

$$g_{ij}(x) = \frac{a^2}{(x^n)^2}\,\delta_{ij}.$$

This Riemannian manifold has constant sectional curvature $K = -\frac{1}{a^2}$. Indeed, using the above lemma, we will show that on $H^n(a)$ there is a field of orthonormal frames

$\{E_1, \ldots, E_n\}$ whose dual field of coframes $\{\omega^1, \ldots, \omega^n\}$ satisfies

$$\Omega_i^j = -K\omega^i \wedge \omega^j \tag{4.9}$$

for $K = -\frac{1}{a^2}$. For that, let us consider the natural coordinate system $x : H^n(a) \to \mathbb{R}^n$ and the corresponding field of coordinate frames $\{X_1, \ldots, X_n\}$ with $X_i = \frac{\partial}{\partial x_i}$. Since

$$\langle X_i, X_j \rangle = \frac{a^2}{(x^n)^2} \delta_{ij},$$

we obtain a field of orthonormal frames $\{E_1, \ldots, E_n\}$ with $E_i = \frac{x^n}{a} X_i$, and the corresponding dual field of coframes $\{\omega^1, \ldots, \omega^n\}$ where $\omega^i = \frac{a}{x^n} dx^i$. Then

$$d\omega^i = \frac{a}{(x^n)^2} dx^i \wedge dx^n = \frac{1}{a} \omega^i \wedge \omega^n = \sum_{j=1}^n \omega^j \wedge \left(-\frac{1}{a}\delta_{jn}\omega^i\right),$$

and so, using the structure equations

$$d\omega^i = \sum_{j=1}^n \omega^j \wedge \omega_j^i$$

$$\omega_i^j + \omega_j^i = 0,$$

we can guess that the connection forms are given by $\omega_j^i = \frac{1}{a}(\delta_{in}\omega^j - \delta_{jn}\omega^i)$. Indeed, we can easily verify that these forms satisfy the above structure equations, and hence must be the connection forms by unicity of solution of these equations. With these forms it is now easy to compute the curvature forms Ω_i^j using the third structure equation

$$d\omega_i^j = \sum_{k=1}^n \omega_i^k \wedge \omega_k^j + \Omega_i^j.$$

We have

$$d\omega_i^j = d\left(\frac{1}{a}(\delta_{jn}\omega^i - \delta_{in}\omega^j)\right) = \frac{1}{a^2}(\delta_{jn}\omega^i \wedge \omega^n - \delta_{in}\omega^j \wedge \omega^n)$$

and

$$\sum_{k=1}^n \omega_i^k \wedge \omega_k^j = \frac{1}{a^2} \sum_{k=1}^n (\delta_{kn}\omega^i - \delta_{in}\omega^k) \wedge (\delta_{jn}\omega^k - \delta_{kn}\omega^j)$$

$$= \frac{1}{a^2} \sum_{k=1}^{n} (\delta_{kn}\delta_{jn}\omega^i \wedge \omega^k - \delta_{kn}\omega^i \wedge \omega^j + \delta_{in}\delta_{kn}\omega^k \wedge \omega^j)$$

$$= \frac{1}{a^2} (\delta_{jn}\omega^i \wedge \omega^n - \omega^i \wedge \omega^j + \delta_{in}\omega^n \wedge \omega^j),$$

and so,

$$\Omega_i^j = d\omega_i^j - \sum_{k=1}^{n} \omega_i^k \wedge \omega_k^j = \frac{1}{a^2}\omega^i \wedge \omega^j.$$

We conclude that $K = -\frac{1}{a^2}$.

The Euclidean spaces $\mathbb{R}^n$ have constant curvature equal to zero. Moreover, we can easily see that the spheres $S^n(r) \subset \mathbb{R}^{n+1}$ of radius r have constant curvature equal to $\frac{1}{r^2}$ [cf. Exercise 5.7(2)]. Therefore we have examples of manifolds with arbitrary constant negative ($H^n(a)$), zero ($\mathbb{R}^n$) or positive ($S^n(r)$) curvature in any dimension. Note that all these examples are simply connected and are geodesically complete. Indeed, the images of the geodesics of the Euclidean space $\mathbb{R}^n$ are straight lines, $S^n(r)$ is compact and the images of the geodesics of $H^n(a)$ are either half circles perpendicular to the plane $x^n = 0$ and centered on this plane, or vertical half lines starting at the plane $x^n = 0$ [cf. Exercise 4.7(4)].

Every simply connected geodesically complete manifold of constant curvature is isometric to one of these examples, as is stated in the following theorem. In general, if the manifold is not simply connected (but still geodesically complete), it is isometric to the quotient of one of the above examples by a free and proper action of a discrete subgroup of the group of isometries (it can be proved that the group of isometries of a Riemannian manifold is always a Lie group).

Theorem 4.3 (Killing-Hopf) *Let M be a connected, geodesically complete n-dimensional Riemannian manifold with constant curvature K.*

(1) If M is simply connected then it is isometric to one of the following: $S^n\left(\frac{1}{\sqrt{K}}\right)$ if $K > 0$, $\mathbb{R}^n$ if $K = 0$, or $H^n\left(\frac{1}{\sqrt{-K}}\right)$ if $K < 0$.

(2) If M is not simply connected then M is isometric to a quotient $\widetilde{M}/\Gamma$, where $\widetilde{M}$ is one of the above simply connected manifolds and Γ is a nontrivial discrete subgroup of the group of isometries of $\widetilde{M}$ acting properly and freely on $\widetilde{M}$.

Proof The proof of this theorem can be found in [dC93]. Here we just give the proof in the case when M is simply connected, $n = 2$ and $K = 0$. In this case, the Cartan–Hadamard theorem [cf. Exercise 5.8(5) in Chap. 3] implies that given $p \in M$ the map $\exp_p : T_pM \to M$ is a diffeomorphism. Let $\{E_1, E_2\}$ be a global orthonormal frame on M (obtained by orthonormalizing the frame associated to global Cartesian coordinates). Since $K = 0$, the corresponding connection form ω_1^2 satisfies $d\omega_1^2 = 0$, and so by the Poincaré Lemma [cf. Exercise 3.8(5) in Chap. 2] we have $\omega_1^2 = df$ for some smooth function $f \in C^\infty(M)$. Let $\{F_1, F_2\}$ be the orthonormal frame with the same orientation as $\{E_1, E_2\}$ such that the angle between E_1 and F_1 is $\theta = -f$. Then its connection form $\overline{\omega}_1^2$ satisfies $\overline{\omega}_1^2 = \omega_1^2 + d\theta = 0$, that is, $\nabla_{F_1} F_1 = \nabla_{F_1} F_2 = \nabla_{F_2} F_1 = \nabla_{F_2} F_2 = 0$. We conclude that $[F_1, F_2] = 0$, and so, by Theorem 6.10 in Chap. 1, their flows commute. We can then introduce local coordinates (x, y) in M by using the parameterization

$$\varphi(x, y) = \psi_{1,x} \circ \psi_{2,x}(p)$$

(where ψ_1, ψ_2 are the flows of F_1, F_2). Using $\frac{\partial}{\partial x} = F_1$ and $\frac{\partial}{\partial y} = F_2$, it is easily shown that $\varphi(x, y) = \exp_p \left(x(F_1)_p + y(F_2)_p \right)$, and so (x, y) are actually global coordinates. Since in these coordinates the metric is written

$$g = dx \otimes dx + dy \otimes dy,$$

we conclude that M is isometric to $\mathbb{R}^2$. □

Example 4.4 Let $\widetilde{M} = \mathbb{R}^2$. Then the subgroup of isometries Γ cannot contain isometries with fixed points (since it acts freely). Hence Γ can only contain translations and gliding reflections (that is, reflections followed by a translation in the direction of the reflection axis). Moreover, Γ is generated by at most two elements, one of which may be assumed to be a translation [cf. Exercise 4.7(6)]. Therefore we have:

(1) if Γ is generated by one translation, then the resulting surface will be a cylinder;
(2) if Γ is generated by two translations we obtain a torus;
(3) if Γ is generated by a gliding reflection we obtain a Möbius band;
(4) if Γ is generated by a translation and a gliding reflection we obtain a Klein bottle.

These are all the possible examples of geodesically complete Euclidean surfaces (2-dimensional manifolds of constant zero curvature).

Example 4.5 The group of orientation-preserving isometries of the hyperbolic plane H^2 is $PSL(2, \mathbb{R}) = SL(2, \mathbb{R})/\{\pm \text{id}\}$, acting on H^2 through

$$\begin{pmatrix} a & b \\ c & d \end{pmatrix} \cdot z := \frac{az + b}{cz + d},$$

where we make the identification $\mathbb{R}^2 \cong \mathbb{C}$ [cf. Exercise 4.7(8) and Sect. 4.6.1]. To find orientable hyperbolic surfaces, that is, surfaces with constant curvature $K = -1$, we have to find discrete subgroups Γ of $PSL(2, \mathbb{R})$ acting properly and freely on H^2. Here there are many more possibilities. As an example, we can consider the group $\Gamma = \langle f \rangle$ generated by the translation $f(z) = z + 2\pi$. The resulting surface is known as a **pseudosphere** and is homeomorphic to a cylinder (cf. Fig. 4.3). However, the width of the end where $y \to +\infty$ converges to zero, while the width of the end where $y \to 0$ converges to $+\infty$. Its height towards both ends is infinite. Note that this surface has geodesics which transversely auto-intersect a finite number of times (cf. Fig. 4.4).

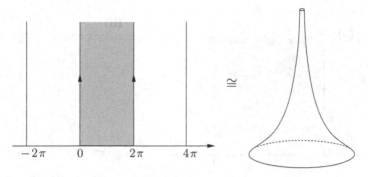

Fig. 4.3 Pseudosphere

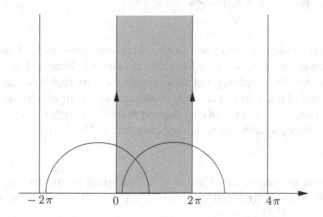

Fig. 4.4 Trajectories of geodesics on the pseudosphere

Other examples can be obtained by considering hyperbolic polygons (bounded by geodesics) and identifying their sides through isometries. For instance, the surface in Fig. 4.5b is obtained by identifying the sides of the polygon in Fig. 4.5a through the isometries $g(z) = z + 2$ and $h(z) = \frac{z}{2z+1}$. Choosing other polygons it is possible to obtain **compact** hyperbolic surfaces. In fact, there exist compact hyperbolic surfaces homeomorphic to any topological 2-manifold with negative Euler characteristic (the Gauss–Bonnet theorem does not allow non-negative Euler characteristics in this case).

Example 4.6 To find Riemannian manifolds of constant positive curvature we have to find discrete subgroups of isometries of the sphere that act properly and freely. Let us consider the case where $K = 1$. Then $\Gamma \subset O(n + 1)$ [cf. Exercise 4.7(11)]. Since it must act freely on S^n, no element of $\Gamma\backslash\{id\}$ can have 1 as an eigenvalue. We will see that, when n is even, S^n and $\mathbb{R}P^n$ are the only geodesically complete manifolds of constant curvature 1. Indeed, if $A \in \Gamma$, then A is an orthogonal $(n+1) \times (n+1)$ matrix and so all its eigenvalues have absolute value equal to 1. Moreover, its characteristic

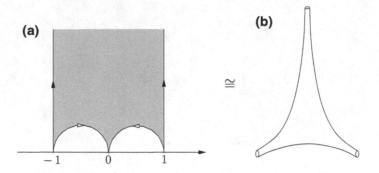

Fig. 4.5 a Hyperbolic polygon, **b** Thrice-punctured sphere

polynomial has odd degree $(n+1)$, and so it has a real root, equal to ± 1. Consequently, A^2 has 1 as an eigenvalue, and so it has to be the identity. Hence, $A = A^{-1} = A^t$, and so A is symmetric, implying that all its eigenvalues are real. The eigenvalues of A are then either all equal to 1 (if $A = $ id) or all equal to -1, in which case $A = -$ id. We conclude that $\Gamma = \{\pm \text{id}\}$ implying that our manifold is either S^n or $\mathbb{R}P^n$. If n is odd there are other possibilities, which are classified in [Wol78].

Exercise 4.7

(1) Show that the metric of $H^n(a)$ is a left-invariant metric for the Lie group structure induced by identifying $(x^1, \ldots, x^n) \in H^n(a)$ with the affine map $g : \mathbb{R}^{n-1} \to \mathbb{R}^{n-1}$ given by

$$g(t^1, \ldots, t^{n-1}) = x^n(t^1, \ldots, t^{n-1}) + (x^1, \ldots, x^{n-1}).$$

(2) Prove that if the forms ω^i in a field of orthonormal coframes satisfy $d\omega^i = \alpha \wedge \omega^i$ (with α a 1-form), then the connection forms ω^j_i are given by $\omega^j_i = \alpha(E_i)\omega^j - \alpha(E_j)\omega^i = -\omega^i_j$. Use this to confirm the results in Example 4.2.

(3) Let K be a real number and let $\rho = 1 + (\frac{K}{4}) \sum_{i=1}^{n}(x^i)^2$. Show that, for the Riemannian metric defined on $\mathbb{R}^n$ by

$$g_{ij}(p) = \frac{1}{\rho^2} \delta_{ij},$$

the sectional curvature is constant equal to K.

(4) Show that any isometry of the Euclidean space $\mathbb{R}^n$ which preserves the coordinate function x^n is an isometry of $H^n(a)$. Use this fact to determine all the geodesics of $H^n(a)$.

(5) (*Schur theorem*) Let M be a connected isotropic Riemannian manifold of dimension $n \geq 3$. Show that M has constant curvature. (**Hint:** Use the structure equations to show that $dK = 0$).

(6) To complete the details in Example 4.4, show that:

  (a) the isometries of $\mathbb{R}^2$ with no fixed points are either translations or gliding reflections;
  (b) any discrete group of isometries of $\mathbb{R}^2$ acting properly and freely is generated by at most two elements, one of which may be assumed to be a translation.

(7) Let $f, g : \mathbb{R}^2 \to \mathbb{R}^2$ be the isometries

$$f(x, y) = (-x, y + 1) \quad \text{and} \quad g(x, y) = (x + 1, y)$$

(thus f is a gliding reflection and g is a translation). Check that $\mathbb{R}^2/\langle f \rangle$ is homeomorphic to a Möbius band (without boundary), and that $\mathbb{R}^2/\langle f, g \rangle$ is homeomorphic to a Klein bottle.

(8) Let H^2 be the hyperbolic plane. Show that:

  (a) the formula

$$\begin{pmatrix} a & b \\ c & d \end{pmatrix} \cdot z := \frac{az + b}{cz + d} \qquad (ad - bc = 1)$$

defines an action of $PSL(2, \mathbb{R}) := SL(2, \mathbb{R})/\{\pm \, \mathrm{id}\}$ on H^2 by orientation-preserving isometries;
  (b) for any two geodesics $c_1, c_2 : \mathbb{R} \to H^2$, parameterized by the arclength, there exists $g \in PSL(2, \mathbb{R})$ such that $c_1(s) = g \cdot c_2(s)$ for all $s \in \mathbb{R}$;
  (c) given $z_1, z_2, z_3, z_4 \in H^2$ with $d(z_1, z_2) = d(z_3, z_4)$, there exists $g \in PSL(2, \mathbb{R})$ such that $g \cdot z_1 = z_3$ and $g \cdot z_2 = z_4$;
  (d) an orientation-preserving isometry of H^2 with two fixed points must be the identity. Conclude that all orientation-preserving isometries are of the form $f(z) = g \cdot z$ for some $g \in PSL(2, \mathbb{R})$.

(9) Check that the isometries $g(z) = z + 2$ and $h(z) = \frac{z}{2z+1}$ of the hyperbolic plane in Example 4.5 identify the sides of the hyperbolic polygon in Fig. 4.5.

(10) A **tractrix** is the curve described parametrically by

$$\begin{cases} x = u - \tanh u \\ y = \mathrm{sech}\, u \end{cases} \qquad (u > 0)$$

(its name derives from the property that the distance between any point in the curve and the x-axis along the tangent is constant equal to 1). Show that the surface of revolution generated by rotating a tractrix about the x-axis (**tractroid**) has constant Gauss curvature $K = -1$. Determine an open subset of the pseudosphere isometric to the tractroid. (Remark: The tractroid is not geodesically complete; in fact, it was proved by Hilbert in 1901 that any surface of constant negative curvature embedded in Euclidean 3-space must be incomplete).

(11) Show that the group of isometries of S^n is $O(n + 1)$.

(12) Let G be a compact Lie group of dimension 2. Show that:

  (a) G is orientable;
  (b) $\chi(G) = 0$;
  (c) any left-invariant metric on G has constant curvature;
  (d) G is the 2-torus T^2.

4.5 Isometric Immersions

Many Riemannian manifolds arise as submanifolds of other Riemannian manifolds, by taking the induced metric (e.g. $S^n \subset \mathbb{R}^{n+1}$). In this section, we will analyze how the curvatures of the two manifolds are related.

Let $f : N \to M$ be an immersion of an n-manifold N on an m-manifold M. We know from Sect. 4.5 of Chap. 1 that for each point $p \in N$ there is a neighborhood $V \subset N$ of p where f is an embedding onto its image. Hence $f(V)$ is a submanifold of M. To simplify notation, we will identify V with $f(V)$, and proceed as if f were the inclusion map. Let $\langle \cdot, \cdot \rangle$ be a Riemannian metric on M and let $\langle\langle \cdot, \cdot \rangle\rangle$ be the metric induced on N by f (which is therefore called an **isometric immersion**). For every $p \in V$, the tangent space $T_p M$ can be decomposed as

$$T_p M = T_p N \oplus (T_p N)^{\perp}.$$

Therefore, every element v of $T_p M$ can be written uniquely as $v = v^{\top} + v^{\perp}$, where $v^{\top} \in T_p N$ is the tangential part of v and $v^{\perp} \in (T_p N)^{\perp}$ is the normal part of v. Let $\widetilde{\nabla}$ and ∇ be the Levi–Civita connections of $(M, \langle \cdot, \cdot \rangle)$ and $(N, \langle\langle \cdot, \cdot \rangle\rangle)$, respectively. Let X, Y be two vector fields in $V \subset N$ and let $\widetilde{X}, \widetilde{Y}$ be two extensions of X, Y to a neighborhood $W \subset M$ of V. Using the Koszul formula, we can easily check that

$$\nabla_X Y = \left(\widetilde{\nabla}_{\widetilde{X}} \widetilde{Y} \right)^{\top}$$

[cf. Exercise 3.3(6) in Chap. 3]. We define the **second fundamental form** of N as

$$B(X, Y) := \widetilde{\nabla}_{\widetilde{X}} \widetilde{Y} - \nabla_X Y.$$

Note that this map is well defined, that is, it does not depend on the extensions $\widetilde{X}, \widetilde{Y}$ of X, Y [cf. Exercise 5.7(1)]. Moreover, it is bilinear, symmetric, and, for each $p \in V$, $B(X, Y)_p \in (T_p N)^{\perp}$ depends only on the values of X_p and Y_p.

Using the second fundamental form, we can define, for each vector $n_p \in (T_p N)^{\perp}$, a symmetric bilinear map $H_{n_p} : T_p N \times T_p N \to \mathbb{R}$ through

$$H_{n_p}(X_p, Y_p) = \langle B(X_p, Y_p), n_p \rangle.$$

The corresponding quadratic form is often called the **second fundamental form of** f **at** p **along the vector** n_p.

Finally, since H_{n_p} is bilinear, there exists a linear map $S_{n_p} : T_pN \to T_pN$ satisfying

$$\langle\langle S_{n_p}(X_p), Y_p\rangle\rangle = H_{n_p}(X_p, Y_p) = \langle B(X_p, Y_p), n_p\rangle$$

for all $X_p, Y_p \in T_pN$. It is easy to check that this linear map is given by

$$S_{n_p}(X_p) = -(\widetilde{\nabla}_{\widetilde{X}} n)_p^\top,$$

where n is a local extension of n_p normal to N. Indeed, since $\langle \widetilde{Y}, n\rangle = 0$ on N and $\widetilde{X}$ is tangent to N, we have on N

$$\begin{aligned}
\langle\langle S_n(X), Y\rangle\rangle &= \langle B(X, Y), n\rangle = \langle\widetilde{\nabla}_{\widetilde{X}}\widetilde{Y} - \nabla_X Y, n\rangle\\
&= \langle\widetilde{\nabla}_{\widetilde{X}}\widetilde{Y}, n\rangle = \widetilde{X}\cdot\langle\widetilde{Y}, n\rangle - \langle\widetilde{Y}, \widetilde{\nabla}_{\widetilde{X}} n\rangle\\
&= \langle -\widetilde{\nabla}_{\widetilde{X}} n, \widetilde{Y}\rangle = \langle\langle -(\widetilde{\nabla}_{\widetilde{X}} n)^\top, Y\rangle\rangle.
\end{aligned}$$

Therefore

$$\langle\langle S_{n_p}(X_p), Y_p\rangle\rangle = \langle\langle -(\widetilde{\nabla}_{\widetilde{X}} n)_p^\top, Y_p\rangle\rangle$$

for all $Y_p \in T_pN$.

Example 5.1 Let N be a **hypersurface** in M, i.e. let $\dim N = n$ and $\dim M = n + 1$. Consider a point $p \in V$ (a neighborhood of N where f is an embedding), and a unit vector n_p normal to N at p. As the linear map $S_{n_p} : T_pN \to T_pN$ is symmetric, there exists an orthonormal basis of T_pN formed by eigenvectors $\{(E_1)_p, \ldots, (E_n)_p\}$ (called **principal directions** at p) corresponding to the set of real eigenvalues $\lambda_1, \ldots, \lambda_n$ of S_{n_p} (called **principal curvatures** at p). The determinant of the map S_{n_p} (equal to the product $\lambda_1 \cdots \lambda_n$) is called the **Gauss curvature of** f and $H := \frac{1}{n} \operatorname{tr} S_{n_p} = \frac{1}{n}(\lambda_1 + \cdots + \lambda_n)$ is called the **mean curvature** of f. When $n = 2$ and $M = \mathbb{R}^3$ with the Euclidean metric, the Gauss curvature of f is in fact the Gauss curvature of N as defined in Sect. 4.1 (cf. Example 5.5).

Example 5.2 If, in the above example, $M = \mathbb{R}^{n+1}$ with the Euclidean metric, we can define the **Gauss map** $g : V \subset N \to S^n$, with values on the unit sphere, which, to each point $p \in V$, assigns the normal unit vector n_p. Since n_p is normal to T_pN, we can identify the tangent spaces T_pN and $T_{g(p)}S^n$ and obtain a well-defined map $(dg)_p : T_pN \to T_pN$. Choosing a curve $c : I \to N$ with $c(0) = p$ and $\dot{c}(0) = X_p \in T_pN$, we have

$$(dg)_p(X_p) = \frac{d}{dt}(g\circ c)_{|t=0} = \frac{d}{dt} n_{c(t)|t=0} = (\widetilde{\nabla}_{\dot{c}} n)_p,$$

where we used the fact $\widetilde{\nabla}$ is the Levi–Civita connection for the Euclidean metric. However, since $\|n\| = 1$, we have

$$0 = \dot{c}(t) \cdot \langle n, n \rangle = 2\langle \widetilde{\nabla}_{\dot{c}}\, n, n \rangle,$$

implying that

$$(dg)_p(X_p) = (\widetilde{\nabla}_{\dot{c}}\, n)_p = (\widetilde{\nabla}_{\dot{c}}\, n)_p^\top = -S_{n_p}(X_p).$$

We conclude that the derivative of the Gauss map at p is $(dg)_p = -S_{n_p}$.

Let us now relate the curvatures of N and M.

Proposition 5.3 *Let p be a point in N, let X_p and Y_p be two linearly independent vectors in $T_pN \subset T_pM$ and let $\Pi \subset T_pN \subset T_pM$ be the 2-dimensional subspace generated by these vectors. Let $K^N(\Pi)$ and $K^M(\Pi)$ denote the corresponding sectional curvatures in N and M, respectively. Then*

$$K^N(\Pi) - K^M(\Pi) = \frac{\langle B(X_p, X_p), B(Y_p, Y_p) \rangle - \|B(X_p, Y_p)\|^2}{\|X_p\|^2\|Y_p\|^2 - \langle X_p, Y_p \rangle^2}.$$

Proof Observing that the right-hand side depends only on Π, we can assume, without loss of generality, that $\{X_p, Y_p\}$ is orthonormal. Let X, Y be local extensions of $X_p, Y_p,$ defined on a neighborhood of p in N and tangent to N, also orthonormal. Let $\widetilde{X}, \widetilde{Y}$ be extensions of X, Y to a neighborhood of p in M. Moreover, consider a field of frames $\{E_1, \ldots, E_{n+k}\}$, also defined on a neighborhood of p in M, such that $E_1, \ldots, E_n$ are tangent to N with $E_1 = X$ and $E_2 = Y$ on N, and $E_{n+1}, \ldots, E_{n+k}$ are normal to N $(n + k = m)$. Then, since $B(X, Y)$ is normal to N,

$$B(X, Y) = \sum_{i=1}^k \langle B(X, Y), E_{n+i} \rangle\, E_{n+i} = \sum_{i=1}^k H_{E_{n+i}}(X, Y)\, E_{n+i}.$$

On the other hand,

$$\begin{aligned} K^N(\Pi) - K^M(\Pi) &= -R^N(X_p, Y_p, X_p, Y_p) + R^M(\widetilde{X}_p, \widetilde{Y}_p, \widetilde{X}_p, \widetilde{Y}_p) \\ &= \langle(-\nabla_X \nabla_Y X + \nabla_Y \nabla_X X + \nabla_{[X,Y]} X \\ &\quad + \widetilde{\nabla}_{\widetilde{X}} \widetilde{\nabla}_{\widetilde{Y}} \widetilde{X} - \widetilde{\nabla}_{\widetilde{Y}} \widetilde{\nabla}_{\widetilde{X}} \widetilde{X} - \widetilde{\nabla}_{[\widetilde{X},\widetilde{Y}]} \widetilde{X})_p, Y_p \rangle \\ &= \langle(-\nabla_X \nabla_Y X + \nabla_Y \nabla_X X + \widetilde{\nabla}_{\widetilde{X}} \widetilde{\nabla}_{\widetilde{Y}} \widetilde{X} - \widetilde{\nabla}_{\widetilde{Y}} \widetilde{\nabla}_{\widetilde{X}} \widetilde{X})_p, Y_p \rangle, \end{aligned}$$

where we have used the fact that $\widetilde{\nabla}_{[\widetilde{X},\widetilde{Y}]}\widetilde{X} - \nabla_{[X,Y]} X$ is normal to N [cf. Exercise 5.7(1)]. However, since on N

$$\widetilde{\nabla}_{\widetilde{Y}} \widetilde{\nabla}_{\widetilde{X}} \widetilde{X} = \widetilde{\nabla}_{\widetilde{Y}}(B(X, X) + \nabla_X X) = \widetilde{\nabla}_{\widetilde{Y}} \left(\sum_{i=1}^{k} H_{E_{n+i}}(X, X) E_{n+i} + \nabla_X X \right)$$

$$= \sum_{i=1}^{k} \left(H_{E_{n+i}}(X, X) \widetilde{\nabla}_{\widetilde{Y}} E_{n+i} + \widetilde{Y} \cdot (H_{E_{n+i}}(X, X)) E_{n+i} \right) + \widetilde{\nabla}_{\widetilde{Y}} \nabla_X X,$$

we have

$$\langle \widetilde{\nabla}_{\widetilde{Y}} \widetilde{\nabla}_{\widetilde{X}} \widetilde{X}, Y \rangle = \sum_{i=1}^{k} H_{E_{n+i}}(X, X) \langle \widetilde{\nabla}_{\widetilde{Y}} E_{n+i}, Y \rangle + \langle \widetilde{\nabla}_{\widetilde{Y}} \nabla_X X, Y \rangle.$$

Moreover,

$$0 = \widetilde{Y} \cdot \langle E_{n+i}, Y \rangle = \langle \widetilde{\nabla}_{\widetilde{Y}} E_{n+i}, Y \rangle + \langle E_{n+i}, \widetilde{\nabla}_{\widetilde{Y}} Y \rangle$$
$$= \langle \widetilde{\nabla}_{\widetilde{Y}} E_{n+i}, Y \rangle + \langle E_{n+i}, B(Y, Y) + \nabla_Y Y \rangle$$
$$= \langle \widetilde{\nabla}_{\widetilde{Y}} E_{n+i}, Y \rangle + \langle E_{n+i}, B(Y, Y) \rangle$$
$$= \langle \widetilde{\nabla}_{\widetilde{Y}} E_{n+i}, Y \rangle + H_{E_{n+i}}(Y, Y),$$

and so

$$\langle \widetilde{\nabla}_{\widetilde{Y}} \widetilde{\nabla}_{\widetilde{X}} \widetilde{X}, Y \rangle = - \sum_{i=1}^{k} H_{E_{n+i}}(X, X) H_{E_{n+i}}(Y, Y) + \langle \widetilde{\nabla}_{\widetilde{Y}} \nabla_X X, Y \rangle$$

$$= - \sum_{i=1}^{k} H_{E_{n+i}}(X, X) H_{E_{n+i}}(Y, Y) + \langle \nabla_Y \nabla_X X, Y \rangle.$$

Similarly, we can conclude that

$$\langle \widetilde{\nabla}_{\widetilde{X}} \widetilde{\nabla}_{\widetilde{Y}} \widetilde{X}, Y \rangle = - \sum_{i=1}^{k} H_{E_{n+i}}(X, Y) H_{E_{n+i}}(X, Y) + \langle \nabla_X \nabla_Y X, Y \rangle,$$

and then

$$K^N(\Pi) - K^M(\Pi)$$

$$= \sum_{i=1}^{k} \left(-(H_{E_{n+i}}(X_p, Y_p))^2 + H_{E_{n+i}}(X_p, X_p) H_{E_{n+i}}(Y_p, Y_p) \right)$$

$$= -\|B(X_p, Y_p)\|^2 + \langle B(X_p, X_p), B(Y_p, Y_p) \rangle. \qquad \square$$

Example 5.4 Again in the case of a hypersurface N, we choose an orthonormal basis $\{(E_1)_p, \ldots, (E_n)_p\}$ of T_pN formed by eigenvectors of S_{n_p}, where

$n_p \in (T_p N)^{\perp}$. Hence, considering a section Π of $T_p N$ generated by two of these vectors $(E_i)_p, (E_j)_p$, and using $B(X_p, Y_p) = \langle\langle S_{n_p}(X_p), Y_p\rangle\rangle n_p$, we have

$$
\begin{aligned}
K^N(\Pi) &- K^M(\Pi) \\
&= -\|B((E_i)_p, (E_j)_p)\|^2 + \langle B((E_i)_p, (E_i)_p), B((E_j)_p, (E_j)_p)\rangle \\
&= -\langle\langle S_{n_p}((E_i)_p), (E_j)_p\rangle\rangle^2 + \langle\langle S_{n_p}((E_i)_p), (E_i)_p\rangle\rangle\langle\langle S_{n_p}((E_j)_p), (E_j)_p\rangle\rangle \\
&= \lambda_i \lambda_j.
\end{aligned}
$$

Example 5.5 In the special case where N is a 2-manifold, and $M = \mathbb{R}^3$ with the Euclidean metric, we have $K^M \equiv 0$ and hence $K^N(p) = \lambda_1 \lambda_2$, as promised in Example 5.1. Therefore, although λ_1 and λ_2 depend on the immersion, their **product** depends only on the intrinsic geometry of N. Gauss was so pleased by this discovery that he called it his **Theorema Egregium** ('remarkable theorem').

Let us now study in detail the particular case where N is a hypersurface in $M = \mathbb{R}^{n+1}$ with the Euclidean metric. Let $c : I \to N$ be a curve in N parameterized by arc length s and such that $c(0) = p$ and $\dot{c}(0) = X_p \in T_p N$. We will identify this curve c with the curve $f \circ c$ in $\mathbb{R}^{n+1}$. Considering the Gauss map $g : V \to S^n$ defined on a neighborhood V of p in N, we take the curve $n(s) := (g \circ c)(s)$ in S^n. Since $\widetilde{\nabla}$ is the Levi–Civita connection corresponding to the Euclidean metric in $\mathbb{R}^3$, we have $\langle \widetilde{\nabla}_{\dot{c}} \dot{c}, n \rangle = \langle \ddot{c}, n \rangle$. On the other hand,

$$
\langle \widetilde{\nabla}_{\dot{c}} \dot{c}, n \rangle = \langle B(\dot{c}, \dot{c}) + \nabla_{\dot{c}} \dot{c}, n \rangle = \langle B(\dot{c}, \dot{c}), n \rangle = H_n(\dot{c}, \dot{c}).
$$

Hence, at $s = 0$, $H_{g(p)}(X_p, X_p) = \langle \ddot{c}(0), n_p \rangle$. This value $k_{n_p} := \langle \ddot{c}(0), n_p \rangle$ is called the **normal curvature** of c at p. Since k_{n_p} is equal to $H_{g(p)}(X_p, X_p)$, it does not depend on the curve, but only on its initial velocity. Because $H_{g(p)}(X_p, X_p) = \langle\langle S_{g(p)}(X_p), X_p\rangle\rangle$, the critical values of these curvatures subject to $\|X_p\| = 1$ are equal to $\lambda_1, \ldots, \lambda_n$, and are called the **principal curvatures**. This is why in Example 5.1 we also called the eigenvalues of S_{n_p} principal curvatures. The Gauss curvature of f is then equal to the product of the principal curvatures, $K = \lambda_1 \ldots \lambda_n$. As the normal curvature does not depend on the choice of curve tangent to X_p at p, we can choose c to take values on the 2-plane generated by X_p and n_p. Then $\ddot{c}(0)$ is parallel to the normal vector n_p, and

$$
|k_n| = |\langle \ddot{c}(0), n \rangle| = \|\ddot{c}(0)\| = k_c,
$$

where $k_c := \|\ddot{c}(0)\|$ is the so-called **curvature** of the curve c at $c(0)$. The same formula holds if c is a geodesic of N [cf. Exercise 5.7(6)].

Example 5.6 Let us consider the following three surfaces: the 2-sphere, the cylinder and the saddle surface $z = xy$.

(1) Let p be any point on the sphere. Intuitively, all points of this surface are on the same side of the tangent plane at p, implying that both principal curvatures have

the same sign (depending on the chosen orientation), and consequently that the
Gauss curvature is positive at all points.

(2) If p is any point on the cylinder, one of the principal curvatures is zero (the
maximum or the minimum, depending on the chosen orientation), and so the
Gauss curvature is zero at all points.

(3) Finally, if p is a point on the saddle surface $z = xy$ then the principal curvatures
at p have opposite signs, and so the Gauss curvature is negative.

Exercise 5.7

(1) Let M be a Riemannian manifold with Levi–Civita connection $\widetilde{\nabla}$, and let N be
a submanifold endowed with the induced metric and Levi–Civita connection ∇.
Let $\widetilde{X}, \widetilde{Y} \in \mathfrak{X}(M)$ be local extensions of $X, Y \in \mathfrak{X}(N)$. Recall that the second
fundamental form of the inclusion of N in M is the map $B : T_pN \times T_pN \to$
$(T_pN)^\perp$ defined at each point $p \in N$ by

$$B(X, Y) := \widetilde{\nabla}_{\widetilde{X}} \widetilde{Y} - \nabla_X Y.$$

Show that:

(a) $B(X, Y)$ does not depend on the choice of the extensions $\widetilde{X}, \widetilde{Y}$;
(b) $B(X, Y)$ is orthogonal to N;
(c) B is symmetric, i.e. $B(X, Y) = B(Y, X)$;
(d) B is bilinear;
(e) $B(X, Y)_p$ depends only on the values of X_p and Y_p;
(f) $\widetilde{\nabla}_{[\widetilde{X}, \widetilde{Y}]} \widetilde{X} - \nabla_{[X,Y]} X$ is orthogonal to N.

(2) Let $S^n(r) \subset \mathbb{R}^{n+1}$ be the n dimensional sphere of radius r.

(a) Choosing at each point the outward pointing normal unit vector, what is the
Gauss map of this inclusion?
(b) What are the eigenvalues of its derivative?
(c) Show that all sectional curvatures are equal to $\frac{1}{r^2}$ (so $S^n(r)$ has constant
curvature $\frac{1}{r^2}$).

(3) Let $(M, \langle \cdot, \cdot \rangle)$ be a Riemannian manifold. A submanifold $N \subset M$ is said to be
totally geodesic if the the geodesics of N are geodesics of M. Show that:

(a) N is totally geodesic if and only if $B \equiv 0$, where B is the second fundamental
form of N;
(b) if N is the set of fixed points of an isometry then N is totally geodesic. Use
this result to give examples of totally geodesic submanifolds of $\mathbb{R}^n$, S^n and
H^n.

(4) Let N be a hypersurface in $\mathbb{R}^{n+1}$ and let p be a point in N. Show that if $K(p) \neq 0$
then

$$|K(p)| = \lim_{D \to p} \frac{\mathrm{vol}(g(D))}{\mathrm{vol}(D)},$$

where $g : V \subset N \to S^n$ is the Gauss map and D is a neighborhood of p whose diameter tends to zero.

(5) Let $(M, \langle \cdot, \cdot \rangle)$ be a Riemannian manifold, p a point in M and Π a section of $T_p M$. For $B_\varepsilon(p) := \exp_p(B_\varepsilon(0))$ a normal ball around p consider the set $N_p := \exp_p(B_\varepsilon(0) \cap \Pi)$. Show that:

   (a) the set N_p is a 2-dimensional submanifold of M formed by the segments of geodesics in $B_\varepsilon(p)$ which are tangent to Π at p;
   (b) if in N_p we use the metric induced by the metric in M, the sectional curvature $K^M(\Pi)$ is equal to the Gauss curvature of the 2-manifold N_p.

(6) Let $(M, \langle \cdot, \cdot \rangle)$ be a Riemannian manifold with Levi–Civita connection $\widetilde{\nabla}$ and let N be a hypersurface in M. The **geodesic curvature** of a curve $c : I \subset \mathbb{R} \to M$, parameterized by arclength, is $k_g(s) = \|\widetilde{\nabla}_{\dot c(s)} \dot c(s)\|$. Show that the absolute values of the principal curvatures are the geodesic curvatures (in M) of the geodesics of N tangent to the principal directions. (**Remark:** In the case of an oriented 2-dimensional Riemannian manifold, k_g is taken to be positive or negative according to the orientation of $\{\dot c(s), \widetilde{\nabla}_{\dot c(s)} \dot c(s)\}$—cf. Sect. 4.2).

(7) Use the Gauss map to compute the Gauss curvature of the following surfaces in $\mathbb{R}^3$:

   (a) the paraboloid $z = \frac{1}{2}(x^2 + y^2)$;
   (b) the saddle surface $z = xy$.

(8) (*Surfaces of revolution*) Consider the map $f : \mathbb{R} \times (0, 2\pi) \to \mathbb{R}^3$ given by

$$f(s, \theta) = (h(s)\cos\theta, h(s)\sin\theta, g(s))$$

with $h > 0$ and g smooth maps such that

$$(h'(s))^2 + (g'(s))^2 = 1.$$

The image of f is the surface of revolution S with axis Oz, obtained by rotating the curve $\alpha(s) = (h(s), g(s))$, parameterized by the arclength s, around that axis.

   (a) Show that f is an immersion.
   (b) Show that $f_s := (df)\left(\frac{\partial}{\partial s}\right)$ and $f_\theta := (df)(\frac{\partial}{\partial \theta})$ are orthogonal.
   (c) Determine the Gauss map and compute the matrix of the second fundamental form of S associated to the frame $\{E_s, E_\theta\}$, where $E_s := f_s$ and $E_\theta := \frac{1}{\|f_\theta\|} f_\theta$.
   (d) Compute the mean curvature H and the Gauss curvature K of S.
   (e) Using these results, give examples of surfaces of revolution with:
       (1) $K \equiv 0$;
       (2) $K \equiv 1$;
       (3) $K \equiv -1$;

(4) $H \equiv 0$ (not a plane).

> (**Remark:** Surfaces with constant zero mean curvature are called **minimal surfaces**; it can be proved
> that if a compact surface with boundary has minimum area among all surfaces with the same boundary
> then it must be a minimal surface).

4.6 Notes

4.6.1 Section 4.4

The isometries of the hyperbolic plane are examples of **linear fractional transfor-mations** (or **Möbius transformations**), i.e. maps $f : \mathbb{C} \to \mathbb{C}$ given by

$$f(z) = \frac{az + b}{cz + d},$$

with $a, b, c, d \in \mathbb{C}$ satisfying $ad - bc \neq 0$. It is easy to see that each of these transformations is a composition of the following types of transformations:

(1) translations: $z \mapsto z + b$;
(2) rotations: $z \mapsto az, |a| = 1$;
(3) homotheties: $z \mapsto rz, r > 0$;
(4) inversions: $z \mapsto 1/z$,

and so it is clear that they carry straight lines and circles to either straight lines or circles.

The special values $f(\infty) = \frac{a}{c}$ and $f(-\frac{d}{c}) = \infty$ can be introduced as limits for $z \to \infty$ and $z \to -d/c$, and so, using the stereographic projection, we can see f as a map from the sphere to itself. Noting that both straight lines and circles in the plane correspond to circles in the sphere, we can say that a Möbius transformation, seen as a map on the sphere, carries circles into circles.

4.6.2 Bibliographical Notes

The material in this chapter can be found in most books on Riemannian geome-try (e.g. [Boo03, dC93, GHL04]). The proof of the Gauss–Bonnet theorem (due to S. Chern) follows [dC93, CCL00] closely. See [KN96, Jos02] to see how this theo-rem fits within the general theory of characteristic classes of fiber bundles. A more elementary discussion of isometric immersions of surfaces in $\mathbb{R}^3$ (including a proof of the Gauss–Bonnet theorem) can be found in [dC76, Mor98].

References

[Boo03] Boothby, W.: An Introduction to Differentiable Manifolds and Riemannian Geometry. Academic Press, New York (2003)

[CCL00] Chern, S., Chen, W., Lam, K.: Lectures on Differential Geometry. World Scientific, Singapore (2000)

[dC76]   do Carmo, M.: Differential Geometry of Curves and Surfaces. Prentice-Hall, Singapore (1976)

[dC93]   do Carmo, M.: Riemannian Geometry. Birkhäuser, Boston (1993)

[GHL04] Gallot, S., Hulin, D., Lafontaine, J.: Riemannian Geometry. Springer, Berlin (2004)

[Jos02] Jost, J.: Riemannian Geometry and Geometric Analysis. Springer, Berlin (2002)

[KN96]   Kobayashi, S., Nomizu, K.: Foundations of Differential Geometry, Vol I and II. Wiley, New York (1996)

[Mor98] Morgan, F.: Riemannian Geometry. A K Peters, Wellesley (1998)

[Wol78] Wolf, J.A.: Spaces of Constant Curvature. Publish or Perish, Berkeley (1978)

Chapter 5
Geometric Mechanics

Mechanics, the science of motion, was basically started by Galileo and his revolutionary empirical approach. The first precise mathematical formulation was laid down by Newton in the **Philosophiae Naturalis Principia Mathematica**, first published in 1687, which contained, among many other things, an explanation for the elliptical orbits of the planets around the Sun. Newton's ideas were developed and extended by a number of mathematicians, including Euler, Lagrange, Laplace, Jacobi, Poisson and Hamilton. Celestial mechanics, in particular, reached an exquisite level of precision: the 1846 discovery of planet Neptune, for instance, was triggered by the need to explain a mismatch between the observed orbit of planet Uranus and its theoretical prediction.

This chapter uses Riemannian geometry to give a geometric formulation of Newtonian mechanics. As explained in Sect. 5.1, this is made possible by the fact that the kinetic energy of any **mechanical system** yields a Riemannian metric on its **configuration space**, that is, the differentiable manifold whose points represent the possible configurations of the system. Section 5.2 describes how **holonomic constraints**, which force the system to move along submanifolds of the configuration space, yield nontrivial mechanical systems. A particularly important example of this, the **rigid body**, is studied in detail in Sect. 5.3. **Non-holonomic constraints**, which restrict velocities rather than configurations, are considered in Sect. 5.4.

Section 5.5 presents the **Lagrangian formulation** of mechanics, where the trajectories are obtained as curves extremizing the **action** integral. Also treated is the **Noether theorem**, which associates conservation laws to symmetries. The dual **Hamiltonian formulation** of mechanics, where the trajectories are obtained from special flows in the cotangent bundle, is described in Sect. 5.6, and used in Sect. 5.7 to formulate the theory of **completely integrable systems**, whose dynamics are particularly simple. Section 5.8 generalizes the Hamiltonian formalism to **symplectic** and **Poisson manifolds**, and discusses **reduction** of these manifolds under appropriate symmetries.

© Springer International Publishing Switzerland 2014

L. Godinho and J. Natário, *An Introduction to Riemannian Geometry*, Universitext, DOI 10.1007/978-3-319-08666-8_5

5.1 Mechanical Systems

In mechanics one studies the motions of particles or systems of particles subject to known forces.

Example 1.1 The motion of a single particle in n-dimensional space is described by a curve $x : I \subset \mathbb{R} \to \mathbb{R}^n$. It is generally assumed that the force acting on the particle depends only on its position and velocity. **Newton's Second Law** requires that the particle's motion satisfies the second-order ordinary differential equation

$$m\ddot{x} = F(x, \dot{x}),$$

where $F : \mathbb{R}^n \times \mathbb{R}^n \to \mathbb{R}^n$ is the force acting on the particle and $m > 0$ is the particle's mass. Therefore the solutions of this equation describe the possible motions of the particle.

It will prove advantageous to make the following generalization:

Definition 1.2 A **mechanical system** is a triple $(M, \langle \cdot, \cdot \rangle, \mathcal{F})$, where:

(i) M is a differentiable manifold, called the **configuration space**;
(ii) $\langle \cdot, \cdot \rangle$ is a Riemannian metric on M yielding the **mass operator** $\mu : TM \to T^*M$, defined by

$$\mu(v)(w) = \langle v, w \rangle$$

for all $v, w \in T_pM$ and $p \in M$;
(iii) $\mathcal{F} : TM \to T^*M$ is a differentiable map satisfying $\mathcal{F}(T_pM) \subset T_p^*M$ for all $p \in M$, called the **external force**.

A **motion** of the mechanical system is a solution $c : I \subset \mathbb{R} \to M$ of the **Newton equation**

$$\mu\left(\frac{D\dot{c}}{dt}\right) = \mathcal{F}(\dot{c}).$$

Remark 1.3 In particular, the geodesics of a Riemannian manifold $(M, \langle \cdot, \cdot \rangle)$ are the motions of the mechanical system $(M, \langle \cdot, \cdot \rangle, 0)$ (describing a **free particle** on M).

Example 1.4 For the mechanical system comprising a single particle moving in n-dimensional space, the configuration space is clearly $\mathbb{R}^n$. If we set

$$\langle\langle v, w \rangle\rangle := m \langle v, w \rangle$$

for all $v, w \in \mathbb{R}^n$, where $\langle \cdot, \cdot \rangle$ is the Euclidean inner product in $\mathbb{R}^n$, then the Levi–Civita connection of $\langle\langle \cdot, \cdot \rangle\rangle$ will still be the trivial connection, and

$$\frac{D\dot{x}}{dt} = \ddot{x}.$$

Setting

$$\mathcal{F}(x, v)(w) := \langle F(x, v), w \rangle \tag{5.1}$$

for all $v, w \in \mathbb{R}^n$, we see that

$$\mu\left(\frac{D\dot{x}}{dt}\right) = \mathcal{F}(x, \dot{x}) \Leftrightarrow \mu\left(\frac{D\dot{x}}{dt}\right)(v) = \mathcal{F}(x, \dot{x})(v) \text{ for all } v \in \mathbb{R}^n$$

$$\Leftrightarrow m\langle \ddot{x}, v \rangle = \langle F(x, \dot{x}), v \rangle \text{ for all } v \in \mathbb{R}^n$$

$$\Leftrightarrow m\ddot{x} = F(x, \dot{x}).$$

Hence the motions of the particle are the motions of the mechanical system $(\mathbb{R}^n, \langle\langle \cdot, \cdot \rangle\rangle, \mathcal{F})$ with $\mathcal{F}$ defined by (5.1).

Definition 1.5 Let $(M, \langle \cdot, \cdot \rangle, \mathcal{F})$ be a mechanical system. The external force $\mathcal{F}$ is said to be:

(i) **positional** if $\mathcal{F}(v)$ depends only on $\pi(v)$, where $\pi : TM \to M$ is the natural projection;
(ii) **conservative** if there exists $U : M \to \mathbb{R}$ such that $\mathcal{F}(v) = -(dU)_{\pi(v)}$ for all $v \in TM$ (the function U is called the **potential energy**).

Remark 1.6 In particular any conservative force is positional. A mechanical system whose exterior force is conservative is called a **conservative mechanical system**.

Definition 1.7 Let $(M, \langle \cdot, \cdot \rangle, \mathcal{F})$ be a mechanical system. The **kinetic energy** is the differentiable map $K : TM \to \mathbb{R}$ given by

$$K(v) = \frac{1}{2}\langle v, v \rangle$$

for all $v \in TM$.

Example 1.8 For the mechanical system comprising a single particle moving in n-dimensional space, one has

$$K(v) := \frac{1}{2}m\langle v, v \rangle.$$

Theorem 1.9 (Conservation of energy) *In a conservative mechanical system $(M, \langle \cdot, \cdot \rangle, -dU)$, the **mechanical energy** $E(t) = K(\dot{c}(t)) + U(c(t))$ is constant along any motion $c : I \subset \mathbb{R} \to M$.*

Proof

$$\frac{dE}{dt}(t) = \frac{d}{dt}\left(\frac{1}{2}\langle \dot{c}(t), \dot{c}(t)\rangle + U(c(t))\right) = \left\langle \frac{D\dot{c}}{dt}(t), \dot{c}(t)\right\rangle + (dU)_{c(t)}\dot{c}(t)$$

$$= \mu\left(\frac{D\dot{c}}{dt}\right)(\dot{c}) - \mathcal{F}(\dot{c})(\dot{c}) = 0. \qquad \square$$

A particularly simple example of a conservative mechanical system is $(M, \langle \cdot, \cdot\rangle, 0)$, whose motions are the geodesics of $(M, \langle \cdot, \cdot\rangle)$. In fact, the motions of **any** conservative system can be suitably reinterpreted as the geodesics of a certain metric.

Definition 1.10 Let $(M, \langle \cdot, \cdot\rangle, -dU)$ be a conservative mechanical system and $h \in \mathbb{R}$ such that

$$M_h := \{p \in M \mid U(p) < h\} \neq \varnothing.$$

The **Jacobi metric** on the manifold M_h is given by

$$\langle\langle v, w\rangle\rangle := 2\left[h - U(p)\right]\langle v, w\rangle$$

for all $v, w \in T_p M_h$ and $p \in M_h$.

Theorem 1.11 (Jacobi) *The motions of a conservative mechanical system $(M, \langle \cdot, \cdot\rangle, -dU)$ with mechanical energy h are, up to reparameterization, geodesics of the Jacobi metric on M_h.*

Proof We shall need the two following lemmas, whose proofs are left as exercises.

Lemma 1.12 *Let $(M, \langle \cdot, \cdot\rangle)$ be a Riemannian manifold with Levi–Civita connection ∇ and let $\langle\langle \cdot, \cdot\rangle\rangle = e^{2\rho}\langle \cdot, \cdot\rangle$ be a metric conformally related to $\langle \cdot, \cdot\rangle$ (where $\rho \in C^\infty(M)$). Then the Levi–Civita connection $\widetilde{\nabla}$ of $\langle\langle \cdot, \cdot\rangle\rangle$ is given by*

$$\widetilde{\nabla}_X Y = \nabla_X Y + d\rho(X)Y + d\rho(Y)X - \langle X, Y\rangle \operatorname{grad} \rho$$

for all $X, Y \in \mathfrak{X}(M)$ (where the gradient is taken with respect to $\langle \cdot, \cdot\rangle$). $\qquad \square$

Lemma 1.13 *A curve $c : I \subset \mathbb{R} \to M$ is a reparameterized geodesic of a Riemannian manifold $(M, \langle \cdot, \cdot\rangle)$ if and only if it satisfies*

$$\frac{D\dot{c}}{dt} = f(t)\dot{c}$$

for some differentiable function $f : I \to \mathbb{R}$. $\qquad \square$

We now prove the Jacobi theorem. Let $c : I \subset \mathbb{R} \to M$ be a motion of $(M, \langle \cdot, \cdot\rangle, -dU)$ with mechanical energy h. Then Lemma 1.12 yields

$$\frac{\tilde{D}\dot{c}}{dt} = \frac{D\dot{c}}{dt} + 2d\rho(\dot{c})\,\dot{c} - \langle \dot{c}, \dot{c} \rangle \, \text{grad}\, \rho,$$

where $\frac{\tilde{D}}{dt}$ is the covariant derivative along c with respect to the Jacobi metric and $e^{2\rho} = 2(h - U)$. The Newton equation yields

$$\mu\left(\frac{D\dot{c}}{dt}\right) = -dU \Leftrightarrow \frac{D\dot{c}}{dt} = -\,\text{grad}\, U = e^{2\rho}\,\text{grad}\,\rho,$$

and by conservation of energy

$$\langle \dot{c}, \dot{c} \rangle = 2K = 2(h - U) = e^{2\rho}.$$

Consequently we have

$$\frac{\tilde{D}\dot{c}}{dt} = 2d\rho(\dot{c})\,\dot{c},$$

which by Lemma 1.13 means that c is a reparameterized geodesic of the Jacobi metric. $\qquad\square$

A very useful expression for writing the Newton equation in local coordinates is the following.

Proposition 1.14 *Let* $(M, \langle \cdot, \cdot \rangle, \mathcal{F})$ *be a mechanical system. If* $(x^1, \ldots, x^n)$ *are local coordinates on* M *and* $(x^1, \ldots, x^n, v^1, \ldots, v^n)$ *are the local coordinates induced on* TM *then*

$$\mu\left(\frac{D\dot{c}}{dt}(t)\right) = \sum_{i=1}^{n} \left[\frac{d}{dt}\left(\frac{\partial K}{\partial v^i}(x(t), \dot{x}(t))\right) - \frac{\partial K}{\partial x^i}(x(t), \dot{x}(t)) \right] dx^i.$$

In particular, if $\mathcal{F} = -dU$ *is conservative then the equations of motion are*

$$\frac{d}{dt}\left(\frac{\partial K}{\partial v^i}(x(t), \dot{x}(t))\right) - \frac{\partial K}{\partial x^i}(x(t), \dot{x}(t)) = -\frac{\partial U}{\partial x^i}(x(t))$$

$(i = 1, \ldots, n)$.

Proof Exercise 1.16(8). $\qquad\square$

Example 1.15

(1) (*Particle in a central field*) Consider a particle of mass $m > 0$ moving in $\mathbb{R}^2$ under the influence of a conservative force whose potential energy U depends only on the distance $r = \sqrt{x^2 + y^2}$ to the origin, $U = u(r)$. The equations of motion are most easily solved when written in **polar coordinates** (r, θ), defined by

$$\begin{cases} x = r\cos\theta \\ y = r\sin\theta \end{cases}.$$

Since

$$dx = \cos\theta dr - r\sin\theta d\theta,$$
$$dy = \sin\theta dr + r\cos\theta d\theta,$$

it is easily seen that the Euclidean metric is written in these coordinates as

$$\langle\cdot,\cdot\rangle = dx \otimes dx + dy \otimes dy = dr \otimes dr + r^2 d\theta \otimes d\theta,$$

and hence

$$K\left(r,\theta,v^r,v^\theta\right) = \frac{1}{2}m\left[\left(v^r\right)^2 + r^2\left(v^\theta\right)^2\right].$$

Therefore we have

$$\frac{\partial K}{\partial v^r} = mv^r, \quad \frac{\partial K}{\partial v^\theta} = mr^2 v^\theta, \quad \frac{\partial K}{\partial r} = mr\left(v^\theta\right)^2, \quad \frac{\partial K}{\partial \theta} = 0,$$

and consequently the Newton equations are written

$$\frac{d}{dt}(m\dot{r}) - mr\dot{\theta}^2 = -u'(r),$$

$$\frac{d}{dt}\left(mr^2\dot{\theta}\right) = 0.$$

Notice that the **angular momentum**

$$p_\theta := mr^2\dot{\theta}$$

is constant along the motion. This conservation law can be traced back to the fact that neither K nor U depend on θ.

(2) (*Christoffel symbols for the 2-sphere*) The metric for the 2-sphere $S^2 \subset \mathbb{R}^3$ is written as

$$\langle\cdot,\cdot\rangle = d\theta \otimes d\theta + \sin^2\theta \, d\varphi \otimes d\varphi$$

in the usual local coordinates (θ,φ) defined by the parameterization

$$\phi(\theta,\varphi) = (\sin\theta\cos\varphi, \sin\theta\sin\varphi, \cos\theta)$$

[cf. Exercise 3.3(4) in Chap. 3]. A quick way to obtain the Christoffel symbols in this coordinate system is to write out the Newton equations for a free particle (of mass $m = 1$, say) on S^2. We have

$$K\left(\theta, \varphi, v^\theta, v^\varphi\right) = \frac{1}{2}\left[\left(v^\theta\right)^2 + \sin^2\theta \left(v^\varphi\right)^2\right]$$

and hence

$$\frac{\partial K}{\partial v^\theta} = v^\theta, \quad \frac{\partial K}{\partial v^\varphi} = \sin^2\theta\, v^\varphi, \quad \frac{\partial K}{\partial \theta} = \sin\theta\cos\theta \left(v^\varphi\right)^2, \quad \frac{\partial K}{\partial \varphi} = 0.$$

Consequently the Newton equations are written

$$\frac{d}{dt}\left(\dot\theta\right) - \sin\theta\cos\theta\,\dot\varphi^2 = 0 \Leftrightarrow \ddot\theta - \sin\theta\cos\theta\,\dot\varphi^2 = 0,$$

$$\frac{d}{dt}\left(\sin^2\theta\dot\varphi\right) = 0 \Leftrightarrow \ddot\varphi + 2\cot\theta\,\dot\theta\,\dot\varphi = 0.$$

Since these must be the equations for a geodesic on S^2, by comparing with the geodesic equations

$$\ddot{x}^i + \sum_{j,k=1}^{2} \Gamma^i_{jk}\dot{x}^j\dot{x}^k = 0 \quad (i = 1, 2),$$

one immediately reads off the nonvanishing Christoffel symbols:

$$\Gamma^\theta_{\varphi\varphi} = -\sin\theta\cos\theta, \qquad \Gamma^\varphi_{\theta\varphi} = \Gamma^\varphi_{\varphi\theta} = \cot\theta.$$

Exercise 1.16

(1) Generalize Examples 1.1, 1.4, and 1.8 to a system of k particles moving in $\mathbb{R}^n$.

(2) Let $(M, \langle\cdot, \cdot\rangle, \mathcal{F})$ be a mechanical system. Show that the Newton equation defines a flow on TM, generated by the vector field $X \in \mathfrak{X}(TM)$ whose local expression is

$$X = v^i\frac{\partial}{\partial x^i} + \left(\sum_{j=1}^{n} g^{ij}(x)F_j(x, v) - \sum_{j,k=1}^{n} \Gamma^i_{jk}(x)v^j v^k\right)\frac{\partial}{\partial v^i},$$

where $(x^1, \ldots, x^n)$ are local coordinates on M, $(x^1, \ldots, x^n, v^1, \ldots, v^n)$ are the local coordinates induced on TM, and

$$\mathcal{F} = \sum_{i=1}^{n} F_i(x, v)dx^i$$

on these coordinates. What are the fixed points of this flow?

(3) (*Harmonic oscillator*) The **harmonic oscillator** (in appropriate units) is the conservative mechanical system $(\mathbb{R}, dx \otimes dx, -dU)$, where $U : \mathbb{R} \to \mathbb{R}$ is

given by

$$U(x) := \frac{1}{2}\omega^2 x^2.$$

(a) Write the equation of motion and its general solution.
(b) Friction can be included in this model by considering the external force

$$\mathcal{F}\left(u\frac{d}{dx}\right) = -dU - 2ku\,dx$$

(where $k > 0$ is a constant). Write the equation of motion of this new mechanical system and its general solution.
(c) Generalize (a) to the n-dimensional harmonic oscillator, whose potential energy $U : \mathbb{R}^n \to \mathbb{R}$ is given by

$$U(x^1, \ldots, x^n) := \frac{1}{2}\omega^2 \left(\left(x^1\right)^2 + \cdots + \left(x^n\right)^2\right).$$

(4) Consider the conservative mechanical system $(\mathbb{R}, dx \otimes dx, -dU)$. Show that:

(a) the flow determined by the Newton equation on $T\mathbb{R} \cong \mathbb{R}^2$ is generated by the vector field

$$X = v\frac{\partial}{\partial x} - U'(x)\frac{\partial}{\partial v} \in \mathfrak{X}(\mathbb{R}^2);$$

(b) the fixed points of the flow are the points of the form $(x_0, 0)$, where x_0 is a critical point of U;
(c) if x_0 is a maximum of U with $U''(x_0) < 0$ then $(x_0, 0)$ is an unstable fixed point;
(d) if x_0 is a minimum of U with $U''(x_0) > 0$ then $(x_0, 0)$ is a stable fixed point, with arbitrarily small neighborhoods formed by periodic orbits.
(e) the periods of these orbits converge to $2\pi U''(x_0)^{-\frac{1}{2}}$ as they approach $(x_0, 0)$;
(f) locally, any conservative mechanical system $(M, \langle \cdot, \cdot \rangle, -dU)$ with dim $M = 1$ is of the form above.

(5) Prove Lemma 1.12. (**Hint:** Use the Koszul formula).
(6) Prove Lemma 1.13.
(7) If $(M, \langle \cdot, \cdot \rangle)$ is a compact Riemannian manifold, it is known that there exists a nontrivial periodic geodesic. Use this fact to show that if M is compact then any conservative mechanical system $(M, \langle \cdot, \cdot \rangle, -dU)$ admits a nontrivial periodic motion.
(8) Prove Proposition 1.14.
(9) Recall that the hyperbolic plane is the upper half plane

$$H = \left\{(x, y) \in \mathbb{R}^2 \mid y > 0\right\}$$

with the Riemannian metric

$$\langle \cdot, \cdot \rangle = \frac{1}{y^2} (dx \otimes dx + dy \otimes dy)$$

[cf. Exercise 3.3(5) in Chap. 3]. Use Proposition 1.14 to compute the Christoffel symbols for the Levi–Civita connection of $(H, \langle \cdot, \cdot \rangle)$ in the coordinates (x, y).

(10) (*Kepler problem*) The **Kepler problem** (in appropriate units) consists in determining the motion of a particle of mass $m = 1$ in the central potential

$$U = -\frac{1}{r}.$$

(a) Show that the equations of motion can be integrated to

$$r^2 \dot{\theta} = p_\theta,$$

$$\frac{\dot{r}^2}{2} + \frac{p_\theta{}^2}{2r^2} - \frac{1}{r} = E,$$

where E and p_θ are integration constants.

(b) Use these equations to show that $u = \frac{1}{r}$ satisfies the linear ODE

$$\frac{d^2 u}{d\theta^2} + u = \frac{1}{p_\theta{}^2}.$$

(c) Assuming that the **pericenter** (i.e. the point in the particle's orbit closer to the center of attraction $r = 0$) occurs at $\theta = 0$, show that the equation of the particle's trajectory is

$$r = \frac{p_\theta{}^2}{1 + \varepsilon \cos \theta},$$

where

$$\varepsilon = \sqrt{1 + 2p_\theta{}^2 E}.$$

(**Remark:** This is the equation of a conic section with eccentricity ε in polar coordinates).

(d) Characterize all geodesics of $\mathbb{R}^2 \setminus \{(0, 0)\}$ with the Riemannian metric

$$\langle \cdot, \cdot \rangle = \frac{1}{\sqrt{x^2 + y^2}} (dx \otimes dx + dy \otimes dy).$$

Show that this manifold is isometric to the surface of a cone with aperture $\frac{\pi}{3}$.

5.2 Holonomic Constraints

Many mechanical systems involve particles or systems of particles whose positions are constrained (for example, a simple pendulum, a particle moving on a given surface, or a rigid system of particles connected by massless rods). To account for these we introduce the following definition:

Definition 2.1 A **holonomic constraint** on a mechanical system $(M, \langle \cdot, \cdot \rangle, \mathcal{F})$ is a submanifold $N \subset M$ with $\dim N < \dim M$. A curve $c : I \subset \mathbb{R} \to M$ is said to be **compatible** with N if $c(t) \in N$ for all $t \in I$.

Example 2.2

(1) A particle of mass $m > 0$ moving in $\mathbb{R}^2$ subject to a constant gravitational accel-
     eration g is modeled by the mechanical system $(\mathbb{R}^2, \langle\langle \cdot, \cdot \rangle\rangle, -mg \, dy)$, where

$$\langle\langle v, w \rangle\rangle := m \langle v, w \rangle$$

($\langle \cdot, \cdot \rangle$ being the Euclidean inner product on $\mathbb{R}^2$). A **simple pendulum** is obtained
by connecting the particle to a fixed pivoting point by an ideal massless rod of
length $l > 0$ (cf. Fig. 5.1). Assuming the pivoting point to be the origin, this
corresponds to the holonomic constraint

$$N = \{(x, y) \in \mathbb{R}^2 \mid x^2 + y^2 = l^2\}$$

(diffeomorphic to S^1).

(2) Similarly, a particle of mass $m > 0$ moving in $\mathbb{R}^3$ subject to a constant gravita-
     tional acceleration g is modeled by the mechanical system $(\mathbb{R}^3, \langle\langle \cdot, \cdot \rangle\rangle, -mg \, dz)$,
     where

$$\langle\langle v, w \rangle\rangle := m \langle v, w \rangle$$

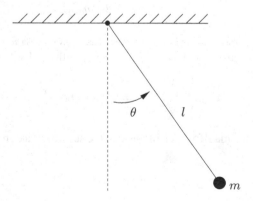

Fig. 5.1 Simple pendulum

($\langle \cdot, \cdot \rangle$ being the Euclidean inner product on $\mathbb{R}^3$). Requiring the particle to move on a surface of equation $z = f(x, y)$ yields the holonomic constraint

$$N = \{(x, y, z) \in \mathbb{R}^3 \mid z = f(x, y)\}.$$

(3) A system of k particles of masses $m_1, \ldots, m_k$ moving freely in $\mathbb{R}^3$ is model-led by the mechanical system $(\mathbb{R}^{3k}, \langle\langle \cdot, \cdot \rangle\rangle, 0)$, where

$$\langle\langle (v_1, \ldots, v_k), (w_1, \ldots, w_k) \rangle\rangle := \sum_{i=1}^{k} m_i \langle v_i, w_i \rangle$$

($\langle \cdot, \cdot \rangle$ being the Euclidean inner product on $\mathbb{R}^3$). A **rigid body** is obtained by connecting all particles by ideal massless rods, and corresponds to the holonomic constraint

$$N = \left\{ (x_1, \ldots, x_k) \in \mathbb{R}^{3k} \mid \|x_i - x_j\| = d_{ij} \text{ for } 1 \le i < j \le k \right\}.$$

If at least three particles are not collinear, N is easily seen to be diffeomorphic to $\mathbb{R}^3 \times O(3)$.

Keeping the particles on the holonomic constraint requires an additional external force (provided by the rods or by the surface in the examples above).

Definition 2.3 A **reaction force** on a mechanical system with holonomic constraint $(M, \langle \cdot, \cdot \rangle, \mathcal{F}, N)$ is a map $\mathcal{R} : TN \to T^*M$ satisfying $\mathcal{R}(T_pN) \subset T_p^*M$ for all $p \in N$ such that, for each $v \in TN$, there is a solution $c : I \subset \mathbb{R} \to N$ of the **generalized Newton equation**

$$\mu\left(\frac{D\dot{c}}{dt}\right) = (\mathcal{F} + \mathcal{R})(\dot{c})$$

with initial condition $\dot{c}(0) = v$.

For any holonomic constraint there exist in general infinite possible choices of reaction forces. The following definition yields a particularly useful criterion for selecting reaction forces.

Definition 2.4 A reaction force in a mechanical system with holonomic constraint $(M, \langle \cdot, \cdot \rangle, \mathcal{F}, N)$ is said to be **perfect**, or to satisfy the **d'Alembert principle**, if

$$\mu^{-1}(\mathcal{R}(v)) \in (T_pN)^{\perp}$$

for all $v \in T_pN$ and $p \in N$.

Remark 2.5 The variation of the kinetic energy of a solution of the generalized Newton equation is

$$\frac{dK}{dt} = \left\langle \frac{D\dot{c}}{dt}, \dot{c} \right\rangle = \mathcal{F}(\dot{c})(\dot{c}) + \mathcal{R}(\dot{c})(\dot{c}) = \mathcal{F}(\dot{c})(\dot{c}) + \left\langle \mu^{-1}(\mathcal{R}(\dot{c})), \dot{c} \right\rangle.$$

Therefore, a reaction force is perfect if and only if it neither creates nor dissipates energy along any motion compatible with the constraint.

Example 2.6 In each of the examples above, requiring the reaction force to be perfect amounts to the following assumptions.

(1) **Simple pendulum** : The force transmitted by the rod is purely radial (i.e. there is no damping);
(2) **Particle on a surface** : The force exerted by the surface is orthogonal to it (i.e. the surface is frictionless);
(3) **Rigid body** : The cohesive forces do not dissipate energy.

The next result establishes the existence and uniqueness of perfect reaction forces.

Theorem 2.7 *Given any mechanical system with holonomic constraint* $(M, \langle \cdot, \cdot \rangle, \mathcal{F}, N)$, *there exists a unique reaction force* $\mathcal{R} : TN \to T^*M$ *satisfying the d'Alembert principle. The solutions of the generalized Newton equation*

$$\mu \left(\frac{D\dot{c}}{dt} \right) = (\mathcal{F} + \mathcal{R})(\dot{c})$$

are exactly the motions of the mechanical system $(N, \langle\langle \cdot, \cdot \rangle\rangle, \mathcal{F}_N)$, *where* $\langle\langle \cdot, \cdot \rangle\rangle$ *is the metric induced on* N *by* $\langle \cdot, \cdot \rangle$ *and* $\mathcal{F}_N$ *is the restriction of* $\mathcal{F}$ *to* N. *In particular, if* $\mathcal{F} = -dU$ *is conservative then* $\mathcal{F}_N = -d(U|_N)$.

Proof Recall from Sect. 4.5 of Chap. 4 that if $\widetilde{\nabla}$ is the Levi–Civita connection of $(M, \langle \cdot, \cdot \rangle)$ and ∇ is the Levi–Civita connection of $(N, \langle\langle \cdot, \cdot \rangle\rangle)$ then

$$\nabla_X Y = \left(\widetilde{\nabla}_{\widetilde{X}} \widetilde{Y} \right)^\top$$

for all $X, Y \in \mathfrak{X}(N)$, where $\widetilde{X}, \widetilde{Y}$ are any extensions of X, Y to $\mathfrak{X}(M)$ (as usual, $v = v^\top + v^\perp$ designates the unique decomposition arising from the splitting $T_p M = T_p N \oplus (T_p N)^\perp$ for each $p \in N$). Moreover, the second fundamental form of N,

$$B(X, Y) = \widetilde{\nabla}_{\widetilde{X}} \widetilde{Y} - \nabla_X Y = \left(\widetilde{\nabla}_{\widetilde{X}} \widetilde{Y} \right)^\perp,$$

is well defined, and $B(X, Y)_p \in (T_p N)^\perp$ is a symmetric bilinear function of X_p, Y_p for all $p \in N$.

Assume that a perfect reaction force $\mathcal{R}$ exists; then the solutions of the generalized Newton equation satisfy

$$\widetilde{\nabla}_{\dot{c}} \dot{c} = \mu^{-1}(\mathcal{F}(\dot{c})) + \mu^{-1}(\mathcal{R}(\dot{c})).$$

Since by hypothesis $\mu^{-1}\mathcal{R}$ is orthogonal to N, the component of this equation tangent to N yields

$$\nabla_{\dot{c}}\dot{c} = \mu_N^{-1}(\mathcal{F}_N(\dot{c}))$$

(where $\mu_N : TN \to T^*N$ is the mass operator on N) as for any $v \in TN$ one has

$$\left\langle\!\left\langle \left(\mu^{-1}(\mathcal{F}(\dot{c}))\right)^{\top}, v \right\rangle\!\right\rangle = \left\langle \mu^{-1}(\mathcal{F}(\dot{c})), v \right\rangle = \mathcal{F}(\dot{c})(v) = \mathcal{F}_N(\dot{c})(v) = \left\langle\!\left\langle \mu_N^{-1}(\mathcal{F}_N(\dot{c})), v \right\rangle\!\right\rangle.$$

Hence c is a motion of $(N, \langle\!\langle \cdot, \cdot \rangle\!\rangle, \mathcal{F}_N)$.

On the other hand, the component of the generalized Newton equation orthogonal to N yields

$$B(\dot{c}, \dot{c}) = \left(\mu^{-1}(\mathcal{F}(\dot{c}))\right)^{\perp} + \mu^{-1}(\mathcal{R}(\dot{c})).$$

Therefore, if $\mathcal{R}$ exists then it must satisfy

$$\mathcal{R}(v) = \mu(B(v, v)) - \mu\left[\left(\mu^{-1}(\mathcal{F}(v))\right)^{\perp}\right] \tag{5.2}$$

for all $v \in TN$. This proves uniqueness.

To prove existence, **define** $\mathcal{R}$ through (5.2), which certainly guarantees that $\mu^{-1}(\mathcal{R}(v)) \in (T_pN)^{\perp}$ for all $v \in T_pN$ and $p \in N$. Given $v \in TN$, let $c : I \subset \mathbb{R} \to N$ be the motion of the mechanical system $(N, \langle\!\langle \cdot, \cdot \rangle\!\rangle, \mathcal{F}_N)$ with initial condition v. Then

$$\tilde{\nabla}_{\dot{c}}\dot{c} = \nabla_{\dot{c}}\dot{c} + B(\dot{c}, \dot{c}) = \mu_N^{-1}(\mathcal{F}_N(\dot{c})) + \left(\mu^{-1}(\mathcal{F}(\dot{c}))\right)^{\perp} + \mu^{-1}(\mathcal{R}(\dot{c}))$$

$$= \left(\mu^{-1}(\mathcal{F}(\dot{c}))\right)^{\top} + \left(\mu^{-1}(\mathcal{F}(\dot{c}))\right)^{\perp} + \mu^{-1}(\mathcal{R}(\dot{c})) = \mu^{-1}(\mathcal{F}(\dot{c})) + \mu^{-1}(\mathcal{R}(\dot{c})).$$

$\square$

Example 2.8 To write the equation of motion of a simple pendulum with a perfect reaction force, we parameterize the holonomic constraint N using the map $\varphi : (-\pi, \pi) \to \mathbb{R}^2$ defined by

$$\varphi(\theta) = (l \sin\theta, -l \cos\theta)$$

(so that $\theta = 0$ labels the stable equilibrium position, cf. Fig. 5.1). We have

$$\frac{d}{d\theta} = \frac{dx}{d\theta}\frac{\partial}{\partial x} + \frac{dy}{d\theta}\frac{\partial}{\partial y} = l\cos\theta\frac{\partial}{\partial x} + l\sin\theta\frac{\partial}{\partial y},$$

and hence the kinetic energy of the pendulum is

$$
K\left(v\frac{d}{d\theta}\right) = \frac{1}{2}m\left\langle vl\cos\theta\frac{\partial}{\partial x} + vl\sin\theta\frac{\partial}{\partial y}, vl\cos\theta\frac{\partial}{\partial x} + vl\sin\theta\frac{\partial}{\partial y}\right\rangle
$$
$$
= \frac{1}{2}ml^2v^2.
$$

On the other hand, the potential energy is given by

$$
U(x, y) = mgy,
$$

and hence its restriction to N has the local expression

$$
U(\theta) = -mgl\cos\theta.
$$

Consequently the equation of motion is

$$
\frac{d}{dt}\left(\frac{\partial K}{\partial v}(\theta, \dot\theta)\right) - \frac{\partial K}{\partial \theta}(\theta, \dot\theta) = -\frac{\partial U}{\partial \theta}(\theta)
$$
$$
\Leftrightarrow \frac{d}{dt}\left(ml^2\dot\theta\right) = -mgl\sin\theta
$$
$$
\Leftrightarrow \ddot\theta = -\frac{g}{l}\sin\theta.
$$

Notice that we did not have to compute the reaction force.

Exercise 2.9

(1) Use spherical coordinates to write the equations of motion for the **spherical pendulum** of length l, i.e. a particle of mass $m > 0$ moving in $\mathbb{R}^3$ subject to a constant gravitational acceleration g and the holonomic constraint

$$
N = \left\{(x, y, z) \in \mathbb{R}^3 \mid x^2 + y^2 + z^2 = l^2\right\}.
$$

Which parallels of N are possible trajectories of the particle?

(2) Write the equations of motion for a particle moving on a frictionless surface of revolution with equation $z = f(r)$ (where $r = \sqrt{x^2 + y^2}$) under a constant gravitational acceleration g.

(3) Write and solve the equations of motion for a free dumbbell, i.e. a system of two particles of masses m_1 and m_2 connected by a massless rod of length l, moving in:

(a) $\mathbb{R}^2$;
(b) $\mathbb{R}^3$.

   (**Hint:** Use the coordinates of the **center of mass**, i.e. the point along the rod at a distance $\frac{m_2}{m_1+m_2}l$ from m_1).

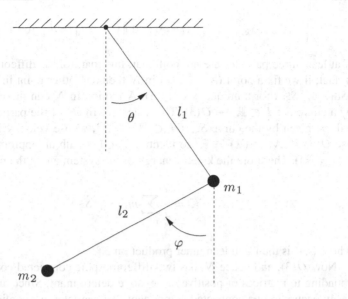

Fig. 5.2 Double pendulum

(4) The **double pendulum** of lengths l_1, l_2 is the mechanical system defined by two particles of masses m_1, m_2 moving in $\mathbb{R}^2$ subject to a constant gravitational acceleration g and the holonomic constraint

$$N = \left\{ (x_1, x_2) \in \mathbb{R}^4 \mid \|x_1\| = l_1 \text{ and } \|x_1 - x_2\| = l_2 \right\}.$$

(diffeomorphic to the 2-torus T^2).

(a) Write the equations of motion for the double pendulum using the parameterization $\phi : (-\pi, \pi) \times (-\pi, \pi) \to N$ given by

$$\phi(\theta, \varphi) = (l_1 \sin\theta, -l_1 \cos\theta, l_1 \sin\theta + l_2 \sin\varphi, -l_1 \cos\theta - l_2 \cos\varphi)$$

(cf. Fig. 5.2).

(b) Linearize the equations of motion around $\theta = \varphi = 0$. Look for solutions of the linearized equations satisfying $\varphi = k\theta$, with $k \in \mathbb{R}$ constant (**normal modes**). What are the periods of the ensuing oscillations?

5.3 Rigid Body

Recall that a rigid body is a system of k particles of masses $m_1, \ldots, m_k$ connected by massless rods in such a way that their mutual distances remain constant. If in addition we assume that a given particle is fixed (at the origin, say) then we obtain the holonomic constraint

$$N = \left\{ (x_1, \ldots, x_k) \in \mathbb{R}^{3k} \mid x_1 = 0 \text{ and } \| x_i - x_j \| = d_{ij} \text{ for } 1 \leq i < j \leq k \right\}.$$

If at least three particles are not collinear, this manifold is diffeomorphic to $O(3)$. In fact, if we fix a point $(\xi_1, \ldots, \xi_k)$ in N then any other point in N is of the form $(S\xi_1, \ldots, S\xi_k)$ for a unique $S \in O(3)$. A motion in N can therefore be specified by a curve $S : I \subset \mathbb{R} \rightarrow O(3)$. The trajectory in $\mathbb{R}^3$ of the particle with mass m_i will be given by the curve $S\xi_i : I \subset \mathbb{R} \rightarrow \mathbb{R}^3$, whose velocity is $\dot{S}\xi_i$ (where we use $O(3) \subset \mathcal{M}_{3\times3}(\mathbb{R}) \cong \mathbb{R}^9$ to identify $T_S O(3)$ with an appropriate subspace of $\mathcal{M}_{3\times3}(\mathbb{R})$). Therefore the kinetic energy of the system along the motion will be

$$K = \frac{1}{2} \sum_{i=1}^{n} m_i \left\langle \dot{S}\xi_i, \dot{S}\xi_i \right\rangle,$$

where $\langle \cdot, \cdot \rangle$ is the Euclidean inner product on $\mathbb{R}^3$.

Now $O(3)$, and hence N, has two diffeomorphic connected components, corresponding to matrices of positive or negative determinant. Since any motion necessarily occurs in one connected component, we can take our configuration space to be simply $SO(3)$. To account for continuum rigid bodies, we make the following generalization:

Definition 3.1 A **rigid body with a fixed point** is any mechanical system of the form $(SO(3), \langle\langle \cdot, \cdot \rangle\rangle, \mathcal{F})$, with

$$\langle\langle V, W \rangle\rangle := \int_{\mathbb{R}^3} \langle V\xi, W\xi \rangle \, dm$$

for all $V, W \in T_S SO(3)$ and all $S \in SO(3)$, where $\langle \cdot, \cdot \rangle$ is the usual Euclidean inner product on $\mathbb{R}^3$ and m (called the **mass distribution of the reference configuration**) is a positive finite measure on $\mathbb{R}^3$, not supported on any straight line through the origin, and satisfying $\int_{\mathbb{R}^3} \| \xi \|^2 dm < +\infty$.

Example 3.2

(1) The rigid body composed by k particles of masses $m_1, \ldots, m_k$ corresponds to the measure

$$m = \sum_{i=1}^{k} m_i \delta_{\xi_i},$$

where δ_{ξ_i} is the Dirac delta centered at the point $\xi_i \in \mathbb{R}^3$.

(2) A continuum rigid body with (say) compactly supported integrable density function $\rho : \mathbb{R}^3 \rightarrow [0, +\infty)$ is described by the measure m defined on the Lebesgue σ-algebra by

$$m(A) := \int_\Lambda \rho(\xi) d^3\xi.$$

Remark 3.3 The rotational motion of a general rigid body can in many cases be reduced to the motion of a rigid body with a fixed point [cf. Exercise 3.20(2)]. Unless otherwise stated, from this point onwards we will take "rigid body" to mean "rigid body with a fixed point".

Proposition 3.4 *The metric* $\langle\langle\cdot,\cdot\rangle\rangle$ *defined on* $SO(3)$ *by a rigid body is left-invariant, that is, any left translation is an isometry.*

Proof Since left multiplication by a fixed matrix $R \in SO(3)$ is a linear map $L_R :$ $\mathcal{M}_{3\times3}(\mathbb{R}) \to \mathcal{M}_{3\times3}(\mathbb{R})$, we have $(dL_R)_S V = RV \in T_{RS}SO(3)$ for any $V \in$ $T_S SO(3)$. Consequently,

$$\langle\langle(dL_R)_S V, (dL_R)_S W\rangle\rangle = \langle\langle RV, RW\rangle\rangle = \int_{\mathbb{R}^3} \langle RV\xi, RW\xi\rangle \, dm$$

$$= \int_{\mathbb{R}^3} \langle V\xi, W\xi\rangle \, dm = \langle\langle V, W\rangle\rangle$$

(as $R \in SO(3)$ preserves the Euclidean inner product). □

Therefore there exist at most as many rigid bodies as inner products on $\mathfrak{so}(3) \cong \mathbb{R}^3$, i.e. as real symmetric positive definite 3×3 matrices [cf. Exercise 1.10(4) in Chap. 3]. In fact, we shall see that any rigid body can be specified by 3 positive numbers.

Proposition 3.5 *The metric* $\langle\langle\cdot,\cdot\rangle\rangle$ *defined on* $SO(3)$ *by a rigid body is given by*

$$\langle\langle V, W\rangle\rangle = \text{tr}\left(VJW^t\right),$$

where

$$J_{ij} = \int_{\mathbb{R}^3} \xi^i \xi^j dm.$$

Proof We just have to notice that

$$\langle\langle V, W\rangle\rangle = \int_{\mathbb{R}^3} \sum_{i=1}^{3} \left(\sum_{j=1}^{3} V_{ij}\xi^j\right)\left(\sum_{k=1}^{3} W_{ik}\xi^k\right) dm$$

$$= \sum_{i,j,k=1}^{3} V_{ij} W_{ik} \int_{\mathbb{R}^3} \xi^j \xi^k \, dm = \sum_{i,j,k=1}^{3} V_{ij} J_{jk} W_{ik}.$$

□

Proposition 3.6 *If* $S : I \subset \mathbb{R} \to SO(3)$ *is a curve and* ∇ *is the Levi–Civita connection on* $(SO(3), \langle\langle \cdot, \cdot \rangle\rangle)$ *then*

$$\langle\langle \nabla_{\dot{S}} \dot{S}, V \rangle\rangle = \int_{\mathbb{R}^3} \langle \ddot{S}\xi, V\xi \rangle \, dm$$

for any $V \in T_S SO(3)$.

Proof We consider first the case in which the rigid body is **non-planar**, i.e. m is not supported in any plane through the origin. In this case, the metric $\langle\langle \cdot, \cdot \rangle\rangle$ can be extended to a flat metric on $\mathcal{M}_{3\times3}(\mathbb{R}) \cong \mathbb{R}^9$ by the same formula

$$\langle\langle\langle V, W \rangle\rangle\rangle = \int_{\mathbb{R}^3} \langle V\xi, W\xi \rangle \, dm$$

for all $V, W \in T_S \mathcal{M}_{3\times3}(\mathbb{R})$ and all $S \in \mathcal{M}_{3\times3}(\mathbb{R})$. Indeed, this formula clearly defines a symmetric 2-tensor on $\mathcal{M}_{3\times3}(\mathbb{R})$. To check positive definiteness, we notice that if $V \in T_S \mathcal{M}_{3\times3}(\mathbb{R})$ is nonzero then its kernel is contained on a plane through the origin. Therefore, the continuous function $\langle V\xi, V\xi \rangle$ is positive on a set of positive measure, and hence

$$\langle\langle\langle V, V \rangle\rangle\rangle = \int_{\mathbb{R}^3} \langle V\xi, V\xi \rangle \, dm > 0.$$

This metric is easily seen to be flat, as the components of the metric on the natural coordinates of $\mathcal{M}_{3\times3}(\mathbb{R})$ are the constant coefficients J_{ij}. Therefore all Christoffel symbols vanish on these coordinates, and the corresponding Levi–Civita connection $\tilde{\nabla}$ is the trivial connection. If $S : I \subset \mathbb{R} \to \mathcal{M}_{3\times3}(\mathbb{R})$ is a curve then

$$\tilde{\nabla}_{\dot{S}} \dot{S} = \ddot{S}.$$

Since $\langle\langle \cdot, \cdot \rangle\rangle$ is the metric induced on $SO(3)$ by $\langle\langle\langle \cdot, \cdot \rangle\rangle\rangle$, we see that for any curve $S : I \subset \mathbb{R} \to SO(3)$ one has

$$\nabla_{\dot{S}} \dot{S} = \left(\tilde{\nabla}_{\dot{S}} \dot{S} \right)^{\top} = \ddot{S}^{\top},$$

and hence

$$\langle\langle \nabla_{\dot{S}} \dot{S}, V \rangle\rangle = \left\langle\left\langle \ddot{S}^{\top}, V \right\rangle\right\rangle = \langle\langle\langle \ddot{S}, V \rangle\rangle\rangle = \int_{\mathbb{R}^3} \langle \ddot{S}\xi, V\xi \rangle \, dm$$

for any $V \in T_S SO(3)$.

For planar rigid bodies the formula can by obtained by a limiting procedure [cf. Exercise 3.20(3)].   □

We can use this result to determine the geodesics of $(SO(3), \langle\langle \cdot, \cdot \rangle\rangle)$. A remarkable shortcut (whose precise nature will be discussed in Sect. 5.5) can be obtained by introducing the following quantity.

Definition 3.7 The **angular momentum** of a rigid body whose motion is described by $S : I \subset \mathbb{R} \to SO(3)$ is the vector

$$p(t) := \int_{\mathbb{R}^3} \left[(S(t)\xi) \times (\dot{S}(t)\xi) \right] dm$$

(where $\times$ is the usual cross product on $\mathbb{R}^3$).

Theorem 3.8 *If $S : I \subset \mathbb{R} \to SO(3)$ is a geodesic of $(SO(3), \langle\langle \cdot, \cdot \rangle\rangle)$ then $p(t)$ is constant.*

Proof We have

$$\dot{p} = \int_{\mathbb{R}^3} \left[(\dot{S}\xi) \times (\dot{S}\xi) + (S\xi) \times (\ddot{S}\xi) \right] dm = \int_{\mathbb{R}^3} \left[(S\xi) \times (\ddot{S}\xi) \right] dm.$$

Take any $v \in \mathbb{R}^3$. Then

$$\langle Sv, \dot{p} \rangle = \left\langle Sv, \int_{\mathbb{R}^3} \left[(S\xi) \times (\ddot{S}\xi) \right] dm \right\rangle = \int_{\mathbb{R}^3} \langle Sv, (S\xi) \times (\ddot{S}\xi) \rangle \, dm$$

$$= \int_{\mathbb{R}^3} \langle \ddot{S}\xi, (Sv) \times (S\xi) \rangle \, dm = \int_{\mathbb{R}^3} \langle \ddot{S}\xi, S(v \times \xi) \rangle \, dm,$$

where we have used the invariance of $\langle \cdot, \cdot \times \cdot \rangle \equiv \det(\cdot, \cdot, \cdot)$ under even permutations of its arguments and the fact that the cross product is equivariant under multiplication by $S \in SO(3)$.

To complete the proof we will need the following lemma, whose proof is left as an exercise.

Lemma 3.9 *There exists a linear isomorphism $\Omega : \mathfrak{so}(3) \to \mathbb{R}^3$ such that*

$$A\xi = \Omega(A) \times \xi$$

for all $\xi \in \mathbb{R}^3$ and $A \in \mathfrak{so}(3)$. Moreover, $\Omega([A, B]) = \Omega(A) \times \Omega(B)$ for all $A, B \in \mathfrak{so}(3)$ (that is, Ω is a Lie algebra isomorphism between $\mathfrak{so}(3)$ and $(\mathbb{R}^3, \times)$). $\square$

Returning to the proof, let $V \in \mathfrak{so}(3)$ be such that $\Omega(V) = v$. Then $SV \in T_S SO(3)$ and

$$\langle Sv, \dot{p}\rangle = \int_{\mathbb{R}^3} \langle \ddot{S}\xi, SV\xi\rangle \, dm = \langle\!\langle \nabla_{\dot{S}}\dot{S}, SV\rangle\!\rangle = 0$$

(as $S : I \subset \mathbb{R} \to SO(3)$ is a geodesic). Since $v \in \mathbb{R}^3$ is arbitrary, we see that $\dot{p} = 0$ along the motion. $\square$

If $S : I \subset \mathbb{R} \to SO(3)$ is a curve then $\dot{S} = SA$ for some $A \in \mathfrak{so}(3)$. Let us define $\Omega := \Omega(A)$. Then

$$p = \int_{\mathbb{R}^3} [(S\xi) \times (SA\xi)] \, dm = \int_{\mathbb{R}^3} S \, [\xi \times (A\xi)] \, dm$$

$$= S \int_{\mathbb{R}^3} [\xi \times (\Omega \times \xi)] \, dm.$$

This suggests the following definition.

Definition 3.10 The linear operator $I : \mathbb{R}^3 \to \mathbb{R}^3$ defined as

$$I(v) := \int_{\mathbb{R}^3} [\xi \times (v \times \xi)] \, dm$$

is called the rigid body's **moment of inertia tensor**.

Proposition 3.11 *The moment of inertia tensor of any given rigid body is a symmetric positive definite linear operator, and the corresponding kinetic energy map* $K : TSO(3) \to \mathbb{R}$ *is given by*

$$K(V) = \frac{1}{2}\langle\!\langle V, V\rangle\!\rangle = \frac{1}{2}\langle\!\langle SA, SA\rangle\!\rangle = \frac{1}{2}\langle I\Omega, \Omega\rangle,$$

for all $V \in T_S SO(3)$ *and all* $S \in SO(3)$, *where* $V = SA$ *and* $\Omega = \Omega(A)$.

Proof We start by checking that I is symmetric:

$$\langle Iv, w\rangle = \left\langle \int_{\mathbb{R}^3} [\xi \times (v \times \xi)] \, dm, w \right\rangle = \int_{\mathbb{R}^3} \langle \xi \times (v \times \xi), w\rangle \, dm$$

$$= \int_{\mathbb{R}^3} \langle v \times \xi, w \times \xi\rangle \, dm = \langle v, Iw\rangle.$$

In particular we have

$$\langle I\Omega, \Omega\rangle = \int_{\mathbb{R}^3} \langle \Omega \times \xi, \Omega \times \xi\rangle \, dm = \int_{\mathbb{R}^3} \langle A\xi, A\xi\rangle \, dm$$

$$= \int_{\mathbb{R}^3} \langle SA\xi, SA\xi \rangle \, dm = 2K(V).$$

The positive definiteness of I is an immediate consequence of this formula. □

Corollary 3.12 *Given any rigid body there exist three positive numbers I_1, I_2, I_3* (**principal moments of inertia**) *and an orthonormal basis of $\mathbb{R}^3$, $\{e_1, e_2, e_3\}$ (**principal axes**), such that $I e_i = I_i e_i$ ($i = 1, 2, 3$).* □

The principal moments of inertia are the three positive numbers which completely specify the rigid body (as they determine the inertia tensor, which in turn yields the kinetic energy). To compute these numbers we must compute the eigenvalues of a matrix representation of the inertia tensor.

Proposition 3.13 *The matrix representation of the inertia tensor in the canonical basis of $\mathbb{R}^3$ is*

$$\begin{pmatrix} \int_{\mathbb{R}^3}(y^2 + z^2)\, dm & -\int_{\mathbb{R}^3} xy\, dm & -\int_{\mathbb{R}^3} xz\, dm \\ -\int_{\mathbb{R}^3} xy\, dm & \int_{\mathbb{R}^3}(x^2 + z^2)\, dm & -\int_{\mathbb{R}^3} yz\, dm \\ -\int_{\mathbb{R}^3} xz\, dm & -\int_{\mathbb{R}^3} yz\, dm & \int_{\mathbb{R}^3}(x^2 + y^2)\, dm \end{pmatrix}.$$

Proof Let $\{u_1, u_2, u_3\}$ be the canonical basis of $\mathbb{R}^3$. Then

$$I_{ij} = \langle I u_i, u_j \rangle = \int_{\mathbb{R}^3} \langle \xi \times (u_i \times \xi), u_j \rangle \, dm.$$

Using the vector identity

$$u \times (v \times w) = \langle u, w \rangle v - \langle u, v \rangle w$$

for all $u, v, w \in \mathbb{R}^3$, we have

$$I_{ij} = \int_{\mathbb{R}^3} \langle \|\xi\|^2 u_i - \langle \xi, u_i \rangle \xi, u_j \rangle \, dm = \int_{\mathbb{R}^3} \left(\|\xi\|^2 \delta_{ij} - \xi^i \xi^j \right) dm. \quad □$$

We can now write the equations for the geodesics of $(SO(3), \langle\langle \cdot, \cdot \rangle\rangle)$, that is, the equations of motion of a rigid body in the absence of external forces. This mechanical system is commonly known as the **Euler top**.

Proposition 3.14 *The equations of motion of the Euler top are given by the **Euler equations***

$$I\dot{\Omega} = (I\Omega) \times \Omega.$$

Proof We just have to notice that

$$p = SI\Omega.$$

Therefore

$$0 = \dot{p} = \dot{S}I\Omega + SI\dot{\Omega} = SAI\Omega + SI\dot{\Omega} = S\left(\Omega \times (I\Omega) + I\dot{\Omega}\right). \qquad \square$$

Remark 3.15 Any point $\xi \in \mathbb{R}^3$ in the rigid body traverses a curve $x(t) = S(t)\xi$ with velocity

$$\dot{x} = \dot{S}\xi = SA\xi = S(\Omega \times \xi) = (S\Omega) \times (S\xi) = (S\Omega) \times x.$$

Therefore $\omega := S\Omega$ is the rigid body's **instantaneous angular velocity**: at each instant, the rigid body rotates about the axis determined by ω with angular speed $\|\omega\|$. Consequently, Ω is the angular velocity as seen in the (accelerated) rigid body's rest frame (cf. Fig. 5.3).

In the basis $\{e_1, e_2, e_3\}$ of the principal axes, the Euler equations are written

$$\begin{cases} I_1\dot{\Omega}^1 = (I_2 - I_3)\Omega^2\Omega^3 \\ I_2\dot{\Omega}^2 = (I_3 - I_1)\Omega^3\Omega^1 \\ I_3\dot{\Omega}^3 = (I_1 - I_2)\Omega^1\Omega^2 \end{cases}.$$

Since I is positive definite (hence invertible), we can change variables to $P := I\Omega$. Notice that $p = SP$, i.e. P is the (constant) angular momentum vector as seen in rigid body's rest frame. In these new variables, the Euler equations are written

$$\dot{P} = P \times \left(I^{-1}P\right).$$

In the basis $\{e_1, e_2, e_3\}$ of the principal axes, these are

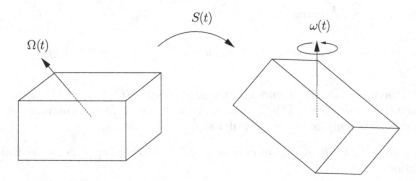

Fig. 5.3 Angular velocities

$$\begin{cases} \dot{P}^1 = \left(\dfrac{1}{I_3} - \dfrac{1}{I_2} \right) P^2 P^3 \\[2mm] \dot{P}^2 = \left(\dfrac{1}{I_1} - \dfrac{1}{I_3} \right) P^3 P^1 \\[2mm] \dot{P}^3 = \left(\dfrac{1}{I_2} - \dfrac{1}{I_1} \right) P^1 P^2 \end{cases}.$$

Proposition 3.16 *If $I_1 > I_2 > I_3$, the stationary points of the Euler equations are given by*

$$P = \lambda e_i \qquad (\lambda \in \mathbb{R}, \ i = 1, 2, 3),$$

and are stable for $i = 1, 3$ and unstable for $i = 2$.

Proof Since there are no external forces, the kinetic energy K, given by

$$2K = \langle I\Omega, \Omega \rangle = \left\langle P, I^{-1}P \right\rangle = \frac{\left(P^1 \right)^2}{I_1} + \frac{\left(P^2 \right)^2}{I_2} + \frac{\left(P^3 \right)^2}{I_3},$$

is conserved. This means that the flow defined by the Euler equations is along ellipsoids with semiaxes of lengths $\sqrt{2KI_1} > \sqrt{2KI_2} > \sqrt{2KI_3}$. On the other hand, since p is constant along the motion, we have a second conserved quantity,

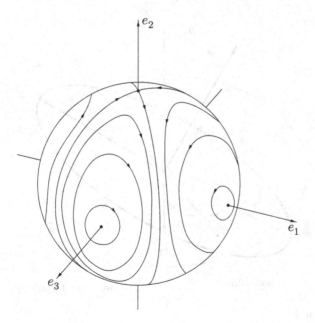

Fig. 5.4 Integral curves of the Euler equations

$$\|p\|^2 = \|P\|^2 = \left(P^1\right)^2 + \left(P^2\right)^2 + \left(P^3\right)^2.$$

Therefore the flow is along spheres. The integral curves on a particular sphere can be found by intersecting it with the ellipsoids corresponding to different values of K, as shown in Fig. 5.4. □

Remark 3.17 Since $\Omega = I^{-1}P$, Proposition 3.16 is still true if we replace P with Ω. The equilibrium points represent rotations about the principal axes with constant angular speed, as they satisfy $\Omega = I_i P$, and hence $\omega = I_i p$ is constant. If the rigid body is placed in a rotation state close to a rotation about the axes e_1 or e_3, P will remain close to these axes, and hence Se_1 or Se_3 will remain close to the fixed vector p. On the other hand, if the rigid body is placed in a rotation state close to a rotation about the axis e_2, then P will drift away from e_2 (approaching $-e_2$ before returning to e_2), and hence Se_2 will drift away from the fixed vector p (approaching $-p$ before returning to p). This can be illustrated by throwing a rigid body (say a brick) in the air, as its rotational motion about the center of mass is that of a rigid body with a fixed point [cf. Exercise 3.20(2)]. When rotating about the smaller or the larger axis [i.e. the principal axes corresponding to the larger or the smaller moments of inertia—cf. Exercise 3.20(6)] it performs a stable rotation, but when rotating about the middle axis it flips in midair.

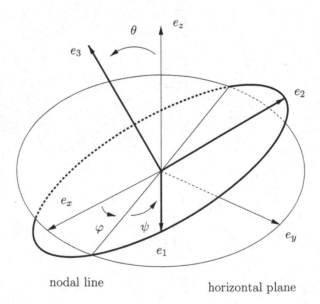

nodal line   horizontal plane

Fig. 5.5 Euler angles

If the rigid body is not free, one must use parameterizations of $SO(3)$.

Definition 3.18 The **Euler angles** correspond to the local coordinates (θ, φ, ψ) : $SO(3) \to (0, \pi) \times (0, 2\pi) \times (0, 2\pi)$ defined by

$$S(\theta, \varphi, \psi) = \begin{pmatrix} \cos\varphi & -\sin\varphi & 0 \\ \sin\varphi & \cos\varphi & 0 \\ 0 & 0 & 1 \end{pmatrix} \begin{pmatrix} 1 & 0 & 0 \\ 0 & \cos\theta & -\sin\theta \\ 0 & \sin\theta & \cos\theta \end{pmatrix} \begin{pmatrix} \cos\psi & -\sin\psi & 0 \\ \sin\psi & \cos\psi & 0 \\ 0 & 0 & 1 \end{pmatrix}.$$

The geometric interpretation of the Euler angles is sketched in Fig. 5.5: if the rotation carries the canonical basis $\{e_x, e_y, e_z\}$ to a new orthonormal basis $\{e_1, e_2, e_3\}$, then θ is the angle between e_3 and e_z, φ is the angle between the line of intersection of the planes spanned by $\{e_1, e_2\}$ and $\{e_x, e_y\}$ (called the **nodal line**) and the x-axis, and ψ is the angle between e_1 and the nodal line.

The general expression of the kinetic energy in the local coordinates of $TSO(3)$ associated to the Euler angles is quite complicated; here we present it only in the simpler case $I_1 = I_2$.

Proposition 3.19 *If $I_1 = I_2$ then the kinetic energy of a rigid body in the local coordinates $(\theta, \varphi, \psi, v^\theta, v^\varphi, v^\psi)$ of $TSO(3)$ is given by*

$$K = \frac{I_1}{2}\left((v^\theta)^2 + (v^\varphi)^2 \sin^2\theta \right) + \frac{I_3}{2} \left(v^\psi + v^\varphi \cos\theta \right)^2.$$

Proof Exercise 3.20(13).   □

A famous model which can be studied using this expression is the so-called **Lagrange top**, corresponding to an axisymmetric rigid body in a constant gravity field g. The potential energy for the corresponding mechanical system is

$$U := g \int_{\mathbb{R}^3} \langle S\xi, e_z \rangle \, dm = Mg \langle S\bar{\xi}, e_z \rangle,$$

where $M = m(\mathbb{R}^3)$ is the total mass and

$$\bar{\xi} := \frac{1}{M} \int_{\mathbb{R}^3} \xi \, dm$$

is the position of the **center of mass** in the rigid body's frame. By axisymmetry, the center of mass satisfies $\bar{\xi} = le_3$ for some $l \in \mathbb{R}$, and so

$$U = Mgl \cos\theta.$$

Exercise 3.20

(1) Show that the bilinear form $\langle\langle \cdot, \cdot \rangle\rangle$ defined on $SO(3)$ by a rigid body is indeed a Riemannian metric.

(2) A general rigid body (i.e. with no fixed points) is any mechanical system of the form $(\mathbb{R}^3 \times SO(3), \langle\langle\langle \cdot, \cdot \rangle\rangle\rangle, \mathcal{F})$, with

$$\langle\langle\langle (v, V), (w, W) \rangle\rangle\rangle := \int_{\mathbb{R}^3} \langle v + V\xi, w + W\xi \rangle \, dm$$

for all $(v, V), (w, W) \in T_{(x,S)}\mathbb{R}^3 \times SO(3)$ and $(x, S) \in \mathbb{R}^3 \times SO(3)$, where $\langle \cdot, \cdot \rangle$ is the usual Euclidean inner product on $\mathbb{R}^3$ and m is a positive finite measure on $\mathbb{R}^3$ not supported on any straight line and satisfying $\int_{\mathbb{R}^3} \|\xi\|^2 dm < +\infty$.

(a) Show that one can always translate m in such a way that

$$\int_{\mathbb{R}^3} \xi \, dm = 0$$

   (i.e. the center of mass of the reference configuration is placed at the origin).
(b) Show that for this choice the kinetic energy of the rigid body is

$$K(v, V) = \frac{1}{2} M \langle v, v \rangle + \frac{1}{2} \langle\langle V, V \rangle\rangle,$$

   where $M = m(\mathbb{R}^3)$ is the total mass of the rigid body and $\langle\langle \cdot, \cdot \rangle\rangle$ is the metric for the rigid body (with a fixed point) determined by m.
(c) Assume that there exists a differentiable function $F : \mathbb{R}^3 \to \mathbb{R}^3$ such that

$$\mathcal{F}(x, S, v, V)(w, W) = \int_{\mathbb{R}^3} \langle F(x + S\xi), w + W\xi \rangle \, dm.$$

   Show that, if

$$\int_{\mathbb{R}^3} (S\xi) \times F(x + S\xi) \, dm = 0$$

   for all $(x, S) \in \mathbb{R}^3 \times SO(3)$, then the projection of any motion on $SO(3)$ is a geodesic of $(SO(3), \langle\langle \cdot, \cdot \rangle\rangle)$.
(d) Describe the motion of a rigid body falling in a constant gravitational field, for which $F = -ge_z$ is constant.

(3) Prove Proposition 3.6 for a planar rigid body. (**Hint:** Include the planar rigid body in a smooth one-parameter family of non-planar rigid bodies).

(4) Prove Lemma 3.9.

(5) Show that $I_1 \leq I_2 + I_3$ (and cyclic permutations). When is $I_1 = I_2 + I_3$?

(6) Determine the principal axes and the corresponding principal moments of inertia of:

  (a) a homogeneous rectangular parallelepiped with mass M, sides $2a, 2b, 2c \in \mathbb{R}^+$ and centered at the origin;

  (b) a homogeneous (solid) ellipsoid with mass M, semiaxes $a, b, c \in \mathbb{R}^+$ and centered at the origin. (Hint: Use the coordinate change $(x, y, z) = (au, bv, cw)$).

(7) A **symmetry** of a rigid body is an isometry $S \in O(3)$ which preserves the mass distribution (i.e. $m(SA) = m(A)$ for any measurable set $A \subset \mathbb{R}^3$). Show that:

  (a) $SIS^t = I$, where I is the matrix representation of the inertia tensor;

  (b) if S is a reflection in a plane then there exists a principal axis orthogonal to the reflection plane;

  (c) if S is a nontrivial rotation about an axis then that axis is principal;

  (d) if moreover the rotation is not by π then all axes orthogonal to the rotation axis are principal.

(8) Consider a rigid body satisfying $I_1 = I_2$. Use the Euler equations to show that:

  (a) the angular velocity satisfies

$$\dot{\omega} = \frac{1}{I_1} p \times \omega;$$

  (b) if $I_1 = I_2 = I_3$ then the rigid body rotates about a fixed axis with constant angular speed (i.e. ω is constant);

  (c) if $I_1 = I_2 \neq I_3$ then ω **precesses** (i.e. rotates) about p with angular velocity

$$\omega_{pr} := \frac{p}{I_1}.$$

(9) Many asteroids have irregular shapes, and hence satisfy $I_1 < I_2 < I_3$. To a very good approximation, their rotational motion about the center of mass is described by the Euler equations. Over very long periods of time, however, their small interactions with the Sun and other planetary bodies tend to decrease their kinetic energy while conserving their angular momentum. Which rotation state do asteroids approach?

(10) Due to its rotation, the Earth is not a perfect sphere, but an oblate spheroid; therefore its moments of inertia are not quite equal, satisfying approximately

$$I_1 = I_2 \neq I_3;$$
$$\frac{I_3 - I_1}{I_1} \simeq \frac{1}{306}.$$

The Earth's rotation axis is very close to e_3, but precesses around it (**Chandler precession**). Find the period of this precession (in the Earth's frame).

(11) Consider a rigid body whose motion is described by the curve $S : \mathbb{R} \to SO(3)$, and let Ω be the corresponding angular velocity. Consider a particle with mass m whose motion **in the rigid body's frame** is given by the curve $\xi : \mathbb{R} \to \mathbb{R}^3$. Let f be the external force on the particle, so that its equation of motion is

$$m \frac{d^2}{dt^2}(S\xi) = f.$$

(a) Show that the equation of motion can be written as

$$m\ddot{\xi} = F - m\Omega \times (\Omega \times \xi) - 2m\Omega \times \dot{\xi} - m\dot{\Omega} \times \xi$$

where $f = SF$. (The terms following F are the so-called **inertial forces**, and are known, respectively, as the **centrifugal force**, the **Coriolis force** and the **Euler force**).

(b) Show that if the rigid body is a homogeneous sphere rotating freely (like the Earth, for instance) then the Euler force vanishes. Why must a long range gun in the Northern hemisphere be aimed at the **left** of the target?

(12) (*Poinsot theorem*) The **inertia ellipsoid** of a rigid body with moment of inertia tensor I is the set
$$E = \left\{ \xi \in \mathbb{R}^3 \mid \langle I\xi, \xi \rangle = 1 \right\}.$$

Show that the inertia ellipsoid of a freely moving rigid body rolls without slipping on a fixed plane orthogonal to p (that is, the contact point has zero velocity at each instant). (**Hint:** Show that any point $S(t)\xi(t)$ where the ellipsoid is tangent to a plane orthogonal to p satisfies $S(t)\xi(t) = \pm\frac{1}{\sqrt{2K}}\omega(t)$).

(13) Prove Proposition 3.19. (**Hint:** Notice that symmetry demands that the expression for K must not depend neither on φ nor on ψ).

(14) Consider the Lagrange top.

(a) Write the equations of motion and determine the equilibrium points.

(b) Show that there exist solutions such that θ, $\dot{\varphi}$ and $\dot{\psi}$ are constant, which in the limit $|\dot{\varphi}| \ll |\dot{\psi}|$ (**fast top**) satisfy

$$\dot{\varphi} \simeq \frac{Mgl}{I_3 \dot{\psi}}.$$

(15) (*Precession of the equinoxes*) Due to its rotation, the Earth is not a perfect sphere, but an oblate ellipsoid; therefore its moments of inertia are not quite equal, satisfying approximately

$$I_1 = I_2 \neq I_3;$$

$$\frac{I_3 - I_1}{I_1} \simeq \frac{1}{306}$$

[cf. Exercise 3.20(10)]. As a consequence, the combined gravitational attraction of the Moon and the Sun disturbs the Earth's rotation motion. This perturbation can be approximately modeled by the potential energy $U : SO(3) \to \mathbb{R}$ given in the Euler angles (θ, φ, ψ) by

$$U = -\frac{\Omega^2}{2}(I_3 - I_1) \cos^2 \theta,$$

where $\dfrac{2\pi}{\Omega} \simeq 168$ days.

(a) Write the equations of motion and determine the equilibrium points.
(b) Show that there exist solutions such that θ, $\dot{\varphi}$ and $\dot{\psi}$ are constant, which in the limit $|\dot{\varphi}| \ll |\dot{\psi}|$ (as is the case with the Earth) satisfy

$$\dot{\varphi} \simeq -\frac{\Omega^2(I_3 - I_1) \cos \theta}{I_3 \dot{\psi}}.$$

Given that for the Earth $\theta \simeq 23°$, determine the approximate value of the period of $\varphi(t)$.

(16) (*Pseudo-rigid body*) Recall that the (non planar) rigid body metric is the restriction to $SO(3)$ of the flat metric on $GL(3)$ given by

$$\langle\langle V, W \rangle\rangle = \mathrm{tr}\left(V J W^t\right),$$

where

$$J_{ij} = \int_{\mathbb{R}^3} \xi^i \xi^j \, dm.$$

(a) What are the geodesics of the Levi–Civita connection for this metric? Is $(GL(3), \langle\langle \cdot, \cdot \rangle\rangle)$ geodesically complete?
(b) The **Euler equation** and the **continuity equation** for an incompressible fluid with velocity field $u : \mathbb{R} \times \mathbb{R}^3 \to \mathbb{R}^3$ and pressure $p : \mathbb{R} \times \mathbb{R}^3 \to \mathbb{R}$ are

$$\frac{\partial u}{\partial t} + (u \cdot \nabla)u = -\nabla p,$$

$$\nabla \cdot u = 0,$$

where

$$\nabla = \left(\frac{\partial}{\partial x^1}, \frac{\partial}{\partial x^2}, \frac{\partial}{\partial x^3} \right)$$

is the usual operator of vector calculus.

Given a geodesic $S : \mathbb{R} \to GL(3)$, we define

$$x(t, \xi) = S(t)\xi,$$
$$u(t, x) = \dot{S}(t)\xi = \dot{S}(t)S^{-1}(t)x.$$

Show that the velocity field u satisfies the Euler equation (with $p = 0$), but not the continuity equation.

(c) Let $f : GL(3) \to \mathbb{R}$ be given by $f(S) = \det S$. Show that

$$\frac{\partial f}{\partial S_{ij}} = \mathrm{cof}(S)_{ij}$$

(where $\mathrm{cof}(S)$ is the matrix of the cofactors of S), and consequently

$$\frac{df}{dt} = (\det S) \, \mathrm{tr} \left(\dot{S} S^{-1} \right).$$

So the continuity equation is satisfied if we impose the constraint $\det S(t) = 1$.

(d) Show that the holonomic constraint $SL(3) \subset GL(3)$ satisfies the d'Alembert principle if and only if

$$\begin{cases} \mu \left(\ddot{S} \right) = \lambda(t) df \\ \det S = 1. \end{cases}$$

Assuming that J is invertible, show that the equation of motion can be rewritten as

$$\ddot{S} = \lambda \left(S^{-1} \right)^t J^{-1}.$$

(e) Show that the geodesics of $(SL(3), \langle\langle \cdot, \cdot \rangle\rangle)$ yield solutions of the Euler equation with

$$p = -\frac{\lambda}{2} x^t \left(S^{-1} \right)^t J^{-1} S^{-1} x$$

which also satisfy the continuity equation.

(**Remark:** More generally, it is possible to interpret the Euler equation on an open set $U \subset \mathbb{R}^n$ as a mechanical system on the group of diffeomorphisms of U (which is an infinite-dimensional Lie group); the continuity equation imposes the holonomic constraint corresponding to the subgroup of volume-preserving diffeomorphisms, and the pressure is the perfect reaction force associated to this constraint).

5.4 Non-holonomic Constraints

Some mechanical systems are subject to constraints which force the motions to proceed in certain admissible directions. To handle such constraints we must first introduce the corresponding geometric concept.

Definition 4.1 A **distribution** Σ of dimension m on a differentiable manifold M is a choice of an m-dimensional subspace $\Sigma_p \subset T_p M$ for each $p \in M$. The distribution is said to be **differentiable** if for all $p \in M$ there exists a neighborhood $U \ni p$ and vector fields $X_1, \ldots, X_m \in \mathfrak{X}(U)$ such that

$$\Sigma_q = \text{span} \left\{ (X_1)_q, \ldots, (X_m)_q \right\}$$

for all $q \in U$.

Equivalently, Σ is differentiable if for all $p \in M$ there exists a neighborhood $U \ni p$ and 1-forms $\omega^1, \ldots, \omega^{n-m} \in \Omega^1(U)$ such that

$$\Sigma_q = \ker \left(\omega^1 \right)_q \cap \cdots \cap \ker \left(\omega^{n-m} \right)_q$$

for all $p \in U$ [cf. Exercise 4.15(1)]. We will assume from this point on that all distributions are differentiable.

Definition 4.2 A **non-holonomic constraint** on a mechanical system $(M, \langle \cdot, \cdot \rangle, \mathcal{F})$ is a distribution Σ on M. A curve $c : I \subset \mathbb{R} \to M$ is said to be **compatible** with Σ if $\dot{c}(t) \in \Sigma_{c(t)}$ for all $t \in I$.

Example 4.3

(1) *(Wheel rolling without slipping)* Consider a vertical wheel of radius R rolling without slipping on a plane. Assuming that the motion takes place along a straight line, we can parameterize any position of the wheel by the position x of the contact point and the angle θ between a fixed radius of the wheel and the radius containing the contact point (cf. Fig. 5.6); hence the configuration space is $\mathbb{R} \times S^1$.

If the wheel is to rotate without slipping, we must require that $\dot{x} = R\dot{\theta}$ along any motion; this is equivalent to requiring that the motion be compatible with the distribution defined on $\mathbb{R} \times S^1$ by the vector field

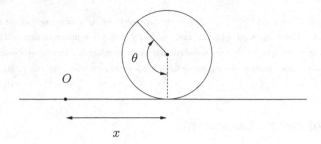

Fig. 5.6 Wheel rolling without slipping

$$X = R\frac{\partial}{\partial x} + \frac{\partial}{\partial \theta},$$

or, equivalently, by the kernel of the 1-form

$$\omega = dx - R d\theta.$$

(2) (*Ice skate*) A simple model for an ice skate is provided by a line segment which can either move along itself or rotate about its middle point. The position of the skate can be specified by the Cartesian coordinates (x, y) of the middle point and the angle θ between the skate and the x-axis (cf. Fig. 5.7); hence the config-uration space is $\mathbb{R}^2 \times S^1$.

If the skate can only move along itself, we must require that $(\dot{x}, \dot{y})$ be proportional to $(\cos\theta, \sin\theta)$; this is equivalent to requiring that the motion be compatible with the distribution defined on $\mathbb{R}^2 \times S^1$ by the vector fields

$$X = \cos\theta\frac{\partial}{\partial x} + \sin\theta\frac{\partial}{\partial y}, \qquad Y = \frac{\partial}{\partial \theta},$$

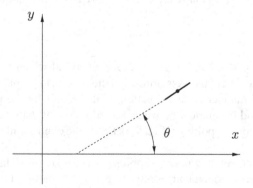

Fig. 5.7 Ice skate

or, equivalently, by the kernel of the 1-form

$$\omega = -\sin\theta\, dx + \cos\theta\, dy.$$

One may wonder whether there exists any connection between holonomic and non-holonomic constraints. To answer this question, we must make a small digression.

Definition 4.4 A **foliation** of dimension m on an n-dimensional differentiable manifold M is a family $\mathcal{F} = \{L_\alpha\}_{\alpha\in A}$ of subsets of M (called **leaves**) satisfying:

(1) $M = \cup_{\alpha\in A} L_\alpha$;
(2) $L_\alpha \cap L_\beta = \varnothing$ if $\alpha \neq \beta$;
(3) each leaf L_α is **pathwise connected**, that is, if $p, q \in L_\alpha$ then there exists a continuous curve $c : [0, 1] \to L_\alpha$ such that $c(0) = p$ and $c(1) = q$;
(4) for each point $p \in M$ there exists an open set $U \ni p$ and local coordinates $(x^1, \ldots, x^n) : U \to \mathbb{R}^n$ such that the connected components of the intersections of the leaves with U are the level sets of $(x^{m+1}, \ldots, x^n) : U \to \mathbb{R}^{n-m}$.

Remark 4.5 The coordinates $(x^1, \ldots, x^m)$ provide local coordinates on the leaves, which are therefore images of injective immersions. In particular, the leaves have well-defined m-dimensional tangent spaces at each point, and consequently any foliation of dimension m defines an m-dimensional distribution. Notice, however, that in general the leaves are not (embedded) submanifolds of M [cf. Exercise 4.15(2)].

Definition 4.6 An m-dimensional distribution Σ on a differential manifold M is said to be **integrable** if there exists an m-dimensional foliation $\mathcal{F} = \{L_\alpha\}_{\alpha\in A}$ on M such that

$$\Sigma_p = T_p L_p$$

for all $p \in M$, where L_p is the leaf containing p. The leaves of $\mathcal{F}$ are called the **integral submanifolds** of the distribution.

Integrable distributions are particularly simple. For instance, the set of points $q \in M$ which are accessible from a given point $p \in M$ by a curve compatible with the distribution is simply the leaf L_p through p. If the leaves are **embedded** submanifolds, then an integrable non-holonomic restriction reduces to a family of holonomic restrictions. For this reason, an integrable distribution is sometimes called a **semi-holonomic constraint**, whereas a non-integrable distribution is called a **true non-holonomic constraint**.

It is therefore important to have a criterion for identifying integrable distributions.

Definition 4.7 Let Σ be a distribution on a differentiable manifold M. A vector field $X \in \mathfrak{X}(M)$ is said to be **compatible** with Σ if $X_p \in \Sigma_p$ for all $p \in M$. We denote by $\mathfrak{X}(\Sigma)$ the linear subspace of $\mathfrak{X}(M)$ formed by all vector fields which are compatible with Σ.

Theorem 4.8 (Frobenius) *A distribution* Σ *is integrable if and only if* $X, Y \in \mathfrak{X}(\Sigma) \Rightarrow [X, Y] \in \mathfrak{X}(\Sigma)$.

Proof The proof of this theorem can be found in [War83] [see Exercise 4.15(3) for the "only if" part]. □

If Σ is locally given by m vector fields $X_1, \ldots, X_m$, then to check integrability it suffices to check whether $[X_i, X_j] = \sum_{k=1}^{m} C_{ij}^k X_k$ for locally defined functions C_{ij}^k [cf. Exercise 4.15(4)]. The next proposition provides an alternative criterion.

Proposition 4.9 *An m-dimensional distribution* Σ *on an n-manifold M is integrable if and only if*

$$d\omega^i \wedge \omega^1 \wedge \cdots \wedge \omega^{n-m} = 0 \qquad (i = 1, \ldots, n - m)$$

for all locally defined sets of differential forms $\{\omega^1, \ldots, \omega^{n-m}\}$ *whose kernels determine* Σ.

Proof Exercise 4.15(5). □

Since the condition of the Frobenius theorem is local, this condition needs to be checked only for sets of differential forms whose domains form an open cover of M.

Example 4.10

(1) (*Wheel rolling without slipping*) Recall that in this case the constraint is given by the kernel of the 1-form
$$\omega = dx - R d\theta.$$

Since $d\omega = 0$, we see that this is a semi-holonomic constraint, corresponding to an integrable distribution. The leaves of the distribution are the submanifolds with equation $x = x_0 + R\theta$.

(2) (*Ice skate*) Recall that in this case the constraint is given by the kernel of the 1-form
$$\omega = -\sin\theta dx + \cos\theta dy.$$

Since

$$d\omega \wedge \omega = (-\cos\theta d\theta \wedge dx - \sin\theta d\theta \wedge dy) \wedge (-\sin\theta dx + \cos\theta dy)$$
$$= -d\theta \wedge dx \wedge dy \neq 0,$$

we see that this is a true non-holonomic constraint.

In a Riemannian manifold $(M, \langle \cdot, \cdot \rangle)$, any distribution Σ determines an **orthogonal distribution** $\Sigma^\perp$, given by

$$\Sigma_p^\perp = (\Sigma_p)^\perp \subset T_p M.$$

Hence we have two orthogonal projections $^\top : TM \to \Sigma$ and $^\perp : TM \to \Sigma^\perp$. The set of all external forces $\mathcal{F} : TM \to T^*M$ satisfying

$$\mathcal{F}(v) = \mathcal{F}\left(v^\top\right)$$

for all $v \in TM$ is denoted by F_Σ.

Definition 4.11 A **reaction force** on a mechanical system with non-holonomic constraints $(M, \langle \cdot, \cdot \rangle, \mathcal{F}, \Sigma)$ is a force $\mathcal{R} \in F_\Sigma$ such that the solutions of the **generalized Newton equation**

$$\mu\left(\frac{D\dot{c}}{dt}\right) = (\mathcal{F} + \mathcal{R})(\dot{c})$$

with initial condition in Σ are compatible with Σ. The reaction force is said to be **perfect**, or to satisfy the **d'Alembert principle**, if

$$\mu^{-1}(\mathcal{R}(v)) \in \Sigma_p^\perp$$

for all $v \in T_p M$, $p \in M$.

Just like in the holonomic case, a reaction force is perfect if and only if it neither creates nor dissipates energy along any motion compatible with the constraint.

Theorem 4.12 *Given a mechanical system with non-holonomic constraints $(M, \langle \cdot, \cdot \rangle, \mathcal{F}, \Sigma)$, there exists a unique reaction force $\mathcal{R} \in F_\Sigma$ satisfying the d'Alembert principle.*

Proof We define the **second fundamental form** of the distribution Σ at a point $p \in M$ as the map $B : T_p M \times \Sigma_p \to \Sigma_p^\perp$ given by

$$B(v, w) = (\nabla_X Y)^\perp,$$

where $X \in \mathfrak{X}(M)$ and $Y \in \mathfrak{X}(\Sigma)$ satisfy $X_p = v$ and $Y_p = w$. To check that B is well defined, let $\{Z_1, \ldots, Z_n\}$ be a local orthonormal frame such that $\{Z_1, \ldots, Z_m\}$ is a basis for Σ and $\{Z_{m+1}, \ldots, Z_n\}$ is a basis for $\Sigma^\perp$. Then

$$\nabla_X Y = \nabla_X \left(\sum_{i=1}^m Y^i Z_i\right) = \sum_{i=1}^m \left((X \cdot Y^i) Z_i + \sum_{j,k=1}^n \Gamma_{ji}^k X^j Y^i Z_k\right),$$

where the functions Γ_{ij}^k are defined by

$$\nabla_{Z_i} Z_j = \sum_{k=1}^n \Gamma_{ij}^k Z_k.$$

Consequently,

$$B(v, w) = (\nabla_X Y)^\perp = \sum_{i=1}^{n} \sum_{j=1}^{m} \sum_{k=m+1}^{n} \Gamma_{ij}^k X^i Y^j Z_k$$

depends only on $v = X_p$ and $w = Y_p$, and is a bilinear map. Incidentally, the restriction of B to $\Sigma_p \times \Sigma_p$ is symmetric for all $p \in M$ if and only if

$$\Gamma_{ij}^k = \Gamma_{ji}^k \Leftrightarrow \langle \nabla_{Z_i} Z_j, Z_k \rangle = \langle \nabla_{Z_j} Z_i, Z_k \rangle \Leftrightarrow \langle [Z_i, Z_j], Z_k \rangle = 0$$

for all $i, j = 1, \ldots, m$ and all $k = m + 1, \ldots, n$, i.e. if and only if Σ is integrable. In this case, B is, of course, the second fundamental form of the leaves.

Let us assume that $\mathcal{R}$ exists. Then any motion $c : I \subset \mathbb{R} \to M$ with initial condition on Σ is compatible with Σ and satisfies

$$\frac{D\dot{c}}{dt} = \mu^{-1} (\mathcal{F}(\dot{c})) + \mu^{-1}(\mathcal{R}(\dot{c})).$$

The projection of this equation on $\Sigma^\perp$ yields

$$B(\dot{c}, \dot{c}) = \left(\mu^{-1}(\mathcal{F}(\dot{c})) \right)^\perp + \mu^{-1}(\mathcal{R}(\dot{c}))$$

(recall that $\frac{D\dot{c}}{dt} = \nabla_{\dot{c}}\dot{c}$). Therefore, if $\mathcal{R}$ exists then it must be given by

$$\mathcal{R}(v) = \mu \left(B(v, v) \right) - \mu \left(\left(\mu^{-1}(\mathcal{F}(v)) \right)^\perp \right)$$

for any $v \in \Sigma$, and by $\mathcal{R}(v) = \mathcal{R}\left(v^\top\right)$ for any $v \in TM$ (as $\mathcal{R} \in F_\Sigma$). This proves the uniqueness of $\mathcal{R}$.

To prove existence, we just have to show that for this choice of $\mathcal{R}$ the solutions of the generalized Newton equation with initial condition on Σ are compatible with Σ. Consider the system

$$\begin{cases} \dot{c} = \sum_{i=1}^{m} v^i Z_i \\ \frac{D\dot{c}}{dt} = \mu^{-1}(\mathcal{F}(\dot{c})) - \left(\mu^{-1}(\mathcal{F}(\dot{c})) \right)^\perp + B(\dot{c}, \dot{c}) \end{cases} . \tag{5.3}$$

When written in local coordinates, this is a system of first-order ODEs with $n + m$ unknowns $x^1(t), \ldots, x^n(t), v^1(t), \ldots, v^m(t)$. Since the second equation is just

$$\frac{D\dot{c}}{dt} = \left(\mu^{-1}(\mathcal{F}(\dot{c})) \right)^\top + \left(\frac{D\dot{c}}{dt} \right)^\perp \Leftrightarrow \left(\frac{D\dot{c}}{dt} \right)^\top = \left(\mu^{-1}(\mathcal{F}(\dot{c})) \right)^\top,$$

we see that this equation has only m nonvanishing components in the local frame $\{Z_1, \ldots, Z_n\}$. Therefore, (5.3) is a system of $(n + m)$ first-order ODEs on $n + m$ unknowns, and has a unique local solution for any initial condition. If $\dot{c}(0) \in \Sigma_{c(0)}$, we can always choose $v^1(0), \ldots, v^m(0)$ such that

$$\dot{c}(0) = \sum_{i=1}^{m} v^i(0) \, (Z_i)_{c(0)} \, .$$

The solution of (5.3) with initial condition $(x^1(0), \ldots, x^n(0), v^1(0), \ldots, v^m(0))$ must then, by uniqueness, be the solution of

$$\frac{D\dot{c}}{dt} = \mu^{-1}(\mathcal{F}(\dot{c})) + \mu^{-1}(\mathcal{R}(\dot{c}))$$

with initial condition $\dot{c}(0)$. On the other hand, it is, by construction, compatible with Σ. $\square$

Example 4.13 (Wheel rolling without slipping) Recall that in this case the constraint is given by the kernel of the 1-form

$$\omega = dx - R \, d\theta.$$

Since $\mu^{-1}\mathcal{R}$ is orthogonal to the constraint for a perfect reaction force $\mathcal{R}$, the constraint must be in the kernel of $\mathcal{R}$, and hence $\mathcal{R} = \lambda \omega$ for some smooth function $\lambda : TM \to \mathbb{R}$.

If the kinetic energy of the wheel is

$$K = \frac{M}{2} \left(v^x\right)^2 + \frac{I}{2} \left(v^\theta\right)^2$$

then

$$\mu\left(\frac{D\dot{c}}{dt}\right) = M\ddot{x} \, dx + I\ddot{\theta} \, d\theta.$$

Just to make things more interesting, consider a constant gravitational acceleration g and suppose that the plane on which the wheel rolls makes an angle α with respect to the horizontal (Fig. 5.8), so that there exists a conservative force with potential energy

$$U = Mgx \sin \alpha.$$

The equation of motion is therefore

$$\mu\left(\frac{D\dot{c}}{dt}\right) = -dU + \mathcal{R}(\dot{c}) \Leftrightarrow M\ddot{x} \, dx + I\ddot{\theta} \, d\theta = -Mg \sin \alpha \, dx + \lambda \, dx - \lambda R \, d\theta.$$

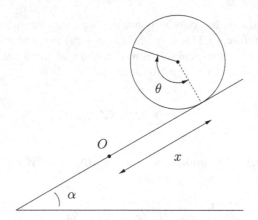

Fig. 5.8 Wheel rolling without slipping on an inclined plane

The motion of the wheel will be given by a solution of this equation which also satisfies the constraint equation, i.e. a solution of the system of ODEs

$$\begin{cases} M\ddot{x} = -Mg\sin\alpha + \lambda \\ I\ddot{\theta} = -R\lambda \\ \dot{x} = R\dot{\theta} \end{cases}.$$

This system is easily solved to yield

$$\begin{cases} x(t) = x_0 + v_0 t - \frac{\gamma}{2}t^2 \\ \theta(t) = \theta_0 + \frac{v_0}{R}t - \frac{\gamma}{2R}t^2 \\ \lambda = \frac{I\gamma}{R^2} \end{cases}$$

where

$$\gamma = \frac{g\sin\alpha}{1 + \frac{I}{MR^2}}$$

and x_0, v_0, θ_0 are integration constants.

Physically, the reaction force can be interpreted as a friction force exerted by the plane on the wheel. This force opposes the translational motion of the wheel but accelerates its spinning motion. Therefore, contrary to intuition, there is no dissipation of energy: all the translational kinetic energy lost by the wheel is restored as rotational kinetic energy.

A perfect reaction force guarantees, as one would expect, conservation of energy.

Theorem 4.14 *Let* $(M, \langle \cdot, \cdot \rangle, -dU, \Sigma)$ *be a conservative mechanical system with non-holonomic constraints. If the reaction force* $\mathcal{R}$ *satisfies the d'Alembert principle then the mechanical energy* $E := K + U$ *is constant along any motion with initial condition in* Σ.

Proof Exercise 4.15(7). $\qquad \square$

Exercise 4.15

(1) Show that an m-dimensional distribution Σ on an n-manifold M is differentiable if and only if for all $p \in M$ there exists a neighborhood $U \ni p$ and 1-forms $\omega^1, \ldots, \omega^{n-m} \in \Omega^1(U)$ such that

$$\Sigma_q = \ker\left(\omega^1\right)_q \cap \cdots \cap \ker\left(\omega^{n-m}\right)_q$$

for all $q \in U$.

(2) Show that the foliation

$$\mathcal{F} = \left\{(x, y) \in \mathbb{R}^2 \mid y = \sqrt{2}x + \alpha\right\}_{\alpha \in \mathbb{R}}$$

of $\mathbb{R}^2$ induces a foliation $\mathcal{F}'$ on $T^2 = \mathbb{R}^2/\mathbb{Z}^2$ whose leaves are not (embedded) submanifolds.

(3) Let Σ be an integrable distribution. Show that $X, Y \in \mathfrak{X}(\Sigma) \Rightarrow [X, Y] \in \mathfrak{X}(\Sigma)$.

(4) Using the Frobenius theorem show that an m-dimensional distribution Σ is integrable if and only if each local basis of vector fields $\{X_1, \ldots, X_m\}$ satisfies $[X_i, X_j] = \sum_{k=1}^{m} C_{ij}^k X_k$ for locally defined functions C_{ij}^k. (**Remark:** Since the condition of the Frobenius theorem is local, this condition needs to be checked only for local bases whose domains form an open cover of M).

(5) Prove Proposition 4.9. (**Hint:** Recall from Exercise 3.8(2) in Chap. 2 that $d\omega(X, Y) = X \cdot \omega(Y) - Y \cdot \omega(X) - \omega([X, Y])$ for any $\omega \in \Omega^1(M)$ and $X, Y \in \mathfrak{X}(M)$).

(6) Let M be an n-dimensional differentiable manifold with an affine connection ∇. Show that the parallel transport of vectors is determined by a distribution Σ on TM, which is integrable if and only if the curvature of ∇ vanishes.

(7) Prove Theorem 4.14.

(8) (*Ice skate*) Recall that our model for an ice skate is given by the non-holonomic constraint Σ defined on $\mathbb{R}^2 \times S^1$ by the kernel of the 1-form $\omega = -\sin\theta dx + \cos\theta dy$.

  (a) Show that the ice skate can access all points in the configuration space: given two points $p, q \in \mathbb{R}^2 \times S^1$ there exists a piecewise smooth curve $c : [0, 1] \to \mathbb{R}^2 \times S^1$ compatible with Σ such that $c(0) = p$ and $c(1) = q$. Why does this show that Σ is non-integrable?

  (b) Assuming that the kinetic energy of the skate is

$$K = \frac{M}{2}\left((v^x)^2 + (v^y)^2\right) + \frac{I}{2}(v^\theta)^2$$

and that the reaction force is perfect, show that the skate moves with constant speed along straight lines or circles. What is the physical interpretation of the reaction force?

(c) Determine the motion of the skate moving on an inclined plane, i.e. subject to a potential energy $U = Mg \sin \alpha \, x$.

(9) Consider a vertical wheel of radius R moving on a plane.

(a) Show that the non-holonomic constraint corresponding to the condition of rolling without slipping or sliding is the distribution determined on the configuration space $\mathbb{R}^2 \times S^1 \times S^1$ by the 1-forms

$$\omega^1 = dx - R \cos \varphi \, d\psi, \qquad \omega^2 = dy - R \sin \varphi \, d\psi,$$

where (x, y, ψ, φ) are the local coordinates indicated in Fig. 5.9.

(b) Assuming that the kinetic energy of the wheel is

$$K = \frac{M}{2} \left((v^x)^2 + (v^y)^2 \right) + \frac{I}{2} (v^\psi)^2 + \frac{J}{2} (v^\varphi)^2$$

and that the reaction force is perfect, show that the wheel moves with constant speed along straight lines or circles. What is the physical interpretation of the reaction force?

(c) Determine the motion of the vertical wheel moving on an inclined plane, i.e. subject to a potential energy $U = Mg \sin \alpha \, x$.

(10) Consider a sphere of radius R and mass M rolling without slipping on a plane.

(a) Show that the condition of rolling without slipping is

$$\dot{x} = R\omega^y, \qquad \dot{y} = -R\omega^x,$$

where (x, y) are the Cartesian coordinates of the contact point on the plane and ω is the angular velocity of the sphere.

(b) Show that if the sphere's mass is symmetrically distributed then its kinetic energy is

$$K = \frac{M}{2} \left(\dot{x}^2 + \dot{y}^2 \right) + \frac{I}{2} \langle \omega, \omega \rangle,$$

where I is the sphere's moment of inertia and $\langle \cdot, \cdot \rangle$ is the Euclidean inner product.

(c) Using ω as coordinates on the fibers of $TSO(3)$, show that

$$\frac{D\dot{c}}{dt} = \ddot{x} \frac{\partial}{\partial x} + \ddot{y} \frac{\partial}{\partial y} + \dot{\omega}.$$

(**Hint:** Recall from Exercise 4.8(3) in Chap. 3 that the integral curves of left-invariant vector fields on a Lie group with a bi-invariant metric are geodesics).

(d) Since we are identifying the fibers of $TSO(3)$ with $\mathbb{R}^3$, we can use the Euclidean inner product to also identify the fibers of $T^*SO(3)$ with $\mathbb{R}^3$.

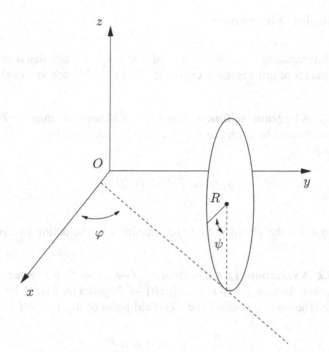

Fig. 5.9 Vertical wheel on a plane

Show that under this identification the non-holonomic constraint yielding the condition of rolling without slipping is the distribution determined by the kernels of the 1-forms

$$\theta^x := dx - R\, e_y, \qquad \theta^y := dy + R\, e_x$$

(where $\{e_x, e_y, e_z\}$ is the canonical basis of $\mathbb{R}^3$). Is this distribution integrable? (**Hint:** Show that any two points of $\mathbb{R}^2 \times SO(3)$ can be connected by a piecewise smooth curve compatible with the distribution).

(e) Show that the sphere moves along straight lines with constant speed and constant angular velocity orthogonal to its motion.

(f) Determine the motion of the sphere moving on an inclined plane, i.e. subject to a potential energy $U = Mg \sin \alpha\, x$.

(11) (*The golfer dilemma*) Show that the center of a symmetric sphere of radius R, mass M and moment of inertia I rolling without slipping inside a vertical cylinder of radius $R + a$ moves with constant angular velocity with respect to the axis of the cylinder while oscillating up and down with a frequency $\sqrt{\frac{I}{I+MR^2}}$ times the frequency of the angular motion.

5.5 Lagrangian Mechanics

Let M be a differentiable manifold, $p, q \in M$ and $a, b \in \mathbb{R}$ such that $a < b$. Let us denote by $\mathcal{C}$ the set of differentiable curves $c : [a, b] \to M$ such that $c(a) = p$ and $c(b) = q$.

Definition 5.1 A **Lagrangian function** on M is a differentiable map $L : TM \to \mathbb{R}$. The **action** determined by L on $\mathcal{C}$ is the map $S : \mathcal{C} \to \mathbb{R}$ given by

$$S(c) := \int_a^b L(\dot{c}(t)) \, dt.$$

We can look for the global minima (or maxima) of the action by considering curves on $\mathcal{C}$.

Definition 5.2 A **variation** of $c \in \mathcal{C}$ is a map $\gamma : (-\varepsilon, \varepsilon) \to \mathcal{C}$ (for some $\varepsilon > 0$) such that $\gamma(0) = c$ and the map $\tilde{\gamma} : (-\varepsilon, \varepsilon) \times [a, b] \to M$ given by $\tilde{\gamma}(s, t) := \gamma(s)(t)$ is differentiable. The curve c is said to be a **critical point** of the action if

$$\frac{d}{ds}_{|s=0} S(\gamma(s)) = 0$$

for any variation γ of c.

Notice that the global minima (or maxima) of S must certainly be attained at critical points. However, a critical point is not necessarily a point of minimum (or maximum). It turns out that the critical points of the action are solutions of second-order ODEs.

Theorem 5.3 *The curve $c \in \mathcal{C}$ is a critical point of the action determined by the Lagrangian $L : TM \to \mathbb{R}$ if and only if it satisfies the **Euler–Lagrange equations***

$$\frac{d}{dt}\left(\frac{\partial L}{\partial v^i}(x(t), \dot{x}(t))\right) - \frac{\partial L}{\partial x^i}(x(t), \dot{x}(t)) = 0 \qquad (i = 1, \dots, n)$$

for any local chart $(x^1, \dots, x^n)$ on M, where $(x^1, \dots, x^n, v^1, \dots, v^n)$ is the corresponding local chart induced on TM.

Proof Assume first that the image of c is contained on the domain of a local chart $(x^1, \dots, x^n)$. Let $\gamma : (-\varepsilon, \varepsilon) \to \mathcal{C}$ be a variation of c. Setting $x(s, t) := (x \circ \tilde{\gamma})(s, t)$, we have

$$S(\gamma(s)) = \int_a^b L\left(x(s, t), \frac{\partial x}{\partial t}(s, t)\right) dt,$$

and hence

$$\frac{d}{ds}\Big|_{s=0} S(\gamma(s)) = \int_a^b \sum_{i=1}^n \frac{\partial L}{\partial x^i}\left(x(0,t), \frac{\partial x}{\partial t}(0,t)\right) \frac{\partial x^i}{\partial s}(0,t)\,dt$$

$$+ \int_a^b \sum_{i=1}^n \frac{\partial L}{\partial v^i}\left(x(0,t), \frac{\partial x}{\partial t}(0,t)\right) \frac{\partial^2 x^i}{\partial s\,\partial t}(0,t)\,dt.$$

Differentiating the relations $x(s,a) = x(p)$, $x(s,b) = x(q)$ with respect to s one obtains

$$\frac{\partial x}{\partial s}(0,a) = \frac{\partial x}{\partial s}(0,b) = 0.$$

Consequently, the second integral above can be integrated by parts to yield

$$-\int_a^b \sum_{i=1}^n \frac{d}{dt}\left(\frac{\partial L}{\partial v^i}\left(x(0,t), \frac{\partial x}{\partial t}(0,t)\right)\right) \frac{\partial x^i}{\partial s}(0,t)\,dt,$$

and hence

$$\frac{d}{ds}\Big|_{s=0} S(\gamma(s)) = \int_a^b \sum_{i=1}^n \left(\frac{\partial L}{\partial x^i}(x(t), \dot{x}(t)) - \frac{d}{dt}\left(\frac{\partial L}{\partial v^i}(x(t), \dot{x}(t))\right)\right) w^i(t)\,dt,$$

where we have set $x(t) := (x \circ c)(t)$ and $w(t) := \frac{\partial x}{\partial s}(0,t)$. This shows that if c satisfies the Euler–Lagrange equations then c is a critical point of the action.

To show the converse, we notice that any smooth function $w : [a,b] \to \mathbb{R}^n$ satisfying $w(a) = w(b) = 0$ determines a variation $\gamma : (-\varepsilon, \varepsilon) \to C$ given in local coordinates by $x(s,t) = x(t) + sw(t)$, satisfying $\frac{\partial x}{\partial s}(0,t) = w(t)$. In particular, if $\rho : [a,b] \to \mathbb{R}$ is a smooth positive function with $\rho(a) = \rho(b) = 0$, we can take

$$w^i(t) := \rho(t)\left(\frac{\partial L}{\partial x^i}(x(t), \dot{x}(t)) - \frac{d}{dt}\left(\frac{\partial L}{\partial v^i}(x(t), \dot{x}(t))\right)\right).$$

Therefore if c is a critical point of the action we must have

$$\int_a^b \sum_{i=1}^n \left(\frac{\partial L}{\partial x^i}(x(t), \dot{x}(t)) - \frac{d}{dt}\left(\frac{\partial L}{\partial v^i}(x(t), \dot{x}(t))\right)\right)^2 \rho(t)\,dt = 0,$$

and hence c must satisfy the Euler–Lagrange equations.

The general case (in which the image of c is not contained in the domain of the local chart) is left as an exercise. $\qquad\square$

Corollary 5.4 *The motions of any conservative mechanical system* $(M, \langle \cdot, \cdot \rangle, -dU)$ *are the critical points of the action determined by the Lagrangian* $L := K - U$.   □

Therefore we can find motions of conservative systems by looking for minima, say, of the action. This variational approach is often very useful.

The energy conservation in a conservative system is, in fact, a particular case of a more general conservation law, which holds for any Lagrangian. Before we state it we need the following definitions.

Definition 5.5 The **fiber derivative** of a Lagrangian function $L : TM \to \mathbb{R}$ at $v \in T_pM$ is the linear map $(\mathbb{F}L)_v : T_pM \to \mathbb{R}$ given by

$$(\mathbb{F}L)_v(w) := \frac{d}{dt}\Big|_{t=0} L(v + tw)$$

for all $w \in T_pM$.

Definition 5.6 If $L : TM \to \mathbb{R}$ is a Lagrangian function then its associated **Hamiltonian function** $H : TM \to \mathbb{R}$ is defined as

$$H(v) := (\mathbb{F}L)_v(v) - L(v).$$

Theorem 5.7 *The Hamiltonian function is constant along the solutions of the Euler–Lagrange equations.*

Proof In local coordinates we have

$$H(x, v) = \sum_{i=1}^{n} v^i \frac{\partial L}{\partial v^i}(x, v) - L(x, v).$$

Consequently, if $c : I \subset \mathbb{R} \to M$ is a solution of the Euler–Lagrange equations, given in local coordinates by $x = x(t)$, then

$$\frac{d}{dt}(H(\dot{c}(t))) = \frac{d}{dt}\left(\sum_{i=1}^{n} \dot{x}^i(t) \frac{\partial L}{\partial v^i}(x(t), \dot{x}(t)) - L(x(t), \dot{x}(t)) \right)$$

$$= \sum_{i,j=1}^{n} \ddot{x}^i(t) \frac{\partial L}{\partial v^i}(x(t), \dot{x}(t)) + \sum_{i=1}^{n} \dot{x}^i(t) \frac{d}{dt}\left(\frac{\partial L}{\partial v^i}(x(t), \dot{x}(t)) \right)$$

$$- \sum_{i=1}^{n} \dot{x}^i(t) \frac{\partial L}{\partial x^i}(x(t), \dot{x}(t)) - \sum_{i=1}^{n} \ddot{x}^i(t) \frac{\partial L}{\partial v^i}(x(t), \dot{x}(t)) = 0.$$

                                                                                            □

Example 5.8 If $(M, \langle \cdot, \cdot \rangle, -dU)$ is a conservative mechanical system then its motions are the solutions of the Euler–Lagrange equations for the Lagrangian $L : TM \to \mathbb{R}$ given by

$$L(v) = \frac{1}{2}\langle v, v \rangle - U(\pi(v))$$

(where $\pi : TM \to M$ is the canonical projection). Clearly,

$$(\mathbb{F}L)_v(w) = \frac{1}{2}\frac{d}{dt}\Big|_{t=0} \langle v + tw, v + tw \rangle = \langle v, w \rangle,$$

and hence

$$H(v) = \langle v, v \rangle - \frac{1}{2}\langle v, v \rangle + U(\pi(v)) = \frac{1}{2}\langle v, v \rangle + U(\pi(v))$$

is the mechanical energy.

The Lagrangian formulation is particularly useful for exploring the relation between symmetry and conservation laws.

Definition 5.9 Let G be a Lie group acting on a manifold M. The Lagrangian $L : TM \to \mathbb{R}$ is said to be **G-invariant** if

$$L\left((dg)_p v\right) = L(v)$$

for all $v \in T_pM$, $p \in M$ and $g \in G$ (where $g : M \to M$ is the map $p \mapsto g \cdot p$).

We will now show that if a Lagrangian is G-invariant then to each element $V \in \mathfrak{g}$ there corresponds a conserved quantity. To do so, we need the following definitions.

Definition 5.10 Let G be a Lie group acting on a manifold M. The **infinitesimal action** of $V \in \mathfrak{g}$ on M is the vector field $X^V \in \mathfrak{X}(M)$ defined as

$$X_p^V := \frac{d}{dt}\Big|_{t=0} (\exp(tV) \cdot p) = \left(dA_p\right)_e V,$$

where $A_p : G \to M$ is the map $A_p(g) = g \cdot p$.

Theorem 5.11 (Noether) *Let G be a Lie group acting on a manifold M. If $L : TM \to \mathbb{R}$ is G-invariant then $J^V : TM \to \mathbb{R}$ defined as $J^V(v) := (\mathbb{F}L)_v\left(X^V\right)$ is constant along the solutions of the Euler–Lagrange equations for all $V \in \mathfrak{g}$.*

Proof Choose local coordinates $(x^1, \ldots, x^n)$ on M and let $(y^1, \ldots, y^m)$ be local coordinates centered at $e \in G$. Let $A : G \times M \to M$ be the action of G on M, written in these local coordinates as

$$\left(A^1\left(x^1, \ldots, x^n, y^1, \ldots, y^m\right), \ldots, A^n\left(x^1, \ldots, x^n, y^1, \ldots, y^m\right)\right).$$

Then the infinitesimal action of $V = \sum_{a=1}^m V^a \frac{\partial}{\partial y^a}$ has components

$$X^i(x) = \sum_{a=1}^{m} \frac{\partial A^i}{\partial y^a}(x, 0)V^a.$$

Since L is G-invariant, we have

$$L\left(A^1(x, y), \ldots, A^n(x, y), \sum_{i=1}^{n} \frac{\partial A^1}{\partial x^i}(x, y)v^i, \ldots, \sum_{i=1}^{n} \frac{\partial A^n}{\partial x^i}(x, y)v^i\right)$$

$$= L\left(x^1, \ldots, x^n, v^1, \ldots, v^n\right).$$

Setting $y = y(t)$ in the above identity, where $\left(y^1(t), \ldots, y^m(t)\right)$ is the expression of the curve $\exp(tV)$ in local coordinates, and differentiating with respect to t at $t = 0$, we obtain

$$\sum_{i=1}^{n}\sum_{a=1}^{m} \frac{\partial L}{\partial x^i}(x, v)\frac{\partial A^i}{\partial y^a}(x, 0)V^a + \sum_{i,j=1}^{n}\sum_{a=1}^{m} \frac{\partial L}{\partial v^i}(x, v)\frac{\partial^2 A^i}{\partial y^a \partial x^j}(x, 0)v^j V^a = 0$$

$$\Leftrightarrow \sum_{i=1}^{n} \frac{\partial L}{\partial x^i}(x, v)X^i(x) + \sum_{i,j=1}^{n} \frac{\partial L}{\partial v^i}(x, v)\frac{\partial X^i}{\partial x^j}(x)v^j = 0.$$

In these coordinates,

$$J^V(x, v) = \sum_{i=1}^{n} \frac{\partial L}{\partial v^i}(x, v)X^i(x).$$

Therefore, if $c : I \subset \mathbb{R} \to M$ is a solution of the Euler–Lagrange equations, given in local coordinates by $x = x(t)$, we have

$$\frac{d}{dt}\left(J^V(\dot{c}(t))\right) = \frac{d}{dt}\left(\sum_{i=1}^{n} \frac{\partial L}{\partial v^i}(x(t), \dot{x}(t))X^i(x(t))\right)$$

$$= \sum_{i=1}^{n} \frac{d}{dt}\left(\frac{\partial L}{\partial v^i}(x(t), \dot{x}(t))\right) X^i(x(t)) + \sum_{i,j=1}^{n} \frac{\partial L}{\partial v^i}(x(t), \dot{x}(t))\frac{\partial X^i}{\partial x^j}(x(t))\dot{x}^j(t)$$

$$= \sum_{i=1}^{n} \frac{d}{dt}\left(\frac{\partial L}{\partial v^i}(x(t), \dot{x}(t))\right) X^i(x(t)) - \sum_{i=1}^{n} \frac{\partial L}{\partial x^i}(x(t), \dot{x}(t))X^i(x(t)) = 0. \qquad \square$$

Remark 5.12 Notice that the map $\mathfrak{g} \ni V \mapsto X^V \in \mathfrak{X}(M)$ is linear. Since $(\mathbb{F}L)_v$ is also linear, we can see J^V as a linear map $\mathfrak{g} \ni V \mapsto J^V \in C^\infty(TM)$. Therefore the Noether theorem yields $m = \dim \mathfrak{g}$ independent conserved quantities.

Example 5.13 Consider a conservative mechanical system consisting of k particles with masses $m_1, \ldots, m_k$ moving in $\mathbb{R}^3$ under a potential energy $U : \mathbb{R}^{3k} \to \mathbb{R}$ which depends only on the distances between them. The motions of the system are the solutions of the Euler–Lagrange equations obtained from the Lagrangian $L : T\mathbb{R}^{3k} \to \mathbb{R}$ given by

$$L(x_1, \ldots, x_k, v_1, \ldots, v_k) = \frac{1}{2} \sum_{i=1}^{k} m_i \langle v_i, v_i \rangle - U(x_1, \ldots, x_k).$$

This Lagrangian is clearly $SO(3)$-invariant, where the action of $SO(3)$ on $\mathbb{R}^{3k}$ is defined through

$$S \cdot (x_1, \ldots, x_k) = (Sx_1, \ldots, Sx_k).$$

The infinitesimal action of $V \in \mathfrak{so}(3)$ is the vector field

$$X^V_{(x_1, \ldots, x_k)} = (Vx_1, \ldots, Vx_k) = (\Omega(V) \times x_1, \ldots, \Omega(V) \times x_k),$$

where $\Omega : \mathfrak{so}(3) \to \mathbb{R}^3$ is the isomorphism in Lemma 3.9. On the other hand,

$$(\mathbb{F}L)_{(v_1, \ldots, v_k)}(w_1, \ldots, w_k) = \sum_{i=1}^{k} m_i \langle v_i, w_i \rangle.$$

Therefore, the Noether theorem guarantees that the quantity

$$J^V = \sum_{i=1}^{k} m_i \langle \dot{x}_i, \Omega(V) \times x_i \rangle = \sum_{i=1}^{k} m_i \langle \Omega(V), x_i \times \dot{x}_i \rangle = \left\langle \Omega(V), \sum_{i=1}^{k} m_i x_i \times \dot{x}_i \right\rangle$$

is conserved along the motion of the system for any $V \in \mathfrak{so}(3)$. In other words, the system's **total angular momentum**

$$Q := \sum_{i=1}^{k} m_i x_i \times \dot{x}_i$$

is conserved.

Exercise 5.14

(1) Complete the proof of Theorem 5.3.
(2) Let $(M, \langle \cdot, \cdot \rangle)$ be a Riemannian manifold. Show that the critical points of the arclength, i.e. of the action determined by the Lagrangian $L : TM \to \mathbb{R}$ given by

$$L(v) = \langle v, v \rangle^{\frac{1}{2}}$$

(where we must restrict the action to curves with nonvanishing velocity) are reparameterized geodesics.
(3) (*Brachistochrone curve*) A particle with mass m moves on a curve $y = y(x)$ under the action of a constant gravitational field, corresponding to the potential energy $U = mgy$. The curve satisfies $y(0) = y(d) = 0$ and $y(x) < 0$ for $0 < x < d$.

(a) Assuming that the particle is set free at the origin with zero velocity, show that its speed at each point is

$$v = \sqrt{-2gy},$$

and that therefore the travel time between the origin and point $(d, 0)$ is

$$S = (2g)^{-\frac{1}{2}} \int\limits_0^d \left(1 + y'^2\right)^{\frac{1}{2}} (-y)^{-\frac{1}{2}} dx,$$

where $y' = \frac{dy}{dx}$.

(b) Show that the curve $y = y(x)$ which corresponds to the minimum travel time satisfies the differential equation

$$\frac{d}{dx}\left[\left(1 + y'^2\right) y\right] = 0.$$

(c) Check that the solution of this equation satisfying $y(0) = y(d) = 0$ is given parametrically by

$$\begin{cases} x = R\theta - R\sin\theta \\ y = -R + R\cos\theta \end{cases}$$

where $d = 2\pi R$. (**Remark:** This curve is called a **cycloid**, because it is the curved traced out by a point on a circle which rolls without slipping on the xx-axis).

(4) (*Charged particle in a stationary electromagnetic field*) The motion of a particle with mass $m > 0$ and charge $e \in \mathbb{R}$ in a stationary electromagnetic field is determined by the Lagrangian $L : T\mathbb{R}^3 \to \mathbb{R}$ given by

$$L = \frac{1}{2}m\langle v, v\rangle + e\langle A, v\rangle - e\,\Phi,$$

where $\langle \cdot, \cdot \rangle$ is the Euclidean inner product, $\Phi \in C^\infty(\mathbb{R}^3)$ is the **electric potential** and $A \in \mathfrak{X}(\mathbb{R}^3)$ is the **magnetic vector potential**.

(a) Show that the equations of motion are

$$m\ddot{x} = eE + e\dot{x} \times B,$$

where $E = -\operatorname{grad}\Phi$ is the **electric field** and $B = \operatorname{curl} A$ is the **magnetic field**.

(b) Write an expression for the Hamiltonian function and use the equations of motion to check that it is constant along any motion.

(5) (*Restricted 3-body problem*) Consider two gravitating particles moving in circular orbit around their common center of mass. We choose our units so that the masses of the particles are $0 < \mu < 1$ and $1 - \mu$, the distance between them is 1 and the orbital angular velocity is also 1. Identifying the plane of the orbit with $\mathbb{R}^2$, with the center of mass at the origin, we can choose fixed positions $p_1 = (1 - \mu, 0)$ and $p_2 = (-\mu, 0)$ for the particles in the rotating frame where they are at rest.

(a) Use Exercise 3.20(11) to show that in this frame the equations of motion of a third particle with negligible mass m moving in the plane of the orbit are

$$\begin{cases} \ddot{x} = \dfrac{F_x}{m} + x + 2\dot{y} \\[4mm] \ddot{y} = \dfrac{F_y}{m} + y - 2\dot{x} \end{cases},$$

where (F_x, F_y) is the force on m as measured in the rotating frame.

(b) Assume that the only forces on m are the gravitational forces produced by μ and $1 - \mu$, so that

$$\begin{cases} \dfrac{F_x}{m} = -\dfrac{\mu}{r_1{}^3}(x - 1 + \mu) - \dfrac{1 - \mu}{r_2{}^3}(x + \mu) \\[4mm] \dfrac{F_y}{m} = -\dfrac{\mu}{r_1{}^3}y - \dfrac{1 - \mu}{r_2{}^3}y = 0 \end{cases},$$

where $r_1, r_2 : \mathbb{R}^2 \to \mathbb{R}$ are the Euclidean distances to p_1, p_2. Show that the equations of motion are the Euler–Lagrange equations for the Lagrangian $L : T\left(\mathbb{R}^2 \setminus \{p_1, p_2\}\right) \to \mathbb{R}$ given by

$$L\left(x, y, v^x, v^y\right) = \frac{1}{2}\left(\left(v^x\right)^2 + \left(v^y\right)^2\right) + xv^y - yv^x$$
$$+ \frac{1}{2}\left(x^2 + y^2\right) + \frac{\mu}{r_1} + \frac{1 - \mu}{r_2}.$$

(c) Find the Hamiltonian function. (**Remark:** The fact that this function remains constant gives the so-called **Tisserand criterion** for identifying the same comet before and after a close encounter with Jupiter).

(d) Compute the equilibrium points (i.e. the points corresponding to stationary solutions) which are not on the x-axis. How many equilibrium points are there in the x-axis?

(e) Show that the linearization of the system around the equilibrium points not in the x-axis is

$$
\begin{cases}
\ddot{\xi} - 2\dot{\eta} = \dfrac{3}{4}\xi \pm \dfrac{3\sqrt{3}}{4}(1 - 2\mu)\eta \\[4mm]
\ddot{\eta} + 2\dot{\xi} = \pm \dfrac{3\sqrt{3}}{4}(1 - 2\mu)\xi + \dfrac{9}{4}\eta
\end{cases},
$$

and show that these equilibrium points are unstable for

$$
\frac{1}{2}\left(1 - \frac{\sqrt{69}}{9}\right) < \mu < \frac{1}{2}\left(1 + \frac{\sqrt{69}}{9}\right).
$$

(6) Consider the mechanical system in Example 5.13.

   (a) Use the Noether theorem to prove that the **total linear momentum**

$$
P := \sum_{i=1}^{k} m_i \dot{x}_i
$$

   is conserved along the motion.

   (b) Show that the system's **center of mass**, defined as the point

$$
X = \frac{\sum_{i=1}^{k} m_i x_i}{\sum_{i=1}^{k} m_i},
$$

   moves with constant velocity.

(7) Generalize Example 5.13 to the case in which the particles move in an arbitrary Riemannian manifold $(M, \langle \cdot, \cdot \rangle)$, by showing that given any Killing vector field $X \in \mathfrak{X}(M)$ [cf. Exercise 3.3(8) in Chap. 3] the quantity

$$
J^X = \sum_{i=1}^{k} m_i \langle \dot{c}_i, X \rangle
$$

   is conserved, where $c_i : I \subset \mathbb{R} \to M$ is the motion of the particle with mass m_i.

(8) Consider the action of $SO(3)$ on itself by left multiplication.

   (a) Show that the infinitesimal action of $B \in \mathfrak{so}(3)$ is the **right**-invariant vector field determined by B.

   (b) Use the Noether theorem to show that the angular momentum of the free rigid body is constant.

(9) Consider a satellite equipped with a small rotor, i.e. a cylinder which can spin freely about its axis. When the rotor is locked the satellite can be modeled by a free rigid body with inertia tensor I. The rotor's axis passes through the satellite's

center of mass, and its direction is given by the unit vector e. The rotor's mass is symmetrically distributed around the axis, producing a moment of inertia J.

(a) Show that the configuration space for the satellite with unlocked rotor is the Lie group $SO(3) \times S^1$, and that its motion is a geodesic of the left-invariant metric corresponding to the kinetic energy

$$K = \frac{1}{2}\langle I\Omega, \Omega \rangle + \frac{1}{2}J\varpi^2 + J\varpi\langle \Omega, e \rangle,$$

where the $\Omega \in \mathbb{R}^3$ is the satellite's angular velocity as seen on the satellite's frame and $\varpi \in \mathbb{R}$ is the rotor's angular speed around its axis.

(b) Use the Noether theorem to show that $l = J(\varpi + \langle \Omega, e \rangle) \in \mathbb{R}$ and $p = S(I\Omega + J\varpi e) \in \mathbb{R}^3$ are conserved along the motion of the satellite with unlocked rotor, where $S : \mathbb{R} \to SO(3)$ describes the satellite's orientation.

5.6 Hamiltonian Mechanics

We will now see that under certain conditions it is possible to study the Euler–Lagrange equations as a flow on the cotangent bundle with special geometric properties.

Let M be an n-dimensional manifold. The set

$$TM \oplus T^*M := \bigcup_{q \in M} T_q M \times T_q^* M$$

has an obvious differentiable structure: if $(x^1, \ldots, x^n)$ are local coordinates on M then $(x^1, \ldots, x^n, v^1, \ldots, v^n, p_1, \ldots, p_n)$ are the local coordinates on $TM \oplus T^*M$ which label the pair $(v, \omega) \in T_q M \times T_q^* M$, where

$$v = \sum_{i=1}^n v^i \frac{\partial}{\partial x^i}, \qquad \omega = \sum_{i=1}^n p_i dx^i,$$

and $q \in M$ is the point with coordinates $(x^1, \ldots, x^n)$. For this differentiable structure, the maps $\pi_1 : TM \oplus T^*M \to TM$ and $\pi_2 : TM \oplus T^*M \to T^*M$ given by $\pi_1(v, \omega) = v$ and $\pi_2(v, \omega) = \omega$ are submersions.

Definition 6.1 The **extended Hamiltonian function** corresponding to a Lagrangian $L : TM \to \mathbb{R}$ is the map $\widetilde{H} : TM \oplus T^*M \to \mathbb{R}$ given by

$$\widetilde{H}(v, \omega) := \omega(v) - L(v).$$

In local coordinates, we have

$$\widetilde{H}\left(x^1,\dots,x^n,v^1,\dots,v^n,p_1,\dots,p_n\right) = \sum_{i=1}^{n} p_i v^i - L\left(x^1,\dots,x^n,v^1,\dots,v^n\right),$$

and hence

$$d\widetilde{H} = \sum_{i=1}^{n}\left(p_i - \frac{\partial L}{\partial v^i}\right)dv^i + \sum_{i=1}^{n} v^i dp_i - \sum_{i=1}^{n}\frac{\partial L}{\partial x^i}dx^i.$$

Thus any critical point of any restriction of $\widetilde{H}$ to a submanifold of the form $T_q M \times \{\omega\}$ (for fixed $q \in M$ and $\omega \in T_q^* M$) must satisfy

$$p_i = \frac{\partial L}{\partial v^i}\left(x^1,\dots,x^n,v^1,\dots,v^n\right) \qquad (i = 1,\dots,n).$$

It follows that the set of all such critical points is naturally a $2n$-dimensional submanifold $S \subset TM \oplus T^*M$ such that $\pi_{1|_S} : S \to TM$ is a diffeomorphism. If $\pi_{2|_S} : S \to T^*M$ is also a diffeomorphism then the Lagrangian is said to be **hyper-regular**. In this case, $\pi_{2|_S} \circ \pi_{1|_S}^{-1} : TM \to T^*M$ is a fiber-preserving diffeomorphism, called the **Legendre transformation**.

Given a hyper-regular Lagrangian, we can use the maps $\pi_{1|_S}$ and $\pi_{2|_S}$ to make the identifications $TM \cong S \cong T^*M$. Since the Hamiltonian function $H : TM \to \mathbb{R}$ is clearly related to the extended Hamiltonian function through $H = \widetilde{H} \circ \pi_{1|_S}^{-1}$, we can under these identifications simply write $H = \widetilde{H}|_S$. Therefore

$$dH = \sum_{i=1}^{n} v^i dp_i - \sum_{i=1}^{n}\frac{\partial L}{\partial x^i}dx^i$$

(here we must think of $\left(x^1,\dots,x^n,v^1,\dots,v^n,p_1,\dots,p_n\right)$ as local functions on S such that **both** $\left(x^1,\dots,x^n,v^1,\dots,v^n\right)$ and $\left(x^1,\dots,x^n,p_1,\dots,p_n\right)$ are local coordinates). On the other hand, thinking of H as a function on the cotangent bundle, we obtain

$$dH = \sum_{i=1}^{n}\frac{\partial H}{\partial x^i}dx^i + \sum_{i=1}^{n}\frac{\partial H}{\partial p_i}dp_i.$$

Therefore we must have

$$\begin{cases} \dfrac{\partial H}{\partial x^i} = -\dfrac{\partial L}{\partial x^i} \\[2mm] \dfrac{\partial H}{\partial p_i} = v^i \end{cases} \qquad (i = 1,\dots,n),$$

where the partial derivatives of the Hamiltonian must be computed with respect to the local coordinates $(x^1, \ldots, x^n, p_1, \ldots, p_n)$ and the partial derivatives of the Lagrangian must be computed with respect to the local coordinates $(x^1, \ldots, x^n, v^1, \ldots, v^n)$.

Proposition 6.2 *The Euler–Lagrange equations for a hyper-regular Lagrangian $L : TM \to \mathbb{R}$ define a flow on TM. This flow is carried by the Legendre transformation to the flow defined on T^*M by the* **Hamilton equations**

$$
\begin{cases}
\dot{x}^i = \dfrac{\partial H}{\partial p_i} \\[2mm]
\dot{p}_i = -\dfrac{\partial H}{\partial x^i}
\end{cases}
\qquad (i = 1, \ldots, n).
$$

Proof The Euler–Lagrange equations can be cast as a system of first-order ordinary differential equations on TM as follows.

$$
\begin{cases}
\dot{x}^i = v^i \\[2mm]
\dfrac{d}{dt}\left(\dfrac{\partial L}{\partial v^i}\right) = \dfrac{\partial L}{\partial x^i}
\end{cases}
\qquad (i = 1, \ldots, n).
$$

Since on S one has

$$
p_i = \frac{\partial L}{\partial v^i}, \qquad v^i = \frac{\partial H}{\partial p_i}, \qquad \frac{\partial L}{\partial x^i} = -\frac{\partial H}{\partial x^i},
$$

we see that this system reduces to the Hamilton equations in the local coordinates $(x^1, \ldots, x^n, p_1, \ldots, p_n)$. Since the Hamilton equations clearly define a flow on T^*M, the Euler–Lagrange equations must define a flow on TM. $\qquad\qquad\square$

Example 6.3 The Lagrangian for a conservative mechanical system $(M, \langle \cdot, \cdot \rangle, -dU)$ is written in local coordinates as

$$
L\left(x^1, \ldots, x^n, v^1, \ldots, v^n\right) = \frac{1}{2} \sum_{i,j=1}^{n} g_{ij}\left(x^1, \ldots, x^n\right) v^i v^j - U\left(x^1, \ldots, x^n\right).
$$

The Legendre transformation is given in these coordinates by

$$
p_i = \frac{\partial L}{\partial v^i} = \sum_{j=1}^{n} g_{ij} v^j \qquad (i = 1, \ldots, n),
$$

and is indeed a fiber-preserving diffeomorphism, whose inverse is given by

$$v^i = \sum_{j=1}^{n} g^{ij} p_j \qquad (i = 1, \ldots, n).$$

As a function on the tangent bundle, the Hamiltonian is (cf. Example 5.8)

$$H = \frac{1}{2} \sum_{i,j=1}^{n} g_{ij} v^i v^j + U.$$

Using the Legendre transformation, we can see the Hamiltonian as the following function on the cotangent bundle.

$$H = \frac{1}{2} \sum_{i,j,k,l=1}^{n} g_{ij} g^{ik} p_k g^{jl} p_l + U = \frac{1}{2} \sum_{k,l=1}^{n} g^{kl} p_k p_l + U.$$

Therefore the Hamilton equations for a conservative mechanical system are

$$\begin{cases} \dot{x}^i = \displaystyle\sum_{j=1}^{n} g^{ij} p_j \\[2em] \dot{p}_i = -\dfrac{1}{2} \displaystyle\sum_{k,l=1}^{n} \dfrac{\partial g^{kl}}{\partial x^i} p_k p_l - \dfrac{\partial U}{\partial x^i} \end{cases} \qquad (i = 1, \ldots, n).$$

The flow defined by the Hamilton equations has remarkable geometric properties, which are better understood by introducing the following definition.

Definition 6.4 The **canonical symplectic potential** (or **Liouville form**) is the 1-form $\theta \in \Omega^1(T^*M)$ given by

$$\theta_\alpha(v) := \alpha\left((d\pi)_\alpha(v)\right)$$

for all $v \in T_\alpha(T^*M)$ and all $\alpha \in T^*M$, where $\pi : T^*M \to M$ is the natural projection. The **canonical symplectic form** on T^*M is the 2-form $\omega \in \Omega^2(T^*M)$ given by $\omega = d\theta$.

In local coordinates, we have

$$\pi\left(x^1, \ldots, x^n, p_1, \ldots, p_n\right) = \left(x^1, \ldots, x^n\right)$$

and

$$v = \sum_{i=1}^{n} dx^i(v) \frac{\partial}{\partial x^i} + \sum_{i=1}^{n} dp_i(v) \frac{\partial}{\partial p_i}.$$

Consequently,

$$(d\pi)_\alpha(v) = \sum_{i=1}^{n} dx^i(v) \frac{\partial}{\partial x^i},$$

and hence

$$\theta_\alpha(v) = \alpha\left((d\pi)_\alpha(v)\right) = \sum_{i=1}^{n} p_i dx^i \left(\sum_{j=1}^{n} dx^j(v) \frac{\partial}{\partial x^j}\right) = \sum_{i=1}^{n} p_i dx^i(v).$$

We conclude that

$$\theta = \sum_{i=1}^{n} p_i dx^i,$$

and consequently

$$\omega = \sum_{i=1}^{n} dp_i \wedge dx^i.$$

Proposition 6.5 *The canonical symplectic form ω is closed ($d\omega = 0$) and nondegenerate. Moreover, $\omega^n = \omega \wedge \cdots \wedge \omega$ is a volume form (in particular T^*M is always orientable, even if M itself is not).*

Proof Exercise 6.15(1). □

Recall from Exercise 1.15(8) in Chap. 2 that if $v \in T_pM$ then $\iota(v)\omega \in T_p^*M$ is the covector given by

$$(\iota(v)\omega)(w) = \omega(v, w)$$

for all $w \in T_pM$. Therefore the first statement in Proposition 6.5 is equivalent to saying that the map $T_pM \ni v \mapsto \iota(v)\omega \in T_p^*M$ is a linear isomorphism for all $p \in M$.

The key to the geometric meaning of the Hamilton equations is contained in the following result.

Proposition 6.6 *The Hamilton equations are the equations for the flow of the vector field X_H satisfying*

$$\iota(X_H)\omega = -dH.$$

Proof The Hamilton equations yield the flow of the vector field

$$X_H = \sum_{i=1}^{n} \left(\frac{\partial H}{\partial p_i} \frac{\partial}{\partial x^i} - \frac{\partial H}{\partial x^i} \frac{\partial}{\partial p_i}\right).$$

Therefore

$$\iota(X_H)\omega = \iota(X_H) \sum_{i=1}^{n} \left(dp_i \otimes dx^i - dx^i \otimes dp_i \right)$$

$$= \sum_{i=1}^{n} \left(-\frac{\partial H}{\partial x^i} dx^i - \frac{\partial H}{\partial p_i} dp_i \right) = -dH.$$

$\square$

Remark 6.7 Notice that H completely determines X_H, as ω is nondegenerate. By analogy with the Riemannian case, $-X_H$ is sometimes called the **symplectic gradient** of H. The vector field X_H is usually referred to as the **Hamiltonian vector field** determined by H.

Definition 6.8 The **Hamiltonian flow** generated by $F \in C^\infty(T^*M)$ is the flow of the unique vector field $X_F \in \mathfrak{X}(T^*M)$ such that

$$\iota(X_F)\omega = -dF.$$

The flow determined on T^*M by a hyper-regular Lagrangian is therefore a particular case of a Hamiltonian flow (in which the generating function is the Hamiltonian function). We will now discuss the geometric properties of general Hamiltonian flows, starting with the Hamiltonian version of energy conservation.

Proposition 6.9 *Hamiltonian flows preserve their generating functions.*

Proof We have

$$X_F \cdot F = dF(X_F) = (-\iota(X_F)\omega)(X_F) = -\omega(X_F, X_F) = 0,$$

as ω is alternating.

$\square$

Proposition 6.10 *Hamiltonian flows preserve the canonical symplectic form: if $\psi_t :$ $T^*M \to T^*M$ is a Hamiltonian flow then $\psi_t^*\omega = \omega$.*

Proof Let $F \in C^\infty(T^*M)$ be the function whose Hamiltonian flow is ψ_t. Recall from Exercise 3.8(7) in Chap. 2 that the Lie derivative of ω along $X_F \in \mathfrak{X}(T^*M)$,

$$L_{X_F}\omega = \frac{d}{dt}\Big|_{t=0} \psi_t^*\omega,$$

can be computed by the Cartan formula:

$$L_{X_F}\omega = \iota(X_F)d\omega + d(\iota(X_F)\omega) = d(-dF) = 0.$$

Therefore

$$\frac{d}{dt} \psi_t{}^* \omega = \frac{d}{ds}\Big|_{s=0} (\psi_{t+s})^* \omega = \frac{d}{ds}\Big|_{s=0} (\psi_s \circ \psi_t)^* \omega = \frac{d}{ds}\Big|_{s=0} \psi_t{}^* (\psi_s{}^* \omega)$$

$$= \psi_t{}^* \frac{d}{ds}\Big|_{s=0} \psi_s{}^* \omega = \psi_t{}^* L_{X_F} \omega = 0.$$

We conclude that

$$\psi_t{}^* \omega = (\psi_0)^* \omega = \omega. \qquad \Box$$

Theorem 6.11 (Liouville) *Hamiltonian flows preserve the integral with respect to the symplectic volume form: if $\psi_t : T^*M \to T^*M$ is a Hamiltonian flow and $F \in C^\infty(T^*M)$ is a compactly supported function then*

$$\int_{T^*M} F \circ \psi_t = \int_{T^*M} F.$$

Proof This is a simple consequence of the fact that ψ_t preserves the symplectic volume form, since

$$\psi_t{}^* (\omega^n) = (\psi_t{}^* \omega)^n = \omega^n.$$

Therefore

$$\int_{T^*M} F \circ \psi_t = \int_{T^*M} (F \circ \psi_t) \omega^n = \int_{T^*M} (F \circ \psi_t) \psi_t{}^* (\omega^n)$$

$$= \int_{T^*M} \psi_t{}^* (F \omega^n) = \int_{T^*M} F \omega^n = \int_{T^*M} F$$

[cf. Exercise 4.2(4) in Chap. 2]. $\qquad \Box$

Theorem 6.12 (Poincaré recurrence) *Let $\psi_t : T^*M \to T^*M$ be a Hamiltonian flow and $K \subset T^*M$ a compact set invariant under ψ_t. Then for each open set $U \subset K$ and each $T > 0$ there exist $\alpha \in U$ and $t \geq T$ such that $\psi_t(\alpha) \in U$.*

Proof Let $F \in C^\infty(T^*M)$ be a compactly supported smooth function with values in $[0, 1]$ such that $F(\alpha) = 1$ for all $\alpha \in K$ (this is easily constructed using, for instance, a partition of unity). Let $G \in C^\infty(T^*M)$ be a smooth function with values in $[0, 1]$ and compact support contained in U such that

$$\int_{T^*M} G > 0.$$

Consider the open sets $U_n := \psi_{nT}(U)$. If these sets were all disjoint then one could define functions $\widetilde{G}_N \in C^\infty(M)$ for each $N \in \mathbb{N}$ as

$$\widetilde{G}_N(\alpha) = \begin{cases} (G \circ \psi_{-nT})(\alpha) & \text{if } \alpha \in U_n \text{ and } n \le N \\ 0 & \text{otherwise} \end{cases}.$$

These functions would have compact support contained in K (K is invariant under ψ_t) and values in $[0, 1]$, and hence would satisfy $\widetilde{G}_N \le F$. Therefore we would have

$$\int_{T^*M} F \ge \int_{T^*M} \widetilde{G}_N = \sum_{n=1}^{N} \int_{T^*M} G \circ \psi_{-nT} = N \int_{T^*M} G$$

for all $N \in \mathbb{N}$, which is absurd. We conclude that there must exist $m, n \in \mathbb{N}$ (with, say, $n > m$) such that

$$U_m \cap U_n \ne \varnothing \Leftrightarrow \psi_{mT}(U) \cap \psi_{nT}(U) \ne \varnothing \Leftrightarrow U \cap \psi_{(n-m)T}(U) \ne \varnothing.$$

Choosing $t = (n - m)T$ and $\alpha \in \psi_{-t}(U \cap \psi_t(U)) = \psi_{-t}(U) \cap U$ yields the result. $\qquad\square$

We can use the symplectic structure of the cotangent bundle to define a new binary operation on the set of differentiable functions on T^*M.

Definition 6.13 The **Poisson bracket** of two differentiable functions $F, G \in C^\infty(T^*M)$ is $\{F, G\} := X_F \cdot G$.

Proposition 6.14 $(C^\infty(T^*M), \{\cdot, \cdot\})$ *is a Lie algebra, and the map that associates to a function $F \in C^\infty(T^*M)$ its Hamiltonian vector field $X_F \in \mathfrak{X}(T^*M)$ is a Lie algebra homomorphism, i.e.*

(i) $\{F, G\} = -\{G, F\}$;
(ii) $\{\alpha F + \beta G, H\} = \alpha\{F, H\} + \beta\{G, H\}$;
(iii) $\{F, \{G, H\}\} + \{G, \{H, F\}\} + \{H, \{F, G\}\} = 0$;
(iv) $X_{\{F,G\}} = [X_F, X_G]$

for any $F, G, H \in C^\infty(T^*M)$ and any $\alpha, \beta \in \mathbb{R}$.

Proof We have

$$\{F, G\} = X_F \cdot G = dG(X_F) = (-\iota(X_G)\omega)(X_F)$$
$$= -\omega(X_G, X_F) = \omega(X_F, X_G),$$

which proves the anti-symmetry and bilinearity of the Poisson bracket. On the other hand,

$$\iota(X_{\{F,G\}})\omega = -d\{F, G\} = -d(X_F \cdot G) = -d(dG(X_F)) = -d(\iota(X_F)dG)$$
$$= -L_{X_F}dG = L_{X_F}(\iota(X_G)\omega) = \iota(L_{X_F}X_G)\omega + \iota(X_G)L_{X_F}\omega$$
$$= \iota([X_F, X_G])\omega$$

[cf. Exercise 3.8(7) in Chap. 2]. Since ω is nondegenerate, we have

$$X_{\{F,G\}} = [X_F, X_G].$$

Finally,

$$\{F, \{G, H\}\} + \{G, \{H, F\}\} + \{H, \{F, G\}\}$$
$$= \{F, X_G \cdot H\} - \{G, X_F \cdot H\} - X_{\{F,G\}} \cdot H$$
$$= X_F \cdot (X_G \cdot H) - X_G \cdot (X_F \cdot H) - [X_F, X_G] \cdot H = 0. \qquad \square$$

Exercise 6.15

(1) Prove Proposition 6.5.

(2) Let $(M, \langle \cdot, \cdot \rangle)$ be a Riemannian manifold, $\alpha \in \Omega^1(M)$ a 1-form and $U \in C^\infty(M)$ a differentiable function.

   (a) Show that the Euler–Lagrange equations for the Lagrangian $L : TM \to \mathbb{R}$ given by

$$L(v) = \frac{1}{2}\langle v, v \rangle + \iota(v)\alpha_p - U(p)$$

   for $v \in T_pM$ yield the motions of the mechanical system $(M, \langle \cdot, \cdot \rangle, \mathcal{F})$, where

$$\mathcal{F}(v) = -(dU)_p - \iota(v)(d\alpha)_p$$

   for $v \in T_pM$.

   (b) Show that the mechanical energy $E = K + U$ is conserved along the motions of $(M, \langle \cdot, \cdot \rangle, \mathcal{F})$ (which is therefore called a **conservative mechanical system with magnetic term**).

   (c) Show that L is hyper-regular and compute the Legendre transformation.

   (d) Find the Hamiltonian $H : T^*M \to \mathbb{R}$ and write the Hamilton equations.

(3) Let $c > 0$ be a positive number, representing the speed of light, and consider the open set $U := \{v \in T\mathbb{R}^n \mid \|v\| < c\}$, where $\| \cdot \|$ is the Euclidean norm. The motion of a free relativistic particle of mass $m > 0$ is determined by the Lagrangian $L : U \to \mathbb{R}$ given by

$$L(v) := -mc^2\sqrt{1 - \frac{\|v\|^2}{c^2}}.$$

(a) Show that L is hyper-regular and compute the Legendre transformation.

(b) Find the Hamiltonian $H : T^*\mathbb{R}^n \to \mathbb{R}$ and write the Hamilton equations.

(4) Show that in the Poincaré recurrence theorem the set of points $\alpha \in U$ such that $\psi_t(\alpha) \in U$ for some $t \geq T$ is dense in U. (**Remark:** It can be shown that this set has full measure).

(5) Let $(M, \langle \cdot, \cdot \rangle)$ be a compact Riemannian manifold. Show that for each normal ball $B \subset M$ and each $T > 0$ there exist geodesics $c : \mathbb{R} \to M$ with $\|\dot{c}(t)\| = 1$ such that $c(0) \in B$ and $c(t) \in B$ for some $t \geq T$.

(6) Let $\left(x^1, \ldots, x^n, p_1, \ldots, p_n\right)$ be the usual local coordinates on T^*M. Compute $X_{x^i}, X_{p_i}, \{x^i, x^j\}, \{p_i, p_j\}$ and $\{p_i, x^j\}$.

(7) Show that the Poisson bracket satisfies the **Leibniz rule**

$$\{F, GH\} = \{F, G\}H + \{F, H\}G$$

for all $F, G, H \in C^\infty(T^*M)$.

5.7 Completely Integrable Systems

We now concentrate on studying the Hamiltonian flow of a Hamiltonian function $H \in C^\infty(T^*M)$. We already know that H is constant along its Hamiltonian flow, so that it suffices to study this flow along the level sets of H. This can be further simplified if there exist additional nontrivial functions $F \in C^\infty(T^*M)$ such that

$$X_H \cdot F = 0 \Leftrightarrow \{H, F\} = 0.$$

Definition 7.1 A function $F \in C^\infty(T^*M)$ is said to be a **first integral** of H if $\{H, F\} = 0$.

In general, there is no reason to expect that there should exist nontrivial first integrals other than H itself. In the special cases when these exist, they often satisfy additional conditions.

Definition 7.2 The functions $F_1, \ldots, F_m \in C^\infty(T^*M)$ are said to be

(i) **in involution** if $\{F_i, F_j\} = 0$ $(i, j = 1, \ldots, m)$;

(ii) **independent** at $\alpha \in T^*M$ if $(dF_1)_\alpha, \ldots, (dF_m)_\alpha \in T^*_\alpha(T^*M)$ are linearly independent covectors.

Proposition 7.3 *If $F_1, \ldots, F_m \in C^\infty(T^*M)$ are in involution and are independent at some point $\alpha \in T^*M$ then $m \leq n$.*

Proof Exercise 7.17(2).   □

The maximal case $m = n$ is especially interesting.

Definition 7.4 The Hamiltonian H is said to be **completely integrable** if there exist n first integrals $F_1, \ldots, F_n$ in involution which are independent on a dense open set $U \subset T^*M$.

Example 7.5

(1) If M is 1-dimensional and $dH \neq 0$ on a dense open set of T^*M then H is completely integrable.
(2) *(Particle in a central field)* Recall Example 1.15 where a particle of mass $m > 0$ moves in a central field. The corresponding Lagrangian function is

$$L\left(r, \theta, v^r, v^\theta\right) = \frac{1}{2}m \left[\left(v^r\right)^2 + r^2 \left(v^\theta\right)^2 \right] - u(r),$$

and so the Legendre transformation is given by

$$p_r = \frac{\partial L}{\partial v^r} = mv^r \quad \text{and} \quad p_r = \frac{\partial L}{\partial v^\theta} = mr^2 v^\theta.$$

The Hamiltonian function is then

$$H\left(r, \theta, p_r, p_\theta\right) = \frac{p_r^2}{2m} + \frac{p_\theta^2}{2mr^2} + u(r).$$

By the Hamilton equations,

$$\dot{p}_\theta = -\frac{\partial H}{\partial \theta} = 0,$$

and hence p_θ is a first integral. Since

$$dH = \left(-\frac{p_\theta^2}{mr^3} + u'(r)\right) dr + \frac{p_r}{m} dp_r + \frac{p_\theta}{mr^2} dp_\theta,$$

we see that dH and dp_θ are independent on the dense open set of $T^*\mathbb{R}^2$ formed by the points whose polar coordinates $(r, \theta, p_r, p_\theta)$ are well defined and do not satisfy

$$u'(r) - \frac{p_\theta^2}{mr^3} = p_r = 0$$

(i.e. are not on a circular orbit – cf. Exercise 7.17(4). Therefore this Hamiltonian is completely integrable.

Proposition 7.6 *Let H be a completely integrable Hamiltonian with first integrals $F_1, \ldots, F_n$ in involution, independent in the dense open set $U \subset T^*M$, and such that $X_{F_1}, \ldots, X_{F_n}$ are complete on U. Then each nonempty level set*

$$L_f := \{\alpha \in U \mid F_1(\alpha) = f_1, \ldots, F_n(\alpha) = f_n\}$$

*is a submanifold of dimension n, invariant for the Hamiltonian flow of H, admitting
a locally free action of $\mathbb{R}^n$ which is transitive on each connected component.*

Proof All points in U are regular points of the map $F : U \to \mathbb{R}^n$ given by
$F(\alpha) = (F_1(\alpha), \ldots, F_n(\alpha))$; therefore all nonempty level sets $L_f := F^{-1}(f)$
are submanifolds of dimension n.

Since $X_H \cdot F_i = 0$ for $i = 1, \ldots, n$, the level sets L_f are invariant for the flow
of X_H. In addition, we have $X_{F_i} \cdot F_j = \{F_i, F_j\} = 0$, and hence these level sets
are invariant for the flow of X_{F_i}. Moreover, these flows commute, as $[X_{F_i}, X_{F_j}] =
X_{\{F_i, F_j\}} = 0$ (cf. Theorem 6.10 in Chap. 1).

Consider the map $A : \mathbb{R}^n \times L_f \to L_f$ given by

$$A(t_1, \ldots, t_n, \alpha) = (\psi_{1,t_1} \circ \cdots \circ \psi_{n,t_n})(\alpha),$$

where $\psi_{i,t} : L_f \to L_f$ is the flow of X_{F_i}. Since these flows commute, this map
defines an action of $\mathbb{R}^n$ on L_f. On the other hand, for each $\alpha \in L_f$, the map
$A_\alpha : \mathbb{R}^n \to L_f$ given by $A_\alpha(t_1, \ldots, t_n) = A(t_1, \ldots, t_n, \alpha)$ is a local diffeomorphism
at the origin, as

$$(dA_\alpha)_0 (e_i) = \frac{d}{dt}_{|t=0} \psi_{i,t}(\alpha) = (X_{F_i})_\alpha$$

and the vector fields X_{F_i} are linearly independent. Therefore the action is locally
free (that is, for each point $\alpha \in L_f$ there exists an open neighborhood $U \subset \mathbb{R}^n$ of
0 such that $A(t, \alpha) \neq \alpha$ for all $t \in U \setminus \{0\}$), meaning that the isotropy groups are
discrete. Also, the action is locally transitive (i.e. each point $\alpha \in L_f$ admits an open
neighborhood $V \subset L_f$ such that every $\beta \in V$ is of the form $\beta = A(t, \alpha)$ for some
$t \in \mathbb{R}^n$), and hence transitive on each connected component (for given $\alpha \in L_f$ both
the set of points $\beta \in L_f$ which are of the form $\beta = A(t, \alpha)$ for some $t \in \mathbb{R}^n$ and the
set of points which are not are open). $\qquad\qquad\qquad\qquad\qquad\qquad\qquad\square$

The isotropy subgroups of the action above are discrete subgroups of $\mathbb{R}^n$. We next
describe the structure of such subgroups.

Proposition 7.7 *Let* Γ *be a discrete subgroup of* $\mathbb{R}^n$. *Then there exist* $k \in
\{0, 1, \ldots, n\}$ *linearly independent vectors* $e_1, \ldots, e_k$ *such that* $\Gamma = \text{span}_{\mathbb{Z}}
\{e_1, \ldots, e_k\}$.

Proof If $\Gamma = \{0\}$ then we are done. If not, let $e \in \Gamma \setminus \{0\}$. Since Γ is discrete, the set

$$\Gamma \cap \{\lambda e \mid 0 < \lambda \leq 1\}$$

is finite (and nonempty). Let e_1 be the element in this set which is closest to 0. Then

$$\Gamma \cap \text{span}_{\mathbb{R}}\{e\} = \text{span}_{\mathbb{Z}}\{e_1\}$$

(for otherwise e_1 would not be the element in this set closest to 0). If $\Gamma = \text{span}_{\mathbb{Z}}\{e_1\}$
then we are done. If not, let $e \in \Gamma \setminus \text{span}_{\mathbb{Z}}\{e_1\}$. Then the set

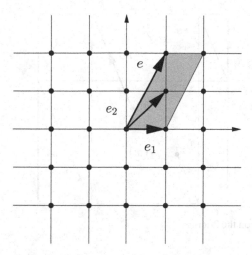

Fig. 5.10 Proof of Proposition 7.7

$$\Gamma \cap \{\lambda e + \lambda_1 e_1 \mid 0 < \lambda, \lambda_1 \le 1\}$$

is finite (and nonempty). Let e_2 be the element in this set which is closest to $\text{span}_{\mathbb{R}}\{e_1\}$ (Fig. 5.10). Then

$$\Gamma \cap \text{span}_{\mathbb{R}}\{e, e_1\} = \text{span}_{\mathbb{Z}}\{e_1, e_2\}.$$

Iterating this procedure yields the result.   □

Proposition 7.8 *Let L_f^α be the connected component of $\alpha \in L_f$. Then L_f^α is diffeomorphic to $T^k \times \mathbb{R}^{n-k}$, where k is the number of generators of the isotropy subgroup Γ_α. In particular, if L_f^α is compact then it is diffeomorphic to the n-dimensional torus T^n.*

Proof Since the action $A : \mathbb{R}^n \times L_f^\alpha \to L_f^\alpha$ is transitive, the local diffeomorphism $A_\alpha : \mathbb{R}^n \to L_f^\alpha$ is surjective. On the other hand, because Γ_α is discrete, the action of Γ_α on $\mathbb{R}^n$ by translation is free and proper, and we can form the quotient $\mathbb{R}^n / \Gamma_\alpha$, which is clearly diffeomorphic to $T^k \times \mathbb{R}^{n-k}$. Finally, it is easily seen that A_α induces a diffeomorphism $\mathbb{R}^n / \Gamma_\alpha \cong L_f^\alpha$.   □

We are now in position to understand the Hamiltonian flow on a completely integrable system. For that we need the following definition (cf. Fig. 5.11).

Definition 7.9 A **linear flow** on the torus $T^n = \mathbb{R}^n / \mathbb{Z}^n$ is the projection of the flow $\psi_t : \mathbb{R}^n \to \mathbb{R}^n$ given by

$$\psi_t(x) = x + vt.$$

The **frequencies** of the linear flow are the components $v^1, \ldots, v^n$ of v.

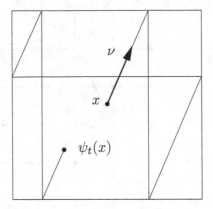

Fig. 5.11 Linear flow on the 2-torus

Theorem 7.10 (Arnold–Liouville) *Let H be a completely integrable Hamiltonian with n first integrals $F_1, \ldots, F_n \in C^\infty(T^*M)$ in involution, independent on the dense open set $U \subset T^*M$. If the connected components of the level sets of the map $(F_1, \ldots, F_n) : U \to \mathbb{R}^n$ are compact then they are n-dimensional tori, invariant for the flow of X_H. The flow of X_H on these tori is a linear flow (for an appropriate choice of coordinates).*

Proof Since the connected components of the level sets of $(F_1, \ldots, F_n)$ are compact, the vector fields X_{F_i} are complete. All that remains to be seen is that the flow of X_H on the invariant tori is a linear flow. It is clear that the flow of each X_{F_i} is linear in the coordinates given by Proposition 7.8. Since X_H is tangent to the invariant tori, we have $X_H = \sum_{i=1}^n f^i X_{F_i}$ for certain functions f^i. Now

$$0 = X_{\{F_i, H\}} = [X_{F_i}, X_H] = \sum_{j=1}^n (X_{F_i} \cdot f^j) X_{F_j},$$

and hence each function f^i is constant on the invariant torus. We conclude that the flow of X_H is linear. □

We next explore in detail the properties of linear flows on the torus.

Definition 7.11 Let $\psi_t : T^n \to T^n$ be a linear flow. The **time average** of a function $f \in C^\infty(T^n)$ along ψ_t is the map

$$\overline{f}(x) := \lim_{T \to +\infty} \frac{1}{T} \int_0^T f(\psi_t(x)) dt$$

(defined on the set of points $x \in T^n$ where the limit exists).

Definition 7.12 The frequencies $v \in \mathbb{R}^n$ of a linear flow $\psi_t : T^n \to T^n$ are said to be **independent** if they are linearly independent over $\mathbb{Q}$, i.e. if $\langle k, v \rangle \neq 0$ for all $k \in \mathbb{Z}^n \setminus \{0\}$.

Theorem 7.13 (Birkhoff ergodicity) *If the frequencies $v \in \mathbb{R}^n$ of a linear flow $\psi_t : T^n \to T^n$ are independent then the time average of any function $f \in C^\infty(T^n)$ exists for all $x \in T^n$ and*

$$\overline{f}(x) = \int_{T^n} f.$$

Proof Since $T^n = \mathbb{R}^n / \mathbb{Z}^n$, the differentiable functions on the torus arise from periodic differentiable functions on $\mathbb{R}^n$, which can be expanded as uniformly convergent Fourier series. Therefore it suffices to show that the theorem holds for $f(x) = e^{2\pi i \langle k, x \rangle}$ with $k \in \mathbb{Z}^n$.

If $k = 0$ then both sides of the equality are 1, and the theorem holds.

If $k \neq 0$ that the right-hand side of the equality is zero, whereas the left-hand side is

$$\overline{f}(x) = \lim_{T \to +\infty} \frac{1}{T} \int_0^T e^{2\pi i \langle k, x + vt \rangle} dt$$

$$= \lim_{T \to +\infty} \frac{1}{T} e^{2\pi i \langle k, x \rangle} \frac{e^{2\pi i \langle k, v \rangle T} - 1}{2\pi i \langle k, v \rangle} = 0$$

(where we used the fact that $\langle k, v \rangle \neq 0$). $\qquad\square$

Corollary 7.14 *If the frequencies of a linear flow $\psi_t : T^n \to T^n$ are independent then $\{\psi_t(x) \mid t \geq 0\}$ is dense on the torus for all $x \in T^n$.*

Proof If $\{\psi_t(x) \mid t \geq 0\}$ were not dense then it would not intersect an open set $U \subset T^n$. Therefore any nonnegative function $f \in C^\infty(T^n)$ with nonempty support contained in U would satisfy $\overline{f}(x) = 0$ and $\int_{T^n} f > 0$, contradicting the Birkhoff ergodicity theorem. $\qquad\square$

Corollary 7.15 *If the frequencies of a linear flow $\psi_t : T^n \to T^n$ are independent and $n \geq 2$ then $\psi_t(x)$ is not periodic.* $\qquad\square$

Remark 7.16 The qualitative behavior of the Hamiltonian flow generated by completely integrable Hamiltonians is completely understood. Complete integrability is however a very strong condition, not satisfied by generic Hamiltonians. The **Kolmogorov–Arnold–Moser (KAM) theorem** guarantees a small measure of genericity by establishing that a large fraction of the invariant tori of a completely integrable Hamiltonians survives under small perturbations of the Hamiltonian, the flow on these tori remaining linear with the same frequencies. On the other hand, many invariant tori, including those whose frequencies are not independent (**resonant tori**), are typically destroyed.

Exercise 7.17

(1) Show that if $F, G \in C^\infty(T^*M)$ are first integrals, then $\{F, G\}$ is also a first integral.
(2) Prove Proposition 7.3.
(3) Consider a surface of revolution $M \subset \mathbb{R}^3$ given in cylindrical coordinates (r, θ, z) by

$$r = f(z),$$

where $f : (a, b) \to (0, +\infty)$ is differentiable.

   (a) Show that the geodesics of M are the critical points of the action determined by the Lagrangian $L : TM \to \mathbb{R}$ given in local coordinates by

$$L\left(\theta, z, v^\theta, v^z\right) = \frac{1}{2}\left((f(z))^2(v^\theta)^2 + \left((f'(z))^2 + 1\right)(v^z)^2\right).$$

   (b) Show that the curves given in local coordinates by $\theta = $ constant or $f'(z) = 0$ are images of geodesics.
   (c) Compute the Legendre transformation, show that L is hyper-regular and write an expression in local coordinates for the Hamiltonian $H : T^*M \to \mathbb{R}$.
   (d) Show that H is completely integrable.
   (e) Show that the projection on M of the invariant set

$$L_{(E,l)} := H^{-1}(E) \cap p_\theta^{-1}(l)$$

   ($E, l > 0$) is given in local coordinates by

$$f(z) \geq \frac{l}{\sqrt{2E}}.$$

   Use this fact to conclude that if f has a strict local maximum at $z = z_0$ then the geodesic whose image is $z = z_0$ is **stable**, i.e. geodesics with initial condition close to the point in TM with coordinates $(\theta_0, z_0, 1, 0)$ stay close to the curve $z = z_0$.

(4) Recall from Example 7.5 that a particle of mass $m > 0$ moving in a central field is described by the completely integrable Hamiltonian function

$$H(r, \theta, p_r, p_\theta) = \frac{p_r^2}{2m} + \frac{p_\theta^2}{2mr^2} + u(r).$$

   (a) Show that there exist circular orbits of radius r_0 whenever $u'(r_0) \geq 0$.
   (b) Verify that the set of points where dH and dp_θ are not independent is the union of these circular orbits.
   (c) Show that the projection of the invariant set

$$L_{(E,l)} := H^{-1}(E) \cap p_\theta^{-1}(l)$$

on $\mathbb{R}^2$ is given in local coordinates by

$$u(r) + \frac{l^2}{2mr^2} \le E.$$

(d) Conclude that if $u'(r_0) \ge 0$ and

$$u''(r_0) + \frac{3u'(r_0)}{r_0} > 0$$

then the circular orbit of radius r_0 is stable.

(5) In general relativity, the motion of a particle in the gravitational field of a point mass $M > 0$ is given by the Lagrangian $L : TU \to \mathbb{R}$ written in cylindrical coordinates (u, r, θ) as

$$L = -\frac{1}{2}\left(1 - \frac{2M}{r}\right)(v^u)^2 + \frac{1}{2}\left(1 - \frac{2M}{r}\right)^{-1}(v^r)^2 + \frac{1}{2}r^2(v^\theta)^2,$$

where $U \subset \mathbb{R}^3$ is the open set given by $r > 2M$ (the coordinate u is called the **time coordinate**, and in general is different from the **proper time** of the particle, i.e. the parameter t of the curve).

(a) Show that L is hyper-regular and compute the corresponding Hamiltonian $H : T^*U \to \mathbb{R}$.

(b) Show that H is completely integrable.

(c) Show that there exist circular orbits of any radius $r_0 > 2M$, with $H < 0$ for $r_0 > 3M$, $H = 0$ for $r_0 = 3M$ and $H > 0$ for $r_0 < 3M$. (**Remark:** The orbits with $H > 0$ are not physical, since they correspond to speeds greater than the speed of light; the orbits with $H = 0$ can only be achieved by massless particles, which move at the speed of light).

(d) Show that the set of points where dH, dp_u and dp_θ are not independent (and $p_u \ne 0$) is the union of these circular orbits.

(e) Show that the projection of the invariant cylinder

$$L_{(E,k,l)} := H^{-1}(E) \cap p_u^{-1}(k) \cap p_\theta^{-1}(l)$$

on U is given in local coordinates by

$$\frac{l^2}{r^2} - \left(1 - \frac{2M}{r}\right)^{-1}k^2 \le 2E.$$

(f) Conclude that if $r_0 > 6M$ then the circular orbit of radius r_0 is stable.

(6) Recall that the Lagrange top is the mechanical system determined by the Lagrangian $L : TSO(3) \to \mathbb{R}$ given in local coordinates by

$$L = \frac{I_1}{2} \left((v^\theta)^2 + (v^\varphi)^2 \sin^2 \theta \right) + \frac{I_3}{2} \left(v^\psi + v^\varphi \cos \theta \right)^2 - Mgl \cos \theta,$$

where (θ, φ, ψ) are the Euler angles, M is the top's mass and l is the distance from the fixed point to the center of mass.

(a) Compute the Legendre transformation, show that L is hyper-regular and write an expression in local coordinates for the Hamiltonian $H : T^*SO(3) \to \mathbb{R}$.

(b) Prove that H is completely integrable.

(c) Show that the solutions found in Exercise 3.20(14) are stable for $|\dot{\varphi}| \ll |\dot{\psi}|$ if $|\dot{\psi}|$ is large enough.

(7) Show that the the Euler top with $I_1 < I_2 < I_3$ defines a completely integrable Hamiltonian on $T^*SO(3)$.

(8) Consider the sequence formed by the first digit of the decimal expansion of each of the integers 2^n for $n \in \mathbb{N}_0$:

$$1, 2, 4, 8, 1, 3, 6, 1, 2, 5, 1, 2, 4, 8, 1, 3, 6, 1, 2, 5, \ldots.$$

The purpose of this exercise is to answer the following question: is there a 7 in this sequence?

(a) Show that if $v \in \mathbb{R} \setminus \mathbb{Q}$ then

$$\lim_{n \to +\infty} \frac{1}{n+1} \sum_{k=0}^n e^{2\pi i v k} = 0.$$

(b) Prove the following discrete version of the Birkhoff ergodicity theorem: if a differentiable function $f : \mathbb{R} \to \mathbb{R}$ is periodic with period 1 and $v \in \mathbb{R} \setminus \mathbb{Q}$ then for all $x \in \mathbb{R}$

$$\lim_{n \to +\infty} \frac{1}{n+1} \sum_{k=0}^n f(x + vk) = \int_0^1 f(x)dx.$$

(c) Show that $\log 2$ is an irrational multiple of $\log 10$.

(d) Is there a 7 in the sequence above?

5.8 Symmetry and Reduction

The symplectic structure on the cotangent bundle can be generalized to arbitrary manifolds.

Definition 8.1 A **symplectic manifold** is a pair (M, ω), where M is a differentiable manifold and $\omega \in \Omega^2(M)$ is nondegenerate and closed.

Example 8.2 If M is an orientable surface and $\omega \in \Omega^2(M)$ is a volume form on M then (M, ω) is a symplectic manifold. In fact, ω is necessarily nondegenerate (if $\iota(v)\omega = 0$ for some nonvanishing $v \in T_p M$ then $\omega_p = 0$), and $d\omega = 0$ trivially.

All definitions and results of Sects. 5.6 and 5.7 (Hamiltonian flow and its properties, Liouville and Poincaré recurrence theorems, Poisson bracket, completely integrable systems and the Arnold–Liouville theorem) are readily extended to arbitrary symplectic manifolds. In fact, all symplectic manifolds are locally the same (i.e. there is no symplectic analogue of the curvature), as we now show.

Theorem 8.3 (Darboux) *Let (M, ω) be a symplectic manifold and $p \in M$. Then there exist local coordinates $(x^1, \ldots, x^n, p_1, \ldots, p_n)$ around p such that*

$$\omega = \sum_{i=1}^{n} dp_i \wedge dx^i$$

(in particular the dimension of M is necessarily even).

Proof We begin by observing that ω is of the form above if and only if $\{x^i, x^j\} = \{p_i, p_j\} = 0$ and $\{p_i, x^j\} = \delta_{ij}$ for $i, j = 1, \ldots, n$ [cf. Exercise 8.23(2)].

Clearly we must have $m := \dim M \geq 2$ (otherwise $\omega = 0$ would be degenerate). Let $P \in C^\infty(M)$ be a function with $(dP)_p \neq 0$, let X_P be the corresponding Hamiltonian vector field and let $T \subset M$ be a hypersurface not tangent to $(X_P)_p$ (cf. Fig. 5.12). Then X_P is not tangent to T on some neighborhood V of p. Possibly reducing V, we can define a smooth function Q on V by the condition that $\psi_{-Q(q)}(q) \in T$ for each $q \in V$, where ψ_t is the flow of X_P. Notice that $T \cap V = Q^{-1}(0)$, implying that X_Q is tangent to T, and so $\{(X_P)_p, (X_Q)_p\}$ is a linearly independent set. This means that $\{(dP)_p, (dQ)_p\}$ is linearly independent, and so, reducing V if necessary, (P, Q) can be extended to a system of local coordinates around p. If $m = 2$ then we are done, because (Q, P) are local coordinates and $\{P, Q\} = X_P \cdot Q = 1$.

If $m > 2$ then, since X_P is not tangent to T, the level set $P^{-1}(P(p))$ intersects T on a $(m-2)$-dimensional manifold $S \subset T$. Since $\{P, Q\} = 1$, we have $X_Q \cdot P = \{Q, P\} = -1$, and so X_Q is not tangent to S. If $q \in S$ and $\{v_1, \ldots, v_{m-2}\}$ is a basis for $T_q S$ then $\{(X_P)_q, (X_Q)_q, v_1, \ldots, v_{m-2}\}$ is a basis for $T_q M$. Moreover, we have $\omega(X_P, v_i) = -dP(v_i) = 0$ (as P is constant in S), and similarly $\omega(X_Q, v_i) = -dQ(v_i) = 0$. We conclude that the matrix $(\omega(v_i, v_j))$ must be nonsingular, that

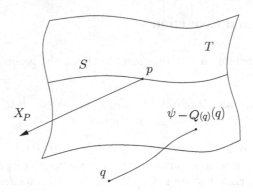

Fig. 5.12 Proof of the Darboux theorem

is, $i^*\omega$ must be nondegenerate, where $i : S \to V$ is the inclusion map. Since $di^*\omega = i^*d\omega = 0$, we see that $(S, i^*\omega)$ is a symplectic manifold. Given any function $F \in C^\infty(S)$, we can extend it to T by making it constant along the flow of X_Q, and then to V by making it constant along the flow of X_P. Since $[X_P, X_Q] = X_{\{P,Q\}} = X_1 = 0$, the flows of X_P and X_Q commute, and so this extension can be done in the reverse order. Consequently, the extended function (which we still call F) satisfies $\{P, F\} = X_P \cdot F = 0$ and $\{Q, F\} = X_Q \cdot F = 0$, that is, $X_F \cdot P = \{F, P\} = 0$ and $X_F \cdot Q = \{F, Q\} = 0$. This implies that X_F is tangent to S, and so X_F coincides on S with the Hamiltonian vector field determined by F on $(S, i^*\omega)$. In the same way, the Poisson bracket $\{F, G\}$ of the extensions to V of two functions $F, G \in C^\infty(S)$ satisfies

$$X_P \cdot \{F, G\} = \{P, \{F, G\}\} = \{\{P, F\}, G\} + \{F, \{P, G\}\} = \{0, G\} + \{F, 0\} = 0,$$

and similarly $X_Q \cdot \{F, G\} = 0$, implying that $\{F, G\}$ is the extension of the Poison bracket on $(S, i^*\omega)$. Therefore, if the Darboux theorem holds for $(S, i^*\omega)$, meaning that we have $m - 2 = 2n - 2$ and local coordinates $(x^1, \ldots, x^{n-1}, p_1, \ldots, p_{n-1})$ with $\{x^i, x^j\} = \{p_i, p_j\} = 0$ and $\{p_i, x^j\} = \delta_{ij}$ for $i, j = 1, \ldots, n - 1$, then, making $x^n = Q$ and $p_n = P$, we have the result for (M, ω).   □

In fact, to have Hamiltonian flows all that is required is the existence of a Poisson bracket. This suggests a further generalization.

Definition 8.4 A **Poisson manifold** is a pair $(M, \{\cdot, \cdot\})$, where M is a differentiable manifold and $\{\cdot, \cdot\}$, called the **Poisson bracket**, is a Lie bracket on $C^\infty(M)$ satisfying the **Leibniz rule**, that is,

   (i) $\{F, G\} = -\{G, F\}$;
   (ii) $\{\alpha F + \beta G, H\} = \alpha\{F, H\} + \beta\{F, H\}$;
   (iii) $\{F, \{G, H\}\} + \{G, \{H, F\}\} + \{H, \{F, G\}\} = 0$;
   (iv) $\{F, GH\} = \{F, G\}H + \{F, H\}G$

for any $\alpha, \beta \in \mathbb{R}$ and $F, G, H \in C^\infty(M)$.

Example 8.5

(1) Any symplectic manifold (M, ω) is naturally a Poisson manifold $(M, \{\cdot, \cdot\})$ [cf. Exercise 6.15(7)].
(2) Any smooth manifold M can be given a Poisson structure, namely the trivial Poisson bracket $\{\cdot, \cdot\} := 0$. This is not true for symplectic structures, even if M is even-dimensional [cf. Exercise 8.23(2)].
(3) If $\langle \cdot, \cdot \rangle$ is the Euclidean inner product on $\mathbb{R}^3$ then the formula

$$\{F, G\}(x) := \langle x, \operatorname{grad} F \times \operatorname{grad} G \rangle$$

defines a nontrivial Poisson bracket on $\mathbb{R}^3$ (cf. Example 8.22).

The bilinearity and Leibniz rule properties of the Poisson bracket imply that $\{F, \cdot\}$ is a derivation (hence a vector field) for any $F \in C^\infty(M)$. This allows us to define Hamiltonian flows.

Definition 8.6 If $(M, \{\cdot, \cdot\})$ is a Poisson manifold and $F \in C^\infty(M)$ then the **Hamiltonian vector field** generated by F is the vector field $X_F \in \mathfrak{X}(M)$ such that

$$X_F \cdot G = \{F, G\}$$

for any function $G \in C^\infty(M)$.

Proposition 8.7 *The map* $C^\infty(M) \ni F \mapsto X_F \in \mathfrak{X}(M)$ *is a Lie algebra homomorphism between* $(C^\infty(M), \{\cdot, \cdot\})$ *and* $(\mathfrak{X}(M), [\cdot, \cdot])$, *that is,*

(i) $X_{\alpha F + \beta G} = \alpha X_F + \beta X_G$;
(ii) $X_{\{F,G\}} = [X_F, X_G]$

for all $\alpha, \beta \in \mathbb{R}$ *and* $F, G, H \in C^\infty(M)$.

Proof Property (i) is immediate from the bilinearity of the Poisson bracket. Property (ii) arises from the Jacobi identity, as

$$\begin{aligned} X_{\{F,G\}} \cdot H &= \{\{F, G\}, H\} = \{F, \{G, H\}\} - \{G, \{F, H\}\} \\ &= X_F \cdot (X_G \cdot H) - X_G \cdot (X_F \cdot H) \\ &= [X_F, X_G] \cdot H \end{aligned}$$

for any $F, G, H \in C^\infty(M)$. $\qquad\qquad\square$

The functions in the kernel of the homomorphism $F \mapsto X_F$ are called the **Casimir functions**, and are simply the functions $F \in C^\infty(M)$ that Poisson commute with all other functions, that is, such that $\{F, G\} = 0$ for all $G \in C^\infty(M)$. Notice that Casimir functions are constant along **any** Hamiltonian flow. The image of the homomorphism $F \mapsto X_F$ is the set of Hamiltonian vector fields, which in particular forms a Lie subalgebra of $(\mathfrak{X}(M), [\cdot, \cdot])$.

Example 8.8

(1) If (M, ω) is a symplectic manifold then the Casimir functions are just the (locally) constant functions.
(2) If $\{\cdot, \cdot\} := 0$ is the trivial Poisson bracket on a smooth manifold M then any function is a Casimir function, and the only Hamiltonian vector field is the zero field.
(3) If $\{\cdot, \cdot\}$ is the Poisson bracket defined on $\mathbb{R}^3$ by the formula

$$\{F, G\}(x) := \langle x, \operatorname{grad} F \times \operatorname{grad} G \rangle$$

then $C(x) := \|x\|^2$ is a Casimir function, as $\operatorname{grad} C = 2x$ and so

$$\{C, F\}(x) = 2\langle x, x \times \operatorname{grad} F \rangle = 2\langle \operatorname{grad} F, x \times x \rangle = 0$$

for any smooth function $F \in C^\infty(\mathbb{R}^3)$. It follows that the Hamiltonian vector fields are necessarily tangent to the spheres of constant C (and in particular must vanish at the origin).

Since the Poisson bracket can be written as

$$\{F, G\} = X_F \cdot G = dG(X_F) = -dF(X_G),$$

we see that $\{F, G\}(p)$ is a linear function of both $(dF)_p$ and $(dG)_p$. Therefore the Poisson bracket determines a bilinear map $B_p : T_p^* M \times T_p^* M \to \mathbb{R}$ for all $p \in M$.

Definition 8.9 The antisymmetric $(0, 2)$-tensor field B satisfying

$$\{F, G\} = B(dF, dG)$$

is called the **Poisson bivector**.

Using the identification $T_p M \cong (T_p^* M)^*$, we have

$$X_F(dG) = dG(X_F) = X_F \cdot G = \{F, G\} = B(dF, dG) = (\iota(dF)B)(dG),$$

where the contraction of a covector with the Poisson bivector is defined in the same way as the contraction of a vector with an alternating tensor [cf. Exercise 1.15(8)] in Chap. 2). Therefore we have

$$X_F = \iota(dF)B,$$

and so the set of all possible values of Hamiltonian vector fields at a given point $p \in M$ is exactly the range of the map $T_p^* M \ni \omega \mapsto \iota(\omega)B \in T_p M$.

Theorem 8.10 (Kirillov) *Let $(M, \{\cdot, \cdot\})$ be a Poisson manifold such that the rank of the map $T_p^* M \ni \omega \mapsto \iota(\omega)B \in T_p M$ is constant (as a function of $p \in M$). Then M is foliated by symplectic submanifolds (S, ω_S) (called **symplectic leaves**) such that*

$$\{F, G\}(p) = \{F|_S, G|_S\}(p)$$

for all $p \in M$, where S is the leaf containing p.

Proof Since the rank r of the map $T_p^* M \ni \omega \mapsto \iota(\omega)B \in T_p M$ is constant, the range Σ_p of this map has dimension r for all $p \in M$, and so determines a distribution Σ of dimension r in M. By construction, all Hamiltonian vector fields are compatible with this distribution, and it is clear that for each $p \in M$ there exist $F_1, \ldots, F_r \in C^\infty(M)$ such that Σ is spanned by $X_{F_1}, \ldots, X_{F_r}$ on a neighborhood of p. Since $[X_{F_i}, X_{F_j}] = X_{\{F_i, F_j\}}$ for $i, j = 1, \ldots, r$, the distribution Σ is integrable, and so M is foliated by r-dimensional leaves S with $T_p S = \Sigma_p$ for all $p \in S$. If $\omega, \eta \in T_p^* M$ then $B(\omega, \eta) = \eta(\iota(\omega)B) = -\omega(\iota(\eta)B)$ depends only on the restrictions of η and ω to Σ_p, that is, B restricts to $\Sigma_p^* \times \Sigma_p^*$. Moreover, this restriction is nondegenerate, since the map $\Sigma_p^* \ni \eta \mapsto \iota(\eta)B \in \Sigma_p$ is surjective. It is then easy to check that the Poisson bracket determined in each leaf S by the restriction of B to $T^* S \times T^* S$ arises from a symplectic form on S [cf. Exercise 8.23(4)].   □

Remark 8.11 Kirillov's theorem still holds in the general case, where the rank of the map $T_p^* M \ni \omega \mapsto \iota(\omega)B \in T_p M$ is not necessarily constant. In this case the symplectic leaves do not necessarily have the same dimension, and form what is called a **singular foliation**.

Example 8.12

(1) If (M, ω) is a symplectic manifold then there is only one symplectic leaf (M itself).
(2) If $\{\cdot, \cdot\} := 0$ is the trivial Poisson bracket on a smooth manifold M then the Poisson bivector vanishes identically and the symplectic leaves are the zero-dimensional points.
(3) If $\{\cdot, \cdot\}$ is the Poisson bracket defined on $\mathbb{R}^3$ by the formula

$$\{F, G\}(x) := \langle x, \operatorname{grad} F \times \operatorname{grad} G \rangle$$

then the Poisson bivector at $x \in \mathbb{R}^3$ is given by

$$B(v, w) = \langle x, v \times w \rangle = \langle w, x \times v \rangle$$

for any $v, w \in \mathbb{R}^3$, where we use the Euclidean inner product $\langle \cdot, \cdot \rangle$ to make the identification $(\mathbb{R}^3)^* \cong \mathbb{R}^3$. Therefore at $x \in \mathbb{R}^3$ we have

$$\iota(v)B = x \times v,$$

and so the range of B at x is the tangent space to the sphere S_x of radius $\|x\|$ centered at the origin. The symplectic leaves are therefore the spheres S_x (including the origin, which is a singular leaf), and if $x \neq 0$ the symplectic form on S_x is given by

$$\omega(v, w) = \frac{1}{\|x\|^2} \langle x, v \times w \rangle$$

for $v, w \in T_x S_x$ (that is, ω is $\frac{1}{\|x\|}$ times the standard volume form). Indeed, if $F \in C^\infty(\mathbb{R}^3)$ and $v \in T_x S_x$ then we have

$$\begin{aligned}
\omega(X_F, v) &= \frac{1}{\|x\|^2} \langle x, X_F \times v \rangle = \frac{1}{\|x\|^2} \langle v, x \times (\iota(\text{grad } F)B) \rangle \\
&= \frac{1}{\|x\|^2} \langle v, x \times (x \times \text{grad } F) \rangle \\
&= \frac{1}{\|x\|^2} \langle v, \langle x, \text{grad } F \rangle x - \|x\|^2 \text{grad } F \rangle \\
&= -\langle v, \text{grad } F \rangle = -dF(v).
\end{aligned}$$

Next we consider the geometric properties of Hamiltonian flows, that is, flows generated by Hamiltonian vector fields. Just like in the symplectic case, we have a Hamiltonian version of energy conservation.

Proposition 8.13 *Hamiltonian flows preserve their generating functions.*

Proof If $F \in C^\infty(M)$ then

$$X_F \cdot F = \{F, F\} = -\{F, F\} = 0. \qquad \square$$

Recall that in the symplectic case Hamiltonian flows preserve the symplectic form. To obtain the analogue of this property in Poisson geometry we make the following definition.

Definition 8.14 A **Poisson map** $f : M \to N$ between two Poisson manifolds $(M, \{\cdot, \cdot\})$ and $(N, \{\cdot, \cdot\})$ is a differentiable map such that

$$\{F, G\} \circ f = \{F \circ f, G \circ f\}$$

for all $F, G \in C^\infty(N)$.

As one would expect, Poisson maps preserve Hamiltonian flows.

Proposition 8.15 *If $(M, \{\cdot, \cdot\})$ and $(N, \{\cdot, \cdot\})$ are Poisson manifolds, $f : M \to N$ is a Poisson map and $F \in C^\infty(N)$ then*

$$f_* X_{F \circ f} = X_F.$$

Proof We just have to notice that given $G \in C^\infty(N)$ we have

$$(f_* X_{F \circ f}) \cdot G = X_{F \circ f} \cdot (G \circ f) = \{F \circ f, G \circ f\} = \{F, G\} = X_F \cdot G. \qquad \square$$

Finally, we show that Hamiltonian flows preserve the Poisson bracket.

Proposition 8.16 *Hamiltonian flows are Poisson maps.*

Proof Let $\psi_t : M \to M$ be the Hamiltonian flow generated by the function $F \in C^\infty(M)$. If $G \in C^\infty(M)$ is another function we have

$$\frac{d}{dt}(G \circ \psi_t) = \frac{d}{dt}(\psi_t{}^*G) = \frac{d}{ds}\Big|_{s=0} ((\psi_{t+s})^*G) = \frac{d}{ds}\Big|_{s=0} ((\psi_t \circ \psi_s)^*G)$$

$$= \frac{d}{ds}\Big|_{s=0} (\psi_s{}^*(\psi_t{}^*G)) = X_F \cdot (\psi_t{}^*G) = \{F, \psi_t{}^*G\}.$$

Given $G, H \in C^\infty(M)$, let $K_t \in C^\infty(M)$ be the function

$$K_t := \{G, H\} \circ \psi_t - \{G \circ \psi_t, H \circ \psi_t\} = \psi_t{}^*\{G, H\} - \{\psi_t{}^*G, \psi_t{}^*H\}.$$

Clearly $K_0 = 0$. Since the Poisson bracket is bilinear, we have

$$\frac{d}{dt}K_t = \frac{d}{dt}(\psi_t{}^*\{G, H\}) - \left\{\frac{d}{dt}(\psi_t{}^*G), \psi_t{}^*H\right\} - \left\{\psi_t{}^*G, \frac{d}{dt}(\psi_t{}^*H)\right\}$$

$$= X_F \cdot (\psi_t{}^*\{G, H\}) - \{\{F, \psi_t{}^*G\}, \psi_t{}^*H\} - \{\psi_t{}^*G, \{F, \psi_t{}^*H\}\}$$

$$= X_F \cdot (\psi_t{}^*\{G, H\}) - \{F, \{\psi_t{}^*G, \psi_t{}^*H\}\} = X_F \cdot K_t.$$

Regarding K_t as a function K defined on $I \times M$, where $I \subset \mathbb{R}$ is the interval of definition of ψ_t, we see that it satisfies

$$\begin{cases} \left(\frac{\partial}{\partial t} - X_F\right) \cdot K = 0 \\ K(0, p) = 0 \text{ for all } p \in M \end{cases}.$$

Integrating from $\{0\} \times M$ along the integral curves of $\frac{\partial}{\partial t} - X_F$ we then obtain $K_t = 0$ for all $t \in I$. □

We are now ready to discuss symmetry and reduction.

Definition 8.17 Let G be a Lie group acting on a Poisson manifold $(M, \{\cdot, \cdot\})$. The action is said to be:

(1) **Poisson** if for each $g \in G$ the map $A_g : M \to M$ given by $A_g(p) := g \cdot p$ is a Poisson map;
(2) **Hamiltonian** if for each $V \in \mathfrak{g}$ there exists a function $J(V) \in C^\infty(M)$ such that the infinitesimal action X^V is the Hamiltonian vector field generated by $J(V)$, that is, $X^V = X_{J(V)}$.

If G is connected then Proposition 8.16 guarantees that a Hamiltonian action is Poisson [cf. Exercise 8.23(6)]. Notice that because X^V is a linear function of V we can take $J(V)$ to be a linear function of V, and thus think of J as a map $J : M \to \mathfrak{g}^*$. This map is called the **momentum map** for the action.

Theorem 8.18 (Noether, Hamiltonian version) *If the action of the Lie group G on the Poisson manifold* $(M, \{\cdot, \cdot\})$ *is Hamiltonian with momentum map* $J : M \to \mathfrak{g}^*$ *and* $H \in C^\infty(M)$ *is G-invariant then J is constant along the Hamiltonian flow of* H.

Proof Since H is G-invariant we have for any $V \in \mathfrak{g}$

$$X^V \cdot H = 0 \Leftrightarrow X_{J(V)} \cdot H = 0 \Leftrightarrow \{J(V), H\} = 0 \Leftrightarrow X_H \cdot J(V) = 0. \quad \square$$

Example 8.19 The relation between the Hamiltonian and the Lagrangian versions of the Noether theorem is made clear by the following important example. Let M be a differentiable manifold, and let G be a Lie group acting on M. We can lift this action to the symplectic (hence Poisson) manifold T^*M by the formula

$$g \cdot \alpha = A^*_{g^{-1}} \alpha$$

for all $\alpha \in T^*M$, where $A_g : M \to M$ is given by $A_g(p) = g \cdot p$ for all $p \in M$. It is easy to check that this formula indeed defines an action of G on T^*M, mapping each cotangent space $T^*_p M$ to $T^*_{g \cdot p} M$.

Let $(x^1, \ldots, x^n)$ be local coordinates on M and let $(x^1, \ldots, x^n, p_1, \ldots, p_n)$ be the corresponding local coordinates on T^*M. Let $(y^1, \ldots, y^m)$ be local coordinates on G centered at the identity $e \in G$ such that $(-y^1, \ldots, -y^m)$ parameterizes the inverse of the element parameterized by $(y^1, \ldots, y^m)$ (this can be easily accomplished by using the exponential map). If in these coordinates the action $A : G \times M \to M$ of G on M is given by

$$(A^1(x^1, \ldots, x^n, y^1, \ldots, y^m), \ldots, A^n(x^1, \ldots, x^n, y^1, \ldots, y^m))$$

then we have

$$A^*_{g^{-1}} \left(\sum_{i=1}^n p_i dx^i \right) = \sum_{i,j=1}^n p_i \frac{\partial A^i}{\partial x^j}(x, -y) dx^j,$$

and so the lift of the action of G to T^*M is written

$$\left(A^1(x, y), \ldots, A^n(x, y), \sum_{i=1}^n \frac{\partial A^i}{\partial x^1}(x, -y) p_i, \ldots, \sum_{i=1}^n \frac{\partial A^i}{\partial x^n}(x, -y) p_i \right).$$

Therefore the infinitesimal action of $V := \sum_{a=1}^m V^a \frac{\partial}{\partial y^a}$ on T^*M is

$$\sum_{i=1}^n X^i(x) \frac{\partial}{\partial x^i} - \sum_{i,j=1}^n \frac{\partial X^j}{\partial x^i}(x) p_j \frac{\partial}{\partial p_i} = \sum_{i=1}^n \frac{\partial J}{\partial p_i} \frac{\partial}{\partial x^i} - \sum_{i=1}^n \frac{\partial J}{\partial x^i} \frac{\partial}{\partial p_i},$$

where

$$X^i(x) = \sum_{a=1}^{m} \frac{\partial A^i}{\partial y^a}(x, 0)V^a$$

are the components of the infinitesimal action of V on M and

$$J = \sum_{i=1}^{n} X^i(x)p_i.$$

We conclude that the lift of the action of G to T^*M is Hamiltonian with momentum map $J : T^*M \to \mathfrak{g}^*$ given by

$$J(\alpha)(V) = \alpha(X^V),$$

where $X^V \in \mathfrak{X}(M)$ is the infinitesimal action of V on M. Notice that $J(V)$ is exactly the image by the Legendre transformation of the conserved quantity J^V in the Lagrangian version of the Noether theorem.

Theorem 8.20 (Poisson reduction) *If the action of G on $(M, \{\cdot, \cdot\})$ is free, proper and Poisson then M/G is naturally a Poisson manifold (identifying $C^\infty(M/G)$ with the G-invariant functions in $C^\infty(M)$), and the natural projection $\pi : M \to M/G$ is a Poisson map. In particular, π carries the Hamiltonian flow of G-invariant functions on M to the Hamiltonian flow of the corresponding functions in M/G.*

Proof We just have to observe that the if the action is Poisson then the Poisson bracket of G-invariant functions is G-invariant. $\qquad\square$

If G is a Lie group then G acts on G by left multiplication, and the lift of this action to T^*G is free, proper and Hamiltonian. If moreover G is connected then the action is Poisson, and we have the following result.

Theorem 8.21 (Lie–Poisson reduction) *If G is a connected Lie group then the quotient Poisson bracket on $T^*G/G \simeq \mathfrak{g}^*$ is given by*

$$\{F, H\}(\mu) := \mu([dF, dH])$$

for all $F, H \in C^\infty(\mathfrak{g}^*)$, where $dF, dH \in \mathfrak{g}^{**} \simeq \mathfrak{g}$. If $(p_1, \ldots, p_m)$ are linear coordinates on $\mathfrak{g}^*$ corresponding to the basis $\{\omega^1, \ldots, \omega^m\}$ then

$$\{F, H\} = \sum_{a,b,c=1}^{m} p_a C_{bc}^a \frac{\partial F}{\partial p_b} \frac{\partial H}{\partial p_c},$$

where C_{bc}^a are the structure constants associated to the dual basis $\{X_1, \ldots, X_m\}$ of $\mathfrak{g}$.

Proof If we think of $\{\omega^1, \ldots, \omega^m\}$ as left-invariant 1-forms on G then the canonical symplectic potential on T^*G is

$$\theta = \sum_{a=1}^{m} p_a \omega^a,$$

(for simplicity we identify ω^a and $\pi^*\omega^a$, where $\pi : T^*G \to G$ is the natural projection). Now from Exercise 2.8(1) in Chap. 4 we know that

$$d\omega^a = -\frac{1}{2} \sum_{b,c=1}^{m} C^a_{bc}\, \omega^b \wedge \omega^c,$$

and so the canonical symplectic form is

$$\omega = d\theta = \sum_{a}^{m} dp_a \wedge \omega^a - \frac{1}{2} \sum_{a,b,c=1}^{m} p_a C^a_{bc}\, \omega^b \wedge \omega^c$$

$$= \sum_{a}^{m} dp_a \otimes \omega^a - \sum_{a}^{m} \omega^a \otimes dp_a - \sum_{a,b,c=1}^{m} p_a C^a_{bc}\, \omega^b \otimes \omega^c.$$

If $F \in C^\infty(T^*G)$ is G-invariant then it only depends on the coordinates $(p_1, \ldots, p_m)$ along the fibers, and so

$$dF = \sum_{a=1}^{m} \frac{\partial F}{\partial p_a} dp_a.$$

Setting

$$X_F := \sum_{a=1}^{m} \xi^a X_a + \sum_{a=1}^{m} \eta_a \frac{\partial}{\partial p_a},$$

where $\{X_1, \ldots, X_m\}$ is the dual basis of $\mathfrak{g}$, we then have

$$\iota(X_F)\omega = -\sum_{a=1}^{m} \xi^a dp_a - \sum_{a,b,c=1}^{m} p_a C^a_{bc}\, \xi^b \omega^c + \sum_{a=1}^{m} \eta_a \omega^a.$$

From $\iota(X_F)\omega = -dF$ we then obtain

$$X_F = \sum_{a=1}^{m} \frac{\partial F}{\partial p_a} X_a + \sum_{a,b,c=1}^{m} p_a C^a_{bc}\, \frac{\partial F}{\partial p_b} \frac{\partial}{\partial p_c},$$

implying that if $H \in C^\infty(T^*G)$ is also G-invariant then

$$\{F, H\} = X_F \cdot H = \sum_{a,b,c=1}^{m} p_a C_{bc}^a \frac{\partial F}{\partial p_b} \frac{\partial H}{\partial p_c}.$$

Notice that as covectors on $\mathfrak{g}^*$ we have $dp_a = X_a$, and so, by definition of the structure functions C_{bc}^a,

$$\{F, H\} = \sum_{a,b,c=1}^{m} p_a \omega^a([X_b, X_c]) \frac{\partial F}{\partial p_b} \frac{\partial H}{\partial p_c} = \sum_{a,b,c=1}^{m} p_a \omega^a\left(\left[\frac{\partial F}{\partial p_b} X_b, \frac{\partial H}{\partial p_c} X_c\right]\right)$$

$$= \sum_{a} p_a \omega^a([dF, dH]).$$

$\square$

Example 8.22 Lie–Poisson reduction on $T^*SO(3)$ yields the Poisson bracket

$$\{F, G\}(x) := \langle x, \nabla F \times \nabla G \rangle$$

on $\mathfrak{so}(3)^* \cong (\mathbb{R}^3)^* \cong \mathbb{R}^3$, where we used Lemma 3.9 to identify $\mathfrak{so}(3)$ with $(\mathbb{R}^3, \times)$ and the Euclidean inner product $\langle \cdot, \cdot \rangle$ to make $(\mathbb{R}^3)^* \cong \mathbb{R}^3$.

Exercise 8.23

(1) Consider the symplectic structure on

$$S^2 = \{(x, y, z) \in \mathbb{R}^3 \mid x^2 + y^2 + z^2 = 1\}$$

determined by the usual volume form. Compute the Hamiltonian flow generated by the function $H(x, y, z) = z$.

(2) Let (M, ω) be a symplectic manifold. Show that:

(a) $\omega = \sum_{i=1}^{n} dp_i \wedge dx^i$ if and only if $\{x^i, x^j\} = \{p_i, p_j\} = 0$ and $\{p_i, x^j\} = \delta_{ij}$ for $i, j = 1, \ldots, n$;

(b) M is orientable;

(c) if M is compact then ω cannot be exact. (**Remark:** In particular if M is compact and all closed 2-forms on M are exact then M does not admit a symplectic structure; this is the case for all even-dimensional spheres S^{2n} with $n > 1$).

(3) Let $(M, \langle \cdot, \cdot \rangle)$ be a Riemannian manifold, $\alpha \in \Omega^1(M)$ a 1-form and $U \in C^\infty(M)$ a differentiable function.

(a) Show that $\tilde{\omega} := \omega + \pi^* d\alpha$ is a symplectic form on T^*M, where ω is the canonical symplectic form and $\pi : T^*M \to M$ is the natural projection ($\tilde{\omega}$ is called a **canonical symplectic form with magnetic term**).

(b) Show that the Hamiltonian flow generated by a function $\tilde{H} \in C^\infty(T^*M)$ with respect to the symplectic form $\tilde{\omega}$ is given by the equations

$$\begin{cases} \dot{x}^i = \dfrac{\partial \tilde{H}}{\partial p_i} \\[2em] \dot{p}_i = -\dfrac{\partial \tilde{H}}{\partial x^i} + \displaystyle\sum_{j=1}^{n} \left(\dfrac{\partial \alpha_j}{\partial x^i} - \dfrac{\partial \alpha_i}{\partial x^j} \right) \dot{x}^j \end{cases}$$

(c) The map $F : T^*M \to T^*M$ given by

$$F(\xi) := \xi - \alpha_p$$

for $\xi \in T_p^*M$ is a fiber-preserving diffeomorphism. Show that F carries the Hamiltonian flow defined in Exercise 6.15(2) to the Hamiltonian flow of $\tilde{H}$ with respect to the symplectic form $\tilde{\omega}$, where

$$\tilde{H}(\xi) := \frac{1}{2}\langle \xi, \xi \rangle + U(p)$$

for $\xi \in T_p^*M$. (**Remark:** Since the projections of the two flows on M coincide, we see that the magnetic term can be introduced by changing either the Lagrangian or the symplectic form).

(4) Let $(M, \{\cdot, \cdot\})$ be a Poisson manifold, B the Poisson bivector and $(x^1, \ldots, x^n)$ local coordinates on M. Show that:

(a) B can be written in these local coordinates as

$$B = \sum_{i,j=1}^{n} B^{ij} \frac{\partial}{\partial x^i} \otimes \frac{\partial}{\partial x^j},$$

where $B^{ij} = \{x^i, x^j\}$ for $i, j = 1, \ldots, n$;

(b) the Hamiltonian vector field generated by $F \in C^{\infty}(M)$ can be written as

$$X_F = \sum_{i,j=1}^{p} B^{ij} \frac{\partial F}{\partial x^i} \frac{\partial}{\partial x^j};$$

(c) the components of B must satisfy

$$\sum_{l=1}^{n} \left(B^{il} \frac{\partial B^{jk}}{\partial x^l} + B^{jl} \frac{\partial B^{ki}}{\partial x^l} + B^{kl} \frac{\partial B^{ij}}{\partial x^l} \right) = 0$$

for all $i, j, k = 1, \ldots, n$;

(d) if $\{\cdot, \cdot\}$ arises from a symplectic form ω then $(B^{ij}) = -(\omega_{ij})^{-1}$;

(e) if B is nondegenerate then it arises from a symplectic form.

(5) (*Action-angle coordinates*) Let (M, ω) be a symplectic manifold and $F = (F_1, \ldots, F_n) : M \to \mathbb{R}^n$ a set of independent first integrals in involution, with compact level sets (n-tori). Choose an invariant torus T_0, a point $\alpha \in T_0$, and an n-dimensional submanifold $N \subset M$ transverse to T_0 at α (that is, $T_\alpha M = T_\alpha T \oplus T_\alpha N$). We fix the the coordinates $x = (x^1, \ldots, x^n)$ determined on each invariant torus T close to T_0 by the identification $T \cong \mathbb{R}^n/\mathbb{Z}^n$ (which arises from the $\mathbb{R}^n$-action associated to F) by setting $x = 0$ on N. In this way we obtain local coordinates (x, F) in a neighborhood of T. Show that:

(a) In these coordinates the components of the Poisson bivector are

$$(B) = \begin{pmatrix} * & | & -A^t \\ - - - - & + & - - - \\ A & | & 0 \end{pmatrix},$$

and so the components of the symplectic form are

$$(\omega) = \begin{pmatrix} 0 & | & -A^{-1} \\ - - - & + & - - - \\ (A^{-1})^t & | & * \end{pmatrix},$$

where $A = A(F)$ is the matrix $A = (a_{ij})$ defined by

$$X_{F_i} = \sum_{j=1}^{n} a_{ij} \frac{\partial}{\partial x^j};$$

(b) it is possible to choose new coordinates $J = J(F)$ such that $\{J_i, x^j\} = \delta_{ij}$;
(c) $\{x^i, x^j\}$ is a function of J only;
(d) it is possible to choose new coordinates $y = x + z(J)$ such that $\{y^i, y^j\} = 0$ and $\{J_i, y^j\} = \delta_{ij}$;
(e) there exists a 1-form θ in a neighborhood of T_0 such that $\omega = d\theta$, and $J_i = \oint_{\gamma_i} \theta$, where γ_i is the projection of the y^i-axis on each invariant torus $T \cong \mathbb{R}^n/\mathbb{Z}^n$.

(**Hint:** You will need to use the Poincaré Lemma – cf. Exercise 3.8(6) in Chap. 2).

(6) Let G be a connected Lie group and $U \subset G$ a neighborhood of the identity. Show that:

(a) $G = \cup_{n=1}^{+\infty} U^n$, where $U^n = \{g_1 \cdots \cdot g_n \mid g_1, \ldots, g_n \in U\}$;
(b) if G acts on a Poisson manifold $(M, \{\cdot, \cdot\})$ and the action is Hamiltonian then it is Poisson.

(7) Let G be a connected Lie group with a free, proper, Hamiltonian action on a Poisson manifold $(M, \{\cdot, \cdot\})$, and let $H \in C^\infty(M)$ be G-invariant. Show that if $p \in M/G$ is a fixed point of $\pi_* X_H \in \mathfrak{X}(M/G)$ (where $\pi : M \to M/G$ is the

quotient map) then the flow of X_H on $\pi^{-1}(p)$ is given by orbits of 1-parameter subgroups of G.

(8) The Lie group $SO(2) \simeq S^1$ acts on $M = \mathbb{R}^2 \setminus \{(0, 0)\}$ through

$$e^{i\varphi} \cdot (r, \theta) = (r, \theta + \varphi),$$

where (r, θ) are polar coordinates on M and θ, φ should be understood mod 2π.

(a) Write an expression for the infinitesimal action $X^V \in \mathfrak{X}(M)$ of $V \in \mathfrak{so}(2) \cong \mathbb{R}$.

(b) Determine the momentum map for the lift of this action to the cotangent bundle.

(c) Write an expression for the Poisson bivector of T^*M with the canonical symplectic structure in the usual coordinates $(r, \theta, p_r, p_\theta)$.

(d) Calculate the Poisson bivector of the Poisson manifold $Q := T^*M/SO(2) \simeq \mathbb{R}^3$. What are the symplectic leaves of this manifold? Give an example of a nonconstant Casimir function.

(e) Consider the Hamiltonian $H : T^*Q \to \mathbb{R}$ given by

$$H(r, \theta, p_r, p_\theta) = \frac{p_r{}^2}{2} + \frac{p_\theta{}^2}{2r^2} + u(r).$$

Show that H is $SO(2)$-invariant, and determine its Hamiltonian flow on the reduced Poisson manifold Q.

(f) Use the Noether theorem to obtain a quantity conserved by the Hamiltonian flow of H on T^*M.

(9) Recall that the upper half plane $H = \{(x, y) \in \mathbb{R}^2 \mid y > 0\}$ has a Lie group structure, given by the operation

$$(x, y) \cdot (z, w) := (yz + x, yw),$$

and that the hyperbolic plane corresponds to the left-invariant metric

$$g := \frac{1}{y^2} (dx \otimes dx + dy \otimes dy)$$

on H [cf. Exercise 7.17(3) in Chap. 1 and Exercise 3.3(5) in Chap. 3]. The geodesics are therefore determined by the Hamiltonian function $K : T^*H \to \mathbb{R}$ given by

$$K(x, y, p_x, p_y) = \frac{y^2}{2} \left(p_x{}^2 + p_y^2 \right).$$

(a) Determine the lift to T^*H of the action of H on itself by left translation, and check that it preserves the Hamiltonian K.

(b) Show that the functions

$$F(x, y, p_x, p_y) = yp_x \quad \text{and} \quad G(x, y, p_x, p_y) = yp_y$$

are also H-invariant, and use this to obtain the quotient Poisson structure on T^*H/H. Is this a symplectic manifold?

(c) Write an expression for the momentum map for the action of H on T^*H, and use it to obtain a nontrivial first integral I of the geodesic equations. Show that the projection on H of a geodesic for which $K = E$, $p_x = l$ and $I = m$ satisfies the equation

$$l^2 x^2 + l^2 y^2 - 2lmx + m^2 = 2E.$$

Assuming $l \neq 0$, what are these curves?

(10) Recall that the Euler top is the mechanical system determined by the Lagrangian $L : TSO(3) \to \mathbb{R}$ given by

$$L = \frac{1}{2} \langle I\Omega, \Omega \rangle,$$

where Ω are the left-invariant coordinates on the fibers resulting from the identifications

$$T_S SO(3) = dL_S(\mathfrak{so}(3)) \cong \mathfrak{so}(3) \cong \mathbb{R}^3.$$

(a) Show that if we use the Euclidean inner product $\langle \cdot, \cdot \rangle$ to identify $(\mathbb{R}^3)^*$ with $\mathbb{R}^3$ then the Legendre transformation is written

$$P = I\Omega,$$

where P are the corresponding left-invariant coordinates on $T^*SO(3)$.

(b) Write the Hamilton equations on the reduced Poisson manifold $T^*SO(3)/SO(3) \cong \mathbb{R}^3$. What are the symplectic leaves? Give an example of a non-constant Casimir function.

(c) Compute the momentum map for the lift to $T^*SO(3)$ of the action of $SO(3)$ on itself by left translation.

(11) Let (P^1, P^2, P^3) be the usual left-invariant coordinates on the fibers of $T^*SO(3)$, and consider the functions $(\Gamma^1, \Gamma^2, \Gamma^3)$ defined through

$$\gamma = S\Gamma$$

for each $S \in SO(3)$, where $\gamma \in \mathbb{R}^3$ is a fixed vector. Show that for $i, j = 1, 2, 3$:

(a) $\{P^i, P^j\} = \sum_{k=1}^{3} \varepsilon_{ijk} P^k;$

(b) $\{\Gamma^i, \Gamma^j\} = 0;$

(c) $\{P^i, \Gamma^j\} = \sum_{k=1}^{3} \varepsilon_{ijk} \Gamma^k,$

where

$$\varepsilon_{ijk} = \begin{cases} +1 & \text{if } (i, j, k) \text{ is an even permutation of } (1, 2, 3) \\ -1 & \text{if } (i, j, k) \text{ is an odd permutation of } (1, 2, 3) \\ 0 & \text{otherwise.} \end{cases}$$

(**Hint:** Show that $\dot{\Gamma} = \Gamma \times \Omega$ along any motion of the Euler top, where Ω is the angular velocity in the Euler top's frame, and regard $\frac{(P^i)^2}{2}$ as the limit of the Euler top Hamiltonian when $I_i = 1$ and $I_j \to +\infty$ for $j \neq i$).

(12) If in Exercise 8.23(11) we set $\gamma = ge_z$, where g is the (constant) gravitational acceleration, then the motion of a rigid body (with a fixed point) of mass M and moment of inertia I, whose center of mass has position vector $L \in \mathbb{R}^3$ in its frame, is given by the Hamiltonian flow of

$$H = \frac{1}{2}\langle P, I^{-1}P\rangle + M\langle \gamma, SL\rangle.$$

(a) Show that H is S^1-invariant for the lift to $T^*SO(3)$ of the action of S^1 on $SO(3)$ determined by $e^{i\theta} \cdot S = R_\theta S$, where

$$R_\theta := \begin{pmatrix} \cos\theta & -\sin\theta & 0 \\ \sin\theta & \cos\theta & 0 \\ 0 & 0 & 1 \end{pmatrix}$$

(corresponding to rotations about the z-axis).

(b) Determine the momentum map for this action.

(c) Show that the functions P and Γ are S^1-invariant, and that the Poisson bracket on the quotient manifold $T^*SO(3)/S^1 \cong (SO(3)/S^1) \times \mathbb{R}^3 \cong S^2 \times \mathbb{R}^3$ is determined by the Poisson brackets of these functions.

(d) Use the functions P and Γ to write the equations of motion on the quotient, and give an example of a nonconstant Casimir function.

5.9 Notes

5.9.1 Section 5.1

Throughout this chapter, starting at the exercises of Sect. 5.1, we need several definitions and facts related to stability of fixed points of vector fields in $\mathbb{R}^n$ (refer for instance to [Arn92, GH02] for additional details). In order to study nonlinear systems

$$\begin{cases} \dot{x} = f(x) \\ x(0) = x_0 \end{cases} \quad (x \in \mathbb{R}^n) \tag{5.4}$$

one usually starts by finding the zeros of f, called **fixed points**, **equilibria** or **stationary solutions**. A fixed point $\bar{x}$ is called **stable** if for each neighborhood U of $\bar{x}$ there exists another (possibly smaller) neighborhood V of $\bar{x}$ such that if $x_0 \in V$ then $x(t) \in U$ for each $t > 0$ where the solution is defined. The behavior of solutions near $\bar{x}$ can, in many situations, be studied by linearizing (5.4) at $\bar{x}$ and analyzing the resulting (linear) system

$$\begin{cases} \dot{\xi} = A\xi \\ \xi(0) = \xi_0 \end{cases} \quad (\xi \in \mathbb{R}^n) \tag{5.5}$$

where $A := (df)_{\bar{x}}$. This linear system has a global solution

$$\xi(\xi_0, t) = e^{tA}\xi_0,$$

where e^{tA} can be seen as a map from $\mathbb{R}^n$ to $\mathbb{R}^n$ defining the flow of the vector field $A\xi$. If we put A in the Jordan canonical form then it is clear that this flow has the following invariant subspaces:

$$\begin{aligned} E^s &:= \text{span}\{v_1, \ldots, v_{n_s}\} & \text{(stable subspace)}; \\ E^u &:= \text{span}\{u_1, \ldots, u_{n_u}\} & \text{(unstable subspace)}; \\ E^c &:= \text{span}\{w_1, \ldots, w_{n_c}\} & \text{(center subspace)}, \end{aligned}$$

where $v_1, \ldots, v_{n_s}$ are the n_s generalized eigenvectors corresponding to eigenvalues with negative real part, $u_1, \ldots, u_{n_u}$ are the n_u generalized eigenvectors corresponding to eigenvalues with positive real part, and $w_1, \ldots, w_{n_s}$ are the n_c generalized eigenvectors corresponding to eigenvalues with zero real part. If $E^c = \varnothing$ then $\bar{x}$ is called a **hyperbolic** or **nondegenerate** fixed point of f. In this case the **Hartman–Grobman theorem** tells us that there exists a homeomorphism from a neighborhood of $\bar{x}$ in $\mathbb{R}^n$ to a neighborhood of 0 in $\mathbb{R}^n$ which takes the orbits of the non-linear flow of (5.4) to those of the linear flow e^{tA} of (5.5). The asymptotic behavior of solutions near $\bar{x}$, and consequently its stability type, is then determined by the eigenvalues λ of A.

5.9.2 Bibliographical Notes

The material in this chapter follows [Oli02],[Arn97] closely. There are of course many other excellent books on mechanics, both traditional [GPS02] and geometric [AM78, MR99]. Non-holonomic systems (including control theory) are treated in greater detail in [Blo03, BL05]. For more information on completely integrable systems see [CB97, Aud96].

References

[AM78]   Abraham, R., Marsden, J.: Foundations of Mechanics. Addison Wesley, New York (1978)
[Arn92]   Arnold, V.I.: Ordinary Differential Equations. Springer, New York (1992)
[Arn97]   Arnold, V.I.: Mathematical Methods of Classical Mechanics. Springer, New York (1997)
[Aud96]   Audin, M.: Spinning Tops. Cambridge University Press, Cambridge (1996)
[Blo03]   Bloch, A.: Nonholonomic Mechanics and Control. Springer, New York (2003)
[BL05]   Bullo, F., Lewis, A.: Geometric Control of Mechanical Systems. Springer, New York (2005)
[CB97]   Cushmann, R., Bates, L.: Global Aspects of Classical Integrable Systems. Birkhäuser, Basel (1997)
[GPS02]   Goldstein, H., Poole, C., Safko, J.: Classical Mechanics. Addison Wesley, New York (2002)
[GH02]   Guckenheimer, J., Holmes, P.: Nonlinear Oscillations Dynamical Systems and Bifurcations of Vector Fields. Springer, New York (2002)
[MR99]   Marsden, J., Ratiu, T.: Introduction to Mechanics and Symmetry. Springer, New York (1999)
[Oli02]   Oliva, W.: Geometric Mechanics. Springer, New York (2002)
[War83]   Warner, F.: Foundations of Differentiable Manifolds and Lie Groups. Springer, New York (1983)

Chapter 6
Relativity

This chapter studies one of the most important applications of Riemannian geometry: the theory of **general relativity**. This theory, which ultimately superseded the classical mechanics of Galileo and Newton, arose from the seemingly paradoxical experimental fact that the speed of light is the same for every observer, independently of their state of motion. In 1905, after a period of great confusion, Einstein came up with an explanation that was as simple as it was radical: time intervals and length measurements are not the same for all observers, but instead depend on their state of motion. In 1908, Minkowski gave a geometric formulation of Einstein's theory by introducing a pseudo-inner product in the four-dimensional **spacetime** $\mathbb{R}^4$. While initially resisting this "excessive mathematization" of his theory, Einstein soon realized that curving spacetime was actually the key to understanding gravity. In 1915, after a long struggle with the mathematics of Riemannian geometry, he was able to arrive at a complete formulation of the general theory of relativity. The predictions of his theory were first confirmed in 1919 by a British solar eclipse expedition, led by Eddington, and have since been verified in every experimental test ever attempted.

To smooth the transition from classical mechanics to relativity, Sect. 6.1 discusses **Galileo spacetime**, the geometric structure underlying Newtonian mechanics, which hinges on the existence of arbitrarily fast motions. If, however, a maximum speed is assumed to exist, then it must be replaced by **Minkowski spacetime**, whose geometry is studied in **special relativity** (Sect. 6.2).

Section 6.3 shows how to include Newtonian gravity in Galileo spacetime by introducing the symmetric **Cartan connection**. Trying to generalize this procedure leads to general **Lorentzian manifolds** satisfying the **Einstein field equation**, of which Minkowski spacetime is the simplest example (Sect. 6.4).

Other simple solutions are analyzed in the subsequent sections: the **Schwarzschild solution**, modeling the gravitational field outside spherically symmetric bodies or **black holes** (Sect. 6.5), and the **Friedmann–Lemaître–Robertson–Walker** models of **cosmology**, describing the behavior of the universe as a whole (Sect. 6.6).

Finally, Sect. 6.7 discusses of the **causal structure** of a Lorentz manifold, in preparation for the proof of the celebrated **singularity theorems** of **Hawking** (Sect. 6.8) and **Penrose** (Sect. 6.9).

© Springer International Publishing Switzerland 2014
L. Godinho and J. Natário, *An Introduction to Riemannian Geometry*, Universitext,
DOI 10.1007/978-3-319-08666-8_6

6.1 Galileo Spacetime

The set of all physical occurrences can be modeled as a connected 4-dimensional manifold M, which we call **spacetime**, and whose points we refer to as **events**. We assume that M is diffeomorphic to $\mathbb{R}^4$, and that there exists a special class of diffeomorphisms $x : M \to \mathbb{R}^4$, called **inertial frames**. An inertial frame yields global coordinates $(x^0, x^1, x^2, x^3) = (t, x, y, z)$. We call the coordinate $t : M \to \mathbb{R}$ the **time function** associated to a given inertial frame. Two events $p, q \in M$ are said to be **simultaneous** on that frame if $t(p) = t(q)$. The level functions of the time function are therefore called **simultaneity hypersurfaces**. The **distance** between two simultaneous events $p, q \in M$ is given by

$$d(p, q) = \sqrt{\sum_{i=1}^{3} \left(x^i(p) - x^i(q)\right)^2}.$$

The motion of a particle is modeled by a smooth curve $c : I \to M$ such that $dt(\dot{c}) \neq 0$. A special class of motions is formed by the motions of **free particles**, i.e. particles which are not acted upon by any external force. The special property that inertial frames have to satisfy is that the motions of free particles are always represented by straight lines. In other words, free particles move with constant velocity relative to inertial frames (**Newton's law of inertia**). In particular, motions of particles at rest in an inertial frame are motions of free particles.

Inertial frames are not unique: if $x : M \to \mathbb{R}^4$ is an inertial frame and $T : \mathbb{R}^4 \to \mathbb{R}^4$ is an invertible affine transformation then $T \circ x$ is another inertial frame. In fact, any two inertial frames must be related by such an affine transformation [cf. Exercise 1.1(3)].

The **Galileo spacetime**, which underlies Newtonian mechanics, is obtained by further requiring that inertial frames should:

(1) agree on the time interval between any two events (and hence on whether two given events are simultaneous);
(2) agree on the distance between simultaneous events.

Therefore, up to translations and reflections, all coordinate transformations between inertial frames belong to the **Galileo group** $Gal(4)$, the group of linear orientation-preserving maps which preserve time functions and the Euclidean structures of the simultaneity hypersurfaces.

When analyzing problems in which only one space dimension is important, we can use a simpler 2-dimensional Galileo spacetime. If (t, x) are the spacetime coordinates associated to an inertial frame and $T \in Gal(2)$ is a Galileo change of basis to a new inertial frame with global coordinates (t', x'), then

$$\frac{\partial}{\partial t'} := T\left(\frac{\partial}{\partial t}\right) = \frac{\partial}{\partial t} + v\frac{\partial}{\partial x}$$

$$\frac{\partial}{\partial x'} := T\left(\frac{\partial}{\partial x}\right) = \frac{\partial}{\partial x}$$

with $v \in \mathbb{R}$, since we must have $t = t'$, and so

$$dt\left(\frac{\partial}{\partial t'}\right) = dt'\left(\frac{\partial}{\partial t'}\right) = 1,$$

and we want the orientation-preserving map T to be an isometry of the simultaneity hypersurface $\{t = 0\} \equiv \{t' = 0\}$. The change of basis matrix is then

$$S = \begin{pmatrix} 1 & 0 \\ v & 1 \end{pmatrix},$$

with inverse

$$S^{-1} = \begin{pmatrix} 1 & 0 \\ -v & 1 \end{pmatrix}.$$

Therefore the corresponding coordinate transformation is

$$\begin{cases} t' = t \\ x' = x - vt \end{cases} \qquad (v \in \mathbb{R})$$

(**Galileo transformation**), and hence the new frame is moving with velocity v with respect to the old one (as the curve $x' = 0$ is the curve $x = vt$). Notice that S^{-1} is obtained from S simply by reversing the sign of v, as one would expect, as the old frame must be moving relative to the new one with velocity $-v$. We shall call this observation the **relativity principle**.

Exercise 1.1

(1) *(Lucas problem)* By the late 19th century there existed a regular transatlantic service between Le Havre and New York. Every day at noon (GMT) a transatlantic ship would depart Le Havre and another one would depart New York. The journey took exactly seven days, so that arrival would also take place at noon (GMT). Therefore, a transatlantic ship traveling from Le Havre to New York would meet a transatlantic ship just arriving from New York at departure, and another one just leaving New York on arrival. Besides these, how many other ships would it meet? At what times? What was the total number of ships needed for this service? (Hint: Represent the ships' motions as curves in a 2-dimensional Galileo spacetime).

(2) Check that free particles move with constant velocity relative to inertial frames.

(3) Let $f : \mathbb{R}^n \to \mathbb{R}^n$ ($n \geq 2$) be a bijection that takes straight lines to straight lines. Show that f must be an affine function, i.e. that

$$f(x) = Ax + b$$

for all $x \in \mathbb{R}^n$, where $A \in GL(n, \mathbb{R})$ and $b \in \mathbb{R}^n$.

(4) Prove that the Galileo group $Gal(4)$ is the subset of $GL(4, \mathbb{R})$ formed by matrices of the form

$$\begin{pmatrix} 1 & 0 \\ v & R \end{pmatrix}$$

where $v \in \mathbb{R}^3$ and $R \in SO(3)$. Conclude that $Gal(4)$ is isomorphic to the group of orientation-preserving isometries of the Euclidean 3-space $\mathbb{R}^3$.

(5) Show that $Gal(2)$ is a subgroup of $Gal(4)$.

6.2 Special Relativity

The Galileo spacetime requirement that all inertial observers should agree on the time interval between two events is intimately connected with the possibility of synchronizing clocks in different frames using signals of arbitrarily high speeds. Experience reveals that this is actually impossible. Instead, there appears to exist a maximum propagation speed, the speed of light (approximately 300,000 km/s), which is the same at all events and in all directions. A more accurate requirement is then that any two inertial frames should

(1′) agree on whether a given particle is moving at the speed of light.

Notice that we no longer require that different inertial frames should agree on the time interval between two events, or even on whether two given events are simultaneous. However, we still require that any two inertial frames should

(2′) agree on the distance between events which are simultaneous in **both** frames.

It is convenient to choose units such that the speed of light is 1 (for instance measuring time in years and distance in light-years). Fix a particular inertial frame with coordinates (x^0, x^1, x^2, x^3). A free particle moving at the speed of light on an inertial frame $x : \mathbb{R}^4 \to \mathbb{R}$ will be a straight line whose tangent vector

$$v = v^0 \frac{\partial}{\partial x^0} + v^1 \frac{\partial}{\partial x^1} + v^2 \frac{\partial}{\partial x^2} + v^3 \frac{\partial}{\partial x^3}$$

must satisfy

$$(v^0)^2 = (v^1)^2 + (v^2)^2 + (v^3)^2, \tag{6.1}$$

so that the distance traveled equals the elapsed time. In other words, v must satisfy $\langle v, v \rangle = 0$, where

$$\langle v, w \rangle := -v^0 w^0 + v^1 w^1 + v^2 w^2 + v^3 w^3 = \sum_{\mu,\nu=0}^{3} \eta_{\mu\nu} v^\mu w^\nu,$$

with $(\eta_{\mu\nu}) = \text{diag}(-1, 1, 1, 1)$. Notice that $\langle \cdot, \cdot \rangle$ is a symmetric nondegenerate tensor which is not positive definite; we call it the **Minkowski (pseudo) inner product**. The coordinate basis

$$\left\{ \frac{\partial}{\partial x^0}, \frac{\partial}{\partial x^1}, \frac{\partial}{\partial x^2}, \frac{\partial}{\partial x^3} \right\}$$

is an orthonormal basis for this inner product [cf. Exercise 2.2(1)], as

$$\left\langle \frac{\partial}{\partial x^\mu}, \frac{\partial}{\partial x^\nu} \right\rangle = \eta_{\mu\nu}$$

$(\mu, \nu = 0, 1, 2, 3)$.

Since we used a particular inertial frame to define the Minkowski inner product, we must now check that it is well defined (i.e. it is independent of the inertial frame we chose to define it). Let $(x^{0\prime}, x^{1\prime}, x^{2\prime}, x^{3\prime})$ be the coordinates associated to another inertial frame. The analogue of (6.1) on the new inertial frame implies that the vectors

$$\frac{\partial}{\partial x^{0\prime}} \pm \frac{\partial}{\partial x^{i\prime}}$$

$(i = 1, 2, 3)$ must be tangent to a motion at the speed of light. By assumption $(1\prime)$, given a motion of a free particle at the speed of light, all inertial observers must agree that the particle is moving at this (maximum) speed. Therefore we must have

$$\left\langle \frac{\partial}{\partial x^{0\prime}} \pm \frac{\partial}{\partial x^{i\prime}}, \frac{\partial}{\partial x^{0\prime}} \pm \frac{\partial}{\partial x^{i\prime}} \right\rangle = 0.$$

This implies that

$$\left\langle \frac{\partial}{\partial x^{0\prime}}, \frac{\partial}{\partial x^{0\prime}} \right\rangle = -\left\langle \frac{\partial}{\partial x^{i\prime}}, \frac{\partial}{\partial x^{i\prime}} \right\rangle;$$

$$\left\langle \frac{\partial}{\partial x^{0\prime}}, \frac{\partial}{\partial x^{i\prime}} \right\rangle = 0.$$

Similarly, we must have

$$\left\langle \sqrt{2}\frac{\partial}{\partial x^{0\prime}} + \frac{\partial}{\partial x^{i\prime}} + \frac{\partial}{\partial x^{j\prime}}, \sqrt{2}\frac{\partial}{\partial x^{0\prime}} + \frac{\partial}{\partial x^{i\prime}} + \frac{\partial}{\partial x^{j\prime}} \right\rangle = 0$$

$(i \neq j)$, and hence

$$\left\langle \frac{\partial}{\partial x^{i\prime}}, \frac{\partial}{\partial x^{j\prime}} \right\rangle = 0.$$

Since $\langle \cdot, \cdot \rangle$ is nondegenerate, we conclude that there must exist $k \neq 0$ such that

$$\left\langle \frac{\partial}{\partial x^{\mu\prime}}, \frac{\partial}{\partial x^{\nu\prime}} \right\rangle = k\eta_{\mu\nu}$$

($\mu, \nu = 0, 1, 2, 3$).

The simultaneity hypersurfaces $\{x^0 = \text{const.}\}$ and $\{x^{0\prime} = \text{const.}\}$ are 3-planes in $\mathbb{R}^4$. If they are parallel, they will coincide for appropriate values of the constants; otherwise, they must intersect along 2-planes of events which are simultaneous in both frames. In either case there exist events which are simultaneous in both frames. Let $v \neq 0$ be a vector connecting two such events. Then $dx^0(v) = dx^{0\prime}(v) = 0$, and hence

$$v = \sum_{i=1}^{3} v^i \frac{\partial}{\partial x^i} = \sum_{i=1}^{3} v^{i\prime} \frac{\partial}{\partial x^{i\prime}}.$$

By assumption (2$\prime$), we must have

$$\sum_{i=1}^{3} \left(v^i\right)^2 = \sum_{i=1}^{3} \left(v^{i\prime}\right)^2.$$

Consequently, from

$$\sum_{i=1}^{3} \left(v^i\right)^2 = \langle v, v \rangle = \left\langle \sum_{i=1}^{3} v^{i\prime} \frac{\partial}{\partial x^{i\prime}}, \sum_{i=1}^{3} v^{i\prime} \frac{\partial}{\partial x^{i\prime}} \right\rangle = k \sum_{i=1}^{3} \left(v^{i\prime}\right)^2,$$

we conclude that we must have $k = 1$. Therefore the coordinate basis

$$\left\{ \frac{\partial}{\partial x^{0\prime}}, \frac{\partial}{\partial x^{1\prime}}, \frac{\partial}{\partial x^{2\prime}}, \frac{\partial}{\partial x^{3\prime}} \right\}$$

must also be an orthonormal basis. In particular, this means that the Minkowski inner product $\langle \cdot, \cdot \rangle$ is well defined (i.e. it is independent of the inertial frame we choose to define it), and that we can identify inertial frames with orthonormal bases of $(\mathbb{R}^4, \langle \cdot, \cdot \rangle)$.

Definition 2.1 $(\mathbb{R}^4, \langle \cdot, \cdot \rangle)$ is said to be the **Minkowski spacetime**. The **length** of a vector $v \in \mathbb{R}^4$ is $|v| = |\langle v, v \rangle|^{\frac{1}{2}}$.

The study of the geometry of Minkowski spacetime is usually called **special relativity**. A vector $v \in \mathbb{R}^4$ is said to be:

(1) **timelike** if $\langle v, v \rangle < 0$; in this case, there exists an inertial frame $(x^{0\prime}, x^{1\prime}, x^{2\prime}, x^{3\prime})$ such that

$$v = |v| \frac{\partial}{\partial x^{0\prime}}$$

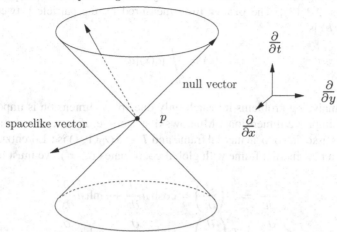

Fig. 6.1 Minkowski geometry (traditionally represented with the t-axis pointing upwards)

[cf. Exercise 2.2(1)], and consequently any two events p and $p + v$ occur on the same spatial location in this frame, separated by a time interval $|v|$;
(2) **spacelike** if $\langle v, v \rangle > 0$; in this case, there exists an inertial frame $(x^{0'}, x^{1'}, x^{2'}, x^{3'})$ such that

$$v = |v| \frac{\partial}{\partial x^{1'}}$$

[cf. Exercise 2.2(1)], and consequently any two events p and $p + v$ occur simultaneously in this frame, a distance $|v|$ apart;
(3) **lightlike**, or **null**, if $\langle v, v \rangle = 0$; in this case any two events p and $p + v$ are connected by a motion at the speed of light in **any** inertial frame.

The set of all null vectors is called the **light cone**, and it is in a way the structure that replaces the absolute simultaneity hypersurfaces of Galileo spacetime. It is the boundary of the set of all timelike vectors, which has two connected components; we represent by $C(v)$ the connected component that contains a given timelike vector v. A **time orientation** for Minkowski spacetime is a choice of one of these components, whose elements are said to be **future-pointing**; this is easily extended to nonzero null vectors (Fig. 6.1).

An inertial frame (x^0, x^1, x^2, x^3) determines a time orientation, namely that for which the future-pointing timelike vectors are the elements of $C\left(\frac{\partial}{\partial x^0}\right)$. Up to translations and reflections, all coordinate transformations between inertial frames belong to the **(proper) Lorentz group** $SO_0(3, 1)$, the group of linear maps which preserve orientation, time orientation and the Minkowski inner product (hence the light cone).

A curve $c : I \subset \mathbb{R} \to \mathbb{R}^4$ is said to be **timelike** if $\langle \dot{c}, \dot{c} \rangle < 0$. Timelike curves represent motions of particles with nonzero mass, since only for these curves it is possible to find an inertial frame in which the particle is instantaneously at rest.

In other words, massive particles must always move at less than the speed of light [cf. Exercise 2.2(13)]. The **proper time** measured by the particle between events $c(a)$ and $c(b)$ is

$$\tau(c) := \int_a^b |\dot{c}(s)| \, ds.$$

When analyzing problems in which only one space dimension is important, we can use a simpler 2-dimensional Minkowski spacetime. If (t, x) are the spacetime coordinates associated to an inertial frame and $T \in SO_0(1, 1)$ is a Lorentzian change of basis to a new inertial frame with global coordinates (t', x'), we must have

$$\frac{\partial}{\partial t'} := T \left(\frac{\partial}{\partial t} \right) = \cosh u \frac{\partial}{\partial t} + \sinh u \frac{\partial}{\partial x}$$

$$\frac{\partial}{\partial x'} := T \left(\frac{\partial}{\partial x} \right) = \sinh u \frac{\partial}{\partial t} + \cosh u \frac{\partial}{\partial x}$$

with $u \in \mathbb{R}$ [cf. Exercise 2.2(3)]. The change of basis matrix is

$$S = \begin{pmatrix} \cosh u & \sinh u \\ \sinh u & \cosh u \end{pmatrix},$$

with inverse

$$S^{-1} = \begin{pmatrix} \cosh u & -\sinh u \\ -\sinh u & \cosh u \end{pmatrix}.$$

Therefore the corresponding coordinate transformation is

$$\begin{cases} t' = t \cosh u - x \sinh u \\ x' = x \cosh u - t \sinh u \end{cases}$$

(**Lorentz transformation**), and hence the new frame is moving with velocity $v = \tanh u$ with respect to the old one (as the curve $x' = 0$ is the curve $x = vt$; notice that $|v| < 1$). The matrix S^{-1} is obtained from S simply by reversing the sign of u, or, equivalently, of v; therefore, the relativity principle still holds for Lorentz transformations.

Moreover, since

$$\begin{cases} \cosh u = \left(1 - v^2\right)^{-\frac{1}{2}} \\ \sinh u = v \left(1 - v^2\right)^{-\frac{1}{2}} \end{cases},$$

one can also write the Lorentz transformation as

$$\begin{cases} t' = \left(1 - v^2\right)^{-\frac{1}{2}} t - v \left(1 - v^2\right)^{-\frac{1}{2}} x \\ x' = \left(1 - v^2\right)^{-\frac{1}{2}} x - v \left(1 - v^2\right)^{-\frac{1}{2}} t. \end{cases}$$

In everyday life situations, we deal with frames whose relative speed is much smaller that the speed of light, $|v| \ll 1$, and with events for which $|x| \ll |t|$ (distances traveled by particles in one second are much smaller that 300,000 km). An approximate expression for the Lorentz transformations in these situations is then

$$\begin{cases} t' = t \\ x' = x - vt \end{cases}$$

which is just a Galileo transformation. In other words, the Galileo group is a convenient low-speed approximation of the Lorentz group.

Suppose that two distinct events p and q occur in the same spatial location in the inertial frame (t', x'),

$$q - p = \Delta t' \frac{\partial}{\partial t'} = \Delta t' \cosh u \frac{\partial}{\partial t} + \Delta t' \sinh u \frac{\partial}{\partial x} = \Delta t \frac{\partial}{\partial t} + \Delta x \frac{\partial}{\partial x}.$$

We see that the time separation between the two events in a different inertial frame (t, x) is **bigger**,

$$\Delta t = \Delta t' \cosh u > \Delta t'.$$

Loosely speaking, moving clocks run slower when compared to stationary ones (**time dilation**).

If, on the other hand, two distinct events p and q occur simultaneously in the inertial frame (t', x'),

$$q - p = \Delta x' \frac{\partial}{\partial x'} = \Delta x' \sinh u \frac{\partial}{\partial t} + \Delta x' \cosh u \frac{\partial}{\partial x} = \Delta t \frac{\partial}{\partial t} + \Delta x \frac{\partial}{\partial x},$$

then they will **not** be simultaneous in the inertial frame (t, x), where the time difference between them is

$$\Delta t = \Delta x' \sinh u \neq 0$$

(**relativity of simultaneity**).

Finally, consider two particles at rest in the inertial frame (t', x'). Their motions are the lines $x' = x_0'$ and $x' = x_0' + l'$. In the inertial frame (t, x), these lines have equations

$$x = \frac{x_0'}{\cosh u} + vt \quad \text{and} \quad x = \frac{x_0' + l'}{\cosh u} + vt,$$

which describe motions of particles moving with velocity v and separated by a distance

$$l = \frac{l'}{\cosh u} < l'.$$

Loosely speaking, moving objects shrink in the direction of their motion (**length contraction**).

Exercise 2.2

(1) Let $\langle \cdot, \cdot \rangle$ be a nondegenerate symmetric 2-tensor on n-dimensional vector space V. Show that there always exists an **orthonormal basis** $\{v_1, \ldots, v_n\}$, i.e. a basis such that $\langle v_i, v_j \rangle = \varepsilon_{ij}$, where $\varepsilon_{ii} = \pm 1$ and $\varepsilon_{ij} = 0$ for $i \neq j$. Moreover, show that $s = \sum_{i=1}^{n} \varepsilon_{ii}$ (known as the **signature** of $\langle \cdot, \cdot \rangle$) does not depend on the choice of orthonormal basis.

(2) Consider the Minkowski inner product $\langle \cdot, \cdot \rangle$ on $\mathbb{R}^4$ with the standard time orientation.

    (a) Let $v \in \mathbb{R}^4$ be timelike and future-pointing. Show that:
       (i) if $w \in \mathbb{R}^4$ is timelike or null and future-pointing then $\langle v, w \rangle < 0$;
      (ii) if $w \in \mathbb{R}^4$ is timelike or null and future-pointing then $v + w$ is timelike and future-pointing;
     (iii) $\{v\}^{\perp} := \{w \in \mathbb{R}^4 \mid \langle v, w \rangle = 0\}$ is a hyperplane containing only spacelike vectors (and the zero vector).

    (b) Let $v \in \mathbb{R}^4$ be null and future-pointing. Show that:
       (i) if $w \in \mathbb{R}^4$ is timelike or null and future-pointing then $\langle v, w \rangle \leq 0$, with equality if and only if $w = \lambda v$ for some $\lambda > 0$;
      (ii) if $w \in \mathbb{R}^4$ is timelike or null and future-pointing then $v + w$ is timelike or null and future-pointing, being null if and only if $w = \lambda v$ for some $\lambda > 0$;
     (iii) $\{v\}^{\perp}$ is a hyperplane containing only spacelike and null vectors, all of which are multiples of v.

    (c) Let $v \in \mathbb{R}^4$ be spacelike. Show that $\{v\}^{\perp}$ is a hyperplane containing timelike, null and spacelike vectors.

(3) Show that if (t, x) are the spacetime coordinates associated to an inertial frame and $T \in SO_0(1, 1)$ is a Lorentzian change of basis to a new inertial frame with global coordinates (t', x'), we must have

$$\frac{\partial}{\partial t'} = T\left(\frac{\partial}{\partial t}\right) = \cosh u \frac{\partial}{\partial t} + \sinh u \frac{\partial}{\partial x}$$

$$\frac{\partial}{\partial x'} = T\left(\frac{\partial}{\partial x}\right) = \sinh u \frac{\partial}{\partial t} + \cosh u \frac{\partial}{\partial x}$$

for some $u \in \mathbb{R}$.

(4) *(Twin paradox)* Twins Alice and Bob part on their 20th birthday: while Alice stays on the Earth (which is approximately an inertial frame), Bob leaves at 80% of the speed of light towards Planet X, 8 light-years away from the Earth, which he therefore reaches 10 years later (as measured in the Earth's frame).

After a short stay, Bob returns to the Earth, again at 80 % of the speed of light. Consequently, Alice is 40 years old when they meet again.

(a) How old is Bob at this meeting?

(b) How do you explain the asymmetry in the twins' ages? Notice that, from Bob's point of view, he is the one who is stationary, while the the Earth moves away and back again.

(c) Imagine that each twin has a very powerful telescope. What does each of them **see**? In particular, how much time elapses for each of them as they see their twin experiencing one year?

(**Hint:** Notice that light rays are represented by null lines, i.e. lines whose tangent vector is null; therefore, if event p in Alice's history is seen by Bob at event q then there must exist a future-directed null line connecting p to q).

(5) (*Car and garage paradox*) A 5-meter long car moves at 80 % of light speed towards a 4-meter long garage with doors at both ends.

(a) Compute the length of the car in the garage's frame, and show that if the garage doors are closed at the right time the car will be completely inside the garage for a few moments.

(b) Compute the garage's length in the car's frame, and show that in this frame the car is never completely inside the garage. How do you explain this apparent contradiction?

(6) Let (t', x') be an inertial frame moving with velocity v with respect to the inertial frame (t, x). Prove the **velocity addition formula**: if a particle moves with velocity w' in the frame (t', x'), the particle's velocity in the frame (t, x) is

$$w = \frac{w' + v}{1 + w'v}.$$

What happens when $w' = \pm 1$?

(7) (*Hyperbolic angle*)

(a) Show that

(i) $\mathfrak{so}(1, 1) = \left\{ \begin{pmatrix} 0 & u \\ u & 0 \end{pmatrix} \mid u \in \mathbb{R} \right\};$

(ii) $\exp \begin{pmatrix} 0 & u \\ u & 0 \end{pmatrix} = \begin{pmatrix} \cosh u & \sinh u \\ \sinh u & \cosh u \end{pmatrix} := S(u);$

(iii) $S(u)S(u') = S(u + u').$

(b) Consider the Minkowski inner product $\langle \cdot, \cdot \rangle$ on $\mathbb{R}^2$ with a given time orientation. If $v, w \in \mathbb{R}^2$ are unit timelike future-pointing vectors then there exists a unique $u \in \mathbb{R}$ such that $w = S(u)v$ (which we call the **hyperbolic angle** between v and w). Show that:

(i) $|u|$ is the length of the curve formed by all unit timelike vectors between v and w;

(ii) $\frac{1}{2}|u|$ is the area of the region swept by the position vector of the curve above;

(iii) hyperbolic angles are additive;

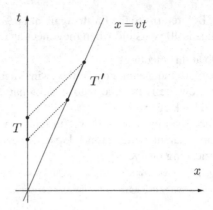

Fig. 6.2 Doppler effect

      (iv) the velocity addition formula of Exercise 2.2(6) is simply the formula
            for the hyperbolic tangent of a sum.

(8) *(Generalized twin paradox)* Let $p, q \in \mathbb{R}^4$ be two events connected by a
    timelike straight line l. Show that the proper time between p and q measured
    along l is bigger than the proper time between p and q measured along any
    other timelike curve connecting these two events. In other words, if an inertial
    observer and a (necessarily) accelerated observer separate at a given event and
    are rejoined at a later event, then the inertial observer always measures a bigger
    (proper) time interval between the two events. In particular, prove the **reversed
    triangle inequality**: if $v, w \in \mathbb{R}^4$ are timelike vectors with $w \in C(v)$ then
    $|v + w| \geq |v| + |w|$.

(9) *(Doppler effect)* Use the spacetime diagram in Fig. 6.2 to show that an observer
    moving with velocity v away from a source of light of period T measures the
    period to be

$$T' = T\sqrt{\frac{1+v}{1-v}}.$$

(**Remark:** This effect allows astronomers to measure the radial velocity of stars and galaxies relative to the
Earth).

(10) *(Aberration)* Suppose that the position in the sky of the star Sirius makes an
     angle θ with the x-axis of a given inertial observer. Show that the angle θ'
     measured by a second inertial observer moving with velocity $v = \tanh u$ along
     the x-axis of the first observer satisfies

$$\tan \theta' = \frac{\sin \theta}{\cosh u \cos \theta + \sinh u}.$$

(11) Minkowski geometry can be used in many contexts. For instance, let $l = \mathbb{R}\frac{\partial}{\partial t}$ represent the motion of an observer at rest in the atmosphere and choose units such that the speed of sound is 1.

    (a) Let $\tau : \mathbb{R}^4 \to \mathbb{R}$ the map such that $\tau(p)$ is the t coordinate of the event in which the observer hears the sound generated at p. Show that the level surfaces of τ are the conical surfaces

$$\tau^{-1}(t_0) = \left\{ p \in \mathbb{R}^4 \mid t_0 \frac{\partial}{\partial t} - p \text{ is null and future-pointing} \right\}.$$

    (b) Show that $c : I \to \mathbb{R}^4$ represents the motion of a supersonic particle iff

$$\left\langle \dot{c}, \frac{\partial}{\partial t} \right\rangle \neq 0 \quad \text{and} \quad \langle \dot{c}, \dot{c} \rangle > 0.$$

    (c) Argue that the observer hears a sonic boom whenever c is tangent to a surface $\tau = $ constant. Assuming that c is a straight line, what does the observer hear before and after the boom?

(12) Let $c : \mathbb{R} \to \mathbb{R}^4$ be the motion of a particle in Minkowski spacetime parameterized by the proper time τ.

    (a) Show that

$$\langle \dot{c}, \dot{c} \rangle = -1$$

    and

$$\langle \dot{c}, \ddot{c} \rangle = 0.$$

    Conclude that $\ddot{c}$ is the particle's acceleration as measured in the particle's **instantaneous rest frame**, i.e. in the inertial frame (t, x, y, z) for which $\dot{c} = \frac{\partial}{\partial t}$. For this reason, $\ddot{c}$ is called the particle's **proper acceleration**, and $|\ddot{c}|$ is interpreted as the acceleration measured by the particle.

    (b) Compute the particle's motion assuming that it is moving along the x-axis and measures a constant acceleration $|\ddot{c}| = a$.

    (c) Consider a spaceship launched from the Earth towards the center of the Galaxy (at a distance of 30,000 light-years) with $a = g$, where g represents the gravitational acceleration at the surface of the Earth. Using the fact that $g \simeq 1$ year^{-1} in units such that $c = 1$, compute the proper time measured aboard the spaceship for this journey. How long would the journey take as measured from the Earth?

(13) (*The faster-than-light missile*) While conducting a surveillance mission on the home planet of the wicked Klingons, the *Enterprise* uncovers their evil plan to build a faster-than-light missile and attack the Earth, 12 light-years away. Captain Kirk immediately orders the *Enterprise* back to the Earth at its top speed ($\frac{12}{13}$ of the speed of light), and at the same time sends out a

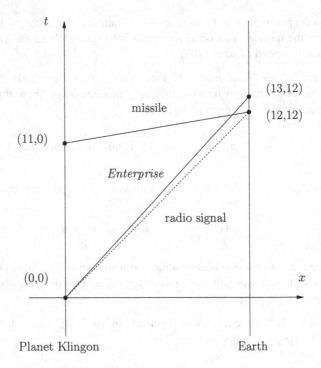

Fig. 6.3 Faster-than-light missile

radio warning. Unfortunately, it is too late: eleven years later (as measured by them), the Klingons launch their missile, moving at 12 times the speed of light. Therefore the radio warning, traveling at the speed of light, reaches the Earth at the same time as the missile, twelve years after its emission, and the *Enterprise* arrives at the ruins of the Earth one year later.

(a) How long does the *Enterprise*'s trip take according to its crew?
(b) On the Earth's frame, let $(0, 0)$ be the (t, x) coordinates of the event in which the *Enterprise* sends the radio warning, $(11, 0)$ the coordinates of the missile's launch, $(12, 12)$ the coordinates of the Earth's destruction and $(13, 12)$ the coordinates of the *Enterprise*'s arrival at the Earth's ruins (cf. Fig. 6.3). Compute the (t', x') coordinates of the same events on the *Enterprise*'s frame.
(c) Plot the motions of the *Enterprise*, the Klingon planet, the Earth, the radio signal and the missile on the *Enterprise*'s frame. Does the missile motion according to the *Enterprise* crew make sense?

(**Remark:** This exercise is based on an exercise in [TW92]).

6.3 The Cartan Connection

Let $(x^0, x^1, x^2, x^3) = (t, x, y, z)$ be an inertial frame on Galileo spacetime, which we can therefore identify with $\mathbb{R}^4$. Recall that Newtonian gravity is described by a **gravitational potential** $\Phi : \mathbb{R}^4 \to \mathbb{R}$. This potential determines the motions of free-falling particles through

$$\frac{d^2 x^i}{dt^2} = -\frac{\partial \Phi}{\partial x^i}$$

$(i = 1, 2, 3)$, and is, in turn, determined by the **matter density function** $\rho : \mathbb{R}^4 \to \mathbb{R}$ through the **Poisson equation**

$$\frac{\partial^2 \Phi}{\partial x^2} + \frac{\partial^2 \Phi}{\partial y^2} + \frac{\partial^2 \Phi}{\partial z^2} = 4\pi\rho$$

(we are using units in which Newton's universal gravitation constant G is set equal to 1). The vacuum Poisson equation (corresponding to the case in which all matter is concentrated on singularities of the gravitational potential) is the well-known **Laplace equation**

$$\frac{\partial^2 \Phi}{\partial x^2} + \frac{\partial^2 \Phi}{\partial y^2} + \frac{\partial^2 \Phi}{\partial z^2} = 0.$$

Notice that the equation of motion is the same for all particles, regardless of their mass. This observation, dating back to Galileo, was made into the so-called **equivalence principle** by Einstein. It implies that a gravitational field determines special curves on Galileo spacetime, namely the motions of free-falling particles. These curves are the geodesics of a symmetric connection, known as the **Cartan connection**, defined through the nonvanishing Christoffel symbols

$$\Gamma^i_{00} = \frac{\partial \Phi}{\partial x^i} \qquad (i = 1, 2, 3)$$

[cf. Exercise 3.1(1)], corresponding to the nonvanishing connection forms

$$\omega^i_0 = \frac{\partial \Phi}{\partial x^i} dt.$$

It is easy to check that the Cartan structure equations

$$\Omega^\mu_\nu = d\omega^\mu_\nu + \sum_{\alpha=0}^{3} \omega^\mu_\alpha \wedge \omega^\alpha_\nu$$

still hold for arbitrary symmetric connections, and hence we have the nonvanishing curvature forms

$$\Omega_0^i = \sum_{j=1}^{3} \frac{\partial^2 \Phi}{\partial x^j \partial x^i} dx^j \wedge dt.$$

The Ricci curvature tensor of this connection is

$$Ric = \left(\frac{\partial^2 \Phi}{\partial x^2} + \frac{\partial^2 \Phi}{\partial y^2} + \frac{\partial^2 \Phi}{\partial z^2} \right) dt \otimes dt$$

[cf. Exercise 3.1(2)], and hence the Poisson equation can be written as

$$Ric = 4\pi\rho \, dt \otimes dt.$$

In particular, the Laplace equation can be written as

$$Ric = 0.$$

Exercise 3.1

(1) Check that the motions of free-falling particles are indeed geodesics of the Cartan connection. What other geodesics are there? How would you interpret them?
(2) Check the formula for the Ricci curvature tensor of the Cartan connection.
(3) Show that the Cartan connection ∇ is compatible with Galileo structure, i.e. show that:

    (a) $\nabla_X dt = 0$ for all $X \in \mathfrak{X}(\mathbb{R}^4)$ [cf. Exercise 2.6(3) in Chap. 3];
    (b) if $E, F \in \mathfrak{X}(\mathbb{R}^4)$ are tangent to the simultaneity hypersurfaces and parallel along some curve $c : \mathbb{R} \to \mathbb{R}^4$, then $\langle E, F \rangle$ is constant.

(4) Show that if the Cartan connection has nonzero curvature then it is not the Levi–Civita connection of any pseudo-Riemannian metric on $\mathbb{R}^4$ (cf. Sect. 6.4).

6.4 General Relativity

Gravity can be introduced in Newtonian mechanics through the symmetric Cartan connection, which preserves Galileo spacetime structure. A natural idea for introducing gravity in special relativity is then to search for symmetric connections preserving the Minkowski inner product. To formalize this, we introduce the following definition.

Definition 4.1 A **pseudo-Riemannian manifold** is a pair (M, g), where M is a connected n-dimensional differentiable manifold and g is a symmetric nondegenerate differentiable 2-tensor field (g is said to be a **pseudo-Riemannian metric** in M). The **signature** of a pseudo-Riemannian manifold is just the signature of g at any tangent space. A **Lorentzian manifold** is a pseudo-Riemannian manifold with signature $n - 2$ [that is, $(g_{\mu\nu}) = \text{diag}(-1, 1, \ldots, 1)$ for appropriate orthonormal frames].

The Minkowski spacetime $(\mathbb{R}^4, \langle \cdot, \cdot \rangle)$ is obviously a Lorentzian manifold. The proof of the Levi–Civita theorem uses the non-degeneracy of the metric, not its positivity. Therefore, the theorem still holds for pseudo-Riemannian manifolds. In other words, given a pseudo-Riemannian manifold (M, g) there exists a unique symmetric connection ∇ which is compatible with g (given by the Koszul formula). Therefore there exists just one symmetric connection preserving the Minkowski metric, which is the trivial connection (obtained in Cartesian coordinates by taking all Christoffel symbols equal to zero). Notice that the geodesics of this connection are straight lines, corresponding to motions of free particles, which in particular do not feel any gravitational field.

To introduce gravity through a symmetric connection we must therefore consider more general 4-dimensional Lorentzian manifolds, which we will still call **spacetimes**. These are no longer required to be diffeomorphic to $\mathbb{R}^4$, nor to have inertial charts. The study of the geometry of these spacetimes is usually called **general relativity**.

Each spacetime comes equipped with its unique Levi–Civita connection, and hence with its geodesics. If $c : I \subset \mathbb{R} \to M$ is a geodesic, then $\langle \dot{c}, \dot{c} \rangle$ is constant, as

$$\frac{d}{ds} \langle \dot{c}(s), \dot{c}(s) \rangle = 2 \left\langle \frac{D\dot{c}}{ds}(s), \dot{c}(s) \right\rangle = 0.$$

A geodesic is called **timelike**, **null**, or **spacelike** according to whether $\langle \dot{c}, \dot{c} \rangle < 0$, $\langle \dot{c}, \dot{c} \rangle = 0$ or $\langle \dot{c}, \dot{c} \rangle > 0$ (i.e. according to whether its tangent vector is timelike, spacelike or null). By analogy with the Cartan connection, we will take timelike geodesics to represent the free-falling motions of massive particles. This ensures that the equivalence principle holds. Null geodesics will be taken to represent the motions of light rays.

In general, any curve $c : I \subset \mathbb{R} \to M$ is said to be **timelike** if $\langle \dot{c}, \dot{c} \rangle < 0$. In this case, c represents the motion of a particle with nonzero mass (which is accelerating unless c is a geodesic). The **proper time** measured by the particle between events $c(a)$ and $c(b)$ is

$$\tau(c) = \int_a^b |\dot{c}(s)| ds,$$

where $|v| = |\langle v, v \rangle|^{\frac{1}{2}}$ for any $v \in TM$.

To select physically relevant spacetimes we must impose some sort of constraint. By analogy with the formulation of the Laplace equation in terms of the Cartan connection, we make the following definition.

Definition 4.2 The Lorentzian manifold (M, g) is said to be a **vacuum solution** of the **Einstein field equation** if its Levi–Civita connection satisfies $Ric = 0$.

The general Einstein field equation is

$$Ric - \frac{S}{2}g = 8\pi E,$$

where $S = \sum_{\mu,\nu=0}^{3} g^{\mu\nu} R_{\mu\nu}$ is the scalar curvature and E is the so-called **energy-momentum tensor** of the matter content of the spacetime. The simplest example of a matter model is that of a **pressureless perfect fluid**, which is described by a **rest density function** $\rho \in C^\infty(M)$ and a unit **velocity vector field** $U \in \mathfrak{X}(M)$ (whose integral lines are the motions of the fluid particles). The energy-momentum tensor for this matter model is

$$E = \rho \nu \otimes \nu,$$

where $\nu \in \Omega^1(M)$ is the 1-form associated to U by the metric g.

The Einstein field equation can be rewritten as

$$Ric = 8\pi T,$$

where

$$T := E - \frac{1}{2}\left(\sum_{\mu,\nu=0}^{3} g^{\mu\nu} E_{\mu\nu} \right) g$$

is the **reduced energy-momentum tensor** [cf. Exercise 4.3(2)]. For a pressureless perfect fluid, the reduced energy-momentum tensor is

$$T = \rho\left(\nu \otimes \nu + \frac{1}{2}g \right),$$

and so Einstein field equation is

$$Ric = 4\pi\rho(2\nu \otimes \nu + g)$$

(compare this with the Poisson equation in terms of the Cartan connection).

It turns out that spacetimes satisfying the Einstein field equation for appropriate choices of T model astronomical phenomena with great accuracy.

Exercise 4.3

(1) Show that the signature of a pseudo-Riemannian manifold (M, g) is well defined, i.e. show that the signature of $g_p \in T^2(T_pM)$ does not depend on $p \in M$.
(2) Show that:

(a) the Einstein field equation can be rewritten as

$$Ric = 8\pi T;$$

(b) the reduced energy-momentum tensor for a pressureless perfect fluid with rest density ρ and unit velocity 1-form ν is

$$T = \rho \left(\nu \otimes \nu + \frac{1}{2} g \right).$$

(3) Let (M, g) be a pseudo-Riemannian manifold and $f : N \to M$ an immersion. Show that f^*g is not necessarily a pseudo-Riemannian metric on N.

(4) Let (M, g) be the $(n + 1)$-dimensional Minkowski spacetime, i.e. $M = \mathbb{R}^{n+1}$ and

$$g = -dx^0 \otimes dx^0 + dx^1 \otimes dx^1 + \cdots + dx^n \otimes dx^n.$$

Let $i : N \to M$ be the inclusion map, where

$$N := \{v \in M \mid \langle v, v \rangle = -1 \text{ and } v^0 > 0\}.$$

Show that (N, i^*g) is the n-dimensional hyperbolic space H^n.

(5) *(Fermi–Walker transport)* Let $c : I \subset \mathbb{R} \to \mathbb{R}^4$ be a timelike curve in Minkowski space parameterized by the proper time, $U := \dot{c}$ the tangent unit vector and $A := \ddot{c}$ the proper acceleration. A vector field $V : I \to \mathbb{R}^4$ is said to be **Fermi–Walker transported** along c if

$$\frac{DV}{d\tau} = \langle V, A \rangle U - \langle V, U \rangle A.$$

(a) Show that U is Fermi–Walker transported along c.
(b) Show that if V and W are Fermi–Walker transported along c then $\langle V, W \rangle$ is constant.
(c) If $\langle V, U \rangle = 0$ then V is tangent at U to the submanifold

$$N := \{v \in \mathbb{R}^4 \mid \langle v, v \rangle = -1 \text{ and } v^0 > 0\},$$

which is isometric to the hyperbolic 3-space [cf. Exercise 4.3(4)]. Show that, in this case, V is Fermi–Walker transported if and only if it is parallel transported along $U : I \to N$.

(d) Assume that c describes a circular motion with constant speed v. Let V be a Fermi–Walker transported vector field, tangent to the plane of the motion, such that $\langle V, U \rangle = 0$. Compute the angle by which V rotates (or **precesses**) after one revolution.

(**Remark:** It is possible to prove that the angular momentum vector of a spinning particle is Fermi–Walker transported along its motion and orthogonal to it; the above precession, which has been observed for spinning particles such as electrons, is called the **Thomas precession**).

(6) *(Twin paradox on a cylinder)* The quotient of Minkowski spacetime by the discrete isometry group generated by the translation $\xi(t, x, y, z) = (t, x +$

8, y, z) is a (flat) vacuum solution of the Einstein field equation. Assume that the Earth's motion is represented by the line $x = y = z = 0$, and that once again Bob departs at 80 % of the speed of light along the x-axis, leaving his twin sister Alice on the Earth, on their 20th birthday [cf. Exercise 2.2(5)]. Because of the topology of space, the two twins meet again after 10 years (as measured on the Earth), *without Bob ever having accelerated.*

(a) Compute the age of each twin in their meeting.
(b) From Bob's viewpoint, it is the Earth which moves away from him. How do you explain the asymmetry in the twins' ages?

(7) *(Rotating frame)*

(a) Show that the metric of Minkowski spacetime can be written as

$$g = -\, dt \otimes dt + dr \otimes dr + r^2 d\theta \otimes d\theta + dz \otimes dz$$

by using cylindrical coordinates (r, θ, z) in $\mathbb{R}^3$.

(b) Let $\omega > 0$ and consider the coordinate change given by $\theta = \theta' + \omega t$. Show that in these coordinates the metric is written as

$$g = -\, (1 - \omega^2 r^2) dt \otimes dt + \omega r^2 dt \otimes d\theta' + \omega r^2 d\theta' \otimes dt$$
$$+ dr \otimes dr + r^2 d\theta' \otimes d\theta' + dz \otimes dz.$$

(c) Show that in the region $U = \{r < \frac{1}{\omega}\}$ the coordinate curves of constant (r, θ', z) are timelike curves corresponding to (accelerated) observers rotating rigidly with respect to the inertial observers of constant (r, θ, z).

(d) The set of the rotating observers is a 3-dimensional smooth manifold Σ with local coordinates (r, θ', z), and there exists a natural projection $\pi : U \to \Sigma$. We introduce a Riemannian metric h on Σ as follows: if $v, w \in T_{\pi(p)}\Sigma$ then

$$h(v, w) = g\left(v^\dagger, w^\dagger\right),$$

where, for each $u \in T_{\pi(p)}\Sigma$, the vector $u^\dagger \in T_p U$ satisfies

$$(d\pi)_p\, u^\dagger = u \quad \text{and} \quad g\left(u^\dagger, \left(\frac{\partial}{\partial t}\right)_p\right) = 0.$$

Show that h is well defined and

$$h = dr \otimes dr + \frac{r^2}{1 - \omega^2 r^2}\, d\theta' \otimes d\theta' + dz \otimes dz.$$

(**Remark:** This is the metric resulting from local distance measurements between the rotating observers; Einstein used the fact that this metric has curvature to argue for the need to use non-Euclidean geometry in the relativistic description of gravity).

(e) The image of a curve $c : I \subset \mathbb{R} \to U$ consists of simultaneous events from the point of view of the rotating observers if $\dot{c}$ is orthogonal to $\frac{\partial}{\partial t}$ at each point. Show that this is equivalent to requiring that $\alpha(\dot{c}) = 0$, where

$$\alpha = dt - \frac{\omega r^2}{1 - \omega^2 r^2} d\theta'.$$

In particular, show that, in general, synchronization of the rotating observers' clocks around closed paths leads to inconsistencies.

(**Remark:** This is the so-called **Sagnac effect**; it must be taken into account when synchronizing the very precise atomic clocks on the GPS system ground stations, because of the Earth's rotation).

(8) (*Static spacetime*) Let (Σ, h) be a 3-dimensional Riemannian manifold and consider the 4-dimensional Lorentzian manifold (M, g) determined by $M := \mathbb{R} \times \Sigma$ and

$$g := -e^{2(\Phi \circ \pi)} dt \otimes dt + \pi^* h,$$

where t is the usual coordinate in $\mathbb{R}$, $\pi : M \to \Sigma$ is the natural projection and $\Phi : \Sigma \to \mathbb{R}$ is a smooth function.

(a) Let $c : I \subset \mathbb{R} \to M$ be a timelike geodesic parameterized by the proper time, and $\gamma := \pi \circ c$. Show that

$$\frac{D\dot{\gamma}}{d\tau} = (1 + h(\dot{\gamma}, \dot{\gamma})) G,$$

where $G = -\operatorname{grad}(\Phi)$ is the vector field associated to $-d\Phi$ by h and can be thought of as the gravitational field. Show that this equation implies that the quantity

$$E^2 := (1 + h(\dot{\gamma}, \dot{\gamma}))e^{2\Phi}$$

is a constant of motion.

(b) Let $c : I \subset \mathbb{R} \to M$ be a null geodesic, $\tilde{c}$ its reparameterization by the coordinate time t, and $\tilde{\gamma} := \pi \circ \tilde{c}$. Show that $\tilde{\gamma}$ is a geodesic of the **Fermat metric**

$$l := e^{-2(\Phi \circ \pi)} h.$$

(**Hint:** Use Lemma 1.12 in Chap. 5).

(c) Show that the vacuum Einstein field equation for g is equivalent to

$$\operatorname{div} G = h(G, G);$$
$$Ric = \nabla d\Phi + d\Phi \otimes d\Phi,$$

where div G is the divergence of G, Ric and ∇ are the Ricci curvature and the Levi–Civita connection of h, and $\nabla d\Phi$ is the tensor defined by $(\nabla d\Phi)(X, Y) := (\nabla_X d\Phi)(Y)$ for all $X, Y \in \mathfrak{X}(\Sigma)$ [cf. Exercises 2.6(3). and 3.3(9). in Chap. 3].

6.5 The Schwarzschild Solution

The vacuum Einstein field equation is nonlinear, and hence much harder to solve than the Laplace equation. One of the first solutions to be discovered was the so-called **Schwarzschild solution**, which can be obtained from the simplifying hypotheses of time independence and spherical symmetry, i.e. by looking for solutions of the form

$$g = -A^2(r)dt \otimes dt + B^2(r)dr \otimes dr + r^2 d\theta \otimes d\theta + r^2 \sin^2 \theta d\varphi \otimes d\varphi$$

for unknown positive smooth functions $A, B : \mathbb{R} \to \mathbb{R}$. Notice that this expression reduces to the Minkowski metric in spherical coordinates for $A \equiv B \equiv 1$).

It is easily seen that the Cartan structure equations still hold for pseudo-Riemannian manifolds. We have

$$g = -\omega^0 \otimes \omega^0 + \omega^r \otimes \omega^r + \omega^\theta \otimes \omega^\theta + \omega^\varphi \otimes \omega^\varphi$$

with

$$\begin{aligned}
\omega^0 &= A(r)dt; \\
\omega^r &= B(r)dr; \\
\omega^\theta &= rd\theta; \\
\omega^\varphi &= r \sin \theta d\varphi,
\end{aligned}$$

and hence $\{\omega^0, \omega^r, \omega^\theta, \omega^\varphi\}$ is an orthonormal coframe. The first structure equations,

$$d\omega^\mu = \sum_{\nu=0}^{3} \omega^\nu \wedge \omega^\mu_\nu;$$

$$dg_{\mu\nu} = \sum_{\alpha=0}^{3} g_{\mu\alpha}\omega^\alpha_\nu + g_{\nu\alpha}\omega^\alpha_\mu,$$

which on an orthonormal frame are written as

$$d\omega^\mu = \sum_{\nu=0}^{3} \omega^\nu \wedge \omega^\mu_\nu;$$

$$\omega_0^0 = \omega_i^i = 0;$$
$$\omega_i^0 = \omega_0^i;$$
$$\omega_j^i = -\omega_i^j$$

$(i, j = 1, 2, 3)$, together with

$$dw^0 = \frac{A'}{B} \omega^r \wedge dt;$$
$$dw^r = 0;$$
$$dw^\theta = \frac{1}{B} \omega^r \wedge d\theta;$$
$$dw^\varphi = \frac{\sin \theta}{B} \omega^r \wedge d\varphi + \cos \theta \, \omega^\theta \wedge d\varphi,$$

yield the nonvanishing connection forms

$$\omega_r^0 = \omega_0^r = \frac{A'}{B} dt;$$
$$\omega_r^\theta = -\omega_\theta^r = \frac{1}{B} d\theta;$$
$$\omega_r^\varphi = -\omega_\varphi^r = \frac{\sin \theta}{B} d\varphi;$$
$$\omega_\theta^\varphi = -\omega_\varphi^\theta = \cos \theta d\varphi.$$

The curvature forms can be computed from the second structure equations

$$\Omega_\nu^\mu = dw_\nu^\mu + \sum_{\alpha=0}^{3} \omega_\alpha^\mu \wedge \omega_\nu^\alpha,$$

and are found to be

$$\Omega_r^0 = \Omega_0^r = \frac{A''B - A'B'}{AB^3} \omega^r \wedge \omega^0;$$
$$\Omega_\theta^0 = \Omega_0^\theta = \frac{A'}{rAB^2} \omega^\theta \wedge \omega^0;$$
$$\Omega_\varphi^0 = \Omega_0^\varphi = \frac{A'}{rAB^2} \omega^\varphi \wedge \omega^0;$$
$$\Omega_r^\theta = -\Omega_\theta^r = \frac{B'}{rB^3} \omega^\theta \wedge \omega^r;$$
$$\Omega_r^\varphi = -\Omega_\varphi^r = \frac{B'}{rB^3} \omega^\varphi \wedge \omega^r;$$

$$\Omega^\varphi_\theta = -\,\Omega^\theta_\varphi = \frac{B^2-1}{r^2B^2}\,\omega^\varphi \wedge \omega^\theta.$$

Thus the components of the curvature tensor on the orthonormal frame can be read off from the curvature forms using

$$\Omega^\mu_\nu = \sum_{\alpha<\beta} R_{\alpha\beta\nu}{}^\mu \omega^\alpha \wedge \omega^\beta.$$

and can in turn be used to compute the components of the Ricci curvature tensor Ric on the same frame. The nonvanishing components of Ric on this frame turn out to be

$$R_{00} = \frac{A''B - A'B'}{AB^3} + \frac{2A'}{rAB^2};$$

$$R_{rr} = -\frac{A''B - A'B'}{AB^3} + \frac{2B'}{rB^3};$$

$$R_{\theta\theta} = R_{\varphi\varphi} = -\frac{A'}{rAB^2} + \frac{B'}{rB^3} + \frac{B^2-1}{r^2B^2}.$$

Thus the vacuum Einstein field equation $Ric = 0$ is equivalent to the ODE system

$$\begin{cases} \dfrac{A''}{A} - \dfrac{A'B'}{AB} + \dfrac{2A'}{rA} = 0 \\[2mm] \dfrac{A''}{A} - \dfrac{A'B'}{AB} - \dfrac{2B'}{rB} = 0 \\[2mm] \dfrac{A'}{A} - \dfrac{B'}{B} - \dfrac{B^2-1}{r} = 0 \end{cases} \Leftrightarrow \begin{cases} \dfrac{A'}{A} + \dfrac{B'}{B} = 0 \\[2mm] \left(\dfrac{A'}{A}\right)' + 2\left(\dfrac{A'}{A}\right)^2 + \dfrac{2A'}{rA} = 0 \\[2mm] \dfrac{2B'}{B} + \dfrac{B^2-1}{r} = 0 \end{cases}.$$

The last equation can be immediately solved to yield

$$B = \left(1 - \frac{2m}{r}\right)^{-\frac{1}{2}},$$

where $m \in \mathbb{R}$ is an integration constant. The first equation implies that $A = \frac{\alpha}{B}$ for some constant $\alpha > 0$. By rescaling the time coordinate t we can assume that $\alpha = 1$. Finally, it is easily checked that the second ODE is identically satisfied. Therefore there exists a one-parameter family of solutions of the vacuum Einstein field equation of the form we sought, given by

$$g = -\left(1 - \frac{2m}{r}\right) dt \otimes dt + \left(1 - \frac{2m}{r}\right)^{-1} dr \otimes dr \tag{6.2}$$
$$+ r^2 d\theta \otimes d\theta + r^2 \sin^2\theta d\varphi \otimes d\varphi.$$

To interpret this family of solutions, we compute the proper acceleration [cf. Exercise 2.2(12)] of the **stationary observers**, whose motions are the integral curves of $\frac{\partial}{\partial t}$. If $\{E_0, E_r, E_\theta, E_\varphi\}$ is the orthonormal frame obtained by normalizing $\left\{\frac{\partial}{\partial t}, \frac{\partial}{\partial r}, \frac{\partial}{\partial \theta}, \frac{\partial}{\partial \varphi}\right\}$ (hence dual to $\{\omega^0, \omega^r, \omega^\theta, \omega^\varphi\}$), we have

$$\nabla_{E_0} E_0 = \sum_{\mu=0}^{3} \omega_0^\mu(E_0) E_\mu = \omega_0^r(E_0) E_r = \frac{A'}{AB} \omega^0(E_0) E_r = \frac{m}{r^2}\left(1 - \frac{2m}{r}\right)^{-\frac{1}{2}} E_r.$$

Therefore, each stationary observer is accelerating with a proper acceleration

$$G(r) = \frac{m}{r^2}\left(1 - \frac{2m}{r}\right)^{-\frac{1}{2}}$$

away from the origin, to prevent falling towards it. In other words, they are experiencing a gravitational field of intensity $G(r)$, directed towards the origin. Since $G(r)$ approaches, for large values of r, the familiar acceleration m/r^2 of the Newtonian gravitational field generated by a point particle of mass m, we interpret the Schwarzschild solution as the general relativistic field of a point particle of mass m. Accordingly, we will assume that $m > 0$ (notice that $m = 0$ corresponds to Minkowski spacetime).

When obtaining the Schwarzschild solution we assumed $A(r) > 0$, and hence $r > 2m$. However, it is easy to check that (6.2) is also a solution of the Einstein vacuum field equation for $r < 2m$. Notice that the coordinate system (t, r, θ, φ) is singular at $r = 2m$, and hence covers only the two disconnected open sets $\{r > 2m\}$ and $\{r < 2m\}$. Both these sets are geodesically incomplete, as for instance radial timelike or null geodesics cannot be extended as they approach $r = 0$ or $r = 2m$ [cf. Exercise 5.1(7)]. While this is to be expected for $r = 0$, as the curvature blows up along geodesics approaching this limit, this is not the case for $r = 2m$. It turns out that it is possible to fit these two open sets together to obtain a solution of the Einstein vacuum field equation regular at $r = 2m$. To do so, we introduce the so-called **Painlevé time coordinate**

$$t' = t + \int \sqrt{\frac{2m}{r}} \left(1 - \frac{2m}{r}\right)^{-1} dr.$$

In the coordinate system (t', r, θ, φ), the Schwarzschild metric is written

$$g = -dt' \otimes dt' + \left(dr + \sqrt{\frac{2m}{r}}\, dt' \right) \otimes \left(dr + \sqrt{\frac{2m}{r}}\, dt' \right) + r^2 d\theta \otimes d\theta + r^2 \sin^2 \theta d\varphi \otimes d\varphi.$$

This expression is nonsingular at $r = 2m$, and is a solution of the Einstein vacuum field equation for $\{r > 2m\}$ and $\{r < 2m\}$. By continuity, it must be a solution also at $r = 2m$.

The submanifold $r = 2m$ is called the **event horizon**, and is ruled by null geodesics. This is easily seen from the fact that $\frac{\partial}{\partial t'} = \frac{\partial}{\partial t}$ becomes null at $r = 2m$, and hence its integral curves are (reparameterizations of) null geodesics.

The causal properties of the Schwarzschild spacetime are best understood by studying the **light cones**, i.e. the set of tangent null vectors at each point. For instance, radial null vectors $v = v^0 \frac{\partial}{\partial t'} + v^r \frac{\partial}{\partial r}$ satisfy

$$-\left(v^0\right)^2 + \left(v^r + \sqrt{\frac{2m}{r}}\, v^0\right)^2 = 0 \Leftrightarrow v^r = \left(\pm 1 - \sqrt{\frac{2m}{r}}\right) v^0.$$

For $r \gg 2m$ we obtain approximately the usual light cones of Minkowski spacetime. As r approaches $2m$, however, the light cones "tip over" towards the origin, becoming tangent to the event horizon at $r = 2m$ (cf. Fig. 6.4). Since the tangent vector to a timelike curve must be inside the light cone, we see that no particle which crosses the event horizon can ever leave the region $r = 2m$ (which for this reason is called a **black hole**). Once inside the black hole, the light cones tip over even more, forcing the particle into the singularity $r = 0$.

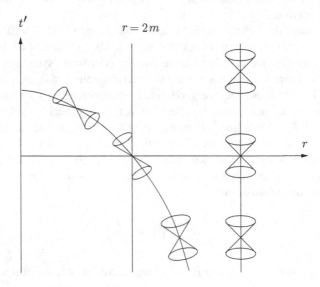

Fig. 6.4 Light cones in Painlevé coordinates

Notice that the Schwarzschild solution in Painlevé coordinates is still not geo-desically complete at the event horizon, as radial timelike and null geodesics are incomplete to the past as they approach $r = 2m$ [cf. Exercise 5.1(7)]. Physically, this is not important: black holes are thought to form through the collapse of (approxi-mately) spherical stars, whose surface follows a radial timelike curve in the spacetime diagram of Fig. 6.4. Since only outside the star is there vacuum, the Schwarzschild solution is expected to hold only above this curve, thereby removing the region of $r = 2m$ leading to incompleteness. Nevertheless, it is possible to glue two copies of the Schwarzschild spacetime in Painlevé coordinates to obtain a solution of the vacuum Einstein field equation which is geodesically incomplete only at the two copies of $r = 0$. This solution, known as the **Kruskal extension**, contains a black hole and its time-reversed version, known as a **white hole**.

For some time it was thought that the curvature singularity at $r = 0$ was an artifact of the high symmetry of Schwarzschild spacetime, and that more realistic models of collapsing stars would be singularity-free. Hawking and Penrose proved that this is not the case: once the collapse has begun, no matter how asymmetric, nothing can prevent a singularity from forming (cf. Sects. 6.8 and 6.9).

Exercise 5.1

(1) Let (M, g) be a 2-dimensional Lorentzian manifold.

   (a) Consider an orthonormal frame $\{E_0, E_1\}$ on an open set $U \subset M$, with associated coframe $\{\omega^0, \omega^1\}$. Check that the Cartan structure equations are

$$\omega^0_1 = \omega^1_0;$$
$$d\omega^0 = \omega^1 \wedge \omega^0_1;$$
$$d\omega^1 = \omega^0 \wedge \omega^0_1;$$
$$\Omega^0_1 = d\omega^0_1.$$

   (b) Let $\{F_0, F_1\}$ be another orthonormal frame such that $F_0 \in C(E_0)$, with associated coframe $\{\overline{\omega}^0, \overline{\omega}^1\}$ and connection form $\overline{\omega}^0_1$. Show that $\sigma = \overline{\omega}^0_1 - \omega^0_1$ is given locally by $\sigma = du$, where u is the hyperbolic angle between F_0 and E_0 [cf. Exercise 2.2(7)].

   (c) Consider a triangle $\Delta \subset U$ whose sides are timelike geodesics, and let α, β and γ be the hyperbolic angles between them (cf. Fig. 6.5). Show that

$$\gamma = \alpha + \beta - \int_\Delta \Omega^0_1,$$

   where, following the usual convention for spacetime diagrams, we orient U so that $\{E_0, E_1\}$ is **negative**.

   (d) Provide a physical interpretation for the formula above in the case in which (M, g) is a totally geodesic submanifold of the Schwarzschild spacetime obtained by fixing (θ, φ) [cf. Exercise 5.7(3) in Chap. 4].

Fig. 6.5 Timelike geodesic
triangle

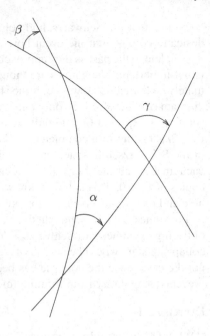

(2) Consider the Schwarzschild spacetime with local coordinates (t, r, θ, φ). An
equatorial circular curve is a curve given in these coordinates by $(t(\tau), r(\tau),$
$\theta(\tau), \varphi(\tau))$ with $\dot{r}(\tau) \equiv 0$ and $\theta(\tau) \equiv \frac{\pi}{2}$.

(a) Show that the conditions for such a curve to be a timelike geodesic parameterized by its proper time are

$$\begin{cases} \ddot{t} = 0 \\ \ddot{\varphi} = 0 \\ r\dot{\varphi}^2 = \frac{m}{r^2}\dot{t}^2 \\ \left(1 - \frac{3m}{r}\right)\dot{t}^2 = 1 \end{cases}$$

Conclude that massive particles can orbit the central mass in circular orbits
for all $r > 3m$.

(b) Show that there exists an equatorial circular null geodesic for $r = 3m$. What
does a stationary observer placed at $r = 3m$, $\theta = \frac{\pi}{2}$ see as he looks along
the direction of this null geodesic?

(c) The angular momentum vector of a free-falling spinning particle is parallel-
transported along its motion, and orthogonal to it [cf. Exercise 4.3(5)]. Consider a spinning particle on a circular orbit around a pointlike mass m. Show
that the angular momentum vector precesses by an angle

$$\delta = 2\pi \left(1 - \left(1 - \frac{3m}{r} \right)^{\frac{1}{2}} \right),$$

after one revolution, if initially aligned with the radial direction.

(**Remark:** The above precession, which has been observed for spinning quartz spheres in orbit around the Earth during the Gravity Probe B experiment, is called the **geodesic precession**).

(3) *(Gravitational redshift)* We consider again the Schwarzschild spacetime with local coordinates (t, r, θ, φ).

(a) Show that the proper time interval $\Delta\tau$ measured by a stationary observer between two events on his history is

$$\Delta\tau = \left(1 - \frac{2m}{r} \right)^{\frac{1}{2}} \Delta t,$$

where Δt is the difference between the time coordinates of the two events.

(**Remark:** This effect has been measured experimentally; loosely speaking, gravity delays time).

(b) Show that if $(t(s), r(s), \theta(s), \varphi(s))$ is a geodesic then so is $(t(s) + \Delta t, r(s), \theta(s), \varphi(s))$ for any $\Delta t \in \mathbb{R}$.

(c) Use the spacetime diagram in Fig. 6.6 to show that if a stationary observer at $r = r_0$ measures a light signal to have period T, a stationary observer at

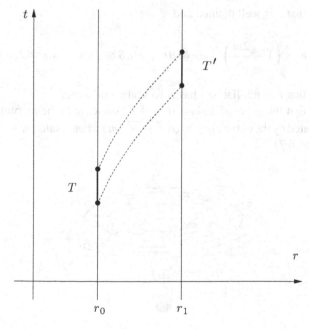

Fig. 6.6 Gravitational redshift

$r = r_1$ measures a period

$$T' = T \sqrt{\frac{1 - \frac{2m}{r_1}}{1 - \frac{2m}{r_0}}}$$

for the same signal.

(**Remark:** This gravitational redshift has been measured experimentally, confirming that spacetime must be curved – in Minkowski spacetime one would necessarily have $T = T'$).

(4) Let (M, g) be the region $r > 2m$ of the Schwarzschild solution with the Schwarzschild metric. The set of all stationary observers in M is a 3-dimensional smooth manifold Σ with local coordinates (r, θ, φ), and there exists a natural projection $\pi : M \to \Sigma$. We introduce a Riemannian metric h on Σ as follows: if $v, w \in T_{\pi(p)}\Sigma$ then

$$h(v, w) = g\left(v^\dagger, w^\dagger\right),$$

where, for each $u \in T_{\pi(p)}\Sigma$, the vector $u^\dagger \in T_pU$ satisfies

$$(d\pi)_p\, u^\dagger = u \quad \text{and} \quad g\left(u^\dagger, \left(\frac{\partial}{\partial t}\right)_p\right) = 0$$

[cf. Exercise 4.3(7)].

(a) Show that h is well defined and

$$h = \left(1 - \frac{2m}{r}\right)^{-1} dr \otimes dr + r^2 d\theta \otimes d\theta + r^2 \sin^2\theta d\varphi \otimes d\varphi.$$

(b) Show that h is not flat, but has zero scalar curvature.

(c) Show that the equatorial plane $\theta = \frac{\pi}{2}$ is isometric to the revolution surface generated by the curve $z(r) = \sqrt{8m(r - 2m)}$ when rotated around the z-axis (cf. Fig. 6.7).

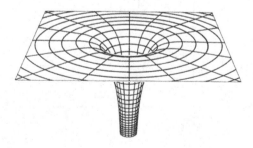

Fig. 6.7 Surface of revolution isometric to the equatorial plane

(**Remark:** This is the metric resulting from local distance measurements between the stationary observers; loosely speaking, gravity deforms space).

(5) In this exercise we study in detail the timelike and null geodesics of the Schwarzschild spacetime. We start by observing that the submanifold $\theta = \frac{\pi}{2}$ is totally geodesic [cf. Exercise 5.7(3) in Chap. 4]. By adequately choosing the angular coordinates (θ, φ), one can always assume that the initial condition of the geodesic is tangent to this submanifold; hence it suffices to study the timelike and null geodesics of the 3-dimensional Lorentzian manifold (M, g), where

$$g = -\left(1 - \frac{2m}{r}\right) dt \otimes dt + \left(1 - \frac{2m}{r}\right)^{-1} dr \otimes dr + r^2 d\varphi \otimes d\varphi.$$

(a) Show that $\frac{\partial}{\partial t}$ and $\frac{\partial}{\partial \varphi}$ are Killing fields [cf. Exercise 3.3(8) in Chap. 3].

(b) Conclude that the equations for a curve $c : \mathbb{R} \to M$ to be a future-directed geodesic (parameterized by proper time if timelike) can be written as

$$\begin{cases} g(\dot{c}, \dot{c}) = -\sigma \\ g\left(\frac{\partial}{\partial t}, \dot{c}\right) = -E \\ g\left(\frac{\partial}{\partial \varphi}, \dot{c}\right) = L \end{cases} \Leftrightarrow \begin{cases} \dot{r}^2 = E^2 - \left(\sigma + \frac{L^2}{r^2}\right)\left(1 - \frac{2m}{r}\right) \\ \left(1 - \frac{2m}{r}\right)\dot{t} = E \\ r^2 \dot{\varphi} = L \end{cases}$$

where $E > 0$ and L are integration constants, $\sigma = 1$ for timelike geodesics and $\sigma = 0$ for null geodesics.

(c) Show that if $L \neq 0$ then $u = \frac{1}{r}$ satisfies

$$\frac{d^2 u}{d\varphi^2} + u = \frac{m\sigma}{L^2} + 3mu^2.$$

(d) For situations where relativistic corrections are small one has $mu \ll 1$, and hence the approximate equation

$$\frac{d^2 u}{d\varphi^2} + u = \frac{m}{L^2}$$

holds for timelike geodesics. Show that the solution to this equation is the conic section given in polar coordinates by

$$u = \frac{m}{L^2}(1 + \varepsilon \cos(\varphi - \varphi_0)),$$

where the integration constants $\varepsilon \geq 0$ and φ_0 are the eccentricity and the argument of the pericenter.

(e) Show that for $\varepsilon \ll 1$ this approximate solution satisfies

$$u^2 \simeq \frac{2m}{L^2} u - \frac{m^2}{L^4}.$$

Argue that timelike geodesics close to circular orbits where relativistic corrections are small yield approximate solutions of the equation

$$\frac{d^2u}{d\varphi^2} + \left(1 - \frac{6m^2}{L^2}\right) u = \frac{m}{L^2}\left(1 - \frac{3m^2}{L^2}\right),$$

and hence the pericenter advances by approximately

$$\frac{6\pi m}{r}$$

radians per revolution.

(Remark: The first success of general relativity was due to this effect, which explained the anomalous precession of Mercury's perihelion—43 arcseconds per century).

(f) Show that if one neglects relativistic corrections then null geodesics satisfy

$$\frac{d^2u}{d\varphi^2} + u = 0.$$

Show that the solution to this equation is the equation for a straight line in polar coordinates,

$$u = \frac{1}{b}\sin(\varphi - \varphi_0),$$

where the integration constants $b > 0$ and φ_0 are the **impact parameter** (distance of closest approach to the center) and the angle between the line and the x-axis.

(g) Assume that $mu \ll 1$. Let us include relativistic corrections by looking for approximate solutions of the form

$$u = \frac{1}{b}\left(\sin\varphi + \frac{m}{b} v\right)$$

(where we take $\varphi_0 = 0$ for simplicity). Show that v is an approximate solution of the equation

$$\frac{d^2v}{d\varphi^2} + v = 3\sin^2\varphi,$$

and hence u is approximately given by

$$u = \frac{1}{b}\left(\sin\varphi + \frac{m}{b}\left(\frac{3}{2} + \frac{1}{2}\cos(2\varphi) + \alpha\cos\varphi + \beta\sin\varphi\right)\right),$$

where α and β are integration constants.

(h) Show that for the incoming part of the null geodesic ($\varphi \simeq 0$) one approximately has

$$u = 0 \Leftrightarrow \varphi = -\frac{m}{b}\,(2 + \alpha)\,.$$

Similarly, show that for the outgoing part of the null geodesic ($\varphi \simeq \pi$) one approximately has

$$u = 0 \Leftrightarrow \varphi = \pi + \frac{m}{b}\,(2 - \alpha)\,.$$

Conclude that φ varies by approximately

$$\Delta\varphi = \pi + \frac{4m}{b}$$

radians along its path, and hence the null geodesic is deflected towards the center by approximately

$$\frac{4m}{b}$$

radians.

(**Remark:** The measurement of this **deflection of light** by the Sun—1.75 arcseconds—was the first experimental confirmation of general relativity, and made Einstein a global celebrity overnight).

(6) (*Birkhoff theorem*) Prove that the only Ricci-flat Lorentzian metric given in local coordinates (t, r, θ, φ) by

$$g = A^2(t, r)dt \otimes dt + B^2(t, r)dr \otimes dr + r^2 d\theta \otimes d\theta + r^2 \sin^2\theta d\varphi \otimes d\varphi$$

is the Schwarzschild metric. Loosely speaking, spherically symmetric mass configurations do not radiate.

(7) (a) Show that the radial timelike or null geodesics in the regions $\{r > 2m\}$ and $\{r < 2m\}$ of the Schwarzschild spacetime cannot be extended as they approach $r = 0$ or $r = 2m$.

   (b) Show that the radial timelike or null geodesics in the Painlevé extension of the Schwarzschild spacetime can be extended to the future, but not to the past, as they approach $r = 2m$.

   (c) Show that radial observers satisfying

$$\frac{dr}{dt'} = -\sqrt{\frac{2m}{r}}$$

in the Painlevé coordinates are free-falling, and that t' is their proper time.

(d) What does a stationary observer see as a particle falls into a black hole?
(8) Show that an observer who crosses the horizon will hit the singularity in a proper
time interval $\Delta\tau \leq \pi m$.

6.6 Cosmology

Cosmology studies the behavior of the universe as a whole. Experimental observations (chiefly that of the cosmic background radiation) suggest that space is isotropic
at the Earth's location. Assuming the **Copernican principle** that the Earth's location in the universe is not in any way special, we take an isotropic (hence constant
curvature) 3-dimensional Riemannian manifold (Σ, h) as our model of space. We
can always find local coordinates (r, θ, φ) on Σ such that

$$h = a^2 \left(\frac{1}{1 - kr^2} dr \otimes dr + r^2 d\theta \otimes d\theta + r^2 \sin^2 \theta d\varphi \otimes d\varphi \right),$$

where $a > 0$ is the "radius" of space and $k = -1, 0, 1$ according to whether the curvature is negative, zero or positive [cf. Exercise 6.1(1)]. Allowing for the possibility
that the "radius" of space may be varying in time, we take our model of the universe
to be (M, g), where $M = \mathbb{R} \times \Sigma$ and

$$g = -dt \otimes dt + a^2(t) \left(\frac{1}{1 - kr^2} dr \otimes dr + r^2 d\theta \otimes d\theta + r^2 \sin^2 \theta d\varphi \otimes d\varphi \right).$$

These are the so-called **Friedmann–Lemaître–Robertson–Walker** (FLRW) models of cosmology.

One can easily compute the Ricci curvature for the metric g. We have

$$g = -\omega^0 \otimes \omega^0 + \omega^r \otimes \omega^r + \omega^\theta \otimes \omega^\theta + \omega^\varphi \otimes \omega^\varphi$$

with

$$\omega^0 = dt;$$
$$\omega^r = a(t) \left(1 - kr^2 \right)^{-\frac{1}{2}} dr;$$
$$\omega^\theta = a(t) r d\theta;$$
$$\omega^\varphi = a(t) r \sin \theta d\varphi,$$

and hence $\{\omega^0, \omega^r, \omega^\theta, \omega^\varphi\}$ is an orthonormal coframe. The first structure equations
yield

$$\omega_r^0 = \omega_0^r = \dot{a}\left(1 - kr^2\right)^{-\frac{1}{2}} dr;$$

$$\omega_\theta^0 = \omega_0^\theta = \dot{a} r d\theta;$$

$$\omega_\varphi^0 = \omega_0^\varphi = \dot{a} r \sin\theta d\varphi;$$

$$\omega_r^\theta = -\omega_\theta^r = \left(1 - kr^2\right)^{\frac{1}{2}} d\theta;$$

$$\omega_r^\varphi = -\omega_\varphi^r = \left(1 - kr^2\right)^{\frac{1}{2}} \sin\theta d\varphi;$$

$$\omega_\theta^\varphi = -\omega_\varphi^\theta = \cos\theta d\varphi.$$

The curvature forms can be computed from the second structure equations, and are found to be

$$\Omega_r^0 = \Omega_0^r = \frac{\ddot{a}}{a}\omega^0 \wedge \omega^r;$$

$$\Omega_\theta^0 = \Omega_0^\theta = \frac{\ddot{a}}{a}\omega^0 \wedge \omega^\theta;$$

$$\Omega_\varphi^0 = \Omega_0^\varphi = \frac{\ddot{a}}{a}\omega^0 \wedge \omega^\varphi;$$

$$\Omega_r^\theta = -\Omega_\theta^r = \left(\frac{k}{a^2} + \frac{\dot{a}^2}{a^2}\right)\omega^\theta \wedge \omega^r;$$

$$\Omega_r^\varphi = -\Omega_\varphi^r = \left(\frac{k}{a^2} + \frac{\dot{a}^2}{a^2}\right)\omega^\varphi \wedge \omega^r;$$

$$\Omega_\theta^\varphi = -\Omega_\varphi^\theta = \left(\frac{k}{a^2} + \frac{\dot{a}^2}{a^2}\right)\omega^\varphi \wedge \omega^\theta.$$

The components of the curvature tensor on the orthonormal frame can be read off from the curvature forms, and can in turn be used to compute the components of the Ricci curvature tensor Ric on the same frame. The nonvanishing components of Ric on this frame turn out to be

$$R_{00} = -\frac{3\ddot{a}}{a};$$

$$R_{rr} = R_{\theta\theta} = R_{\varphi\varphi} = \frac{\ddot{a}}{a} + \frac{2\dot{a}^2}{a^2} + \frac{2k}{a^2}.$$

At very large scales, galaxies and clusters of galaxies are expected to behave as particles of a pressureless fluid, which we take to be our matter model. By isotropy, the average spatial motion of the galaxies must vanish, and hence their unit velocity vector field must be $\frac{\partial}{\partial t}$ (corresponding to the 1-form $-dt$). Therefore the Einstein field equation is

$$Ric = 4\pi\rho(2dt \otimes dt + g),$$

which is equivalent to the ODE system

$$\begin{cases} -\dfrac{3\ddot{a}}{a} = 4\pi\rho \\[2mm] \dfrac{\ddot{a}}{a} + \dfrac{2\dot{a}^2}{a^2} + \dfrac{2k}{a^2} = 4\pi\rho \end{cases} \Leftrightarrow \begin{cases} \ddot{a} + \dfrac{\dot{a}^2}{2a} + \dfrac{k}{2a} = 0 \\[2mm] \rho = -\dfrac{3\ddot{a}}{4\pi a} \end{cases}.$$

The first equation allows us to determine the function $a(t)$, and the second yields ρ (which in particular must be a function of the t coordinate only; this is to be taken to mean that the average density of matter at cosmological scales is spatially constant). On the other hand, the quantity

$$\frac{4\pi\rho a^3}{3} = -\ddot{a}a^2$$

is constant, since

$$\frac{d}{dt}\left(-\ddot{a}a^2\right) = \frac{d}{dt}\left(\frac{a\dot{a}^2}{2} + \frac{ka}{2}\right) = a\dot{a}\ddot{a} + \frac{\dot{a}^3}{2} + \frac{k\dot{a}}{2} = 0.$$

Hence we have

$$\ddot{a} = -\frac{\alpha}{a^2}$$

for some integration constant α (we take $\alpha > 0$ so that $\rho > 0$). Substituting in the equation for $a(t)$ we get the first-order ODE

$$\frac{\dot{a}^2}{2} - \frac{\alpha}{a} = -\frac{k}{2}.$$

This can be used to show that $a(t)$ is bounded if and only if $k = 1$ [cf. Exercise 6.1(4)]. Moreover, in all cases $a(t)$ vanishes (and hence $\dot{a}(t)$, $\ddot{a}(t)$ and $\rho(t)$ blow up) for some value of t, usually taken to be $t = 0$. This singularity is called the **big bang** of the solution defined for $t > 0$. It was once thought to be a consequence of the high degree of symmetry of the FLRW models. Hawking and Penrose, however, showed that the big bang is actually a generic feature of cosmological models (cf. Sects. 6.8 and 6.9).

The function

$$H(t) = \frac{\dot{a}}{a}$$

is (somewhat confusingly) called the **Hubble constant**. It is easy to see from the above equations that

$$H^2 + \frac{k}{a^2} = \frac{8\pi}{3}\rho.$$

Therefore, in these models one has $k = -1$, $k = 0$ or $k = 1$ according to whether the average density ρ of the universe is smaller than, equal to or bigger than the so-called **critical density**

$$\rho_c = \frac{3H^2}{8\pi}.$$

These models were the standard models for cosmology for a long time. Currently, however, things are thought to be slightly more complicated [cf. Exercise 6.1(7)].

Exercise 6.1

(1) Show that the Riemannian metric h given in local coordinates (r, θ, φ) by

$$h = a^2 \left(\frac{1}{1 - kr^2} dr \otimes dr + r^2 d\theta \otimes d\theta + r^2 \sin^2 \theta d\varphi \otimes d\varphi \right)$$

has constant curvature $K = \frac{k}{a^2}$.

(2) The motions of galaxies and groups of galaxies in the FLRW models are the integral curves of $\frac{\partial}{\partial t}$. Show that these are timelike geodesics, and that the time coordinate t is the proper time of such observers.

(3) Consider two galaxies in a FLRW model, whose spatial locations can be assumed to be $r = 0$ and $(r, \theta, \varphi) = (r_1, \theta_1, \varphi_1)$. Show that:

(a) the spatial distance $d(t)$ between the two galaxies along the spatial Riemannian manifold of constant t satisfies the **Hubble law**

$$\dot{d} = Hd,$$

where $H = \dot{a}/a$ is the Hubble constant;

(b) the family (reparameterized) null geodesics connecting the first galaxy to the second galaxy can be written as

$$(t, r, \theta, \varphi) = (t(r, t_0), r, \theta_1, \varphi_1) \qquad (0 < r < r_1),$$

where $(t(r, t_0))$ is the solution of

$$\begin{cases} \dfrac{dt}{dr} = \dfrac{a(t)}{\sqrt{1 - kr^2}} \\[2mm] t(0, t_0) = t_0 \end{cases} ;$$

(c) $\dfrac{\partial t}{\partial t_0}(r_1, t_0) = \dfrac{a(t_1)}{a(t_0)}$, where $t_1 = t(r_1, t_0)$.

(**Remark:** This means that light emitted by the first galaxy with period T is measured by the second galaxy to have period $T' = \frac{a(t_1)}{a(t_0)} T$).

(4) Recall that in an FLRW model the "radius" of space, $a(t)$, evolves according to the ODE

$$\frac{\dot{a}^2}{2} - \frac{\alpha}{a} = -\frac{k}{2} \quad \Rightarrow \quad \ddot{a} = -\frac{\alpha}{a^2}.$$

Show that:

(a) $a(t)$ vanishes in finite time (assume that this happens at $t = 0$);
(b) if $k = -1$ or $k = 0$ then the solution can be extended to all values of $t > 0$;
(c) if $k = 1$ then the solution cannot be extended past $t = 2\pi\alpha$ (**big crunch**);
(d) if $k = 1$ then no observer can circumnavigate the universe, no matter how fast he moves;
(e) the solution can be given parametrically by:
   (i) $k = 1$:

$$\begin{cases} a = \alpha(1 - \cos u) \\ t = \alpha(u - \sin u) \end{cases} ;$$

   (ii) $k = 0$:

$$\begin{cases} a = \frac{\alpha}{2}u^2 \\ t = \frac{\alpha}{6}u^3 \end{cases} ;$$

   (iii) $k = -1$:

$$\begin{cases} a = \alpha(\cosh u - 1) \\ t = \alpha(\sinh u - u) \end{cases} .$$

(5) Show that the FLRW model with $k = 1$ is isometric to the hypersurface with equation

$$\sqrt{x^2 + y^2 + z^2 + w^2} = 2\alpha - \frac{t^2}{8\alpha}$$

in the 5-dimensional Minkowski spacetime $(\mathbb{R}^5, g)$ with metric

$$g = -dt \otimes dt + dx \otimes dx + dy \otimes dy + dz \otimes dz + dw \otimes dw.$$

(6) (*A model of collapse*) Show that the radius of a spherical shell $r = r_0$ in a FLRW model changes with proper time in exactly the same fashion as the radius of a radially free-falling spherical shell in a Schwarzschild spacetime of mass parameter m moving with energy parameter E [cf. Exercise 5.1(5)], provided that

$$\begin{cases} m = \alpha r_0^3 \\ E^2 - 1 = -k r_0^2 \end{cases} .$$

Therefore these two spacetimes can be matched along the 3-dimensional hyper-surface determined by the spherical shell's motion to yield a model of collapsing matter. Can you give a physical interpretation of this model?

(7) Show that if we allow for a **cosmological constant** $\Lambda \in \mathbb{R}$, i.e. for an Einstein equation of the form

$$Ric = 4\pi\rho(2\nu \otimes \nu + g) + \Lambda g$$

then the equations for the FLRW models become

$$\begin{cases} \dfrac{\dot{a}^2}{2} - \dfrac{\alpha}{a} - \dfrac{\Lambda}{6}a^2 = -\dfrac{k}{2} \\[3mm] \dfrac{4\pi}{3}a^3\rho = \alpha \end{cases}$$

Analyze the possible behaviors of the function $a(t)$.

(**Remark:** It is currently thought that there indeed exists a positive cosmological constant, also known as **dark energy**. The model favored by experimental observations seems to be $k = 0$, $\Lambda > 0$).

(8) Consider the 5-dimensional Minkowski spacetime $(\mathbb{R}^5, g)$ with metric

$$g = -dt \otimes dt + dx \otimes dx + dy \otimes dy + dz \otimes dz + dw \otimes dw.$$

Show that the induced metric on each of the following hypersurfaces determines FLRW models with the indicated parameters.

(a) **Einstein universe**: the "cylinder" of equation

$$x^2 + y^2 + z^2 + w^2 = \frac{1}{\Lambda},$$

satisfies $k = 1$, $\Lambda > 0$ and $\rho = \frac{\Lambda}{4\pi}$.

(b) **de Sitter universe**: the "sphere" of equation

$$-t^2 + x^2 + y^2 + z^2 + w^2 = \frac{3}{\Lambda}$$

satisfies $k = 1$, $\Lambda > 0$ and $\rho = 0$.

(9) A light signal emitted with period T and received with period T' is said to have suffered a **redshift**

$$z = \frac{T'}{T} - 1$$

[so that in the case of the Doppler effect one has $z \simeq v$ for small velocities, cf. Exercise 2.2(9)]. If the light is emitted by a galaxy at $r = 0$ at time $t = t_0$ and received by a galaxy at $r = r_1$ at time $t = t_1$ then its redshift is

$$z = \frac{a(t_1)}{a(t_0)} - 1$$

[cf. Exercise 6.1(3)]. This light is spread over a sphere of radius $R = a(t_1)r_1$, and so its brightness is inversely proportional to R^2. Compute R as a function of z for the following FLRW models:

(a) **Milne universe** ($k = -1, \alpha = \Lambda = 0$), for which $a(t) = t$;
(b) **Flat de Sitter universe** ($k = \alpha = 0, \Lambda = 3H^2$), for which $a(t) = e^{Ht}$;
(c) **Einstein–de Sitter universe** ($k = \Lambda = 0, \alpha = 2/9t_1{}^2$), for which $a(t) = (t/t_1)^{2/3}$.

(**Remark:** The brightness of distant galaxies is further reduced by a factor of $(1 + z)^2$, since each photon has frequency, hence energy, $(1 + z)$ times smaller at reception, and the rate of detection of photons is $(1 + z)$ times smaller than the rate of emission; with this correction, R can be deduced from the observed brightness for galaxies of known luminosity, and the correct FLRW model chosen as the one whose curve $R = R(z)$ best fits observations).

6.7 Causality

In this section we will study the causal features of spacetimes. This is a subject which has no parallel in Riemannian geometry, where the metric is positive definite. Although we will focus on 4-dimensional Lorentzian manifolds, the discussion can be easily generalized to any dimension $n \geq 2$.

A spacetime (M, g) is said to be **time-orientable** if there exists a vector field $X \in \mathfrak{X}(M)$ such that $\langle X, X \rangle < 0$. In this case, we can define a time orientation on each tangent space $T_p M$ (which is, of course, isometric to Minkowski spacetime) by choosing the timelike vectors in the connected component $C(X_p)$ to be future-pointing.

Assume that (M, g) is **time-oriented** (i.e. time-orientable with a definite choice of time orientation). A timelike curve $c : I \subset \mathbb{R} \to M$ is said to be **future-directed** if $\dot{c}$ is future-pointing. The **chronological future** of $p \in M$ is the set $I^+(p)$ of all points to which p can be connected by a future-directed timelike curve. A **future-directed causal curve** is a curve $c : I \subset \mathbb{R} \to M$ such that $\dot{c}$ is timelike or null and future-pointing (if nonzero). The **causal future** of $p \in M$ is the set $J^+(p)$ of all points to which p can be connected by a future-directed causal curve. Notice that $I^+(p)$ is simply the set of all events which are accessible to a particle with nonzero mass at p, whereas $J^+(p)$ is the set of events which can be causally influenced by p (as this causal influence cannot propagate faster than the speed of light). Analogously, the **chronological past** of $p \in M$ is the set $I^-(p)$ of all points which can be connected to p by a future-directed timelike curve, and the **causal past** of $p \in M$ is the set $J^-(p)$ of all points which can be connected to p by a future-directed causal curve.

In general, the chronological and causal pasts and futures can be quite complicated sets, because of global features of the spacetime. Locally, however, causal properties are similar to those of Minkowski spacetime. More precisely, we have the following statement:

Proposition 7.1 *Let (M, g) be a time-oriented spacetime. Then each point $p_0 \in M$ has an open neighborhood $V \subset M$ such that the spacetime (V, g) obtained by restricting g to V satisfies:*

(1) V is a normal neighborhood of each of its points, and given $p, q \in V$ there exists a unique geodesic (up to reparameterization) joining p to q (i.e. V is **geodesically convex***);*

(2) $q \in I^+(p)$ if and only if there exists a future-directed timelike geodesic connecting p to q;

(3) $J^+(p) = \overline{I^+(p)}$;

(4) $q \in J^+(p) \setminus I^+(p)$ if and only if there exists a future-directed null geodesic connecting p to q.

Proof Let U be a normal neighborhood of p_0 and choose **normal coordinates** (x^0, x^1, x^2, x^3) on U, given by the parameterization

$$\varphi(x^0, x^1, x^2, x^3) = \exp_{p_0}(x^0 v_0 + x^1 v_1 + x^2 v_2 + x^3 v_3),$$

where $\{v_0, v_1, v_2, v_3\}$ is a basis of $T_{p_0}(M)$ [cf. Exercise 4.8(2) in Chap. 3]. Let $D : U \to \mathbb{R}$ be the differentiable function

$$D(p) := \sum_{\alpha=0}^{3} \left(x^\alpha(p) \right)^2,$$

and let us define for each $\varepsilon > 0$ the set

$$B_\varepsilon := \{ p \in U \mid D(p) < \varepsilon \},$$

which for sufficiently small ε is diffeomorphic to an open ball in $T_{p_0} M$. Assume, for simplicity, that U is of this form.

Let us show that there exists $\rho > 0$ such that if $c : I \subset \mathbb{R} \to B_\rho$ is a geodesic then all critical points of $D(t) := D(c(t))$ are strict local minima. In fact, setting $x^\mu(t) := x^\mu(c(t))$, we have

$$\dot{D}(t) = 2 \sum_{\alpha=0}^{3} x^\alpha(t) \dot{x}^\alpha(t);$$

$$\ddot{D}(t) = 2 \sum_{\alpha=0}^{3} \left(\dot{x}^\alpha(t) \right)^2 + 2 \sum_{\alpha=0}^{3} x^\alpha(t) \ddot{x}^\alpha(t)$$

$$= 2 \sum_{\mu,\nu=0}^{3} \left(\delta_{\mu\nu} - \sum_{\alpha=0}^{3} \Gamma^\alpha_{\mu\nu}(c(t)) x^\alpha(t) \right) \dot{x}^\mu(t) \dot{x}^\nu(t),$$

and for ρ sufficiently small the matrix

$$\delta_{\mu\nu} - \sum_{\alpha=0}^{3} \Gamma_{\mu\nu}^{\alpha} x^{\alpha}$$

is positive definite on B_ρ.

Consider the map $F : W \subset TM \to M \times M$, defined on some open neighborhood W of $0 \in T_{p_0} M$ by

$$F(v) = (\pi(v), \exp(v)).$$

As we saw in the Riemannian case (cf. Chap. 3, Sect. 3.4), this map is a local diffeomorphism at $0 \in T_{p_0} M$. Choosing $\delta > 0$ sufficiently small and reducing W, we can assume that F maps W diffeomorphically to $B_\delta \times B_\delta$, and that $\exp(tv) \in B_\rho$ for all $t \in [0, 1]$ and $v \in W$ (as otherwise it would be possible to construct a sequence $v_n \to 0 \in T_{p_0} M$ such that $\exp(v_n) \not\to p_0$).

Finally, set $V = B_\delta$. If $p, q \in V$ and $v = F^{-1}(p, q)$, then $c(t) = \exp_p(tv)$ is a geodesic connecting p to q whose image is contained in B_ρ. If its image were not contained in V, there would necessarily exist a point of local maximum of $D(t)$, which cannot occur. Therefore, there is a geodesic in V connecting p to q. Since $\exp_p$ is a diffeomorphism onto V, this geodesic is unique (up to reparameterization). This proves (1).

To prove assertion (2), we start by noticing that if there exists a future-directed timelike geodesic connecting p to q then it is obvious that $q \in I^+(p)$. Suppose now that $q \in I^+(p)$; then there exists a future-directed timelike curve $c : [0, 1] \to V$ such that $c(0) = p$ and $c(1) = q$. Choose normal coordinates (x^0, x^1, x^2, x^3) given by the parameterization

$$\varphi(x^0, x^1, x^2, x^3) = \exp_p(x^0 E_0 + x^1 E_1 + x^2 E_2 + x^3 E_3),$$

where $\{E_0, E_1, E_2, E_3\}$ is an orthonormal basis of $T_p M$ with E_0 timelike and future-pointing. These are global coordinates in V, since $F : W \to V \times V$ is a diffeomorphism. Defining

$$W_p(q) := - \left(x^0(q)\right)^2 + \left(x^1(q)\right)^2 + \left(x^2(q)\right)^2 + \left(x^3(q)\right)^2$$

$$= \sum_{\mu,\nu=0}^{3} \eta_{\mu\nu} x^\mu(q) x^\nu(q),$$

with $(\eta_{\mu\nu}) = \mathrm{diag}(-1, 1, 1, 1)$, we have to show that $W_p(q) < 0$. Let $W_p(t) := W_p(c(t))$. Since $x^\mu(p) = 0$ ($\mu = 0, 1, 2, 3$), we have $W_p(0) = 0$. Setting $x^\mu(t) := x^\mu(c(t))$, we obtain

$$\dot{W}_p(t) = 2 \sum_{\mu,\nu=0}^{3} \eta_{\mu\nu} x^\mu(t) \dot{x}^\nu(t);$$

$$\ddot{W}_p(t) = 2 \sum_{\mu,\nu=0}^{3} \eta_{\mu\nu} x^\mu(t) \ddot{x}^\nu(t) + 2 \sum_{\mu,\nu=0}^{3} \eta_{\mu\nu} \dot{x}^\mu(t) \dot{x}^\nu(t),$$

and consequently (recalling that $(d \exp_p)_0 = \mathrm{id}$)

$$\dot{W}_p(0) = 0;$$
$$\ddot{W}_p(0) = 2\langle \dot{c}(0), \dot{c}(0) \rangle < 0.$$

Therefore there exists $\varepsilon > 0$ such that $W_p(t) < 0$ for $t \in (0, \varepsilon)$.

Using the same ideas as in the Riemannian case (cf. Chap. 3, Sect. 3.4), it is easy to prove that the level surfaces of W_p are orthogonal to the geodesics through p. Therefore, if $c_v(t) = \exp_p(tv)$ is the geodesic with initial condition $v \in T_p M$, we have

$$(\mathrm{grad}\, W_p)_{c_v(1)} = a(v)\dot{c}_v(1),$$

where the gradient of a function is defined as in the Riemannian case (notice, however, that in the Lorentzian case a smooth function f **decreases** along the direction of grad f if grad f is timelike). Now

$$\langle (\mathrm{grad}\, W_p)_{c_v(t)}, \dot{c}_v(t) \rangle = \frac{d}{dt} W_p(c_v(t)) = \frac{d}{dt} W_p(c_{tv}(1))$$
$$= \frac{d}{dt} \left(t^2 W_p(c_v(1)) \right) = 2t W_p(c_v(1)),$$

and hence

$$\langle (\mathrm{grad}\, W_p)_{c_v(1)}, \dot{c}_v(1) \rangle = 2 W_p(c_v(1)).$$

On the other hand,

$$\langle (\mathrm{grad}\, W_p)_{c_v(1)}, \dot{c}_v(1) \rangle = \langle a(v)\dot{c}_v(1), \dot{c}_v(1) \rangle$$
$$= a(v)\langle v, v \rangle = a(v) W_p(c_v(1)).$$

We conclude that $a(v) = 2$, and therefore

$$(\mathrm{grad}\, W_p)_{c_v(1)} = 2\dot{c}_v(1).$$

Consequently, grad W_p is tangent to geodesics through p, being future-pointing on future-directed geodesics.

Suppose that $W_p(t) < 0$. Then $(\mathrm{grad}\, W_p)_{c(t)}$ is timelike future-pointing, and so

$$\dot{W}(t) = \left\langle \left(\text{grad } W_p\right)_{c(t)}, \dot{c}(t)\right\rangle < 0,$$

as $\dot{c}(t)$ is also timelike future-pointing [cf. Exercise 2.2(2)]. We conclude that we must have $W_p(t) < 0$ for all $t \in [0, 1]$. In particular, $W_p(q) = W_p(1) < 0$, and hence there exists a future-directed timelike geodesic connecting p to q.

To prove assertion (3), let us see first that $I^+(p) \subset J^+(p)$. If $q \in I^+(p)$, then q is the limit of a sequence of points $q_n \in I^+(p)$. By (2), $q_n = \exp_p(v_n)$ with $v_n \in T_pM$ timelike future-pointing. Since $\exp_p$ is a diffeomorphism, v_n converges to a causal future-pointing vector $v \in T_pM$, and so $q = \exp_p(v)$ can be reached from p by a future-directed causal geodesic. The converse inclusion $J^+(p) \subset \overline{I^+(p)}$ holds in general (cf. Proposition 7.2).

Finally, (4) is obvious from (3) and the fact that $\exp_p$ is a diffeomorphism onto V. $\square$

This local behavior can be used to prove the following global result.

Proposition 7.2 *Let (M, g) be a time-oriented spacetime and $p \in M$. Then:*

(1) $I^+(p)$ is open;
(2) $J^+(p) \subset \overline{I^+(p)}$;
(3) $I^+(p) = \text{int } J^+(p)$
(4) if $r \in J^+(p)$ and $q \in I^+(r)$ then $q \in I^+(p)$;
(5) if $r \in I^+(p)$ and $q \in J^+(r)$ then $q \in I^+(p)$.

Proof Exercise 7.10(2). $\square$

The generalized twin paradox [cf. Exercise 2.2(8)] also holds locally for general spacetimes. More precisely, we have the following statement:

Proposition 7.3 *Let (M, g) be a time-oriented spacetime, $p_0 \in M$ and $V \subset M$ a geodesically convex open neighborhood of p_0. The spacetime (V, g) obtained by restricting g to V satisfies the following property: if $p, q \in V$ with $q \in I^+(p)$, c is the timelike geodesic connecting p to q and γ is any timelike curve connecting p to q, then $\tau(\gamma) \leq \tau(c)$, with equality if and only if γ is a reparameterization of c.*

Proof Any timelike curve $\gamma : [0, 1] \to V$ satisfying $\gamma(0) = p$, $\gamma(1) = q$ can be written as

$$\gamma(t) = \exp_p(r(t)n(t)),$$

for $t \in [0, 1]$, where $r(t) \geq 0$ and $\langle n(t), n(t)\rangle = -1$. We have

$$\dot{\gamma}(t) = (\exp_p)_* (\dot{r}(t)n(t) + r(t)\dot{n}(t)).$$

Since $\langle n(t), n(t)\rangle = -1$, we have $\langle \dot{n}(t), n(t)\rangle = 0$, and consequently $\dot{n}(t)$ is tangent to the level surfaces of the function $v \mapsto \langle v, v\rangle$. We conclude that

$$\dot{\gamma}(t) = \dot{r}(t)X_{\gamma(t)} + Y(t),$$

where X is the unit tangent vector field to timelike geodesics through p and $Y(t) = r(t)(\exp_p)_*\dot{n}(t)$ is tangent to the level surfaces of W_p (hence orthogonal to $X_{\gamma(t)}$). Consequently,

$$\tau(\gamma) = \int_0^1 \left| \langle \dot{r}(t)X_{\gamma(t)} + Y(t), \dot{r}(t)X_{\gamma(t)} + Y(t) \rangle \right|^{\frac{1}{2}} dt$$

$$= \int_0^1 \left(\dot{r}(t)^2 - |Y(t)|^2 \right)^{\frac{1}{2}} dt$$

$$\leq \int_0^1 \dot{r}(t)dt = r(1) = \tau(c),$$

where we have used the facts that γ is timelike, $\dot{r}(t) > 0$ for all $t \in [0, 1]$ (as $\dot{\gamma}$ is future-pointing) and $\tau(c) = r(1)$ (as $q = \exp_p(r(1)n(1))$). It should be clear that $\tau(\gamma) = \tau(c)$ if and only if $|Y(t)| \equiv 0 \Leftrightarrow Y(t) \equiv 0$ ($Y(t)$ is spacelike or zero) for all $t \in [0, 1]$, implying that n is constant. In this case, $\gamma(t) = \exp_p(r(t)n)$ is, up to reparameterization, the geodesic through p with initial condition $n \in T_pM$. $\square$

There is also a local property characterizing null geodesics.

Proposition 7.4 *Let (M, g) be a time-oriented spacetime, $p_0 \in M$ and $V \subset M$ a geodesically convex open neighborhood of p_0. The spacetime (V, g) obtained by restricting g to V satisfies the following property: if for $p, q \in V$ there exists a future-directed null geodesic c connecting p to q and γ is a causal curve connecting p to q then γ is a reparameterization of c.*

Proof Since p and q are connected by a null geodesic, we conclude from Proposition 7.1 that $q \in J^+(p) \setminus I^+(p)$. Let $\gamma : [0, 1] \to V$ be a causal curve connecting p to q. Then we must have $\gamma(t) \in J^+(p) \setminus I^+(p)$ for all $t \in [0, 1]$, since $\gamma(t_0) \in I^+(p)$ implies $\gamma(t) \in I^+(p)$ for all $t > t_0$ (see Proposition 7.2). Consequently, we have

$$\left\langle (\operatorname{grad} W_p)_{\gamma(t)}, \dot{\gamma}(t) \right\rangle = 0,$$

where W_p was defined in the proof of Proposition 7.1. The formula

$$(\operatorname{grad} W_p)_{c_v(1)} = 2\dot{c}_v(1),$$

which was proved for timelike geodesics c_v with initial condition $v \in T_pM$, must also hold for null geodesics (by continuity). Hence grad W_p is tangent to the null geodesics ruling $J^+(p) \setminus I^+(p)$ and future-pointing. Since $\dot{\gamma}(t)$ is also future-pointing, we conclude that $\dot{\gamma}$ is proportional to grad W_p [cf. Exercise 2.2(8)], and therefore γ must be a reparameterization of a null geodesic (which must be c). $\square$

Corollary 7.5 *Let (M, g) be a time-oriented spacetime and $p \in M$. If $q \in J^+$ $(p) \setminus I^+(p)$ then any future-directed causal curve connecting p to q must be a reparameterized null geodesic.* $\qquad\square$

For physical applications, it is important to require that the spacetime satisfies reasonable causality conditions. The simplest of these conditions excludes time travel, i.e. the possibility of a particle returning to an event in its past history.

Definition 7.6 A spacetime (M, g) is said to satisfy the **chronology condition** if it does not contain closed timelike curves.

This condition is violated by compact spacetimes:

Proposition 7.7 *Any compact spacetime (M, g) contains closed timelike curves.*

Proof Taking if necessary the time-orientable double covering [cf. Exercise 7.10(1)], we can assume that (M, g) is time-oriented. Since $I^+(p)$ is an open set for any $p \in M$, it is clear that $\{I^+(p)\}_{p \in M}$ is an open cover of M. If M is compact, we can obtain a finite subcover $\{I^+(p_1), \ldots, I^+(p_N)\}$. Now if $p_1 \in I^+(p_i)$ for $i \neq 1$ then $I^+(p_1) \subset I^+(p_i)$, and we can exclude $I^+(p_1)$ from the subcover. Therefore, we can assume without loss of generality that $p_1 \in I^+(p_1)$, and hence there exists a closed timelike curve starting and ending at p_1. $\qquad\square$

A stronger restriction on the causal behavior of the spacetime is the following:

Definition 7.8 A spacetime (M, g) is said to be **stably causal** if there exists a **global time function**, i.e. a smooth function $t : M \to \mathbb{R}$ such that $\text{grad}(t)$ is timelike.

In particular, a stably causal spacetime is time-orientable. We choose the time orientation defined by $-\text{grad}(t)$, so that t increases along future-directed time-like curves. Notice that this implies that no closed timelike curves can exist, i.e. any stably causal spacetime satisfies the chronology condition. In fact, any small perturbation of a stably causal spacetime still satisfies the chronology condition [cf. Exercise 7.10(4)].

Let (M, g) be a time-oriented spacetime. A smooth future-directed causal curve $c : (a, b) \to M$ (with possibly $a = -\infty$ or $b = +\infty$) is said to be **future-inextendible** if $\lim_{t \to b} c(t)$ does not exist. The definition of a **past-inextendible** causal curve is analogous. The **future domain of dependence** of $S \subset M$ is the set $D^+(S)$ of all events $p \in M$ such that any past-inextendible causal curve starting at p intersects S. Therefore any causal influence on an event $p \in D^+(S)$ had to register somewhere in S, and one can expect that what happens at p can be predicted from data on S. Similarly, the **past domain of dependence** of S is the set $D^-(S)$ of all events $p \in M$ such that any future-inextendible causal curve starting at p intersects S. Therefore any causal influence of an event $p \in D^-(S)$ will register somewhere in S, and one can expect that what happened at p can be retrodicted from data on S. The **domain of dependence** of S is simply the set $D(S) = D^+(S) \cup D^-(S)$.

Let (M, g) be a stably causal spacetime with time function $t : M \to \mathbb{R}$. The level sets $S_a = t^{-1}(a)$ are said to be **Cauchy hypersurfaces** if $D(S_a) = M$. Spacetimes for which this happens have particularly good causal properties.

Definition 7.9 A stably causal spacetime possessing a time function whose level sets are Cauchy hypersurfaces is said to be **globally hyperbolic**.

Notice that the future and past domains of dependence of the Cauchy hypersurfaces S_a are $D^+(S_a) = t^{-1}([a, +\infty))$ and $D^-(S_a) = t^{-1}((-\infty, a])$.

Exercise 7.10

(1) *(Time-orientable double covering)* Using ideas similar to those of Exercise 8.6(9) in Chap. 1, show that if (M, g) is a non-time-orientable Lorentzian manifold then there exists a **time-orientable double covering**, i.e. a time-orientable Lorentzian manifold $(\overline{M}, \overline{g})$ and a local isometry $\pi : \overline{M} \to M$ such that every point in M has two preimages by π. Use this to conclude that the only compact surfaces which admit a Lorentzian metric are the torus T^2 and the Klein bottle K^2.

(2) Let (M, g) be a time-oriented spacetime and $p \in M$. Show that:

    (a) $I^+(p)$ is open;

    (b) $J^+(p)$ is not necessarily closed;

    (c) $J^+(p) \subset \overline{I^+(p)}$;

    (d) $I^+(p) = \text{int } J^+(p)$

    (e) if $r \in J^+(p)$ and $q \in I^+(r)$ then $q \in I^+(p)$;

    (f) if $r \in I^+(p)$ and $q \in J^+(r)$ then $q \in I^+(p)$;

    (g) it may happen that $I^+(p) = M$.

(3) Consider the 3-dimensional Minkowski spacetime $(\mathbb{R}^3, g)$, where

$$g = -dt \otimes dt + dx \otimes dx + dy \otimes dy.$$

Let $c : \mathbb{R} \to \mathbb{R}^3$ be the curve $c(t) = (t, \cos t, \sin t)$. Show that although $\dot{c}(t)$ is null for all $t \in \mathbb{R}$ we have $c(t) \in I^+(c(0))$ for all $t > 0$. What kind of motion does this curve represent?

(4) Let (M, g) be a stably causal spacetime and h an arbitrary symmetric $(2, 0)$-tensor field with compact support. Show that for sufficiently small $|\varepsilon|$ the tensor field $g_\varepsilon := g + \varepsilon h$ is still a Lorentzian metric on M, and (M, g_ε) satisfies the chronology condition.

(5) Let (M, g) be the quotient of the 2-dimensional Minkowski spacetime by the discrete group of isometries generated by the map $f(t, x) = (t + 1, x + 1)$. Show that (M, g) satisfies the chronology condition, but there exist arbitrarily small perturbations of (M, g) [in the sense of Exercise 7.10(4)] which do not.

(6) Let (M, g) be a time-oriented spacetime and $S \subset M$. Show that:

    (a) $S \subset D^+(S)$;

    (b) $D^+(S)$ is not necessarily open;

    (c) $D^+(S)$ is not necessarily closed.

(7) Let (M, g) be the 2-dimensional spacetime obtained by removing the positive x-semi-axis of Minkowski 2-dimensional spacetime (cf. Fig. 6.8). Show that:

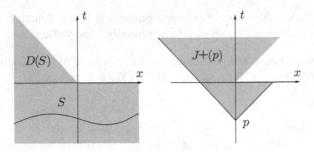

Fig. 6.8 Stably causal but not globally hyperbolic spacetime

 (a) (M, g) is stably causal but not globally hyperbolic;
 (b) there exist points $p, q \in M$ such that $J^+(p) \cap J^-(q)$ is not compact;
 (c) there exist points $p, q \in M$ with $q \in I^+(p)$ such that the supremum of the
      lengths of timelike curves connecting p to q is not attained by any timelike
      curve.

 (8) Let (Σ, h) be a 3-dimensional Riemannian manifold. Show that the spacetime
      $(M, g) = (\mathbb{R} \times \Sigma, -dt \otimes dt + h)$ is globally hyperbolic if and only if (Σ, h)
      is complete.
 (9) Show that the following spacetimes are globally hyperbolic:

     (a) the Minkowski spacetime;
     (b) the FLRW spacetimes;
     (c) the region $\{r > 2m\}$ of Schwarzschild spacetime;
     (d) the region $\{r < 2m\}$ of Schwarzschild spacetime.

 (10) Let (M, g) be a global hyperbolic spacetime with Cauchy hypersurface S. Show
       that M is diffeomorphic to $\mathbb{R} \times S$.

6.8 Hawking Singularity Theorem

As we have seen in Sects. 6.5 and 6.6, both the Schwarzschild solution and the FLRW
cosmological models display singularities, beyond which timelike and null geodesics
cannot be continued.

Definition 8.1 A spacetime (M, g) is said to be **singular** if it is not geodesically
complete.

It was once thought that the examples above were singular due to their high degree
of symmetry, and that more realistic spacetimes would be nonsingular. Following
Hawking and Penrose [Pen65, Haw67, HP70], we will show that this is not the case:
any sufficiently small perturbation of these solutions will still be singular.

The question of whether a given Riemannian manifold is geodesically complete is settled by the Hopf–Rinow theorem. Unfortunately, this theorem does not hold in Lorentzian geometry (essentially because one cannot use the metric to define a distance function). For instance, compact manifolds are not necessarily geodesically complete [cf. Exercise 8.12(1)], and the exponential map is not necessarily surjective in geodesically complete manifolds [cf. Exercise 8.12(2)].

Let (M, g) be a globally hyperbolic spacetime and S a Cauchy hypersurface with future-pointing unit normal vector field n. Let c_p be the timelike geodesic with initial condition n_p for each point $p \in S$. We define a smooth map $\exp : U \to M$ on an open set $U \subset \mathbb{R} \times S$ containing $\{0\} \times S$ as $\exp(t, p) = c_p(t)$.

Definition 8.2 The critical values of $\exp$ are said to be **conjugate points** to S.

Loosely speaking, conjugate points are points where geodesics starting orthogonally at nearby points of S intersect.

Let $q = \exp(t_0, p)$ be a point not conjugate to S, and let (x^1, x^2, x^3) be local coordinates on S around p. Then (t, x^1, x^2, x^3) are local coordinates on some open set $V \ni q$. Since $\frac{\partial}{\partial t}$ is the unit tangent field to the geodesics orthogonal to S, we have $g_{00} = \left\langle \frac{\partial}{\partial t}, \frac{\partial}{\partial t} \right\rangle = -1$. On the other hand, we have

$$\frac{\partial g_{0i}}{\partial t} = \frac{\partial}{\partial t} \left\langle \frac{\partial}{\partial t}, \frac{\partial}{\partial x^i} \right\rangle = \left\langle \frac{\partial}{\partial t}, \nabla_{\frac{\partial}{\partial t}} \frac{\partial}{\partial x^i} \right\rangle$$

$$= \left\langle \frac{\partial}{\partial t}, \nabla_{\frac{\partial}{\partial x^i}} \frac{\partial}{\partial t} \right\rangle = \frac{1}{2} \frac{\partial}{\partial x^i} \left\langle \frac{\partial}{\partial t}, \frac{\partial}{\partial t} \right\rangle = 0$$

for $i = 1, 2, 3$, and, since $g_{0i} = 0$ on S, we have $g_{0i} = 0$ on V. Therefore the surfaces of constant t are orthogonal to the geodesics tangent to $\frac{\partial}{\partial t}$. For this reason, (t, x^1, x^2, x^3) is said to be a **synchronized** coordinate system. On this coordinate system we have

$$g = -dt \otimes dt + \sum_{i,j=1}^{3} \gamma_{ij} dx^i \otimes dx^j,$$

where the functions

$$\gamma_{ij} := \left\langle \frac{\partial}{\partial x^i}, \frac{\partial}{\partial x^j} \right\rangle$$

form a positive definite matrix. Since the vector fields $\frac{\partial}{\partial x^i}$ can always be defined along c_p, the matrix (γ_{ij}) is also well defined along c_p, even at points where the synchronized coordinate system breaks down, i.e. at points which are conjugate to S. These are the points for which $\gamma(t) := \det\left(\gamma_{ij}(t)\right)$ vanishes, since only then will $\left\{ \frac{\partial}{\partial t}, \frac{\partial}{\partial x^1}, \frac{\partial}{\partial x^2}, \frac{\partial}{\partial x^3} \right\}$ fail to be linearly independent. (In fact the vector fields $\frac{\partial}{\partial x^i}$ are Jacobi fields along c_p—see Exercise 4.8(6) in Chap. 3).

It is easy to see that

$$\Gamma_{00}^0 = \Gamma_{00}^i = 0 \quad \text{and} \quad \Gamma_{0j}^i = \sum_{k=1}^{3} \gamma^{ik} \beta_{kj},$$

where $(\gamma^{ij}) = (\gamma_{ij})^{-1}$ and $\beta_{ij} = \frac{1}{2} \frac{\partial \gamma_{ij}}{\partial t}$ [cf. Exercise 8.12(4)]. Consequently,

$$R_{00} = \sum_{i=1}^{3} R_{i00}{}^i = \sum_{i=1}^{3} \left(\frac{\partial \Gamma_{00}^i}{\partial x^i} - \frac{\partial \Gamma_{i0}^i}{\partial t} + \sum_{j=1}^{3} \Gamma_{00}^j \Gamma_{ij}^i - \sum_{j=1}^{3} \Gamma_{i0}^j \Gamma_{0j}^i \right)$$

$$= -\frac{\partial}{\partial t} \left(\sum_{i,j=1}^{3} \gamma^{ij} \beta_{ij} \right) - \sum_{i,j,k,l=1}^{3} \gamma^{jk} \gamma^{il} \beta_{ki} \beta_{lj}.$$

(cf. Chap. 4, Sect. 6.1). The quantity

$$\theta := \sum_{i,j=1}^{3} \gamma^{ij} \beta_{ij}$$

appearing in this expression is called the **expansion** of the synchronized observers, and has an important geometric meaning:

$$\theta = \frac{1}{2} \operatorname{tr} \left((\gamma_{ij})^{-1} \frac{\partial}{\partial t} (\gamma_{ij}) \right) = \frac{1}{2} \frac{\partial}{\partial t} \log \gamma = \frac{\partial}{\partial t} \log \gamma^{\frac{1}{2}}.$$

Here we have used the formula

$$(\log(\det A))' = \operatorname{tr} \left(A^{-1} A' \right)$$

which holds for any smooth matrix function $A : \mathbb{R} \to GL(n)$ [cf. Example 7.1(4) in Chap. 1]. Therefore the expansion yields the variation of the 3-dimensional volume element measured by synchronized observers. More importantly for our purposes, we see that a singularity of the expansion indicates a zero of γ, i.e. a conjugate point to S.

Definition 8.3 A spacetime (M, g) is said to satisfy the **strong energy condition** if $Ric(V, V) \geq 0$ for any timelike vector field $V \in \mathfrak{X}(M)$.

By the Einstein equation, this is equivalent to requiring that the reduced energy-momentum tensor T satisfies $T(V, V) \geq 0$ for any timelike vector field $V \in \mathfrak{X}(M)$. In the case of a pressureless fluid with rest density function $\rho \in C^\infty(M)$ and unit velocity vector field $U \in \mathfrak{X}(M)$, this requirement becomes

$$\rho \left(\langle U, V \rangle^2 + \frac{1}{2} \langle V, V \rangle \right) \geq 0,$$

or, since the term in brackets is always positive [cf. Exercis 8.12(5)], simply $\rho \geq 0$. For more complicated matter models, the strong energy condition produces equally reasonable restrictions.

Proposition 8.4 *Let (M, g) be a globally hyperbolic spacetime satisfying the strong energy condition, $S \subset M$ a Cauchy hypersurface and $p \in S$ a point where $\theta = \theta_0 < 0$. Then the geodesic c_p contains at least a point conjugate to S, at a distance of at most $-\frac{3}{\theta_0}$ to the future of S (assuming that it can be extended that far).*

Proof Since (M, g) satisfies the strong energy condition, we have $R_{00} = Ric \left(\frac{\partial}{\partial t}, \frac{\partial}{\partial t} \right) \geq 0$ on any synchronized frame. Consequently,

$$\frac{\partial \theta}{\partial t} + \sum_{i,j,k,l=1}^{3} \gamma^{jk} \gamma^{il} \beta_{ki} \beta_{lj} \leq 0$$

on such a frame. Choosing an orthonormal basis (where $\gamma^{ij} = \delta_{ij}$) and using the inequality

$$(\text{tr } A)^2 \leq n \, \text{tr}(A^t A),$$

which holds for square $n \times n$ matrices (as a simple consequence of the Cauchy–Schwarz inequality), it is easy to show that

$$\sum_{i,j,k,l=1}^{3} \gamma^{jk} \gamma^{il} \beta_{ki} \beta_{lj} = \sum_{i,j=1}^{3} \beta_{ji} \beta_{ij} = \text{tr} \left((\beta_{ij})(\beta_{ij})^t \right) \geq \frac{1}{3} \theta^2.$$

Consequently θ must satisfy

$$\frac{\partial \theta}{\partial t} + \frac{1}{3} \theta^2 \leq 0.$$

Integrating this inequality yields

$$\frac{1}{\theta} \geq \frac{1}{\theta_0} + \frac{t}{3},$$

and hence θ must blow up at a value of t no greater than $-\frac{3}{\theta_0}$. $\square$

Proposition 8.5 *Let (M, g) be a globally hyperbolic spacetime, S a Cauchy hypersurface, $p \in M$ and c a timelike geodesic through p orthogonal to S. If there exists a conjugate point between S and p then c does not maximize length (among the timelike curves connecting S to p).*

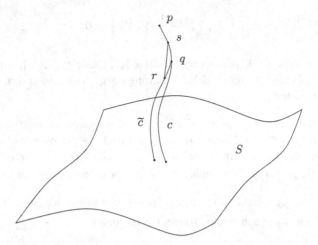

Fig. 6.9 Proof of Proposition 8.5

Proof We will offer only a sketch of the proof. Let q be the first conjugate point along c between S and p. Then we can use a synchronized coordinate system around the portion of c between S and q. Since q is conjugate to S, there exists another geodesic $\tilde{c}$, orthogonal to S, with the same length $t(q)$, which (approximately) intersects c at q. Let V be a geodesically convex neighborhood of q, let $r \in V$ be a point along $\tilde{c}$ between S and q, and let $s \in V$ be a point along c between q and p (cf. Fig.6.9). Then the piecewise smooth timelike curve obtained by following $\tilde{c}$ between S and r, the unique geodesic in V between r and s, and c between s and p, connects S to p and has strictly bigger length than c (by the generalized twin paradox). This curve can be easily smoothed while retaining bigger length than c. $\square$

Proposition 8.6 *Let (M, g) be a globally hyperbolic spacetime, S a Cauchy hypersurface and $p \in D^+(S)$. Then $D^+(S) \cap J^-(p)$ is compact.*

Proof Let us define a **simple neighborhood** $U \subset M$ to be a geodesically convex open set diffeomorphic to an open ball whose boundary is a compact submanifold of a larger geodesically convex open set (therefore ∂U is diffeomorphic to S^3 and $\overline{U}$ is compact). It is clear that simple neighborhoods form a basis for the topology of M. Also, it is easy to show that any open cover $\{V_\alpha\}_{\alpha \in A}$ has a countable, locally finite refinement $\{U_n\}_{n \in \mathbb{N}}$ by simple neighborhoods [cf. Exercise 8.12(7)].

If $A = D^+(S) \cap J^-(p)$ were not compact, there would exist a countable, locally finite open cover $\{U_n\}_{n \in \mathbb{N}}$ of A by simple neighborhoods not admitting any finite subcover. Take $q_n \in A \cap U_n$ such that $q_m \neq q_n$ for $m \neq n$. The sequence $\{q_n\}_{n \in \mathbb{N}}$ cannot have accumulation points, since any point in M has a neighborhood intersecting only finite simple neighborhoods U_n. In particular, each simple neighborhood U_n contains only a finite number of points in the sequence (as $\overline{U}_n$ is compact).

Set $p_1 = p$. Since $p_1 \in A$, we have $p_1 \in U_{n_1}$ for some $n_1 \in \mathbb{N}$. Let $q_n \notin U_{n_1}$. Since $q_n \in J^-(p_1)$, there exists a future-directed causal curve c_n connecting q_n to

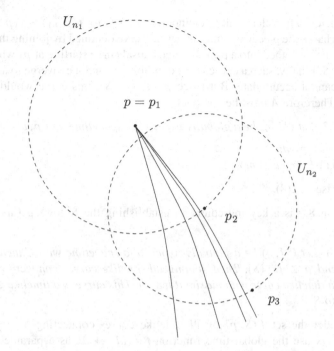

Fig. 6.10 Proof of Proposition 8.6

p_1. This curve will necessarily intersect ∂U_{n_1}. Let $r_{1,n}$ be an intersection point. Since U_{n_1} contains only a finite number of points in the sequence $\{q_n\}_{n\in\mathbb{N}}$, there will exist infinite intersection points $r_{1,n}$. As ∂U_{n_1} is compact, these will accumulate to some point $p_2 \in \partial U_{n_1}$ (cf. Fig. 6.10).

Because $\overline{U}_{n_1}$ is contained in a geodesically convex open set V, which can be chosen so that $v \mapsto (\pi(v), \exp(v))$ is a diffeomorphism onto $V \times V$, we have $p_2 \in J^-(p_1)$: if $\gamma_{1,n}$ is the unique causal geodesic connecting p_1 to $r_{1,n}$, parameterized by the global time function $t : M \to \mathbb{R}$, then the subsequence of $\{\gamma_{1,n}\}$ corresponding to a convergent subsequence of $\{r_{1,n}\}$ will converge to a causal geodesic γ_1 connecting p_1 to p_2. If $S = t^{-1}(0)$ then we have $t(r_{1,n}) \geq 0$, implying that $t(p_2) \geq 0$ and hence $p_2 \in A$. Since $p_2 \notin U_{n_1}$, there must exist $n_2 \in \mathbb{N}$ such that $p_2 \in U_{n_2}$.

Since U_{n_2} contains only a finite number of points in the sequence $\{q_n\}_{n\in\mathbb{N}}$, an infinite number of curves c_n must intersect ∂U_{n_2} to the past of $r_{1,n}$. Let $r_{2,n}$ be the intersection points. As ∂U_{n_2} is compact, $\{r_{2,n}\}$ must accumulate to some point $p_3 \in \partial U_{n_2}$. Because $\overline{U}_{n_2}$ is contained in a geodesically convex open set, $p_3 \in J^-(p_2)$: if $\gamma_{2,n}$ is the unique causal geodesic connecting $r_{1,n}$ to $r_{2,n}$, parameterized by the global time function, then the subsequence of $\{\gamma_{2,n}\}$ corresponding to convergent subsequences of both $\{r_{1,n}\}$ and $\{r_{2,n}\}$ will converge to a causal geodesic connecting p_2 to p_3. Since $J^-(p_2) \subset J^-(p_1)$ and $t(r_{2,n}) \geq 0 \Rightarrow t(p_3) \geq 0$, we have $p_3 \in A$.

Iterating the procedure above, we can construct a sequence $\{p_i\}_{i\in\mathbb{N}}$ of points in A satisfying $p_i \in U_{n_i}$ with $n_i \neq n_j$ if $i \neq j$, such that p_i is connected to p_{i+1} by a causal geodesic γ_i. It is clear that γ_i cannot intersect S, for $t(p_{i+1}) > t(p_{i+2}) \geq 0$. On the other hand, the piecewise smooth causal curve obtained by joining the curves γ_i can easily be smoothed into a past-directed causal curve starting at p_1 which does not intersect S. Finally, such a curve is inextendible: it cannot converge to any point, as $\{p_i\}_{i\in\mathbb{N}}$ cannot accumulate. But since $p_1 \in D^+(S)$, this curve would have to intersect S. Therefore A must be compact. $\square$

Corollary 8.7 *Let (M, g) be a globally hyperbolic spacetime and $p, q \in M$. Then*

(i) *$J^+(p)$ is closed;*
(ii) *$J^+(p) \cap J^-(q)$ is compact.*

Proof Exercise 8.12(8). $\square$

Proposition 8.6 is a key ingredient in establishing the following fundamental result.

Theorem 8.8 *Let (M, g) be a globally hyperbolic spacetime with Cauchy hypersurface S, and $p \in D^+(S)$. Then, among all timelike curves connecting p to S, there exists a timelike curve with maximal length. This curve is a timelike geodesic, orthogonal to S.*

Proof Consider the set $T(S, p)$ of all timelike curves connecting S to p. Since we can always use the global time function $t : M \to \mathbb{R}$ as a parameter, these curves are determined by their images, which are compact subsets of the compact set $A = D^+(S) \cap J^-(p)$. As is well known (cf. [Mun00]), the set $C(A)$ of all compact subsets of A is a compact metric space for the **Hausdorff metric** d_H, defined as follows: if $d : M \times M \to \mathbb{R}$ is a metric yielding the topology of M,

$$d_H(K, L) = \inf\{\varepsilon > 0 \mid K \subset U_\varepsilon(L) \text{ and } L \subset U_\varepsilon(K)\},$$

where $U_\varepsilon(K)$ is a ε-neighborhood of K for the metric d. Therefore, the closure $C(S, p) := \overline{T(S, p)}$ is a compact subset of $C(A)$. It is not difficult to show that $C(S, p)$ can be identified with the set of **continuous causal curves** connecting S to p (a continuous curve $c : [0, t(p)] \to M$ is said to be **causal** if $c(t_2) \in J^+(c(t_1))$ whenever $t_2 > t_1$).

The length function $\tau : T(S, p) \to \mathbb{R}$ is defined by

$$\tau(c) := \int_0^{t(p)} |\dot{c}(t)| dt.$$

This function is **upper semicontinuous**, i.e. continuous for the topology

$$\mathcal{O} = \{(-\infty, a) \mid -\infty \leq a \leq +\infty\}$$

in $\mathbb{R}$. Indeed, let $c \in T(S, p)$ be parameterized by its arclength u. For a sufficiently small $\varepsilon > 0$, the function u can be extended to the ε-neighborhood $U_\varepsilon(c)$ in such a

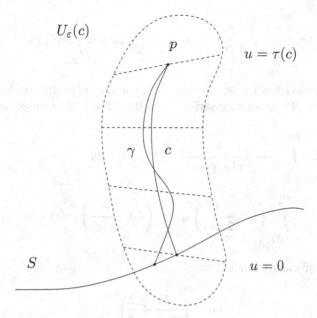

Fig. 6.11 Proof of Theorem 8.8

way that its level hypersurfaces are spacelike and orthogonal to c, that is, $-\operatorname{grad} u$ is timelike and coincides with $\dot{c}$ on c (cf. Fig. 6.11). If $\gamma \in T(S, p)$ is in the open ball $B_\varepsilon(c) \subset C(A)$ for the Hausdorff metric d_H then we can use u as a parameter, thus obtaining

$$du(\dot{\gamma}) = 1 \Leftrightarrow \langle \dot{\gamma}, \operatorname{grad} u \rangle = 1.$$

Therefore $\dot{\gamma}$ can be decomposed as

$$\dot{\gamma} = \frac{1}{\langle \operatorname{grad} u, \operatorname{grad} u \rangle} \operatorname{grad} u + X,$$

where X is spacelike and orthogonal to $\operatorname{grad} u$, and so

$$|\dot{\gamma}| = \left| \frac{1}{\langle \operatorname{grad} u, \operatorname{grad} u \rangle} + \langle X, X \rangle \right|^{\frac{1}{2}}.$$

Given $\delta > 0$, we can choose $\varepsilon > 0$ sufficiently small so that

$$-\frac{1}{\langle \operatorname{grad} u, \operatorname{grad} u \rangle} < \left(1 + \frac{\delta}{2\tau(c)}\right)^2$$

on the ε-neighborhood $U_\varepsilon(c)$ (as $\langle \operatorname{grad} u, \operatorname{grad} u \rangle = -1$ along c). We have

$$\tau(\gamma) = \int\limits_{0}^{t(p)} \left| \frac{d\gamma}{dt} \right| dt = \int\limits_{0}^{t(p)} |\dot{\gamma}| \frac{du}{dt} dt = \int\limits_{u(\gamma \cap S)}^{\tau(c)} |\dot{\gamma}| du,$$

where we have to allow for the fact that c is not necessarily orthogonal to S, and so the initial point of γ is not necessarily at $u = 0$ (cf. Fig. 6.11). Consequently,

$$\tau(\gamma) = \int\limits_{u(\gamma \cap S)}^{\tau(c)} \left| -\frac{1}{\langle \text{grad } u, \text{grad } u \rangle} - \langle X, X \rangle \right|^{\frac{1}{2}} du$$

$$< \int\limits_{u(\gamma \cap S)}^{\tau(c)} \left(1 + \frac{\delta}{2\tau(c)} \right) du = \left(1 + \frac{\delta}{2\tau(c)} \right) (\tau(c) - u(\gamma \cap S)).$$

Choosing ε sufficiently small so that

$$|u| < \left(\frac{1}{\tau(c)} + \frac{2}{\delta} \right)^{-1}$$

on $S \cap U_\varepsilon(c)$, we obtain $\tau(\gamma) < \tau(c) + \delta$, proving upper semicontinuity in $T(S, p)$. As a consequence, the length function can be extended to $C(S, p)$ through

$$\tau(c) = \lim\limits_{\varepsilon \to 0} \sup\{\tau(\gamma) \mid \gamma \in B_\varepsilon(c) \cap T(S, p)\}$$

(as for $\varepsilon > 0$ sufficiently small the supremum will be finite). Also, it is clear that if $c \in T(S, p)$ then the upper semicontinuity of the length forces the two definitions of $\tau(c)$ to coincide. The extension of the length function to $C(S, p)$ is trivially upper semicontinuous: given $c \in C(S, p)$ and $\delta > 0$, let $\varepsilon > 0$ be such that $\tau(\gamma) < \tau(c) + \frac{\delta}{2}$ for any $\gamma \in B_{2\varepsilon}(c) \cap T(S, p)$. Then it is clear that $\tau(c') \le \tau(c) + \frac{\delta}{2} < \tau(c) + \delta$ for any $c' \in B_\varepsilon(c)$.

Finally, we notice that the compact sets of $\mathbb{R}$ for the topology $\mathcal{O}$ are the sets with a maximum. Therefore, the length function attains a maximum at some point $c \in C(S, p)$. All that remains to be seen is that the maximum is also attained at a smooth timelike curve γ. To do so, cover c with finitely many geodesically convex neighborhoods and choose points $p_1, \ldots, p_k$ in c such that $p_1 \in S$, $p_k = p$ and the portion of c between p_{i-1} and p_i is contained in a geodesically convex neighborhood for all $i = 2, \ldots, k$. It is clear that there exists a sequence $c_n \in T(S, p)$ such that $c_n \to c$ and $\tau(c_n) \to \tau(c)$. Let $t_i = t(p_i)$ and $p_{i,n}$ be the intersection of c_n with $t^{-1}(t_i)$. Replace c_n by the sectionally geodesic curve γ_n obtained by joining $p_{i-1,n}$ to $p_{i,n}$ in the corresponding geodesically convex neighborhood. Then $\tau(\gamma_n) \ge \tau(c_n)$, and therefore $\tau(\gamma_n) \to \tau(c)$. Since each sequence $p_{i,n}$ converges to p_i, γ_n converges to the sectionally geodesic curve γ obtained by joining p_{i-1} to p_i ($i = 2, \ldots, k$), and it is clear that $\tau(\gamma_n) \to \tau(\gamma) = \tau(c)$. Therefore γ is a point of maximum for the length. Finally, we notice that γ must be smooth at the points p_i, for otherwise we could increase its length by using the generalized twin paradox. Therefore γ must be a timelike geodesic. Using a synchronized coordinate system around $\gamma(0)$, it is clear that γ must be orthogonal to S, for otherwise it would be possible to increase its length. $\qquad\square$

We have now all the necessary ingredients to prove the Hawking singularity theorem:

Theorem 8.9 (Hawking) *Let (M, g) be a globally hyperbolic spacetime satisfying the strong energy condition, and suppose that the expansion satisfies $\theta \leq \theta_0 < 0$ on a Cauchy hypersurface S. Then (M, g) is singular.*

Proof We will show that no future-directed timelike geodesic orthogonal to S can be extended to proper time greater than $\tau_0 = -\frac{3}{\theta_0}$ to the future of S. Suppose that this was not so. Then there would exist a future-directed timelike geodesic c orthogonal to S, parameterized by proper time, defined in an interval $[0, \tau_0 + \varepsilon]$ for some $\varepsilon > 0$. Let $p = c(\tau_0 + \varepsilon)$. According to Theorem 8.8, there would exist a timelike geodesic γ with maximal length connecting S to p, orthogonal to S. Because $\tau(c) = \tau_0 + \varepsilon$, we would necessarily have $\tau(\gamma) \geq \tau_0 + \varepsilon$. Proposition 8.4 guarantees that γ would develop a conjugate point at a distance of at most τ_0 to the future of S, and Proposition 8.5 states that γ would cease to be maximizing beyond this point. Therefore we arrive at a contradiction. $\qquad\square$

Remark 8.10 It should be clear that (M, g) is singular if the condition $\theta \leq \theta_0 < 0$ on a Cauchy hypersurface S is replaced by the condition $\theta \geq \theta_0 > 0$ on S. In this case, no **past-directed** timelike geodesic orthogonal to S can be extended to proper time greater than $\tau_0 = \frac{3}{\theta_0}$ to the **past** of S.

Example 8.11

(1) The FLRW models are globally hyperbolic [cf. Exercise 7.10(9)], and satisfy the strong energy condition (as $\rho > 0$). Moreover,

$$\beta_{ij} = \frac{\dot{a}}{a}\gamma_{ij} \Rightarrow \theta = \frac{3\dot{a}}{a}.$$

Assume that the model is expanding at time t_0. Then $\theta = \theta_0 = \frac{3\dot{a}(t_0)}{a(t_0)} > 0$ on the Cauchy hypersurface $S = \{t = t_0\}$, and hence Theorem 8.9 guarantees that this model is singular to the past of S (i.e. there exists a big bang). Moreover, Theorem 8.9 implies that this singularity is generic: any sufficiently small perturbation of an expanding FLRW model satisfying the strong energy condition will also be singular. Loosely speaking, any expanding universe must have begun at a big bang.

(2) The region $\{r < 2m\}$ of the Schwarzschild solution is globally hyperbolic [cf. Exercise 7.10(9)], and satisfies the strong energy condition (as $Ric = 0$). The metric can be written in this region as

$$g = -d\tau \otimes d\tau + \left(\frac{2m}{r} - 1\right) dt \otimes dt + r^2 d\theta \otimes d\theta + r^2 \sin^2 \theta d\varphi \otimes d\varphi,$$

where

$$\tau = \int_r^{2m} \left(\frac{2m}{u} - 1\right)^{-\frac{1}{2}} du.$$

Therefore the inside of the black hole can be pictured as a cylinder $\mathbb{R} \times S^2$ whose shape is evolving in time. As $r \to 0$, the S^2 contracts to a singularity, with the t-direction expanding. Since

$$\sum_{i,j=1}^{3} \beta_{ij} dx^i \otimes dx^j = \frac{dr}{d\tau} \left(-\frac{m}{r^2} dt \otimes dt + r d\theta \otimes d\theta + r \sin^2 \theta d\varphi \otimes d\varphi\right),$$

we have

$$\theta = \left(\frac{2m}{r} - 1\right)^{-\frac{1}{2}} \left(\frac{2}{r} - \frac{3m}{r^2}\right).$$

Therefore we have $\theta = \theta_0 < 0$ on any Cauchy hypersurface $S = \{r = r_0\}$ with $r_0 < \frac{3m}{2}$, and hence Theorem 8.9 guarantees that the Schwarzschild solution is singular to the future of S. Moreover, Theorem 8.9 implies that this singularity is generic: any sufficiently small perturbation of the Schwarzschild solution satisfying the strong energy condition will also be singular. Loosely speaking, once the collapse has advanced long enough, nothing can prevent the formation of a singularity.

Exercise 8.12

(1) *(Clifton–Pohl torus)* Consider the Lorentzian metric

$$\bar{g} := \frac{1}{u^2 + v^2} (du \otimes dv + dv \otimes du)$$

on $\overline{M} = \mathbb{R}^2 \setminus \{0\}$. The Lie group $\mathbb{Z}$ acts freely and properly on $\overline{M}$ by isometries through

$$n \cdot (u, v) = (2^n u, 2^n v),$$

and this determines a Lorentzian metric g on $M = \overline{M}/\mathbb{Z} \cong T^2$. Show that (M, g) is not geodesically complete (although M is compact).
(**Hint:** Look for null geodesics with $v \equiv 0$.)

(2) *(2-dimensional Anti-de Sitter universe)* Consider $\mathbb{R}^3$ with the pseudo-Riemannian metric

$$\bar{g} = -du \otimes du - dv \otimes dv + dw \otimes dw,$$

and let (M, g) be the universal covering of the submanifold

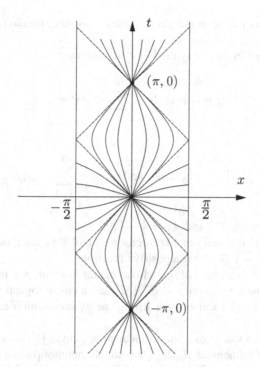

Fig. 6.12 The exponential map is not surjective in the 2-dimensional Anti-de Sitter universe

$$H = \{(u, v, w) \in \mathbb{R}^3 \mid u^2 + v^2 - w^2 = 1)\}$$

with the induced metric. Show that:

(a) a model for (M, g) is $M = \mathbb{R} \times \left(-\frac{\pi}{2}, \frac{\pi}{2}\right)$ and

$$g = \frac{1}{\cos^2 x}(-dt \otimes dt + dx \otimes dx)$$

(Fig. 6.12; hence (M, g) is not globally hyperbolic);

(b) (M, g) is geodesically complete, but $\exp_p$ is not surjective for any $p \in M$
(**Hint:** Notice that each isometry of $(\mathbb{R}^3, \bar{g})$ determines an isometry of (M, g));

(c) there exist points $p, q \in M$ connected by arbitrarily long timelike curves [cf. Exercise 8.12(10)].

(3) By analogy with Exercise 3.3(5) in Chap. 3, we can define a left-invariant Lorentzian metric on the Lie group $H = \mathbb{R} \times \mathbb{R}^+$ of Exercise 7.17(3) in Chap. 1 as

$$g := \frac{1}{x^2}(-dt \otimes dt + dx \otimes dx).$$

Show that this metric is not geodesically complete. (**Remark:** This cannot happen in Riemannian geometry – cf. Exercise 5.8(4) in Chap. 3).

(4) Show that the Christoffel symbols for the metric

$$g = -dt \otimes dt + \sum_{i,j=1}^{3} \gamma_{ij} dx^i \otimes dx^j,$$

satisfy

$$\Gamma_{00}^0 = \Gamma_{00}^i = 0 \quad \text{and} \quad \Gamma_{0j}^i = \sum_{k=1}^{3} \gamma^{ik} \beta_{kj},$$

where $(\gamma^{ij}) = (\gamma_{ij})^{-1}$ and $\beta_{ij} = \frac{1}{2} \frac{\partial \gamma_{ij}}{\partial t}$.

(5) Show that if U is a unit timelike vector field and V is any timelike vector field then $\langle U, V \rangle^2 + \frac{1}{2} \langle V, V \rangle$ is a positive function.

(6) Show that a spacetime (M, g) whose matter content is a pressureless fluid with rest density function $\rho \in C^\infty(M)$ and a cosmological constant $\Lambda \in \mathbb{R}$ [cf. Exercise 6.1(7)] satisfies the strong energy condition if and only if $\rho \geq 0$ and $\rho \geq \frac{\Lambda}{4\pi}$.

(7) Let (M, g) be a spacetime. Show that any open cover $\{V_\alpha\}_{\alpha \in A}$ has a countable, locally finite refinement $\{U_n\}_{n \in \mathbb{N}}$ by simple neighborhoods (i.e. $\cup_{n \in \mathbb{N}} U_n = \cup_{\alpha \in A} V_\alpha$, for each $n \in \mathbb{N}$ there exists $\alpha \in A$ such that $U_n \subset V_\alpha$, and each point $p \in M$ has a neighborhood which intersects only finite simple neighborhoods U_n).

(8) Prove Corollary 8.7.

(9) Let (M, g) be a globally hyperbolic spacetime, $t : M \to \mathbb{R}$ a global time function, $S = t^{-1}(0)$ a Cauchy hypersurface, $p \in D^+(S)$ and $A = D^+(S) \cap J^-(p)$. Show that the closure $C(S, p) := \overline{T(S, p)}$ in the space $C(A)$ of all compact subsets of A with the Hausdorff metric can be identified with the set of continuous causal curves connecting S to p (parameterized by t).

(10) Let (M, g) be a globally hyperbolic spacetime and $p, q \in M$ with $q \in I^+(p)$. Show that among all timelike curves connecting p to q there exists a timelike curve with maximal length, which is a timelike geodesic.

(11) Consider two events p and q on Schwarzschild spacetime corresponding to the beginning and the end of a complete circular orbit of radius r [cf. Exercise 5.1(2)]. Show that the corresponding timelike geodesic is not maximal.

(12) (*Myers theorem*) Use ideas similar to those leading to the proof of Theorem 8.9 to prove the **Myers theorem**: if $(M, \langle \cdot, \cdot \rangle)$ is a complete Riemannian manifold whose Ricci curvature satisfies $Ric(X, X) \geq \varepsilon \langle X, X \rangle$ for some $\varepsilon > 0$ then M is compact. Can these ideas be used to prove a singularity theorem in Riemannian geometry?

(13) Explain why the Hawking singularity theorem does not apply to each of the following spacetimes:

(a) Minkowski spacetime;
(b) Einstein universe [cf. Exercise 6.1(8)];
(c) de Sitter universe [cf. Exercise 6.1(8)];
(d) 2-dimensional Anti-de Sitter universe [cf. Exercise 8.12(2)].

6.9 Penrose Singularity Theorem

Let (M, g) be a globally hyperbolic spacetime, S a Cauchy hypersurface with future-pointing unit normal vector field n, and $\Sigma \subset S$ a compact 2-dimensional submanifold with unit normal vector field ν in S. Let c_p be the null geodesic with initial condition $n_p + \nu_p$ for each point $p \in \Sigma$. We define a smooth map $\exp : (-\varepsilon, \varepsilon) \times \Sigma \to M$ for some $\varepsilon > 0$ as $\exp(r, p) = c_p(r)$.

Definition 9.1 The critical values of exp are said to be **conjugate points** to Σ.

Loosely speaking, conjugate points are points where geodesics starting orthogonally at nearby points of Σ intersect [see also Exercise 4.8(6) in Chap. 3].
Let $q = \exp(r_0, p)$ be a point not conjugate to Σ. If φ is a local parameterization of Σ around p, then we can construct a system of local coordinates (u, r, x^2, x^3) on some open set $V \ni q$ by using the map

$$(u, r, x^2, x^3) \mapsto \exp(r, \psi_u(\varphi(x^2, x^3))),$$

where ψ_u is the flow along the timelike geodesics orthogonal to S and the map $\exp : (-\varepsilon, \varepsilon) \times \psi_u(\Sigma) \to M$ is defined as above.
Since $\frac{\partial}{\partial r}$ is tangent to null geodesics, we have $g_{rr} = \left\langle \frac{\partial}{\partial r}, \frac{\partial}{\partial r} \right\rangle = 0$. On the other hand, we have

$$\frac{\partial g_{r\mu}}{\partial r} = \frac{\partial}{\partial r} \left\langle \frac{\partial}{\partial r}, \frac{\partial}{\partial x^\mu} \right\rangle = \left\langle \frac{\partial}{\partial r}, \nabla_{\frac{\partial}{\partial r}} \frac{\partial}{\partial x^\mu} \right\rangle$$
$$= \left\langle \frac{\partial}{\partial r}, \nabla_{\frac{\partial}{\partial x^\mu}} \frac{\partial}{\partial r} \right\rangle = \frac{1}{2} \frac{\partial}{\partial x^\mu} \left\langle \frac{\partial}{\partial r}, \frac{\partial}{\partial r} \right\rangle = 0,$$

for $\mu = 0, 1, 2, 3$. Since $g_{ru} = -1$ and $g_{r2} = g_{r3} = 0$ on $\psi_u(\Sigma)$, we have $g_{ru} = -1$ and $g_{r2} = g_{r3} = 0$ on V. Therefore the metric is written in this coordinate system as

$$g = \alpha du \otimes du - du \otimes dr - dr \otimes du + \sum_{i=2}^{3} \beta_i \left(du \otimes dx^i + dx^i \otimes du \right) + \sum_{i,j=2}^{3} \gamma_{ij} dx^i \otimes dx^j.$$

Since

$$\det \begin{pmatrix} \alpha & -1 & \beta_2 & \beta_3 \\ -1 & 0 & 0 & 0 \\ \beta_2 & 0 & \gamma_{22} & \gamma_{23} \\ \beta_3 & 0 & \gamma_{32} & \gamma_{33} \end{pmatrix} = -\det \begin{pmatrix} \gamma_{22} & \gamma_{23} \\ \gamma_{32} & \gamma_{33} \end{pmatrix},$$

we see that the functions

$$\gamma_{ij} := \left\langle \frac{\partial}{\partial x^i}, \frac{\partial}{\partial x^j} \right\rangle$$

form a positive definite matrix, and so g induces a Riemannian metric on the 2-dimensional surfaces $\exp(r, \psi_u(\Sigma))$, which are then spacelike. Since the vector fields $\frac{\partial}{\partial x^i}$ can always be defined along c_p, the matrix (γ_{ij}) is also well defined along c_p, even at points where the coordinate system breaks down, i.e. at points which are conjugate to Σ. These are the points for which $\gamma := \det(\gamma_{ij})$ vanishes, since only then will $\left\{ \frac{\partial}{\partial u}, \frac{\partial}{\partial r}, \frac{\partial}{\partial x^2}, \frac{\partial}{\partial x^3} \right\}$ fail to be linearly independent. (In fact the vector fields $\frac{\partial}{\partial x^i}$ are Jacobi fields along c_p – see Exercise 4.8(6) in Chap. 3).

It is easy to see that

$$\Gamma^u_{ur} = \Gamma^u_{rr} = \Gamma^u_{ri} = \Gamma^r_{rr} = \Gamma^i_{rr} = 0 \quad \text{and} \quad \Gamma^i_{rj} = \sum_{k=2}^{3} \gamma^{ik} \beta_{kj},$$

where $(\gamma^{ij}) = (\gamma_{ij})^{-1}$ and $\beta_{ij} = \frac{1}{2} \frac{\partial \gamma_{ij}}{\partial r}$ [cf. Exercise 9.9(1)]. Consequently,

$$R_{rr} = R_{urr}{}^u + \sum_{i=2}^{3} R_{irr}{}^i = \sum_{i=2}^{3} \left(-\frac{\partial \Gamma^i_{ir}}{\partial r} - \sum_{j=2}^{3} \Gamma^j_{ir} \Gamma^i_{rj} \right)$$

$$= -\frac{\partial}{\partial r} \left(\sum_{i,j=2}^{3} \gamma^{ij} \beta_{ij} \right) - \sum_{i,j,k,l=2}^{3} \gamma^{jk} \gamma^{il} \beta_{ki} \beta_{lj}.$$

(cf. Chap. 4, Sect. 4.1). The quantity

$$\theta := \sum_{i,j=2}^{3} \gamma^{ij} \beta_{ij}$$

appearing in this expression is called the **expansion** of the null geodesics, and has an important geometric meaning:

$$\theta = \frac{1}{2} \operatorname{tr} \left((\gamma_{ij})^{-1} \frac{\partial}{\partial r} (\gamma_{ij}) \right) = \frac{1}{2} \frac{\partial}{\partial r} \log \gamma = \frac{\partial}{\partial r} \log \gamma^{\frac{1}{2}},$$

where $\gamma := \det\left(\gamma_{ij}\right)$. Therefore the expansion yields the variation of the area element of the spacelike 2-dimensional surfaces $\exp(r, \psi_u(\Sigma))$. More importantly for our purposes, we see that a singularity of the expansion indicates a zero of γ, i.e. a conjugate point to $\psi_u(\Sigma)$.

Definition 9.2 A spacetime (M, g) is said to satisfy the **null energy condition** if $Ric(V, V) \geq 0$ for any null vector field $V \in \mathfrak{X}(M)$.

It is easily seen that this condition is implied by (but weaker than) the strong energy condition. By the Einstein equation, it is equivalent to requiring that the reduced energy-momentum tensor T satisfies $T(V, V) \geq 0$ for any null vector field $V \in \mathfrak{X}(M)$. In the case of a pressureless fluid with rest density function $\rho \in C^\infty(M)$ and unit velocity vector field $U \in \mathfrak{X}(M)$, this requirement becomes

$$\rho \langle U, V \rangle^2 \geq 0 \Leftrightarrow \rho \geq 0.$$

For more complicated matter models, the null energy condition produces equally reasonable restrictions.

Proposition 9.3 *Let (M, g) be a globally hyperbolic spacetime satisfying the null energy condition, $S \subset M$ a Cauchy hypersurface, $\Sigma \subset S$ a compact 2-dimensional submanifold with unit normal vector field ν in S and $p \in \Sigma$ a point where $\theta = \theta_0 < 0$. Then the null geodesic c_p contains at least a point conjugate to Σ, at an affine parameter distance of at most $-\frac{2}{\theta_0}$ to the future of Σ (assuming that it can be extended that far).*

Proof Since (M, g) satisfies the null energy condition, we have $R_{rr} = Ric\left(\frac{\partial}{\partial r}, \frac{\partial}{\partial r}\right) \geq 0$. Consequently,

$$\frac{\partial \theta}{\partial r} + \sum_{i,j,k,l=2}^{3} \gamma^{jk} \gamma^{il} \beta_{ki} \beta_{lj} \leq 0.$$

Choosing an orthonormal basis (where $\gamma^{ij} = \delta_{ij}$), and using the inequality

$$(\operatorname{tr} A)^2 \leq n \operatorname{tr}(A^t A)$$

for square $n \times n$ matrices, it is easy to show that

$$\sum_{i,j,k,l=2}^{3} \gamma^{jk} \gamma^{il} \beta_{ki} \beta_{lj} = \sum_{i,j=2}^{3} \beta_{ji} \beta_{ij} = \operatorname{tr}\left((\beta_{ij})(\beta_{ij})^t\right) \geq \frac{1}{2} \theta^2.$$

Consequently θ must satisfy

$$\frac{\partial \theta}{\partial r} + \frac{1}{2} \theta^2 \leq 0.$$

Integrating this inequality yields

$$\frac{1}{\theta} \geq \frac{1}{\theta_0} + \frac{r}{2},$$

and hence θ must blow up at a value of r no greater than $-\frac{2}{\theta_0}$.   □

We define the **chronological future** and the **causal future** of the compact surface Σ as

$$I^+(\Sigma) = \bigcup_{p \in \Sigma} I^+(p) \quad \text{and} \quad J^+(\Sigma) = \bigcup_{p \in \Sigma} J^+(p)$$

(with similar definitions for the **chronological past** and the **causal past** of Σ). It is clear that $I^+(\Sigma)$, being the union of open sets, is itself open, and also that $J^+(\Sigma) \subset \overline{I^+(\Sigma)}$ and $I^+(\Sigma) = \text{int } J^+(\Sigma)$. On the other hand, it is easy to generalize Proposition 8.6 (and consequently Corollary 8.7) to the corresponding statements with compact surfaces replacing points [cf. Exercise 9.9(2)]. In particular, $J^+(\Sigma)$ is closed. Therefore

$$\partial J^+(\Sigma) = \partial I^+(\Sigma) = J^+(\Sigma) \setminus I^+(\Sigma),$$

and so, by a straightforward generalization of Corollary 7.5, every point in this boundary can be reached from a point in Σ by a future-directed null geodesic. Moreover, this geodesic must be orthogonal to Σ. Indeed, at Σ we have

$$\frac{\partial}{\partial u} = n \quad \text{and} \quad \frac{\partial}{\partial r} = n + \nu,$$

and so the metric takes the form

$$g = -du \otimes du - du \otimes dr - dr \otimes du + \sum_{i,j=2}^{3} \gamma_{ij} dx^i \otimes dx^j.$$

If $c : I \subset \mathbb{R} \to M$ is a future-directed null geodesic with $c(0) \in \Sigma$, its initial tangent vector

$$\dot{c}(0) = \dot{u}\frac{\partial}{\partial u} + \dot{r}\frac{\partial}{\partial r} + \sum_{i=2}^{3} \dot{x}^i \frac{\partial}{\partial x^i} = (\dot{u} + \dot{r})n + \dot{r}\nu + \sum_{i=2}^{3} \dot{x}^i \frac{\partial}{\partial x^i}$$

must satisfy

$$\dot{u}(\dot{u} + 2\dot{r}) = \sum_{i,j=2}^{3} \gamma_{ij} \dot{x}^i \dot{x}^j.$$

Since c is future-directed we must have $\dot{u} + \dot{r} > 0$. On the other hand, by choosing the unit normal to Σ on S to be either ν or $-\nu$, we can assume $\dot{r} \geq 0$. If c is not orthogonal to Σ we then have

$$\sum_{i,j=2}^{3} \gamma_{ij} \dot{x}^i \dot{x}^j > 0 \Rightarrow \dot{u}(\dot{u} + 2\dot{r}) > 0 \Rightarrow \dot{u} > 0.$$

Now the region where $u > 0$ and $r \geq 0$ is clearly a subset of $I^+(\Sigma)$, since its points can be reached from Σ by a sectionally smooth curve composed of an arc of timelike geodesic and an arc of null geodesic. Therefore, we see that if c is not orthogonal to Σ then $c(t) \in I^+(\Sigma)$ for all $t > 0$.

Even future-directed null geodesics orthogonal to Σ may eventually enter $I^+(\Sigma)$. A sufficient condition for this to happen is given in the following result.

Proposition 9.4 *Let (M, g) be a globally hyperbolic spacetime, S a Cauchy hypersurface with future-pointing unit normal vector field n, $\Sigma \subset S$ a compact 2-dimensional submanifold with unit normal vector field ν in S, $p \in \Sigma$, c_p the null geodesic through p with initial condition $n_p + \nu_p$ and $q = c_p(r)$ for some $r > 0$. If c_p has a conjugate point between p and q then $q \in I^+(\Sigma)$.*

Proof We will offer only a sketch of the proof. Let s be the first conjugate point along c_p between p and q. Since q is conjugate to p, there exists another null geodesic γ starting at Σ which (approximately) intersects c_p at s. The piecewise smooth null curve obtained by following γ between Σ and s, and c_p between s and q is a causal curve but not a null geodesic. This curve can be easily smoothed while remaining causal and nongeodesic, and so by the generalization of Corollary 7.5 we have $q \in I^+(\Sigma)$ (see Fig. 6.13). $\qquad\square$

Definition 9.5 Let (M, g) be a globally hyperbolic spacetime and S a Cauchy hypersurface with future-pointing unit normal vector field n. A compact 2-dimensional submanifold $\Sigma \subset S$ with unit normal vector field ν in S is said to be **trapped** if the expansions θ^+ and θ^- of the null geodesics with initial conditions $n + \nu$ and $n - \nu$ are both negative everywhere on Σ.

We have now all the necessary ingredients to prove the Penrose singularity theorem.

Theorem 9.6 (Penrose) *Let (M, g) be a connected globally hyperbolic spacetime with a noncompact Cauchy hypersurface S, satisfying the null energy condition. If S contains a trapped surface Σ then (M, g) is singular.*

Proof Let $t : M \to \mathbb{R}$ be a global time function such that $S = t^{-1}(0)$. The integral curves of $\operatorname{grad} t$, being timelike, intersect S exactly once, and $\partial I^+(\Sigma)$ at most once. This defines a continuous injective map $\pi : \partial I^+(\Sigma) \to S$, whose image is open. Indeed, if $q = \pi(p)$, then all points in some neighborhood of q are images of points in $\partial I^+(\Sigma)$, as otherwise there would be a sequence $q_n \in S$ with $q_n \to q$ such that

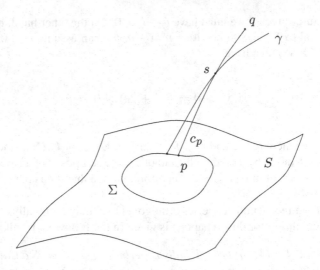

Fig. 6.13 Proof of Proposition 9.4

the integral curves of grad t through q_n would not intersect $\partial I^+(\Sigma)$. Letting r_n be the intersections of these curves with the Cauchy hypersurface $t^{-1}(t(r))$, for some point r to the future of p along the integral line of grad t, we would have $r_n \to r$, and so $r_n \in I^+(\Sigma)$ for sufficiently large n (as $I^+(\Sigma)$ is open), leading to a contradiction.

Since Σ is trapped (and compact), there exists $\theta_0 < 0$ such that the expansions θ^+ and θ^- of the null geodesics orthogonal to Σ both satisfy $\theta^+, \theta^- \leq \theta_0$. We will show that there exists a future-directed null geodesic orthogonal to Σ which cannot be extended to an affine parameter greater than $r_0 = -\frac{2}{\theta_0}$ to the future of Σ. Suppose that this was not so. Then, according to Proposition 9.3, any null geodesic orthogonal to Σ would have a conjugate point at an affine parameter distance of at most r_0 to the future of Σ, after which it would be in $I^+(\Sigma)$, by Proposition 9.4. Consequently, $\partial I^+(\Sigma)$ would be a (closed) subset of the compact set

$$\exp^+([0, r_0] \times \Sigma) \cup \exp^-([0, r_0] \times \Sigma)$$

(where $\exp^+$ and $\exp^-$ refer to the exponential map constructed using the unit normals ν and $-\nu$), hence compact. Therefore the image of π would also be compact, hence closed as well as open. Since M, and therefore S, are connected, the image of π would be S, which would then be homeomorphic to $\partial I^+(\Sigma)$. But S is noncompact by hypothesis, and we reach a contradiction.   □

Remark 9.7 It should be clear that (M, g) is singular if the condition of existence of a trapped surface is replaced by the condition of existence of an **anti-trapped surface**, that is, a compact surface $\Sigma \subset S$ such that the expansions of null geodesics orthogonal to Σ are both positive. In this case, there exists a **past-directed** null geodesic orthogonal to Σ which cannot be extended to an affine parameter time greater than $r_0 = \frac{2}{\theta_0}$ to the **past** of Σ.

Example 9.8

(1) The region $\{r < 2m\}$ of the Schwarzschild solution is globally hyperbolic [cf. Exercise 7.10(9)], and satisfies the null energy condition (as $Ric = 0$). Since r (or $-r$) is clearly a time function (depending on the choice of time orientation), it must increase (or decrease) along any future-pointing null geodesic, and therefore any sphere Σ of constant (t, r) is anti-trapped (or trapped). Since any Cauchy hypersurface is diffeomorphic to $\mathbb{R} \times S^2$, hence noncompact, we conclude from Theorem 9.6 that the Schwarzschild solution is singular to past (or future) of Σ. Moreover, Theorem 8.9 implies that this singularity is generic: any sufficiently small perturbation of the Schwarzschild solution satisfying the null energy condition will also be singular. Loosely speaking, once the collapse has advanced long enough, nothing can prevent the formation of a singularity.

(2) The FLRW models are globally hyperbolic [cf. Exercise 7.10(9)], and satisfy the null energy condition (as $\rho > 0$). Moreover, radial null geodesics satisfy

$$\frac{dr}{dt} = \pm \frac{1}{a} \sqrt{1 - kr^2}.$$

Therefore, if we start with a sphere Σ of constant (t, r) and follow the orthogonal null geodesics along the direction of increasing or decreasing r, we obtain spheres whose radii ar satisfy

$$\frac{d}{dt}(ar) = \dot{a}r + a\dot{r} = \dot{a}r \pm \sqrt{1 - kr^2}.$$

Assume that the model is expanding, with the big bang at $t = 0$, and spatially noncompact (in particular $k \neq 1$). Then, for sufficiently small $t > 0$, the sphere Σ is anti-trapped, and hence Theorem 9.6 guarantees that this model is singular to the past of Σ (i.e. there exists a big bang). Moreover, Theorem 9.6 implies that this singularity is generic: any sufficiently small perturbation of an expanding, spatially noncompact FLRW model satisfying the null energy condition will also be singular. Loosely speaking, any expanding universe must have begun at a big bang.

Exercise 9.9

(1) Show that the Christoffel symbols for the metric

$$g = \alpha du \otimes du - du \otimes dr - dr \otimes du + \sum_{i=2}^{3} \beta_i \left(du \otimes dx^i + dx^i \otimes du \right)$$

$$+ \sum_{i,j=2}^{3} \gamma_{ij} dx^i \otimes dx^j$$

satisfy

$$\Gamma^u_{ur} = \Gamma^u_{rr} = \Gamma^u_{ri} = \Gamma^r_{rr} = \Gamma^i_{ur} = \Gamma^i_{rr} = 0 \quad \text{and} \quad \Gamma^i_{rj} = \sum_{k=2}^{3} \gamma^{ik}\beta_{kj},$$

where $(\gamma^{ij}) = (\gamma_{ij})^{-1}$ and $\beta_{ij} = \frac{1}{2}\frac{\partial \gamma_{ij}}{\partial r}$.

(2) Let (M, g) be a globally hyperbolic spacetime with Cauchy hypersurfaces S_0 and S_1 satisfying $S_1 \subset D^+(S_0)$, and $\Sigma \subset S_1$ a compact surface. Show that:

   (a) $D^+(S_0) \cap J^-(\Sigma)$ is compact;
   (b) $J^-(\Sigma)$ is closed.

(3) Explain why the Penrose singularity theorem does not apply to each of the following spacetimes:

   (a) Minkowski spacetime;
   (b) Einstein universe [cf. Exercise 6.1(8)];
   (c) de Sitter universe [cf. Exercise 6.1(8)];
   (d) 2-dimensional Anti-de Sitter universe [cf. Exercise 8.12(2)].

6.10 Notes

6.10.1 Bibliographical Notes

There are many excellent texts on general relativity, usually containing also the relevant differential and Lorentzian geometry. These range from introductory [Sch02] to more advanced [Wal84] to encyclopedic [MTW73]. A more mathematically oriented treatment can be found in [BEE96, O'N83] ([GHL04] also contains a brief glance at pseudo-Riemannian geometry). For more information on special relativity and the Lorentz group see [Nab92, Oli02]. Causality and the singularity theorems are treated in greater detail in [Pen87, HE95, Nab88], and in the original papers [Pen65, Haw67, HP70].

References

[BEE96]   Beem, J., Ehrlich, P., Easley, K.: Global Lorentzian Geometry. Marcel Dekker, New York (1996)
[GHL04]   Gallot, S., Hulin, D., Lafontaine, J.: Riemannian Geometry. Springer, New York (2004)
[HE95]   Hawking, S., Ellis, G.: The Large Scale Structure of Space-Time. Cambridge University Press, Cambridge (1995)
[HP70]   Hawking, S., Penrose, R.: The singularities of gravitational collapse and cosmology. Proc. Roy. Soc. Lon. A. **314**, 529–548 (1970)

[Haw67]   Hawking, S.: The occurrence of singularities in cosmology. III. Causality and singularities. Proc. Roy. Soc. Lon. 300, 187–201 (1967)
[Mun00]   Munkres, J.: Topology. Prentice-Hall, Upper Saddle River (2000)
[MTW73]   Misner, C., Thorne, K., Wheeler, J.A.: Gravitation. Freeman, New York (1973)
[Nab88]   Naber, G.: Spacetime and Singularities - An Introduction. Cambridge University Press, Cambridge (1988)
[Nab92]   Naber, G.L.: The Geometry of Minkowski Spacetime. Springer, New York (1992)
[Oli02]   Oliva, W.: Geometric Mechanics. Springer, New York (2002)
[O'N83]   O'Neill, B.: Semi-Riemannian Geometry. Academic Press, Waltham (1983)
[Pen65]   Penrose, R.: Gravitational collapse and space-time singularities. Phys. Rev. Lett. **14**, 57–59 (1965)
[Pen87]   Penrose, R.: Techniques of Differential Topology in Relativity. Society for Industrial and Applied Mathematics, Philadelphia (1987)
[Sch02]   Schutz, B.: A First Course in General Relativity. Cambridge University Press, Cambridge (2002)
[TW92]   Taylor, E., Wheeler, J.: Spacetime Physics. Freeman, New York (1992)
[Wal84]   Wald, R.: General Relativity. University of Chicago Press, Chicago (1984)

Chapter 7
Solutions to Selected Exercises

7.1 Chapter 1

7.1.1 Section 1.1

(1) In all these examples conditions (i) and (iii) in the definition of a topological manifold are inherited from the ambient space, and so we just have to worry about (ii).

(a) D^2 is an open subset of $\mathbb{R}^2$ so it is trivially a topological 2-manifold.

(b) $M = S^2 \backslash \{p\}$ is a topological manifold of dimension 2. Treating $\mathbb{R}^2$ as the complex plane, and assuming without loss of generality that p is the north pole of the sphere, we obtain a simple homeomorphism $\varphi : \mathbb{R}^2 \cong \mathbb{C} \to M$ by taking

$$\varphi(z) = \left(\frac{2x}{1 + |z|^2}, \frac{2y}{1 + |z|^2}, \frac{|z|^2 - 1}{1 + |z|^2} \right)$$

where $z = x + iy$. The inverse of this map is called the stereographic projection [cf. Exercise 2.5(7)].

(c) $N = S^2 \backslash \{p, q\}$ is also a topological 2-manifold. We can assume, without loss of generality, that p and q are the north and south poles of the sphere and then φ defined in (b) is a homeomorphism between the open set $\mathbb{R}^2 \backslash \{0\} \cong \mathbb{C} \backslash \{0\}$ and N.

(d) The cylinder $V = \{(x, y) \in \mathbb{R}^3 \mid x^2 + y^2 = 1\}$ is a topological 2-manifold. For each point $p := (x, y, z) \in V$ we take the normal vector $n := (x, y, 0)$ and consider a plane generated by two coordinate axis that are not parallel to n. Then there is a neighborhood of p in V homeomorphic to its projection on this plane.

(e) The cone $S := \{(x, y) \mid x^2 + y^2 = z^2\}$ is not a topological manifold. If that were the case, there would exist a connected open set W in S, containing the origin, homeomorphic to an open subset $U \subset \mathbb{R}^2$. Then $W \backslash \{0\}$ would

© Springer International Publishing Switzerland 2014
L. Godinho and J. Natário, *An Introduction to Riemannian Geometry*, Universitext,
DOI 10.1007/978-3-319-08666-8_7

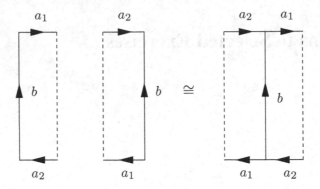

Fig. 7.1 Klein bottle with a Möbius band deleted

be homeomorphic to $U \backslash \{q\}$ (where q is the preimage of the origin). But $W \backslash \{0\}$ is disconnected, while $U \backslash \{q\}$ is connected, and hence they cannot be homeomorphic.

(3) It is easy to show that the Klein bottle with a Möbius band deleted is again homeomorphic to a Möbius band (see Fig. 7.1).

(5) (a) If the new vertex is on a face, then it must be connected to the three vertices of that face. In doing this, the number of vertices has increased by 1, the number of edges has increased by 3 and the number of faces has increased by 2 (as one face has been divided into three faces). Therefore $V - E + F$ has changed by $1 - 3 + 2 = 0$. If the new vertex is on an edge, then it must be connected to the two vertices opposite to that edge. In doing this, the number of vertices has increased by 1, the number of edges has increased by 3 (as two new edges have been created and one edge has been divided into two) and the number of faces has increased by 2 (as two faces have been divided into four faces). Again, $V - E + F$ does not change.

(b) For the triangulation of S^2 determined by the tetrahedron, one has $V = 4$, $E = 6$ and $F = 4$, and so $\chi(S^2) = 4 - 6 + 4 = 2$.

(c) A decomposition of T^2 into triangles can be obtained by adding a diagonal to the square whose sides are identified. This is not exactly a triangulation (because the intersection of the two triangles consists of all three edges), but it can be turned into one by adding vertices, and so, by (a), it can be used to compute the Euler characteristic. Allowing for the identifications, we have $V = 1$, $E = 3$, $F = 2$, and so $\chi(T^2) = 1 - 3 + 2 = 0$.

(d) Same as for T^2.

(e) Same as for T^2, except that now the identifications yield $V = 2$, and so $\chi(\mathbb{R}P^2) = 2 - 3 + 2 = 1$.

(f) Consider triangulations of M and N. Deleting a triangle on each surface and identifying their edges yields $M \# N$ with a triangulation. Since 3 vertices get identified, the total number of vertices goes down by 3, and the same is true for the total number of edges. The total number of faces goes down by two, corresponding to the two faces which were deleted. Therefore $\chi(M\#N) = \chi(M) + \chi(N) - 2$.

7.1.2 Section 1.2

(4) $(\varphi_2^{-1} \circ \varphi_1)(x) = x^{\frac{1}{3}}$, which is not differentiable at the origin. By Exercise 2.5(1), the two atlases are not equivalent, as $\{(\mathbb{R}, \varphi_1), (\mathbb{R}, \varphi_2)\}$ is not an atlas.

(7) Let us write the point $p \in S^n \subset \mathbb{R}^{n+1}$ as $p = (x, x^{n+1})$, with $x \in \mathbb{R}^n$ and $x^{n+1} \in \mathbb{R}$. The line through N and p is given parametrically by

$$c(t) = (tx, 1 + t(x^{n+1} - 1)),$$

and intersects the hyperplane $x^{n+1} = 0$ at $t = \frac{1}{1-x^{n+1}}$. Therefore,

$$y = \pi_N(p) = \frac{x}{1 - x^{n+1}}.$$

The squared norm of this equation yields

$$\|y\|^2 = \frac{\|x\|^2}{(1 - x^{n+1})^2} = \frac{1 - (x^{n+1})^2}{(1 - x^{n+1})^2} = \frac{1 + x^{n+1}}{1 - x^{n+1}}$$

(where we have used $\|p\|^2 = \|x\|^2 + (x^{n+1})^2 = 1$); equivalently,

$$x^{n+1} = \frac{\|y\|^2 - 1}{\|y\|^2 + 1}.$$

Therefore the relation between x and y can be written as

$$x = \frac{2y}{1 + \|y\|^2},$$

and consequently

$$p = \pi_N^{-1}(y) = \left(\frac{2y}{1 + \|y\|^2}, \frac{\|y\|^2 - 1}{\|y\|^2 + 1} \right).$$

Analogously, we have

$$y = \pi_S(p) = \frac{x}{1 + x^{n+1}}.$$

and

$$p = \pi_S^{-1}(y) = \left(\frac{2y}{1 + \|y\|^2}, \frac{1 - \|y\|^2}{1 + \|y\|^2} \right).$$

Thus the map $\pi_S \circ \pi_N^{-1}$, which maps $\pi_N(S^n \setminus \{N, S\}) = \mathbb{R}^n \setminus \{0\}$ to $\pi_S(S^n \setminus \{N, S\}) = \mathbb{R}^n \setminus \{0\}$, is given by

$$\pi_S \circ \pi_N^{-1}(y) = \frac{y}{\|y\|^2},$$

and hence is differentiable on its domain. The same is true for

$$\pi_N \circ \pi_S^{-1}(y) = \frac{y}{\|y\|^2}.$$

Therefore $\{(\mathbb{R}^n, \pi_N^{-1}), (\mathbb{R}^n, \pi_S^{-1})\}$ is an atlas for S^n.

To see that this atlas is equivalent to the atlas on Example 2.3(5). we have to compute $\pi_N \circ \varphi_i^+$, $\pi_N \circ \varphi_i^-$, $\pi_S \circ \varphi_i^+$, $\pi_S \circ \varphi_i^-$ and their inverses for $i = 1, \ldots, n + 1$. There are essentially two different cases, corresponding to $i = n + 1$ and $i \neq n + 1$. As an example, we have

$$\pi_N \circ \varphi_{n+1}^+(x) = \pi_N(x, g(x)) = \frac{x}{1 - g(x)} = \frac{x}{1 - (1 - \|x\|^2)^{\frac{1}{2}}},$$

which is differentiable on its domain $U \setminus \{0\}$. The other case is done similarly.

(9) (a) It is clear that $\mathbb{R}P^n = \bigcup_{1=1}^{n+1} V_i$. Moreover, if $\varphi_i(x^1, \ldots, x^n) \in V_j$ then

$$\varphi_i(x^1, \ldots, x^n) = [x^1, \ldots, x^j, \ldots, x^{i-1}, 1, x^i, \ldots, x^n]$$
$$= \left[\frac{x^1}{x^j}, \ldots, 1, \ldots, \frac{x^{i-1}}{x^j}, \frac{1}{x^j}, \frac{x^i}{x^j}, \ldots, \frac{x^n}{x^j} \right],$$

and hence

$$\varphi_j^{-1} \circ \varphi_i(x^1, \ldots, x^n) = \left(\frac{x^1}{x^j}, \ldots, \frac{x^{i-1}}{x^j}, \frac{1}{x^j}, \frac{x^i}{x^j}, \ldots, \frac{x^n}{x^j} \right)$$

is differentiable on its domain.

(b) The quotients $(\mathbb{R}^{n+1} \setminus \{0\}) / \sim$ and $S^n / \sim$ are in bijection because any line through the origin in $\mathbb{R}^{n+1}$ intersects S^n in two antipodal points. The two topologies are also the same, because an open set of $\mathbb{R}^{n+1} \setminus \{0\}$ formed by lines through the origin intersects S^n on an open set. To check that the two atlases are equivalent we need to check that the maps $\varphi_j^{-1} \circ (\pi \circ \varphi_i^+)$ and their inverses are differentiable for $i, j = 1, \ldots, n + 1$. As an example, we have

$$\varphi_{n+1}^{-1} \circ (\pi \circ \varphi_{n+1}^{+})(x) = \varphi_{n+1}^{-1}(\pi(x, g(x))) = \varphi_{n+1}^{-1}([x, g(x)])$$

$$= \varphi_{n+1}^{-1}\left(\left[\frac{x}{g(x)}, 1\right]\right) = \frac{x}{g(x)},$$

which is differentiable on its domain U. The other cases are done similarly.

7.1.3 Section 1.3

(2) Given an open set W in N we want to show that $f^{-1}(W) \subset M$ is open. By Exercise 2.5(2), we just need to show that, for every parameterization (U, φ) of M, the set $\tilde{U} := \varphi^{-1}\left(f^{-1}(W)\right) \subset \mathbb{R}^m$ is open (with $m = \dim M$). Considering an atlas $\{(V_\beta, \psi_\beta)\}$ for N, we have

$$\tilde{U} = \varphi^{-1}\left(f^{-1}(W)\right) = \varphi^{-1}\left(\bigcup_\beta f^{-1}(W \cap \psi_\beta(V_\beta))\right)$$

$$= \bigcup_\beta \varphi^{-1}\left(f^{-1}\left(\psi_\beta\left(\psi_\beta^{-1}(W)\right)\right)\right)$$

$$= \bigcup_\beta (\psi_\beta^{-1} \circ f \circ \varphi)^{-1}\left(\psi_\beta^{-1}(W)\right).$$

Since W is open we know that $\psi_\beta^{-1}(W)$ is an open set in $\mathbb{R}^n$ ($n = \dim N$). Then, since the map $\psi_\beta^{-1} \circ f \circ \varphi$ is C^∞ and hence continuous, the set $(\psi_\beta^{-1} \circ f \circ \varphi)^{-1}\left(\psi_\beta^{-1}(W)\right)$ is also open, implying that $\tilde{U}$ is open.

(4) We use the atlas $\{(\mathbb{R}^n, \pi_N^{-1}), (\mathbb{R}^n, \pi_S^{-1})\}$ for S^n [cf. Exercise 2.5(7)]. To check that f is differentiable we must show that the four maps $\pi_N \circ f \circ \pi_N^{-1}, \pi_S \circ f \circ \pi_N^{-1}, \pi_N \circ f \circ \pi_S^{-1}$ and $\pi_S \circ f \circ \pi_S^{-1}$ are differentiable. Since $f(S) = N$, we see that the first map is defined on $\pi_N(f^{-1}(\pi_N^{-1}(\mathbb{R}^n))) = \pi_N(f^{-1}(S^n\backslash\{N\})) = \pi_N(S^n\backslash\{S\}) = \mathbb{R}^n\backslash\{0\}$ (where we have slightly abused the notation in the interest of clarity). We have

$$\pi_N \circ f \circ \pi_N^{-1}(y) = \pi_N \circ f\left(\frac{2y}{1 + \|y\|^2}, \frac{\|y\|^2 - 1}{\|y\|^2 + 1}\right)$$

$$= \pi_N\left(-\frac{2y}{1 + \|y\|^2}, \frac{1 - \|y\|^2}{\|y\|^2 + 1}\right) = -\frac{y}{\|y\|^2},$$

which is differentiable on $\mathbb{R}^n\backslash\{0\}$. The other three maps are similarly shown to be differentiable on their domains.

(6) (a) The identity map is not a diffeomorphism because $\widehat{\mathrm{id}}(x) := \varphi_2^{-1} \circ \mathrm{id} \circ \varphi_1(x) = x^{\frac{1}{3}}$ is not differentiable at the origin.

(b) The map f is a diffeomorphism since $\hat{f} := \varphi_2^{-1} \circ f \circ \varphi_1 = \mathrm{id}$ and $\widehat{f^{-1}} := \varphi_1^{-1} \circ f^{-1} \circ \varphi_2 = \mathrm{id}$ are C^∞.

7.1.4 Section 1.4

(1) Consider a parameterization $\varphi : U \subset \mathbb{R}^n \to M$ around p and take the operators $\left(\frac{\partial}{\partial x^i}\right)_p$ defined in the text. If $\sum_{i=1}^n \alpha^i \left(\frac{\partial}{\partial x^i}\right)_p = 0$ for some $\alpha^1, \ldots, \alpha^n \in \mathbb{R}$ then

$$\left(\sum_{i=1}^n \alpha^i \left(\frac{\partial}{\partial x^i}\right)_p\right)(f) = 0$$

for every function $f : M \to \mathbb{R}$ differentiable at p. If, in particular, we take the coordinate functions of φ^{-1}, i.e. the functions $f_j := (\varphi^{-1})^j : \varphi(U) \to \mathbb{R}$, then $\hat{f}_j(x^1, \ldots, x^n) = x^j$ and so

$$0 = \left(\sum_{i=1}^n \alpha^i \left(\frac{\partial}{\partial x^i}\right)_p\right)(f_j) = \sum_{i=1}^n \alpha^i \left(\frac{\partial x^j}{\partial x^i}\right)(\varphi^{-1}(p)) = \alpha^j,$$

implying that the α^j are all equal to 0.

(4) Let $v \in T_p M$ be given by $v = \dot{c}(0)$ for some curve $c : (-\varepsilon, \varepsilon) \to M$. Then

$$(d(g \circ f))_p(v) = \left.\frac{d}{dt}\right|_{t=0} (g \circ f)(c(t))$$

$$= \left.\frac{d}{dt}\right|_{t=0} g((f \circ c)(t)) = (dg)_{f(p)}(w),$$

where $w \in T_{f(p)} N$ is the tangent vector to the differentiable curve $(f \circ c) : (-\varepsilon, \varepsilon) \to N$ at $t = 0$, i.e.

$$w = \left.\frac{d}{dt}\right|_{t=0} f(c(t)) = (df)_p(v).$$

Therefore

$$(d(g \circ f))_p(v) = (dg)_{f(p)}((df)_p(v))$$

for all $v \in T_p M$.

(6) Identifying $T_N S^n$ and $T_S S^n$ with the subspace of $\mathbb{R}^{n+1}$ given by $x^{n+1} = 0$, we have

$$(df)_N(v) = \left.\frac{d}{dt}\right|_{t=0} f(c(t)) = \left.\frac{d}{dt}\right|_{t=0} (-c(t)) = -\dot{c}(0) = -v$$

(where $c : (-\varepsilon, \varepsilon) \to S^n$ satisfies $\dot{c}(0) = v$), i.e. $(df)_N$ is simply multiplication by -1. Alternatively, using the parameterizations $(\mathbb{R}^n, \pi_S^{-1})$ and $(\mathbb{R}^n, \pi_N^{-1})$ at N and S, we have

$$\hat{f}(y) = \pi_N \circ f \circ \pi_S^{-1}(y) = \pi_N \circ f\left(\frac{2y}{1 + \|y\|^2}, \frac{1 - \|y\|^2}{1 + \|y\|^2}\right)$$

$$= \pi_N\left(-\frac{2y}{1 + \|y\|^2}, \frac{\|y\|^2 - 1}{1 + \|y\|^2}\right) = -y.$$

The Jacobian matrix of this map at $\pi_S(N) = 0$ is $(d\hat{f})_0 = -\,\mathrm{id}$. Therefore if y are the local coordinates corresponding to the first parameterization and z are the local coordinates corresponding to the second parameterization, we have

$$(df)_N\left(v^1\frac{\partial}{\partial y^1} + \cdots + v^n\frac{\partial}{\partial y^n}\right) = -v^1\frac{\partial}{\partial z^1} - \cdots - v^n\frac{\partial}{\partial z^n}.$$

(7) Let $c : (-\varepsilon, \varepsilon) \to W$ be a curve in $W \subset M$ such that $c(0) = p$ and let $v := \dot{c}(0)$. Then on the coordinate chart $x : W \to \mathbb{R}^n$,

$$(df)_p v = \frac{d(f \circ c)}{dt}(0) = \sum_{i=1}^n \dot{x}^i(0)\frac{\partial \hat{f}}{\partial x^i}(x(p)),$$

where in local coordinates we write $\hat{c}(t) = (x^1(t), \ldots, x^n(t))$. On the other hand,

$$(dx^j)_p v = \frac{d(x^j \circ c)}{dt}(0) = \sum_{i=1}^n \dot{x}^i(0)\frac{\partial x^j}{\partial x^i}(x(p)) = \dot{x}^j(0)$$

and the result follows.

(8) Clearly

$$\bigcup_\alpha \Phi_\alpha(U_\alpha \times \mathbb{R}^n) = \bigcup_{p \in \bigcup_\alpha \varphi_\alpha(U_\alpha)} T_pM = TM.$$

Moreover, the topology defined on TM by these parameterizations is easily seen to be Hausdorff and second countable. Finally, for $W = \Phi_\alpha(U_\alpha) \cap \Phi_\beta(U_\beta) \neq \varnothing$, the overlap maps

$$\Phi_\beta^{-1} \circ \Phi_\alpha : \Phi_\alpha^{-1}(W) \to \Phi_\beta^{-1}(W)$$

and

$$\Phi_\alpha^{-1} \circ \Phi_\beta : \Phi_\beta^{-1}(W) \to \Phi_\alpha^{-1}(W)$$

are given by

$$(\Phi_\beta^{-1} \circ \Phi_\alpha)(x, v) = \left((\varphi_\beta^{-1} \circ \varphi_\alpha)(x), (d(\varphi_\beta^{-1} \circ \varphi_\alpha))_x(v)\right)$$

and

$$(\Phi_\alpha^{-1} \circ \Phi_\beta)(x, v) = \left((\varphi_\alpha^{-1} \circ \varphi_\beta)(x), (d(\varphi_\alpha^{-1} \circ \varphi_\beta))_x(v)\right)$$

and so they are differentiable (In the formulae above we use the standard identification $T_x\mathbb{R}^n \cong \mathbb{R}^n$).

7.1.5 Section 1.5

(3) If M is a submanifold of N of dimension m then the inclusion map $i : M \hookrightarrow N$ is an embedding. In particular, the map i is an immersion and then, by the local immersion theorem, for every point $p \in W$ there are parameterizations (U, φ) and (V, ψ) around p on M and $f(p)$ on N for which

$$\hat{i}(x^1, \ldots, x^m) := (\psi^{-1} \circ i \circ \varphi)(x^1, \ldots, x^m) = (x^1, \ldots, x^m, 0, \ldots, 0) \in \mathbb{R}^n.$$

Taking an open set $W \subset N$ contained in $\psi((U \times \mathbb{R}^{n-m}) \cap V)$ and the coordinate system $x : W \to \mathbb{R}^n$ given by $x = \psi^{-1}$, we have

$$M \cap W = \{q \in W \mid x^{m+1}(q) = \cdots = x^n(q) = 0\}.$$

Conversely, if for every $p \in M$ there is a coordinate system $x_p : W_p \to \mathbb{R}^n$ around p on N such that

$$M \cap W_p = \{q \in W_p \mid x_p^{m+1}(q) = \cdots = x_p^n(q) = 0\},$$

then, taking the standard projection π onto the first m factors, the map

$$\widetilde{x}_p : M \cap W_p \to \mathbb{R}^m := \pi \circ x_p$$

is a coordinate system around p on M for the subspace topology on M. Indeed, $\widetilde{x}_p$ is a homeomorphism onto its image:

  (i) if A is an open subset of $\widetilde{x}_p(M \cap W_p)$ then $(A \times \mathbb{R}^{n-m}) \cap x_p(W_p)$ is an open subset of $x_p(W_p)$ and so $\widetilde{x}_p^{-1}(A) = x_p^{-1}((A \times \mathbb{R}^{n-m}) \cap x_p(W_p)) \cap M$ is an open set of M for the subspace topology;
 (ii) if B is an open subset of $M \cap W_p$ then there is an open set of N, $B' \subset W_p$ such that $B = B' \cap M$ and so

$$\widetilde{x}_p(B) = \pi\left(x_p(B') \cap \{x \in \mathbb{R}^n \mid x^{m+1} = \cdots = x^n = 0\}\right)$$

is an open subset of $\mathbb{R}^m$.

Moreover, $\{(\tilde{x}_p(W_p), \tilde{x}_p^{-1})\}_{p \in M}$ forms an atlas for M: if, for instance, $L := W_p \cap W_q \cap M \neq \varnothing$ and $f_{p,q} := x_q \circ x_p^{-1} : x_p(W_p \cap W_q) \to x_q(W_p \cap W_q)$ is the corresponding overlap map on N, then the overlap map on M

$$\tilde{x}_q \circ \tilde{x}_p^{-1} : \tilde{x}_p(L) \to \tilde{x}_q(L)$$

is given by

$$(\tilde{x}_q \circ \tilde{x}_p^{-1})(x^1, \ldots, x^m) = \pi(f_{p,q}(x^1, \ldots, x^m, 0, \ldots, 0)).$$

It is clear from the choice of coordinates that the inclusion map is an embedding.
(4) Consider the map $f : \mathbb{R}^{n+1} \to \mathbb{R}$ given by

$$f(x^1, \ldots, x^n) = (x^1)^2 + \cdots + (x^{n+1})^2.$$

Its derivative

$$(df)_x = 2x^1 dx^1 + \cdots + 2x^{n+1} dx^{n+1}$$

is clearly injective for $x \neq 0$, as it is represented by the nonvanishing matrix

$$\left(2x^1 \mid \cdots \mid 2x^{n+1}\right).$$

Therefore, 1 is a regular value of f, and so $S^n = f^{-1}(1)$ is an n-dimensional manifold (cf. Theorem 5.6). Moreover, we have

$$\begin{aligned} T_x S^n = \ker(df)_x &= \{v \in T_x \mathbb{R}^{n+1} \mid (df)_x(v) = 0\} \\ &= \{v \in \mathbb{R}^{n+1} \mid x^1 v^1 + \cdots + x^{n+1} v^{n+1} = 0\} \\ &= \{v \in \mathbb{R}^{n+1} \mid \langle x, v \rangle = 0\}, \end{aligned}$$

where we have used the identification $T_x \mathbb{R}^{n+1} \cong \mathbb{R}^{n+1}$.
(5) Let $i : V \to M$ be the inclusion map. Then $f \circ i : V \to N$ is differentiable. For each point $p \in W$ let $x_p : U_p \to \mathbb{R}^n$ ($n = \dim N$) be a local chart on N such that

$$W \cap U_p = \{q \in U_p \mid x_p^{k+1}(q) = \cdots = x_p^n(q) = 0\},$$

where $k = \dim W$ [cf. Exercise 5.9(3)]. The maps $\tilde{x}_p : W \cap U_p \to \mathbb{R}^k$ defined by $\tilde{x}_p(q) := (x_p^1(q), \ldots, x_p^k(q))$ are local charts defining an atlas for W. Hence, for every $p \in W$ the maps

$$\tilde{x}_p \circ f \circ i : (f \circ i)^{-1}\left(W \cap U_p\right) \to \mathbb{R}^k$$

are differentiable, implying that $f : V \to W$ is differentiable.

7.1.6 Section 1.6

(2) Let $X, Y, Z \in \mathfrak{X}(M)$.

(a) Let $\alpha, \beta \in \mathbb{R}$. Then

$$
\begin{aligned}
[\alpha X + \beta Y, Z] &= (\alpha X + \beta Y) \circ Z - Z \circ (\alpha X + \beta Y) \\
&= \alpha(X \circ Z - Z \circ X) + \beta(Y \circ Z - Z \circ Y) \\
&= \alpha[X, Z] + \beta[Y, Z].
\end{aligned}
$$

Similarly, $[X, \alpha Y + \beta Z] = \alpha[X, Y] + \beta[X, Z]$.

(b) We have

$$
[X, Y] = X \circ Y - Y \circ X = -(Y \circ X - X \circ Y) = -[Y, X].
$$

(c) The Jacobi identity can be proved as follows:

$$
\begin{aligned}
[[X, Y], Z] &+ [[Y, Z], X] + [[Z, X], Y] \\
&= (X \circ Y - Y \circ X) \circ Z - Z \circ (X \circ Y - Y \circ X) \\
&+ (Y \circ Z - Z \circ Y) \circ X - X \circ (Y \circ Z - Z \circ Y) \\
&+ (Z \circ X - X \circ Z) \circ Y - Y \circ (Z \circ X - X \circ Z) = 0.
\end{aligned}
$$

(d) Let $f, g \in C^\infty(M)$. Then,

$$
\begin{aligned}
[fX, gY] &= (fX) \circ (gY) - (gY) \circ (fX) \\
&= fgX \circ Y + f(X \cdot g)Y - gfY \circ X - g(Y \cdot f)X \\
&= fg[X, Y] + f(X \cdot g)Y - g(Y \cdot f)X,
\end{aligned}
$$

where we used Exercise 6.11(1).

(5) (a) Let $f : \mathbb{R}^3 \to \mathbb{R}$ be a smooth function.

$$
\begin{aligned}
[X_1, X_2] \cdot f &= (X_1 \circ X_2 - X_2 \circ X_1) \cdot f \\
&= X_1 \cdot \left(z\frac{\partial f}{\partial x} - x\frac{\partial f}{\partial z} \right) - X_2 \cdot \left(y\frac{\partial f}{\partial z} - z\frac{\partial f}{\partial y} \right) \\
&= y\frac{\partial f}{\partial x} - x\frac{\partial f}{\partial y} = -X_3 \cdot f
\end{aligned}
$$

and so $[X_1, X_2] = -X_3$. Similarly, we conclude that $[X_1, X_3] = X_2$ and $[X_2, X_3] = -X_1$.

(b) Let $V := \text{span}\{X_1, X_2, X_3\}$. From (a) we know that the Lie bracket determines a bilinear map $[\cdot, \cdot] : V \times V \to V$ so we conclude that $V = \text{span}\{X_1, X_2, X_3\}$ is a Lie subalgebra of $\mathfrak{X}(\mathbb{R}^3)$. To show that it is isomorphic to $(\mathbb{R}^3, \times)$ we will use that map $F : V \to \mathbb{R}^3$ given by

$$F(a_1 X_1 + a_2 X_2 + a_3 X_3) = (a_1, -a_2, a_3).$$

This map is clearly bijective so we just have to show that it is also a Lie algebra homomorphism. For that we see that

$$F([X_1, X_2]) = F(-X_3) = (0, 0, -1) = (1, 0, 0) \times (0, -1, 0) = F(X_1) \times F(X_2);$$
$$F([X_1, X_3]) = F(X_2) = (0, -1, 0) = (1, 0, 0) \times (0, 0, 1) = F(X_1) \times F(X_3);$$
$$F([X_2, X_3]) = F(-X_1) = (-1, 0, 0) = (0, -1, 0) \times (0, 0, 1) = F(X_2) \times F(X_3).$$

(c) Let $q \in \mathbb{R}^3$. We know that $\psi_{1,t}(q)$ is an integral curve of X_1 at q. Hence,

$$\frac{d\psi_{1,t}(q)}{dt} = (X_1)_{\psi_{1,t}(q)} \quad \text{and} \quad \psi_{1,0}(q) = q.$$

Consequently,

$$\dot\psi_{1,t}^1(q) = X_1^1(\psi_{1,t}(q)) = 0;$$
$$\dot\psi_{1,t}^2(q) = X_1^2(\psi_{1,t}(q)) = -\psi_{1,t}^3(q);$$
$$\dot\psi_{1,t}^3(q) = X_1^3(\psi_{1,t}(q)) = \psi_{1,t}^2(q).$$

Hence, $\psi_{1,t}^1 = C$, $\ddot\psi_{1,t}^2 = -\dot\psi_{1,t}^3 = -\psi_{1,t}^2$, and so

$$\psi_{1,t}^2 = A \cos t + B \sin t \quad \text{and} \quad \psi_{1,t}^3 = A \sin t - B \cos t,$$

where A, B and C are functions of $q = (x, y, z)$. Since $\psi_{1,0}(x, y, z) = (x, y, z)$ we have $C = x$, $A = y$ and $B = -z$ and we conclude that

$$\psi_{1,t}(x, y, z) = (x, y \cos t - z \sin t, y \sin t + z \cos t).$$

Similarly, we see that

$$\psi_{2,t}(x, y, z) = (x \cos t + z \sin t, y, -x \sin t + z \cos t)$$

and

$$\psi_{3,t}(x, y, z) = (x \cos t - y \sin t, x \sin t + y \cos t, z).$$

(d) We will show that $\psi_{1,\frac{\pi}{2}} \circ \psi_{2,\frac{\pi}{2}} \neq \psi_{2,\frac{\pi}{2}} \circ \psi_{1,\frac{\pi}{2}}$. The other combinations are similar.

$$\left(\psi_{1,\frac{\pi}{2}} \circ \psi_{2,\frac{\pi}{2}}\right)(x, y, z) = \psi_{1,\frac{\pi}{2}}(z, y, -x) = (z, x, y)$$

while

$$\left(\psi_{2,\frac{\pi}{2}} \circ \psi_{1,\frac{\pi}{2}}\right)(x, y, z) = \psi_{2,\frac{\pi}{2}}(x, -z, y) = (y, -z, -x).$$

(6) Let $X \in \mathfrak{X}(\mathbb{R})$ be given by $X = x^2 \frac{d}{dx}$. The equation defining its integral curves is

$$\dot{x} = x^2$$

and so the integral curve at $x_0 \in \mathbb{R}$ is $c_{x_0}(t) = -\frac{1}{t+C}$, where C is a function of x_0. Since $c_{x_0}(0) = -\frac{1}{C}$, we conclude that $C = -\frac{1}{x_0}$, implying that the local flow of X at y is the map

$$F : W \times (-\varepsilon, \varepsilon) \to \mathbb{R}$$

$$(x, t) \mapsto \frac{x}{1 - tx},$$

where $W = (a, b)$ is an open interval containing y. For instance if $a > 0$, the local flow can only be extended to $W \times (-\infty, \frac{1}{b})$. We conclude that X is not a complete vector field since we can never extend the local flow to $\mathbb{R} \times \mathbb{R}$.

(10) (a) If $c : I \to M$ is an integral curve of X then $\dot{c}(t) = X_{c(t)}$ for all $t \in I$. Therefore, the curve $\gamma : I \to N$ defined by $\gamma(t) := f(c(t))$ satisfies

$$\dot{\gamma}(t) = (df)_{c(t)}\dot{c}(t) = (df)_{c(t)}X_{c(t)} = Y_{f(c(t))} = Y_{\gamma(t)},$$

i.e. γ is an integral curve of Y.

(b) We just showed that if X and Y are f-related then $f(F_X(p, t)) = F_Y(f(p), t)$ for all $(p, t) \in M \times \mathbb{R}$ for which both sides are defined. On the other hand, if this relation holds then differentiating at $t = 0$ yields $(df)_p X_p = Y_{f(p)}$ for each $p \in M$.

(12) (a) We have

$$(L_X Y)_p = \frac{d}{dt}\left((d\psi_{-t})_{\psi_t(p)}Y_{\psi_t(p)}\right)_{|t=0},$$

and hence

$$(L_X Y)_p \cdot f = \frac{d}{dt}\left(Y_{\psi_t(p)} \cdot (f \circ \psi_{-t})\right)_{|t=0}$$

for any differentiable function $f \in C^\infty(M)$. Let us define

$$H(v, v) := Y_{\psi_v(p)} \cdot (f \circ \psi_{-v}).$$

We have

$$\frac{\partial H}{\partial v}(0, 0) = \frac{d}{dv}(Y_{\psi_v(p)} \cdot f)_{|v=0} = \frac{d}{dv}(Y \cdot f)(\psi_v(p))_{|v=0}$$

$$= (X \cdot (Y \cdot f))(p)$$

and

$$\frac{\partial H}{\partial v}(0,0) = \frac{d}{dv}(Y_p \cdot (f \circ \psi_{-v}))_{|v=0} = Y_p \cdot \left(\frac{d}{dv}(f \circ \psi_{-v})_{|v=0}\right)$$
$$= Y_p \cdot (-X \cdot f) = -(Y \cdot (X \cdot f))(p).$$

Consequently,

$$(L_X Y)_p \cdot f = \frac{d}{dt}(H(t,t))_{|t=0} = \frac{\partial H}{\partial v}(0,0) + \frac{\partial H}{\partial v}(0,0)$$
$$= (X \cdot (Y \cdot f) - Y \cdot (X \cdot f))(p) = [X,Y]_p \cdot f$$

for any differentiable function $f \in C^\infty(M)$, implying that $L_X Y = [X,Y]$.
(b) We have

$$L_X[Y,Z] = \frac{d}{dt}((\psi_{-t})_*[Y,Z])_{|t=0} = \frac{d}{dt}[(\psi_{-t})_*Y, (\psi_{-t})_*Z]_{|t=0}$$
$$= \left[\frac{d}{dt}((\psi_{-t})_*Y)_{|t=0}, Z\right] + \left[Y, \frac{d}{dt}((\psi_{-t})_*Z)_{|t=0}\right]$$
$$= [L_X Y, Z] + [Y, L_X Z].$$

Notice that using (a) this formula can be written as

$$[X,[Y,Z]] = [[X,Y],Z] + [Y,[X,Z]]$$

(i.e. it is just the Jacobi identity).
(c) We have

$$L_X(L_Y Z) = [X,[Y,Z]] \quad \text{and} \quad L_Y(L_X Z) = [Y,[X,Z]].$$

Therefore,

$$(L_X \circ L_Y - L_Y \circ L_X)Z = [X,[Y,Z]] - [Y,[X,Z]]$$
$$= [X,[Y,Z]] + [Y,[Z,X]]$$
$$= -[Z,[X,Y]] = [[X,Y],Z] = L_{[X,Y]}Z,$$

where we have used the Jacobi identity.
(13) (a) This is an immediate consequence of Exercise 6.11(10).
(b) If $\psi_t \circ \phi_s = \phi_s \circ \psi_t$ for all $s, t \in \mathbb{R}$ then, by (a), $(\psi_t)_*Y = Y$ for all $t \in \mathbb{R}$.
Therefore,

$$[X,Y] = L_X Y = \frac{d}{dt}((\psi_{-t})_*Y)_{|t=0} = \frac{d}{dt}(Y)_{|t=0} = 0.$$

If, on the other hand, $[X, Y] = 0$ then

$$\frac{d}{dt}((\psi_t)_*Y) = \frac{d}{d\varepsilon}((\psi_{t+\varepsilon})_*Y)\Big|_{\varepsilon=0} = \frac{d}{d\varepsilon}((\psi_t)_*(\psi_\varepsilon)_*Y)\Big|_{\varepsilon=0}$$

$$= (\psi_t)_*\frac{d}{d\varepsilon}((\psi_\varepsilon)_*Y)\Big|_{\varepsilon=0} = -(\psi_t)_*L_XY = 0.$$

Since $(\psi_0)_*Y = Y$, we conclude that $(\psi_t)_*Y = Y$ for all $t \in \mathbb{R}$. Therefore $\psi_t \circ \phi_s = \phi_s \circ \psi_t$ for all $s, t \in \mathbb{R}$.

7.1.7 Section 1.7

(3) (a) Given two affine maps $g(t) = yt + x$ and $h(t) = wt + z$, we have

$$(g \circ h)(t) = g(h(t)) = g(wt + z) = ywt + yz + x.$$

Therefore the group operation is given by

$$(x, y) \cdot (z, w) = (yz + x, yw).$$

The identity element is clearly $e = (0, 1)$ (corresponding to the identity map), and hence

$$(z, w) = (x, y)^{-1} \Leftrightarrow (yz + x, yw) = (0, 1)$$

$$\Leftrightarrow (z, w) = \left(-\frac{x}{y}, \frac{1}{y}\right).$$

Therefore the maps $H \times H \ni (g, h) \mapsto g \cdot h \in H$ and $H \ni g \mapsto g^{-1} \in H$ are smooth, and hence H is a Lie group.

(b) Because

$$L_{(x,y)}(z, w) = (yz + x, yw),$$

the matrix representation of $(dL_{(x,y)})_{(z,w)}$ is

$$(dL_{(x,y)})_{(z,w)} = \begin{pmatrix} y & 0 \\ 0 & y \end{pmatrix}.$$

Therefore $X^V_{(x,y)}$ has components

$$\begin{pmatrix} y & 0 \\ 0 & y \end{pmatrix} \begin{pmatrix} \xi \\ \eta \end{pmatrix} = \begin{pmatrix} y\xi \\ y\eta \end{pmatrix}.$$

(c) If

$$V = \xi \frac{\partial}{\partial x} + \eta \frac{\partial}{\partial y} \quad \text{and} \quad W = \zeta \frac{\partial}{\partial x} + \omega \frac{\partial}{\partial y}$$

then

$$\left[X^V, X^W \right] = \left[\xi y \frac{\partial}{\partial x} + \eta y \frac{\partial}{\partial y}, \zeta y \frac{\partial}{\partial x} + \omega y \frac{\partial}{\partial y} \right]$$

$$= (\eta \zeta - \omega \xi) y \frac{\partial}{\partial x}.$$

Therefore

$$[V, W] = \left[X^V, X^W \right]_{(0,1)} = (\eta \zeta - \omega \xi) \frac{\partial}{\partial x}.$$

(d) The flow of X^V is given by the solution of the system of ODEs

$$\begin{cases} \dot{x} = \xi y \\ \dot{y} = \eta y \end{cases}$$

which is

$$\begin{cases} x = x_0 + \frac{y_0 \xi (e^{\eta t} - 1)}{\eta} \\ y = y_0 e^{\eta t} \end{cases}$$

for $\eta \neq 0$ and

$$\begin{cases} x = x_0 + y_0 \xi t \\ y = y_0 \end{cases}$$

for $\eta = 0$. The exponential map is obtained by setting $(x_0, y_0) = e = (0, 1)$ and $t = 1$:

$$\exp(V) = \left(\frac{\xi (e^{\eta} - 1)}{\eta}, e^{\eta} \right)$$

for $\eta \neq 0$ and

$$\exp(V) = (\xi, 1)$$

for $\eta = 0$.

(e) The multiplication of two such matrices is

$$\begin{pmatrix} y & x \\ 0 & 1 \end{pmatrix} \begin{pmatrix} w & z \\ 0 & 1 \end{pmatrix} = \begin{pmatrix} yw & yz + x \\ 0 & 1 \end{pmatrix},$$

which reproduces the group operation on H. Therefore H can be identified with the corresponding subgroup of $GL(2)$. A curve $c : (-\varepsilon, \varepsilon) \to H$ with $c(0) = I$ is then given by

$$c(t) = \begin{pmatrix} y(t) & x(t) \\ 0 & 1 \end{pmatrix}$$

with $x(0) = 0$ and $y(0) = 1$, and its derivative at $t = 0$ is

$$\dot{c}(0) = \begin{pmatrix} \dot{y}(0) & \dot{x}(0) \\ 0 & 0 \end{pmatrix}.$$

We conclude that $\mathfrak{h}$ can be identified with the vector space of matrices of the form

$$\begin{pmatrix} \eta & \xi \\ 0 & 0 \end{pmatrix}.$$

The Lie bracket must then be given by

$$\left[\begin{pmatrix} \eta & \xi \\ 0 & 0 \end{pmatrix}, \begin{pmatrix} \omega & \zeta \\ 0 & 0 \end{pmatrix} \right] = \begin{pmatrix} \eta & \xi \\ 0 & 0 \end{pmatrix} \begin{pmatrix} \omega & \zeta \\ 0 & 0 \end{pmatrix} - \begin{pmatrix} \omega & \zeta \\ 0 & 0 \end{pmatrix} \begin{pmatrix} \eta & \xi \\ 0 & 0 \end{pmatrix}$$

$$= \begin{pmatrix} 0 & \eta\zeta - \omega\xi \\ 0 & 0 \end{pmatrix},$$

which agrees with (c). Moreover, the exponential map must be given by

$$\exp \begin{pmatrix} \eta & \xi \\ 0 & 0 \end{pmatrix} = \sum_{k=0}^{+\infty} \frac{1}{k!} \begin{pmatrix} \eta & \xi \\ 0 & 0 \end{pmatrix}^k$$

$$= \begin{pmatrix} 1 & 0 \\ 0 & 1 \end{pmatrix} + \begin{pmatrix} \eta & \xi \\ 0 & 0 \end{pmatrix} + \frac{1}{2} \begin{pmatrix} \eta^2 & \eta\xi \\ 0 & 0 \end{pmatrix} + \cdots,$$

yielding

$$\exp \begin{pmatrix} \eta & \xi \\ 0 & 0 \end{pmatrix} = \begin{pmatrix} e^\eta & \frac{\xi(e^\eta - 1)}{\eta} \\ 0 & 1 \end{pmatrix}$$

for $\eta \neq 0$ and

$$\exp \begin{pmatrix} \eta & \xi \\ 0 & 0 \end{pmatrix} = \begin{pmatrix} 1 & \xi \\ 0 & 1 \end{pmatrix},$$

for $\eta = 0$, which agrees with (d).

(6) (a) Clearly $h(\mathbb{R}) = \mathbb{R} \setminus \{0\}$ and

$$h(x + y) = \det e^{A(x+y)} = \det (e^{Ax} e^{Ay})$$

$$= (\det e^{Ax})(\det e^{Ay}) = h(x) \cdot h(y).$$

(b) Note that $h(t) = f(e^{At})$ where $f : GL(n) \to \mathbb{R}$ is given by $f(B) = \det(B)$. Hence, since $(e^{At})' = Ae^{At}$, we have, by Example 7.1(4), that

$$h'(0) = (df)_I(A) = \text{tr}(A).$$

(c) Again using Example 7.1.4 we have

$$h'(t) = (df)_{e^{At}}(Ae^{At}) = \det(e^{At})\,\text{tr}\,(e^{-At}Ae^{At}) = h(t)\,\text{tr}(A).$$

Hence, $h(t) = ke^{\text{tr}(A)t}$ for a constant $k \in \mathbb{R}$. Since we know that $h(0) = \det I = 1$ we conclude that $h(t) = e^{\text{tr}(A)t}$ and so $\det(e^A) = h(1) = e^{\text{tr}(A)}$.

(12) (a) Let $V \in \mathfrak{g}$. We begin by showing that X^V is F-related to $X^{(dF)_eV}$ [cf. Exercise 6.11(8)]. Indeed,

$$\begin{aligned} (dF)_g X_g^V &= (dF)_g(dL_g)_e V = d(F \circ L_g)_e V \\ &= d(L_{F(g)} \circ F)_e V = (dL_{F(g)})_e (dF)_e V \\ &= X_{F(g)}^{(dF)_e V}, \end{aligned}$$

where $F \circ L_g = L_{F(g)} \circ F$ follows from the fact that F is a Lie group homomorphism.

Let $V, W \in \mathfrak{g}$. Then

$$\begin{aligned} (dF)_e[V, W] &= (dF)_e[X^V, X^W]_e = (F_*[X^V, X^W])_e \\ &= [F_*X^V, F_*X^W]_e = [X^{(dF)_eV}, X^{(dF)_eW}]_e \\ &= [(dF)_eV, (dF)_eW]. \end{aligned}$$

Here we have used the result of Exercise 6.11(9), which is easily extended to arbitrary differentiable maps.

(b) Given $g \in G$ we have $F = L_{F(g)} \circ F \circ L_{g^{-1}}$. Consequently,

$$(dF)_g = (dL_{F(g)})_e \circ (dF)_e \circ (dL_{g^{-1}})_g.$$

Since the left multiplication map is a diffeomorphism, we conclude that if $(dF)_e$ is an isomorphism then $(dF)_g$ is an isomorphism for any $g \in G$. The inverse function theorem then guarantees that F is a local diffeomorphism.

(c) Let $U \ni e$ be an open set such that the restriction of F to U is a diffeomorphism onto $F(U)$. Then for any $g \in G$ the restriction of $F = L_{F(g)} \circ F \circ L_{g^{-1}}$ to $L_g(U)$ is a diffeomorphism onto its image. Moreover, it is easily seen that $F^{-1}(F(L_g(U))) = L_g(F^{-1}(F(U)))$. Therefore we just have to check that $F^{-1}(F(U))$ is a disjoint union of open sets diffeomorphic to $F(U)$. Now $F(h) \in F(U)$ if and only if $F(h) = F(h_0)$ for some $h_0 \in U$, i.e. if and only if $hh_0^{-1} = g \in \ker(F)$. We conclude that

$$F^{-1}(F(U)) = \bigcup_{g \in \ker(F)} L_g(U).$$

On the other hand, the open sets $L_g(U)$ are clearly disjoint, because if $g \in \ker(F)$ and $h_0 \in U$ then $F(gh_0) = F(h_0)$, and hence gh_0 can only be on U (where F is injective) if $g = e$.

7.1.8 Section 1.8

(2) Let us assume that M is orientable and let us fix an orientation on M. Then, for each $p \in M$, there exists a parameterization (U, φ) around p such that $(d\varphi)_x$ preserves the standard orientation of $\mathbb{R}^n$ at each point $x \in U$. Taking the atlas $\mathcal{A}$ formed by these parameterizations we have that the overlap maps are orientation-preserving and $\det d(\varphi_\beta^{-1} \circ \varphi_\alpha) > 0$ on $W := \varphi_\alpha(U_\alpha) \cap \varphi_\beta(U_\beta)$ for every $(U_\alpha, \varphi_\alpha), (U_\beta, \varphi_\beta) \in \mathcal{A}$. Conversely, if there exists an atlas $\mathcal{A}$ for which the overlap maps are orientation-preserving, then we can choose an orientation on M in the following way. For each $p \in M$ we take a parameterization $(U, \varphi) \in \mathcal{A}$ around p and we assign a positive sign to the ordered bases of $T_p M$ that are equivalent to the ordered basis $\{(d\varphi)_x e_i\}_{i=1}^n$, where $x := \varphi^{-1}(p)$ and $\{e_i\}_{i=1}^n$ is the standard basis of $\mathbb{R}^n$. This choice of orientation does not depend on the parameterization φ as all overlap maps are orientation-preserving.

(4) Let $t_0 \in I$ and let (U, φ) be a parameterization around $c(t_0)$ such that $(d\varphi)_x$ is orientation-preserving for all $x \in U$. In the corresponding local coordinates $x := \varphi^{-1}$ we have

$$V_i(t) = \sum_{j=1}^n V_i^j(t) \left(\frac{\partial}{\partial x^j} \right)_{c(t)},$$

where the functions $V_i^j : J \to \mathbb{R}$, defined on a neighborhood $J \subset I$ of t_0, are smooth. Therefore the map $d : J \to \mathbb{R}$ defined by $d(t) = \det(V_i^j(t))$ is also smooth. Moreover, since $\{V_1(t_0), \ldots, V_n(t_0)\}$ is a basis of $T_{c(t_0)} M$, we either have $d(t_0) > 0$ or $d(t_0) < 0$. Consequently, we will also have $d(t) > 0$ or $d(t) < 0$ for t in a neighborhood of t_0 in J. We conclude that the set of points $t \in I$ where $\{V_1(t), \ldots, V_n(t)\}$ has positive orientation is an open subset of I, and so is the set of points where $\{V_1(t), \ldots, V_n(t)\}$ has negative orientation. Since I is connected (it is an interval), we conclude that $\{V_1(t), \ldots, V_n(t)\}$ has either positive orientation for all $t \in I$ or negative orientation for all $t \in I$.

(5) Let $c : [0, 2\pi] \to M$ be the curve defined on the Möbius band M by

$$c(\varphi) := g(0, \varphi) = (\cos \varphi, \sin \varphi, 0),$$

and consider the smooth vector fields along c defined by

$$V_1(\varphi) = \frac{\partial g}{\partial t}(0, \varphi) = \left(\cos \left(\frac{\varphi}{2} \right) \cos \varphi, \cos \left(\frac{\varphi}{2} \right) \sin \varphi, \sin \left(\frac{\varphi}{2} \right) \right);$$

$$V_2(\varphi) = \frac{\partial g}{\partial \varphi}(0, \varphi) = (-\sin \varphi, \cos \varphi, 0).$$

Since
$$\|V_1(\varphi)\| = \|V_2(\varphi)\| = 1 \quad \text{and} \quad \langle V_1(\varphi), V_2(\varphi) \rangle = 0$$

for all $\varphi \in [0, 2\pi]$, these vector fields form a basis of $T_{c(\varphi)}M$ for all $\varphi \in [0, 2\pi]$. Moreover, since $c(0) = c(2\pi) = (1, 0, 0)$, we know that

$$\{V_1(0), V_2(0)\} = \{(1, 0, 0); (0, 1, 0)\}$$

and

$$\{V_1(2\pi), V_2(2\pi)\} = \{(-1, 0, 0); (0, 1, 0)\}$$

are two bases for $T_{(1,0,0)}M$. However, the change-of-basis matrix from the first basis to the second basis,

$$S = \begin{pmatrix} -1 & 0 \\ 0 & 1 \end{pmatrix},$$

has negative determinant. Therefore M cannot be orientable.

(9) (a) We begin with the observation that for each parameterization (U, φ) of M there exists a parameterization $(\widetilde{U}, \widetilde{\varphi})$ which induces the opposite orientation on $T_p M$ for every $p \in \varphi(U)$ (one just has to reverse the order of two of the variables in $\mathbb{R}^n$).

The maps $\overline{\varphi} : U \to \overline{M}$ define a topology on $\overline{M}$, given by the basis

$$\{\overline{\varphi}(U) \mid (U, \varphi) \text{ is a parameterization of } M\}.$$

That this is indeed a basis for a topology and that such topology is Hausdorff and second countable follows from the fact that

$$\{\varphi(U) \mid (U, \varphi) \text{ is a parameterization of } M\}$$

is a basis for the topology of M with the same properties.

Given two parameterizations (U, φ) and (V, ψ) of M, the map $\overline{\psi}^{-1} \circ \overline{\varphi}$ is defined on the image by φ^{-1} of the connected components of $\varphi(U) \cap \psi(V)$ where the orientations induced by φ and ψ agree. Therefore $\overline{\psi}^{-1} \circ \overline{\varphi}$ is not only differentiable $(\overline{\psi}^{-1} \circ \overline{\varphi} = \psi^{-1} \circ \varphi$ on the points where it is defined) but also orientation-preserving. We conclude that

$$\{(U, \overline{\varphi}) \mid (U, \varphi) \text{ is a parameterization of } M\}$$

is an atlas for $\overline{M}$ whose overlap maps are orientation-preserving. Therefore $\overline{M}$ is an **orientable** n-dimensional manifold [cf. Exercise 8.6(2)].

(b) This is immediate from the fact that, for the parameterizations above, $\varphi^{-1} \circ \pi \circ \overline{\varphi}$ is the identity map and from the above observation that for each parameterization (U, φ) of M there exists a parameterization $(\widetilde{U}, \widetilde{\varphi})$ which induces the opposite orientation on $T_p M$ for every $p \in \varphi(U)$.

(c) If $\overline{W}$ is a connected component of $\overline{M}$ then $\pi : \overline{W} \to M$ is still a covering map. Therefore the fibers $\pi^{-1}(p) \subset \overline{W}$ can either have one point or two points. In the former case $\pi : \overline{W} \to M$ is a diffeomorphism and M would be orientable (because $\overline{W}$ is). Since this is not the case, we conclude that $\overline{W} = \overline{M}$, and so M is connected.

(d) The identities $\pi \circ \sigma = \pi$ and $\sigma \circ \sigma = \sigma$ are immediate. That σ is a diffeomorphism is clear from $\pi \circ \sigma = \pi$ and π being a local diffeomorphism.

(e) This is immediate from the fact that the only covering map admitted by a simply connected manifold is the trivial covering map.

7.1.9 Section 1.9

(2) Let $\mathcal{A} = \{(U_\alpha, \varphi_\alpha)\}$ and $\mathcal{B} = \{(V_\beta, \psi_\beta)\}$ be atlases respectively for M and N. Note that, since M is a manifold without boundary and N is a manifold with boundary, the sets U_α are open subsets of $\mathbb{R}^m$ while the sets V_β are open subsets of $\mathbb{H}^n$ (where $m = \dim M$ and $n = \dim N$). We will show that $\{(U_\alpha \times V_\beta, \varphi_\alpha \times \psi_\beta)\}$ is an atlas for the product manifold $M \times N$ and that this set is a manifold with boundary. Indeed,

(a) the sets $U_\alpha \times V_\beta$ are open subsets of $\mathbb{R}^m \times \mathbb{H}^n \cong \mathbb{H}^{m+n}$;

(b) the sets $\varphi_\alpha(U_\alpha) \times \psi_\beta(V_\beta)$ are open subsets of $M \times N$;

(c) the maps $\theta_{\alpha,\beta} := \varphi_\alpha \times \psi_\beta : U_\alpha \times V_\beta \to M \times N$ are homeomorphisms;

(d) for $W := \theta_{\alpha_1,\beta_1}(U_{\alpha_1} \times V_{\beta_1}) \cap \theta_{\alpha_2,\beta_2}(U_{\alpha_2} \times V_{\beta_2}) \neq \varnothing$ the overlap maps

$$\theta_{\alpha_1,\beta_1}^{-1} \circ \theta_{\alpha_2,\beta_2} : \theta_{\alpha_2,\beta_2}^{-1}(W) \to \theta_{\alpha_1,\beta_1}^{-1}(W)$$

are differentiable since

$$(\theta_{\alpha_1,\beta_1}^{-1} \circ \theta_{\alpha_2,\beta_2})(p, q) = \theta_{\alpha_1,\beta_1}^{-1}(\varphi_{\alpha_2}(p), \psi_{\beta_2}(q))$$

$$= \left((\varphi_{\alpha_1}^{-1} \circ \varphi_{\alpha_2})(p), (\psi_{\beta_1}^{-1} \circ \psi_{\beta_2})(q)\right)$$

and $\varphi_{\alpha_1}^{-1} \circ \varphi_{\alpha_2}$ and $\psi_{\beta_1}^{-1} \circ \psi_{\beta_2}$ are differentiable;

(e) we have

$$\bigcup_{\alpha,\beta} \theta_{\alpha,\beta}(U_\alpha \times V_\beta) = \bigcup_{\alpha,\beta} \varphi_\alpha(U_\alpha) \times \psi_\beta(V_\beta) = M \times N.$$

We will now show that $\partial(M \times N) = M \times \partial N$. For that we see that $(p, q) \in \partial(M \times N)$ if and only if there is a pair (α, β) for which

$$(p, q) \in \theta_{\alpha,\beta} \left(\partial \mathbb{H}^{m+n} \cap (U_\alpha \times V_\beta)\right)$$
$$= \theta_{\alpha,\beta} \left(\{(x, y) \in (U_\alpha \times V_\beta) \mid y^n = 0\}\right)$$
$$= \theta_{\alpha,\beta} \left(U_\alpha \times (V_\beta \cap \partial \mathbb{H}^n)\right) = \varphi_\alpha(U_\alpha) \times \psi_\beta(V_\beta \cap \partial \mathbb{H}^n).$$

Hence, $(p, q) \in \partial(M \times N)$ if and only if $(p, q) \in M \times \partial N$ and we conclude that $\partial(M \times N) = M \times \partial N$.

7.2 Chapter 2

7.2.1 Section 2.1

(3) Let $T \in T^k(V^*)$.

(a) Consider the tensor $\text{Alt}(T) := \frac{1}{k!} \sum_{\sigma \in S_k} (\text{sgn } \sigma)(T \circ \sigma)$. Then, if σ_0 is a permutation that interchanges two indices and keeps the others fixed, we have

$$\text{Alt}(T) \circ \sigma_0 = \frac{1}{k!} \sum_{\sigma \in S_k} (\text{sgn } \sigma)(T \circ \sigma \circ \sigma_0)$$

$$= \frac{1}{k!} (\text{sgn } \sigma_0) \sum_{\sigma \in S_k} (\text{sgn}(\sigma \circ \sigma_0))(T \circ (\sigma \circ \sigma_0))$$

$$= \frac{1}{k!} (\text{sgn } \sigma_0) \sum_{\tau \in S_k} (\text{sgn } \tau)(T \circ \tau)$$

$$= (\text{sgn } \sigma_0) \text{Alt}(T) = -\text{Alt}(T)$$

and we conclude that $\text{Alt}(T)$ is an alternating tensor.

(b) If T is alternating then $T \circ \sigma = (\text{sgn } \sigma)T$ for any permutation $\sigma \in S_k$. Hence,

$$\text{Alt}(T) = \frac{1}{k!} \sum_{\sigma \in S_k} (\text{sgn } \sigma)(T \circ \sigma) = \frac{1}{k!} \sum_{\sigma \in S_k} (\text{sgn } \sigma)^2 T = T,$$

where we used the fact that S_k has $k!$ elements.

(c) Since we know from (a) that $\text{Alt}(T)$ is an alternating tensor we conclude from (b) that $\text{Alt}(\text{Alt}(T)) = \text{Alt}(T)$.

(4) Let $\{T_1, \ldots, T_n\}$ be a basis of V^*. We have

$$T = \sum_{i_1 < \ldots < i_k} a_{i_1 \ldots i_k} T_{i_1} \wedge \ldots \wedge T_{i_k}$$

and
$$S = \sum_{j_1 < \ldots < j_m} b_{j_1 \ldots j_m} T_{j_1} \wedge \ldots \wedge T_{j_m}.$$

Therefore

$$T \wedge S = \sum_{\substack{i_1 < \ldots < i_k \\ j_1 < \ldots < j_m}} a_{i_1 \ldots i_k} b_{j_1 \ldots j_m} T_{i_1} \wedge \ldots \wedge T_{i_k} \wedge T_{j_1} \wedge \ldots \wedge T_{j_m}$$

$$= \sum_{\substack{i_1 < \ldots < i_k \\ j_1 < \ldots < j_m}} a_{i_1 \ldots i_k} b_{j_1 \ldots j_m} (-1)^k T_{j_1} \wedge T_{i_1} \wedge \ldots \wedge T_{i_k} \wedge T_{j_2} \wedge \ldots \wedge T_{j_m}$$

$$= \sum_{\substack{i_1 < \ldots < i_k \\ j_1 < \ldots < j_m}} a_{i_1 \ldots i_k} b_{j_1 \ldots j_m} (-1)^{km} T_{j_1} \wedge \ldots \wedge T_{j_m} \wedge T_{i_1} \wedge \ldots \wedge T_{i_k}$$

$$= (-1)^{km} S \wedge T.$$

(5) (1) If $v_1, \ldots, v_{k+m} \in V$ then

$$\big(F^*(T \otimes S)\big)(v_1, \ldots, v_{k+m}) = (T \otimes S)(F(v_1), \ldots, F(v_{k+m}))$$
$$= T(F(v_1), \ldots, F(v_k)) S(F(v_{k+1}), \ldots, F(v_{k+m}))$$
$$= (F^*T(v_1, \ldots, v_k))(F^*S(v_{k+1}, \ldots, v_{k+m}))$$
$$= ((F^*T) \otimes (F^*S))(v_1, \ldots, v_{k+m}).$$

(2) Obvious.

(3) If $v_1, \ldots, v_{k+m} \in V$ then

$$\big(F^*(T \wedge S)\big)(v_1, \ldots, v_{k+m}) = (T \wedge S)(F(v_1), \ldots, F(v_{k+m}))$$
$$= \frac{(k+m)!}{k!m!} (\mathrm{Alt}(T \otimes S))(F(v_1), \ldots, F(v_{k+m}))$$
$$= \frac{1}{k!m!} \sum_{\sigma \in S_{k+m}} (\mathrm{sgn}\,\sigma)(T \otimes S)(F(v_{\sigma(1)}), \ldots, F(v_{\sigma(k+m)}))$$
$$= \frac{1}{k!m!} \sum_{\sigma \in S_{k+m}} (\mathrm{sgn}\,\sigma)(F^*(T \otimes S))(v_{\sigma(1)}, \ldots, v_{\sigma(k+m)})$$
$$= \frac{1}{k!m!} \sum_{\sigma \in S_{k+m}} (\mathrm{sgn}\,\sigma)((F^*T) \otimes (F^*S))(v_{\sigma(1)}, \ldots, v_{\sigma(k+m)})$$
$$= \frac{(k+m)!}{k!m!} (\mathrm{Alt}((F^*T) \otimes (F^*S)))(v_1, \ldots, v_{k+m})$$
$$= ((F^*T) \wedge (F^*S))(v_1, \ldots, v_{k+m}).$$

(4) If $v_1, \ldots, v_k \in V$ then

$$((H \circ F)^*(T))(v_1, \ldots, v_k) = T(H(F(v_1)), \ldots, H(F(v_k)))$$
$$= (H^*T)(F(v_1), \ldots, F(v_k)) = (F^*(H^*T))(v_1, \ldots, v_k).$$

7.2.2 Section 2.2

(3) (a) We have

$$L_X(T(Y_1, \ldots, Y_k))(p) = \frac{d}{dt}(T(Y_1, \ldots, Y_k)(\psi_t(p)))|_{t=0}$$
$$= \frac{d}{dt}\left(T_{\psi_t(p)}\left((Y_1)_{\psi_t(p)}, \ldots, (Y_k)_{\psi_t(p)}\right)\right)|_{t=0}$$
$$= \frac{d}{dt}\left((\psi_t^*T)_p\left((d\psi_{-t})_{\psi_t(p)}(Y_1)_{\psi_t(p)}, \ldots, (d\psi_{-t})_{\psi_t(p)}(Y_k)_{\psi_t(p)}\right)\right)|_{t=0}$$
$$= (L_XT)(Y_1, \ldots, Y_k)(p) + T(L_XY_1, \ldots, Y_k)(p) + \ldots + T(Y_1, \ldots, L_XY_k)(p).$$

(b) A possible definition is to set

$$(L_XT)(Y_1, \ldots, Y_k, \omega^1, \ldots, \omega^k) = L_X(T(Y_1, \ldots, Y_k, \omega^1, \ldots, \omega^k))$$
$$- T(L_XY_1, \ldots, Y_k, \omega^1, \ldots, \omega^k) - \cdots - T(Y_1, \ldots, L_XY_k, \omega^1, \ldots, \omega^k)$$
$$- T(Y_1, \ldots, Y_k, L_X\omega^1, \ldots, \omega^k) - \cdots - T(Y_1, \ldots, Y_k, \omega^1, \ldots, L_X\omega^k)$$

for all vector fields $Y_1, \ldots, Y_k$ and all 1-tensor fields $\omega^1, \ldots, \omega^k$.

7.2.3 Section 2.3

(1) Property (i) is trivially true, and property (iii) is an immediate consequence of Proposition 1.12. To prove (ii), we notice that given $p \in M$ and $v_1, \ldots, v_k \in T_pM$ we have

$$(f^*(g\alpha))_p(v_1, \ldots, v_k) = (g\alpha)_{f(p)}((df)_pv_1, \ldots, (df)_pv_k)$$
$$= g(f(p))\alpha_{f(p)}((df)_pv_1, \ldots, (df)_pv_k)$$
$$= (g \circ f)(p)(f^*\alpha)_p(v_1, \ldots, v_k) = ((f^*g)(f^*\alpha))_p(v_1, \ldots, v_k).$$

Finally, (iv) follows from

$$(g^*f^*\alpha)_p(v_1, \ldots, v_k) = (f^*\alpha)_{g(p)}((dg)_pv_1, \ldots, (dg)_pv_k)$$
$$= \alpha_{f(g(p))}((df)_{g(p)}(dg)_pv_1, \ldots, (df)_{g(p)}(dg)_pv_k)$$

$$= \alpha_{(f \circ g)(p)}((d(f \circ g))_p v_1, \ldots, (d(f \circ g))_p v_k)$$
$$= ((f \circ g)^* \alpha)_p (v_1, \ldots, v_k).$$

(3) Let M be a smooth manifold.

(i) $d(\omega_1 + \omega_2)$ is the form locally represented by $d(\omega_1 + \omega_2)_\alpha$ for each parameterization $\varphi_\alpha : U_\alpha \to M$. Moreover, since

$$(\omega_1 + \omega_2)_\alpha = \varphi_\alpha^*(\omega_1 + \omega_2) = \varphi_\alpha^* \omega_1 + \varphi_\alpha^* \omega_2 = (\omega_1)_\alpha + (\omega_2)_\alpha$$

we have, by Proposition 3.7(i) that

$$d(\omega_1 + \omega_2)_\alpha = d((\omega_1)_\alpha + (\omega_2)_\alpha) = d(\omega_1)_\alpha + d(\omega_2)_\alpha$$
$$= \varphi_\alpha^* d\omega_1 + \varphi_\alpha^* d\omega_2 = \varphi_\alpha^* (d\omega_1 + d\omega_2)$$

and we conclude that $d(\omega_1 + \omega_2) = d\omega_1 + d\omega_2$.

(ii) Let ω be a k-form. Then, $d(\omega \wedge \gamma)$ is the form locally represented by $d(\omega \wedge \gamma)_\alpha$ for each parameterization $\varphi_\alpha : U_\alpha \to M$. Moreover, by Proposition 3.3(iii)

$$(\omega \wedge \gamma)_\alpha = \varphi_\alpha^*(\omega \wedge \gamma) = \varphi_\alpha^* \omega \wedge \varphi_\alpha^* \gamma = \omega_\alpha \wedge \gamma_\alpha.$$

Hence, by Proposition 3.7(ii),

$$d(\omega \wedge \gamma)_\alpha = d(\omega_\alpha \wedge \gamma_\alpha) = d\omega_\alpha \wedge \gamma_\alpha + (-1)^k \omega_\alpha \wedge d\gamma_\alpha$$
$$= \varphi_\alpha^*(d\omega) \wedge \varphi_\alpha^* \gamma + (-1)^k \varphi_\alpha^* \omega \wedge \varphi_\alpha^*(d\gamma)$$
$$= \varphi_\alpha^*(d\omega \wedge \gamma + (-1)^k \omega \wedge d\gamma) = (d\omega \wedge \gamma + (-1)^k \omega \wedge d\gamma)_\alpha$$

and the result follows.

(iii) The form $d(d\omega)$ is locally represented by $d(d\omega)_\alpha = d(d\omega_\alpha)$ for each parameterization $\varphi_\alpha : U_\alpha \to M$. Since $d(d\omega_\alpha) = 0$ by Proposition 3.7(iii), we conclude that $d(d\omega) = 0$.

(iv) Let $f : M \to N$ be a smooth map between two manifolds. Let ω be a form on N. Then $d(f^* \omega)$ is the form locally represented by $d(f^* \omega)_\alpha$ for each parameterization $\varphi_\alpha : U_\alpha \to M$. Moreover, by Proposition 3.3(iv),

$$(f^* \omega)_\alpha = \varphi_\alpha^*(f^* \omega) = (f \circ \varphi_\alpha)^* \omega,$$

and so

$$d(f^* \omega)_\alpha = d((f \circ \varphi_\alpha)^* \omega).$$

On the other hand, $f^*(d\omega)$ is the form locally represented by

$$\varphi_\alpha^*(f^*(d\omega)) = (f \circ \varphi_\alpha)^*(d\omega)$$

for each parameterization $\varphi_\alpha : U_\alpha \to M$. Now, if $\psi_\beta : V \to N$ is a parameterization of N then we have on $\psi_\beta(V)$

$$d\omega = (\psi_\beta^{-1})^* d\omega_\beta = (\psi_\beta^{-1})^* d(\psi_\beta^* \omega),$$

and so by Proposition 3.7

$$\begin{aligned}
\varphi_\alpha^*(f^*(d\omega)) &= (f \circ \varphi_\alpha)^* (\psi_\beta^{-1})^* d\omega_\beta = (\psi_\beta^{-1} \circ f \circ \varphi_\alpha)^* d\omega_\beta \\
&= d((\psi_\beta^{-1} \circ f \circ \varphi_\alpha)^* \omega_\beta) = d((f \circ \varphi_\alpha)^* (\psi_\beta^{-1})^* \omega_\beta) \\
&= d((f \circ \varphi_\alpha)^* \omega)
\end{aligned}$$

on $\varphi_\alpha^{-1}(f^{-1}(\psi_\beta(V)))$. Since $\psi_\beta : V \to N$ is arbitrary, the result follows.

(6) (a) We have

$$\tilde{f}^* \omega = dt \wedge \sum_I (a_I \circ \tilde{f}) df^I + \sum_J (b_J \circ \tilde{f}) df^J,$$

and hence

$$Q(\tilde{f}^* \omega) = \sum_I \left(\int_{t_0}^t (a_I \circ \tilde{f}) ds \right) df^I.$$

On the other hand, since $\tilde{f}$ is the identity on the first coordinate, precomposition with $\tilde{f}$ commutes with integration with respect to t. Therefore

$$\begin{aligned}
\tilde{f}^*(Q(\omega)) &= \sum_I \left(\left(\int_{t_0}^t a_I \, ds \right) \circ \tilde{f} \right) df^I \\
&= \sum_I \left(\int_{t_0}^t (a_I \circ \tilde{f}) \, ds \right) df^I = Q(\tilde{f}^* \omega).
\end{aligned}$$

(b) Each parameterization $\varphi : U \to M$ yields a parameterization $\tilde{\varphi} : \mathbb{R} \times U \to \mathbb{R} \times M$ through $\tilde{\varphi} = \mathrm{id} \times \varphi$. Given a k-form $\omega \in \Omega^k(\mathbb{R} \times M)$ we then define $Q(\omega) \in \Omega^{k-1}(\mathbb{R} \times M)$ as the $(k-1)$-form whose local representation associated to the parameterization $\tilde{\varphi}_\alpha$ is $(Q(\omega))_\alpha = Q(\omega_\alpha)$. e $Q(\omega) = (\tilde{\varphi}_\alpha^{-1})^* (Q(\omega_\alpha))$ on $\mathbb{R} \times \varphi_\alpha(U_\alpha)$. To check that this definition is consistent consider another parameterization $\varphi_\beta : U_\beta \to M$ such that $W := \varphi_\alpha(U_\alpha) \cap \varphi_\beta(U_\beta) \neq \varnothing$. Let $f : \varphi_\alpha^{-1}(W) \to \varphi_\beta^{-1}(W)$ be the diffeomorphism given by $f = \varphi_\beta^{-1} \circ \varphi_\alpha$. Then on $\mathbb{R} \times W$ we have

$$\left(\widetilde{\varphi}_\alpha^{-1}\right)^*(Q(\omega_\alpha)) = \left(\widetilde{\varphi}_\alpha^{-1}\right)^*\left(Q\left(\widetilde{f}^*\omega_\beta\right)\right) = \left(\widetilde{\varphi}_\alpha^{-1}\right)^*\left(\widetilde{f}^*(Q(\omega_\beta))\right)$$
$$= \left(\widetilde{f}\circ\widetilde{\varphi}_\alpha^{-1}\right)^*(Q(\omega_\beta)) = \left(\widetilde{\varphi}_\beta^{-1}\right)^*(Q(\omega_\beta)).$$

To see that $\widetilde{f}^*\circ Q = Q\circ\widetilde{f}^*$ at a given point $(t,p)\in\mathbb{R}\times M$ consider parameterizations $\varphi:U\to M$ and $\psi:V\to N$ around p and $f(p)$, and let $\hat{f} = \psi^{-1}\circ f\circ\varphi$ be the corresponding local representation of f. Then at p we have

$$\widetilde{f}^*(Q(\omega)) = \widetilde{f}^*\left(\widetilde{\psi}^{-1}\right)^*\left(Q\left(\widetilde{\psi}^*\omega\right)\right) = \left(\widetilde{\psi}^{-1}\circ\widetilde{f}\right)^*\left(Q\left(\widetilde{\psi}^*\omega\right)\right)$$
$$= \left(\widetilde{\hat{f}}\circ\widetilde{\varphi}^{-1}\right)^*\left(Q\left(\widetilde{\psi}^*\omega\right)\right) = \left(\widetilde{\varphi}^{-1}\right)^*\widetilde{\hat{f}}^*\left(Q\left(\widetilde{\psi}^*\omega\right)\right)$$
$$= \left(\widetilde{\varphi}^{-1}\right)^*\left(Q\left(\widetilde{\hat{f}}^*\left(\widetilde{\psi}^*\omega\right)\right)\right) = \left(\widetilde{\varphi}^{-1}\right)^*\left(Q\left(\left(\widetilde{\psi}\circ\widetilde{\hat{f}}\right)^*\omega\right)\right)$$
$$= \left(\widetilde{\varphi}^{-1}\right)^*\left(Q\left(\left(\widetilde{f}\circ\widetilde{\varphi}\right)^*\omega\right)\right) = \left(\widetilde{\varphi}^{-1}\right)^*\left(Q\left(\widetilde{\varphi}^*\widetilde{f}^*\omega\right)\right)$$
$$= Q\left(\widetilde{f}^*\omega\right).$$

The linearity of Q on M is similarly obtained from the linearity of Q on $\mathbb{R}^n$.
(c) Notice that in local coordinates the formula for Q reduces to the formula in $\mathbb{R}^n$. If

$$\omega = dt\wedge\sum_I a_I(t,x)\,dx^I + \sum_J b_J(t,x)\,dx^J$$

we have

$$i_{t_0}^*\omega = \sum_J b_J(t_0,x)\,dx^J$$

($dt_0 = 0$ as t_0 is constant), and $\pi^*i_{t_0}^*\omega$ is given by the same expression. Consequently

$$\omega - \pi^*i_{t_0}^*\omega = dt\wedge\sum_I a_I(t,x)\,dx^I + \sum_J (b_J(t,x) - b_J(t_0,x))\,dx^J.$$

On the other hand,

$$d(Q(\omega)) = dt\wedge\sum_I a_I(t,x)\,dx^I + \sum_I\sum_i\left(\int_{t_0}^t\frac{\partial a_I}{\partial x^i}\,ds\right)dx^i\wedge dx^I$$

and

$$d\omega = -dt \wedge \sum_I \sum_i \frac{\partial a_I}{\partial x^i} dx^i \wedge dx^I$$

$$+ dt \wedge \sum_J \frac{\partial b_J}{\partial t} dx^J + \sum_J \sum_i \frac{\partial b_J}{\partial x^i} dx^i \wedge dx^J,$$

leading to

$$\mathcal{Q}(d\omega) = -\sum_I \sum_i \left(\int_{t_0}^t \frac{\partial a_I}{\partial x^i} ds \right) dx^i \wedge dx^I + \sum_J \left(\int_{t_0}^t \frac{\partial b_J}{\partial s} ds \right) dx^J.$$

Consequently,

$$d(\mathcal{Q}(\omega)) + \mathcal{Q}(d\omega) = dt \wedge \sum_I a_I(t, x) dx^I + \sum_J (b(t, x) - b(t_0, x)) dx^J$$

$$= \omega - \pi^* i_{t_0}^* \omega.$$

(d) Since $\pi \circ i_{t_0} = \mathrm{id}_M$, we have $i_{t_0}^* \circ \pi^* = \mathrm{id}$ and hence $i_{t_0}^\sharp \circ \pi^\sharp = \mathrm{id}$. On the other hand, if ω is closed then

$$\omega - \pi^* i_{t_0}^* \omega = d(\mathcal{Q}(\omega)),$$

meaning that ω and $\pi^* i_{t_0}^* \omega$ are in the same cohomology class. Therefore $\pi^\sharp \circ i_{t_0}^\sharp = \mathrm{id}$.

(e) By (d), $H^k(\mathbb{R}^n) = H^k(\mathbb{R}^{n-1}) = \ldots = H^k(\mathbb{R})$. Now we know that $H^k(\mathbb{R}) = 0$ for all $k > 1$. On the other hand, any 1-form on $\mathbb{R}$ is necessarily exact, since

$$a(t) dt = d \left(\int_0^t a(s) ds \right).$$

We conclude that $H^1(\mathbb{R}) = 0$. Therefore $H^k(\mathbb{R}^n) = 0$ for all $k > 0$.

(f) We have $f = H \circ i_{t_0}$ and $g = H \circ i_{t_1}$, and therefore $f^\sharp = i_{t_0}^\sharp \circ H^\sharp$ and $g^\sharp = i_{t_1}^\sharp \circ H^\sharp$. Now $i_{t_0}^\sharp = i_{t_1}^\sharp$, as they are both the inverse map to $\pi^\sharp$. Consequently $f^\sharp = g^\sharp$.

(g) The map $H : \mathbb{R} \times \mathbb{R}^n \to \mathbb{R}^n$ given by $H(t, x) = (1 - t)x$ is clearly a homotopy between the identity map $f(x) = H(0, x) = x$ and the constant map $g(x) = H(1, x) = 0$.

(h) Since M is contractible, the identity map $\mathrm{id} : M \to M$ is smoothly homotopic to a constant map $g : M \to M$, and hence $\mathrm{id}^\sharp = g^\sharp$. Using local coordinates, it is immediate to check that if $\omega \in \Omega^k(M)$ is a k-form with

$k > 0$ then $g^*\omega = 0$, implying that $g^\sharp : H^k(M) \to H^k(M)$ is the zero map. But $g^\sharp = \mathrm{id}^\sharp$ is the identity map on $H^k(M)$, and hence we must have $H^k(M) = 0$.

(7) (a) We have

$$
\begin{aligned}
L_X(\omega_1 \wedge \omega_2) &= \frac{d}{dt}(\psi_t{}^*(\omega_1 \wedge \omega_2))\Big|_{t=0} = \frac{d}{dt}((\psi_t{}^*\omega_1) \wedge (\psi_t{}^*\omega_2))\Big|_{t=0} \\
&= \left(\frac{d}{dt}(\psi_t{}^*\omega_1)\Big|_{t=0}\right) \wedge \omega_2 + \omega_1 \wedge \left(\frac{d}{dt}(\psi_t{}^*\omega_2)\Big|_{t=0}\right) \\
&= (L_X\omega_1) \wedge \omega_2 + \omega_1 \wedge (L_X\omega_2)
\end{aligned}
$$

(where we have used the fact that the wedge product is bilinear).

(b) Similarly,

$$
\begin{aligned}
L_X(d\omega) &= \frac{d}{dt}(\psi_t{}^*(d\omega))\Big|_{t=0} = \frac{d}{dt}(d(\psi_t{}^*\omega))\Big|_{t=0} \\
&= d\left(\frac{d}{dt}(\psi_t{}^*\omega)\Big|_{t=0}\right) = d(L_X\omega)
\end{aligned}
$$

(where we have used the fact that the exterior derivative is linear).

(c) We will prove this formula by induction. We start with the case when $\omega = f\,dg$ for smooth functions $f, g \in \Omega^0(M)$. In this case we can use the properties proved above to obtain

$$
L_X\omega = (L_X f)dg + f\,d(L_X g) = (X \cdot f)dg + f\,d(X \cdot g).
$$

On the other hand,

$$
\begin{aligned}
\iota(X)d\omega + d(\iota(X)\omega) &= \iota(X)(df \wedge dg) + d(f(X \cdot g)) \\
&= \iota(X)(df \otimes dg - dg \otimes df) + (X \cdot g)df + f\,d(X \cdot g) \\
&= (X \cdot f)dg - (X \cdot g)df + (X \cdot g)df + f\,d(X \cdot g) \\
&= L_X\omega.
\end{aligned}
$$

Next we prove that if the Cartan formula holds for ω and η then it holds for $\omega \wedge \eta$. Since locally any form can be obtained by taking wedge products of 1-forms of the type $f\,dg$, this will complete the proof. If ω is a k-form, we have

$$
\begin{aligned}
&\iota(X)d(\omega \wedge \eta) + d(\iota(X)(\omega \wedge \eta)) \\
&= \iota(X)(d\omega \wedge \eta + (-1)^k \omega \wedge d\eta) + d((\iota(X)\omega) \wedge \eta + (-1)^k \omega \wedge (\iota(X)\eta)) \\
&= (\iota(X)d\omega) \wedge \eta + (-1)^{k+1}d\omega \wedge (\iota(X)\eta) + (-1)^k(\iota(X)\omega) \wedge d\eta + \omega \wedge (\iota(X)d\eta) \\
&\quad + d(\iota(X)\omega) \wedge \eta + (-1)^{k-1}(\iota(X)\omega) \wedge d\eta + (-1)^k dw \wedge (\iota(X)\eta) + \omega \wedge d(\iota(X)\eta)
\end{aligned}
$$

$$= (\iota(X)d\omega) \wedge \eta + \omega \wedge (\iota(X)d\eta) + d(\iota(X)\omega) \wedge \eta + \omega \wedge d(\iota(X)\eta)$$
$$= (L_X\omega) \wedge \eta + \omega \wedge (L_X\eta) = L_X(\omega \wedge \eta)$$

(where we have used Exercise 1.15(8)).
(d) We begin by noticing that

$$\left(\psi_t^*(\iota(Y)\omega)\right)_p (v_1, \ldots, v_k) = (\iota(Y)\omega)_{\psi_t(p)}((d\psi_t)_p v_1, \ldots, (d\psi_t)_p v_k)$$
$$= \omega_{\psi_t(p)}(Y_{\psi_t(p)}, (d\psi_t)_p v_1, \ldots, (d\psi_t)_p v_k)$$
$$= \omega_{\psi_t(p)}((d\psi_t)_p (d\psi_{-t})_{\psi_t(p)} Y_{\psi_t(p)}, (d\psi_t)_p v_1, \ldots, (d\psi_t)_p v_k)$$
$$= \left(\psi_t^*\omega\right)_p ((d\psi_{-t})_{\psi_t(p)} Y_{\psi_t(p)}, v_1, \ldots, v_k)$$

i.e.

$$\psi_t^*(\iota(Y)\omega) = \iota((\psi_{-t})_* Y)(\psi_t^*\omega).$$

Taking the derivative with respect to t at $t = 0$ and using the fact that the contraction is a bilinear operation yields the result.

7.2.4 Section 2.4

(3) T^2 divides S^3 into two connected components

$$M_1 := \{(x, y, z, w) \in S^3 \mid x^2 + y^2 < 1\}$$

and

$$M_2 := \{(x, y, z, w) \in S^3 \mid x^2 + y^2 > 1\}.$$

Let us consider the parameterizations $\varphi_i : U_i \to M_i$ $(i = 1, 2)$ defined by

$$\varphi_i(r, v, v) = (r \cos v, r \sin v, \sqrt{2 - r^2} \cos v, \sqrt{2 - r^2} \sin v)$$

on $U_1 := (0, 1) \times (0, 2\pi) \times (0, 2\pi)$ and $U_2 := (1, \sqrt{2}) \times (0, 2\pi) \times (0, 2\pi)$.
Then,

$$\varphi_i^* dx = \cos v \, dr - r \sin v \, dv$$
$$\varphi_i^* dy = \sin v \, dr + r \cos v \, dv$$
$$\varphi_i^* dz = -\frac{r}{\sqrt{2 - r^2}} \cos v \, dr - \sqrt{2 - r^2} \sin v \, dv$$
$$\varphi_i^* dw = -\frac{r}{\sqrt{2 - r^2}} \sin v \, dr + \sqrt{2 - r^2} \cos v \, dv$$

and so $\varphi_i^*\omega = (r(2-r^2)\cos^2 v - r^3 \cos^2 v)\, dr \wedge dv \wedge dv$. Hence, since the sets $M_1\backslash\varphi_1(U_1)$ and $M_2\backslash\varphi_2(U_2)$ have measure zero we have

$$\int_{M_1} \omega = \pm \int_{U_1} \varphi_1^*\omega = \pm \int_0^1 \int_0^{2\pi} \int_0^{2\pi} (r(2-r^2)\cos^2 v - r^3 \cos^2 v)\, dv\, dv\, dr = \pm\pi^2$$

$$\int_{M_2} \omega = \pm \int_{U_2} \varphi_2^*\omega = \pm \int_1^{\sqrt{2}} \int_0^{2\pi} \int_0^{2\pi} (r(2-r^2)\cos^2 v - r^3 \cos^2 v)\, dv\, dv\, dr = \mp\pi^2,$$

where the $\pm$ signs depend on the choice of orientation.

(4) Let us consider an atlas $\{(U_\alpha, \varphi_\alpha)\}$ of orientation-preserving parameterizations on M. Then, since f is an orientation-preserving diffeomorphism $\{(U_\alpha, f \circ \varphi_\alpha)\}$ is an atlas of orientation-preserving parameterizations on N. Let $\{\rho_i\}_{i \in I}$ be a partition of unity subordinate to the cover $\{\widetilde{W}_\alpha\}$ of N where $\widetilde{W}_\alpha := (f \circ \varphi_\alpha)(U_\alpha)$. Since ω is compactly supported we can assume that I is finite. Hence, considering the forms $\omega_i := \rho_i \omega$, we have

$$\int_N \omega := \sum_{i \in I} \int_N \omega_i = \sum_{i \in I} \int_{U_{\alpha_i}} (f \circ \varphi_{\alpha_i})^*\omega_i.$$

Now

$$(f \circ \varphi_{\alpha_i})^*\omega_i = (f \circ \varphi_{\alpha_i})^*(\rho_i \omega) = ((f \circ \varphi_{\alpha_i})^*\rho_i)((f \circ \varphi_{\alpha_i})^*\omega)$$
$$= (\rho_i \circ f \circ \varphi_{\alpha_i})(\varphi_{\alpha_i}^*(f^*\omega)) = \varphi_{\alpha_i}^*((\rho_i \circ f)f^*\omega).$$

Moreover, $\{\rho_i \circ f\}_{i \in I}$ is clearly a partition of unity subordinate to the cover $\{W_\alpha\}$ of M where $W_\alpha := \varphi_\alpha(U_\alpha)$. Hence,

$$\int_N \omega = \sum_{i \in I} \int_{U_{\alpha_i}} (f \circ \varphi_{\alpha_i})^*\omega_i = \sum_{i \in I} \int_{U_{\alpha_i}} \varphi_{\alpha_i}^*((\rho_i \circ f)f^*\omega)$$

$$= \sum_{i \in I} \int_{U_{\alpha_i}} \varphi_{\alpha_i}^*(f^*\omega)_i = \int_M f^*\omega.$$

7.2.5 Section 2.5

(1) First we note that the form ω is exact. Indeed, $\omega = d\alpha$ with $\alpha = xz\, dy \wedge dw$. Then, denoting by M_1 and M_2 the two connected components of $S^3\backslash T^2$, we have $T^2 = \partial M_1 = \partial M_2$ and so, by the Stokes theorem,

$$\int_{M_1} \omega = \int_{M_1} d\alpha = \int_{T^2} i_1^* \alpha,$$

where $i_1 : \partial M_1 \to M_1$ is the inclusion map and T^2 is equipped with the induced orientation. Similarly,

$$\int_{M_2} \omega = \int_{M_2} d\alpha = \int_{T^2} i_2^* \alpha,$$

where $i_2 : \partial M_2 \to M_2$ is the inclusion map and T^2 is equipped with the induced orientation. Note that this orientation on T^2 is the opposite of the one induced by the orientation of M_1.

To compute these integrals we consider the parameterization $\varphi : U \to T^2$ defined by

$$\varphi(v, v) = (\cos v, \sin v, \cos v, \sin v)$$

on $U := (0, 2\pi) \times (0, 2\pi)$. Then,

$$\varphi^*(i_j^* dx) = -\sin v \, dv, \qquad \varphi^*(i_j^* dy) = \cos v \, dv,$$
$$\varphi^*(i_j^* dz) = -\sin v \, dv, \qquad \varphi^*(i_j^* dw) = \cos v \, dv,$$

($j = 1, 2$), and so $\varphi^*(i_j^* \alpha) = \cos^2 v \cos^2 v \, dv \wedge dv$. Hence, since the set $T^2 \backslash \varphi(U)$ has measure zero, we have

$$\int_{T_2} i_1^* \alpha = \pm \int_U \varphi^*(i_1^* \alpha) = \pm \int_0^{2\pi} \int_0^{2\pi} \cos^2 v \cos^2 v \, dv dv = \pm \pi^2$$

$$\int_{T^2} i_2^* \alpha = \mp \int_U \varphi^*(i_2^* \alpha) = \mp \int_0^{2\pi} \int_0^{2\pi} \cos^2 v \cos^2 v \, dv dv = \mp \pi^2,$$

where the $\pm$ signs depend on the choice of orientation on S^3. Note that the sum of the two integrals is zero. This is not surprising since

$$\int_{M_1} \omega + \int_{M_2} \omega = \int_{S^3} \omega = \int_{S^3} d\alpha = \int_{\partial S^3} i^* \alpha = 0$$

as $\partial S^3 = \varnothing$ (here we used the fact that T^2 has measure zero in S^3).

(3) (a) Under this identification, we have

$$\langle p, X_p \rangle = 0$$

for all $p \in S^n$, where $\langle \cdot, \cdot \rangle$ is the Euclidean inner product in $\mathbb{R}^{n+1}$. Therefore,

$$\|H(t, p)\|^2 = \cos^2(\pi t)\|p\|^2 + \sin^2(\pi t) = 1,$$

and H indeed maps $\mathbb{R} \times S^n$ to S^n. Moreover, H is clearly smooth and $H(0, p) = p$ and $H(1, p) = -p$ for all $p \in S^n$. Therefore H is a smooth homotopy between the identity map and the antipodal map. Geometrically, for fixed $p \in S^n$ the curve $H(t, p)$ traverses half of the great circle tangent to X_p at p.

(b) We have
$$d\omega = (n + 1)dx^1 \wedge \cdots \wedge dx^{n+1}.$$

If $B^{n+1} := \{x \in \mathbb{R}^{n+1} \mid \|x\| \leq 1\}$, we have by the Stokes theorem that

$$\int_{S^n} \omega = \int_{B^{n+1}} d\omega = (n + 1) \int_{B^{n+1}} dx^1 \wedge \cdots \wedge dx^{n+1}$$

$$= (n + 1) \int_{B^{n+1}} dx^1 \cdots dx^{n+1} > 0.$$

(c) Suppose that X exists. Then the antipodal map $f : S^n \to S^n$ is homotopic to the identity map. Now it is very easy to check that $f^*\omega = (-1)^{n+1}\omega$. Since ω is closed in S^n (it is an n-form), we have by Exercise 5.3(2). that

$$\int_{S^n} \omega = \int_{S^n} f^*\omega = (-1)^{n+1} \int_{S^n} \omega.$$

As $\int_{S^n} \omega > 0$, we must have $(-1)^{n+1} = 1$, and so n must be odd. For odd n there exist vector fields $X \in \mathfrak{X}(S^n)$ with no zeros. An example is the vector field given by

$$X_{(x^1,\ldots,x^{n+1})} = (-x^2, x^1, \ldots, -x^{n+1}, x^n),$$

which is indeed tangent to S^n ($\langle x, X_x \rangle = 0$) and does not vanish on S^n ($\|X_x\| = \|x\| = 1$).

7.2.6 Section 2.6

(4) (a) If $\omega \in \Omega^k(S^n)$ satisfies $\omega = \pi^*\theta$ for some $\theta \in \Omega^k(\mathbb{R}P^n)$ then

$$f^*\omega = f^*(\pi^*\theta) = (\pi \circ f)^*\theta = \pi^*\theta = \omega.$$

Conversely, assume that $\omega \in \Omega^k(S^n)$ satisfies $f^*\omega = \omega$. Then for each open set $U \subset S^n$ such that $\pi_{|_U}$ is a diffeomorphism we define $\theta_U \in \Omega^k(\pi(U))$ as $\theta_U := (\pi_{|_U}^{-1})^*\omega$. We now show that if $p \in \pi(U) \cap \pi(V)$, for some other open set $V \subset S^n$ such that $\pi_{|_V}$ is a diffeomorphism, then $(\theta_U)_p = (\theta_V)_p$. Indeed, we have $\pi^{-1}(p) = \{p_1, p_2\}$ for some $p_1, p_2 \in S^2$ with $p_2 = f(p_1)$. Assume, without loss of generality, that $p_1 \in U$. Then either $p_1 \in V$ or $p_2 \in V$. If $p_1 \in V$ then $\pi_{|_U}$ and $\pi_{|_V}$ agree on a neighborhood of p_1, and so do θ_U and θ_V on the image of this neighborhood (which contains p). If $p_2 \in V$, then $\pi_{|_U} = \pi_{|_V} \circ f$ on a neighborhood of p_1, and hence

$$\pi_{|_U}^{-1} = f^{-1} \circ \pi_{|_V}^{-1} = f \circ \pi_{|_V}^{-1}$$

on this neighborhood. Therefore,

$$\theta_U = (f \circ \pi_{|_V}^{-1})^*\omega = (\pi_{|_V}^{-1})^* f^*\omega = (\pi_{|_V}^{-1})^*\omega = \theta_V$$

on the image of this neighborhood (and in particular at p). We conclude that $\theta_U = \theta_V$ on $\pi(U) \cap \pi(V)$, which shows that there exists $\theta \in \Omega^k(\mathbb{R}P^n)$ such that $\theta_U = \theta_{|_{\pi(U)}}$ for each U. Finally, since $\omega_{|_U} = \pi_{|_U}^*\theta_U$ for each U, we have $\omega = \pi^*\theta$.

(b) It is easy to check that

$$\omega = \sum_{i=1}^{n+1} (-1)^{i+1} x^i dx^1 \wedge \cdots \wedge dx^{i-1} \wedge dx^{i+1} \wedge \cdots \wedge dx^{n+1}$$

is a volume form for S^n [cf. Exercise 5.3(3)]. Indeed, if $v_1, \ldots, v_n \in \mathbb{R}^{n+1}$ are n linearly independent vectors tangent to S^n at x then

$$\omega(v_1, \ldots, v_n) = \sum_{i=1}^{n+1} (-1)^{i+1} x^i \begin{vmatrix} v_1^1 & \cdots & v_1^{i-1} & v_1^{i+1} & \cdots & v_1^{n+1} \\ \cdots & \cdots & \cdots & \cdots & \cdots & \cdots \\ v_n^1 & \cdots & v_n^{i-1} & v_n^{i+1} & \cdots & v_n^{n+1} \end{vmatrix}$$

$$= \begin{vmatrix} x^1 & \cdots & x^{n+1} \\ v_1^1 & \cdots & v_1^{n+1} \\ \cdots & \cdots & \cdots \\ v_n^1 & \cdots & v_n^{n+1} \end{vmatrix} \neq 0,$$

as $\{x, v_1, \ldots, v_n\}$ is a basis of $\mathbb{R}^{n+1}$. Since

$$f^*\omega = (-1)^{n+1}\omega,$$

we see that for odd n we have $\omega = \pi^*\theta$ for some $\theta \in \Omega^n(\mathbb{R}P^n)$, which must be a volume form (as π is a local diffeomorphism). Therefore $\mathbb{R}P^n$ is orientable for odd n. Assume now that n is even. If $\mathbb{R}P^n$ was orientable, there would exist a volume form $\theta \in \Omega^n(\mathbb{R}P^n)$. Then $\widetilde{\omega} := \pi^*\theta$ would be a volume form for S^n satisfying $f^*\widetilde{\omega} = \widetilde{\omega}$. Since ω is also a volume form for S^n, we would have $\widetilde{\omega} = g\omega$ for a nonvanishing function $g \in C^\infty(S^n)$. But, since $f^*\omega = -\omega$, this function would have to satisfy $g \circ f = -g$, i.e. it would have to assume opposite signs at antipodal points, and hence would have to vanish at some point. We conclude that $\mathbb{R}P^n$ is not orientable for even n. Finally, let n be odd and let $U \subset S^n$ be an open hemisphere. Then $\pi|_U$ is a diffeomorphism and $\mathbb{R}P^n \backslash \pi(U)$ has zero measure. Therefore

$$\int_{S^n} \pi^*\theta = \int_U \pi^*\theta + \int_{f(U)} \pi^*\theta = \int_U \pi^*\theta + \int_U f^*\pi^*\theta$$

$$= 2\int_U \pi^*\theta = 2\int_{\mathbb{R}P^n} \theta.$$

(c) Consider the orientation on S^n defined by the volume form ω of (b). If $\widetilde{\mathbb{R}P^n}$ is the orientable double covering of $\mathbb{R}P^n$ [cf. Exercise 8.6(9) in Chap. 1], we define $g : S^n \to \widetilde{\mathbb{R}P^n}$ as

$$g(p) = (\pi(p), [(d\pi)_p v_1, \ldots, (d\pi)_p v_n]),$$

where $\{v_1, \ldots, v_n\}$ is a positive basis of $T_p S^n$. Using the fact that for even n we have $f^*\omega = -\omega$, i.e. f reverses orientations, it is now very easy to show that g is a diffeomorphism such that $\widetilde{\pi} \circ g = \pi$, where $\widetilde{\pi} : \widetilde{\mathbb{R}P^n} \to \mathbb{R}P^n$ is the natural projection.

(6) (a) Let $\omega \in \Omega^{n-1}(\partial M)$ be a volume form for ∂M compatible with the induced orientation. Since ∂M is compact (because M is) we have

$$\int_{\partial M} \omega > 0.$$

If f existed it would satisfy $f \circ i = \text{id}$, where $i : \partial M \to M$ is the inclusion map. Then, using the Stokes theorem, we would have

$$\int_{\partial M} \omega = \int_{\partial M} (f \circ i)^*\omega = \int_{\partial M} i^* f^*\omega = \int_M df^*\omega$$

$$= \int_M f^*d\omega = \int_M f^*0 = 0$$

($d\omega = 0$ as it is an n-form on the $(n-1)$-dimensional manifold ∂M). Therefore f cannot exist.

(b) Assume that there existed a differentiable map $g : B \to B$ without fixed points. Then for each $x \in B$ there would exist a unique ray r_x starting at $g(x)$

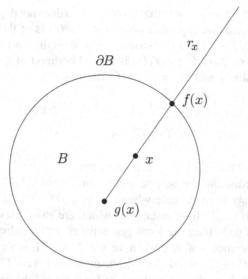

Fig. 7.2 Construction of the map $f : B \to \partial B$

and passing through x, and clearly $r_x \setminus \{g(x)\}$ would intersect ∂B in a unique point $f(x)$ (cf. Fig. 7.2). The map $f : B \to \partial B$ would be differentiable, since

$$f(x) = g(x) + t(x)\frac{x - g(x)}{\|x - g(x)\|}$$

where $t(x)$ is the unique positive root of the equation

$$\|f(x)\|^2 = 1 \Leftrightarrow t^2 + 2t\frac{\langle x, g(x) \rangle - \|g(x)\|^2}{\|x - g(x)\|} + \|g(x)\|^2 = 1.$$

Moreover, we would have $f_{|\partial B} = \mathrm{id}$. Since by (a) the map f cannot exist, neither can g.

7.3 Chapter 3

7.3.1 Section 3.1

(3) (a) In this case the natural projection map $\pi : M \to M/G$ is a covering map (hence a local diffeomorphism). For each point $r \in M/G$ we can select a point $p \in \pi^{-1}(r)$ and a neighborhood $U \ni p$ such that $\pi_{|U}$ is a diffeomorphism onto its image. We then define a metric h on $\pi(U)$ through

$h := ((\pi_{|U})^{-1})^* g$. To show that this definition does not depend on the choice of p we notice that any other point $q \in \pi^{-1}(r)$ is of the form $q = f(p)$, where $f : M \to M$ is an isometry corresponding to the action by some element of G. Now $V := f(U)$ is a neighborhood of q such that $\pi_{|V}$ is a diffeomorphism, and $\pi_{|U} = \pi_{|V} \circ f$. Therefore

$$((\pi_{|U})^{-1})^* g = (f^{-1} \circ (\pi_{|V})^{-1})^* g = ((\pi_{|V})^{-1})^* (f^{-1})^* g$$
$$= ((\pi_{|V})^{-1})^* g,$$

since f (and therefore f^{-1}) is an isometry.

(b) We can define the flat square metric on $T^n = \mathbb{R}^n/\mathbb{Z}^n$ by applying the procedure above to the case when (M, g) is $\mathbb{R}^n$ with the Euclidean metric and $\mathbb{Z}^n$ acts on $\mathbb{R}^n$ by translations (which are isometries of the Euclidean metric). In particular, the local geometry of T^n is indistinguishable from the local geometry of $\mathbb{R}^n$. We have $\operatorname{vol}(T^n) = 1$, since, if $U := (0, 1)^n$ then $\pi_{|U} : U \to T^n$ is a diffeomorphism which covers T^n except for a zero measure set. Note that each choice of basis for $\mathbb{R}^n$ determines an action of $\mathbb{Z}^n$ by translations by integer multiples of the basis vectors, whose quotient is diffeomorphic to T^n. The metrics obtained on T^n from these actions are in general different (in particular the corresponding volumes of T^n do not have to be 1).

(c) We can define the standard metric on $\mathbb{R}P^n = S^n/\mathbb{Z}_2$ by applying the procedure above to the case when (M, g) is S^n with the standard metric and $\mathbb{Z}_2$ acts on S^n by the antipodal map (which is an isometry of the standard metric). In particular, the local geometry of $\mathbb{R}P^n$ is indistinguishable from the local geometry of S^n. Notice also that $\operatorname{vol}(\mathbb{R}P^n) = \frac{1}{2}\operatorname{vol}(S^n)$, since if U is a hemisphere then $\pi_{|U} : U \to \mathbb{R}P^n$ is a diffeomorphism which covers $\mathbb{R}P^n$ except for a zero measure set.

(4) (a) If g is left-invariant, then we must have

$$\langle v, w \rangle_x = \langle (dL_{x^{-1}})_x v, (dL_{x^{-1}})_x w \rangle_e$$

for all $x \in G$ and all $v, w \in T_x G$. Thus we just have to show that this formula indeed defines a left-invariant metric on G. It is easy to check that the smoothness of the map

$$G \times G \ni (x, y) \mapsto x^{-1}y = L_{x^{-1}}y \in G$$

implies the smoothness of the map

$$G \times TG \ni (x, v) \mapsto (dL_{x^{-1}})_x v \in TG,$$

and that therefore the formula above defines a smooth tensor field g on G. It should also be clear that g is symmetric and positive definite. All that remains to be proved is that g is left-invariant, that is,

$$\langle (dL_y)_x v, (dL_y)_x w \rangle_{yx} = \langle v, w \rangle_x$$

for all $v, w \in T_x G$ and all $x, y \in G$. Indeed we have

$$
\begin{aligned}
&\langle (dL_y)_x v, (dL_y)_x w \rangle_{yx} \\
&= \left\langle \left(dL_{(yx)^{-1}}\right)_{yx} (dL_y)_x v, \left(dL_{(yx)^{-1}}\right)_{yx} (dL_y)_x w \right\rangle_e \\
&= \left\langle \left(d\left(L_{x^{-1}y^{-1}} \circ L_y\right)\right)_x v, \left(d\left(L_{x^{-1}y^{-1}} \circ L_y\right)\right)_x w \right\rangle_e \\
&= \langle (dL_{x^{-1}})_x v, (dL_{x^{-1}})_x w \rangle_e = \langle v, w \rangle_x.
\end{aligned}
$$

Thus any inner product on the Lie algebra $\mathfrak{g} = T_e G$ determines a left-invariant metric on G.

(b) Recall that every quaternion $q \in \mathbb{R}SU(2)$ can be written as

$$q = a1 + bi + cj + dk$$

with $a, b, c, d \in \mathbb{R}$, where $1, i, j, k \in SU(2)$ are given in Exercise 7.17(13) of Chap. 1, and that $SU(2)$ is the set of quaternions of Euclidean length 1. Given the identities $i^2 = j^2 = k^2 = ijk = -1$, is easy to check that

$$qq^* = q^*q = (a^2 + b^2 + c^2 + d^2)1.$$

Furthermore, using this basis to identify $\mathbb{R}SU(2)$ with $\mathbb{R}^4$, we can write the Euclidean inner product as

$$\langle q_1, q_2 \rangle = \mathrm{Re}(q_1 q_2^*) = \mathrm{Re}(q_1^* q_2),$$

where $\mathrm{Re}(q) = a$. If $v \in SU(2)$ is a unit quaternion, that is, $v^*v = 1$, then

$$\langle vq_1, vq_2 \rangle = \mathrm{Re}(q_1^* v^* v q_2) = \mathrm{Re}(q_1^* q_2) = \langle q_1, q_2 \rangle.$$

Therefore multiplication by unit quaternions preserves the Euclidean inner product. Restricting to vectors tangent to $SU(2)$ we conclude that the standard metric on $SU(2)$ is left-invariant.

(c) The Euclidean inner product on $\mathcal{M}_{n \times n} \cong \mathbb{R}^{n^2}$ is given by

$$\langle A, B \rangle = \mathrm{tr}(AB^t).$$

Therefore, if $S \in O(n)$ then

$$\langle SA, SB \rangle = \mathrm{tr}(SAB^t S^t) = \mathrm{tr}(S^t SAB^t) = \mathrm{tr}(AB^t) = \langle A, B \rangle.$$

Restricting to vectors tangent to $O(n)$ we conclude that the metric induced on $O(n)$ by the Euclidean metric of $\mathcal{M}_{n \times n} \cong \mathbb{R}^{n^2}$ is left-invariant.

7.3.2 Section 3.2

(1) (a) Let $p \in W$ and $\rho \in C^\infty(M)$ a bump function satisfying svpp $\rho \subset W$ and $\rho(p) > 0$. Thus we have

$$\rho X = \rho \tilde{X} \quad \text{and} \quad \rho Y = \rho \tilde{Y}.$$

Using the properties of an affine connection we obtain

$$\nabla_{\rho X}(\rho Y) = \rho(\rho \nabla_X Y + (X \cdot \rho)Y)$$

and similarly

$$\nabla_{\rho \tilde{X}}(\rho \tilde{Y}) = \rho(\rho \nabla_{\tilde{X}} \tilde{Y} + (\tilde{X} \cdot \rho)\tilde{Y}).$$

Equating the two expressions yields

$$\rho \nabla_X Y + (X \cdot \rho)Y = \rho \nabla_{\tilde{X}} \tilde{Y} + (\tilde{X} \cdot \rho)\tilde{Y},$$

which at p reads

$$\rho(p)(\nabla_X Y)_p + (X_p \cdot \rho)Y_p = \rho(p)(\nabla_{\tilde{X}} \tilde{Y})_p + (\tilde{X}_p \cdot \rho)\tilde{Y}_p.$$

Since $p \in W$, we have $X_p = \tilde{X}_p$ and $Y_p = \tilde{Y}_p$. Therefore

$$\rho(p)(\nabla_X Y)_p = \rho(p)(\nabla_{\tilde{X}} \tilde{Y})_p \Leftrightarrow (\nabla_X Y)_p = (\nabla_{\tilde{X}} \tilde{Y})_p,$$

where we've used the fact that $\rho(p) > 0$. Since $p \in W$ is arbitrary, we conclude that $\nabla_X Y = \nabla_{\tilde{X}} \tilde{Y}$ on W.

(b) We have

$$\nabla_X Y = \nabla_X \left(\sum_{i=1}^n Y^i \frac{\partial}{\partial x^i} \right) = \sum_{i=1}^n (X \cdot Y^i) \frac{\partial}{\partial x^i} + \sum_{i=1}^n Y^i \nabla_X \frac{\partial}{\partial x^i}$$

$$= \sum_{i=1}^n (X \cdot Y^i) \frac{\partial}{\partial x^i} + \sum_{k=1}^n Y^k \nabla_{\left(\sum_{j=1}^n X^j \frac{\partial}{\partial x^j}\right)} \frac{\partial}{\partial x^k}$$

$$= \sum_{i=1}^n (X \cdot Y^i) \frac{\partial}{\partial x^i} + \sum_{j,k=1}^n X^j Y^k \nabla_{\frac{\partial}{\partial x^j}} \frac{\partial}{\partial x^k}$$

$$= \sum_{i=1}^{n} (X \cdot Y^i) \frac{\partial}{\partial x^i} + \sum_{i,j,k=1}^{n} X^j Y^k \Gamma^i_{jk} \frac{\partial}{\partial x^i}.$$

(c) Using

$$\dot{c}(t) = \sum_{i=1}^{n} \dot{x}^i(t) \left(\frac{\partial}{\partial x^i} \right)_{c(t)} \quad \text{and} \quad V(t) = \sum_{i=1}^{n} V^i(t) \left(\frac{\partial}{\partial x^i} \right)_{c(t)}$$

we obtain

$$\frac{DV}{dt}(t) = \nabla_{\dot{c}(t)} V = \sum_{i=1}^{n} \left(\dot{c}(t) \cdot V^i(t) + \sum_{j,k=1}^{n} \Gamma^i_{jk}(c(t)) \dot{x}^j(t) V^k(t) \right) \left(\frac{\partial}{\partial x^i} \right)_{c(t)}$$

$$= \sum_{i=1}^{n} \left(\dot{V}^i(t) + \sum_{j,k=1}^{n} \Gamma^i_{jk}(c(t)) \dot{x}^j(t) V^k(t) \right) \left(\frac{\partial}{\partial x^i} \right)_{c(t)}.$$

Therefore the coordinate equations for the parallel transport law are

$$\dot{V}^i(t) + \sum_{j,k=1}^{n} \Gamma^i_{jk}(c(t)) \dot{x}^j(t) V^k(t) = 0 \quad (i = 1, \ldots, n).$$

(d) Using (7) in the case when $V = \dot{c}$, i.e. $V^i(t) = \dot{x}^i(t)$, yields

$$\ddot{x}^i(t) + \sum_{j,k=1}^{n} \Gamma^i_{jk}(c(t)) \dot{x}^j(t) \dot{x}^k(t) = 0 \quad (i = 1, \ldots, n).$$

(3) (a) Let $X, Y, Z \in \mathfrak{X}(M)$ and $f, g \in C^\infty(M)$. Then

$$(\nabla_X \omega)(f Y + g Z) = X \cdot (\omega(f Y + g Z)) - \omega(\nabla_X(f Y + g Z))$$
$$= X \cdot (f \omega(Y) + g \omega(Z)) - \omega(f \nabla_X Y + (X \cdot f)Y + g \nabla_X Z + (X \cdot g)Z)$$
$$= (X \cdot f)\omega(Y) + (X \cdot g)\omega(Z) + f(X \cdot (\omega(Y)) - \omega(\nabla_X Y))$$
$$+ g(X \cdot (\omega(Z)) - \omega(\nabla_X Z)) - (X \cdot f)\omega(Y) - (X \cdot g)\omega(Z)$$
$$= (f \nabla_X \omega)(Y) + g(\nabla_X \omega)(Z).$$

(b) (i) We have

$$(\nabla_{f X + g Y} \omega)(Z) = (f X + g Y) \cdot (\omega(Z)) - \omega(\nabla_{f X + g Y} Z)$$
$$= f X \cdot (\omega(Z)) + g Y \cdot (\omega(Z)) - f \omega(\nabla_X Z) - g \omega(\nabla_Y Z)$$
$$= f(\nabla_X \omega + g \nabla_Y \omega)(Z);$$

(ii) Also,

$$(\nabla_X(\omega + \eta))(Y) = X \cdot ((\omega + \eta)(Y)) - (\omega + \eta)(\nabla_X Y)$$
$$= X \cdot (\omega(Y)) + X \cdot (\eta(Y)) - \omega(\nabla_X Y) - \eta(\nabla_X Y)$$
$$= (\nabla_X \omega + \nabla_X \eta)(Y);$$

(iii) Finally,

$$(\nabla_X(f\omega))(Y) = X \cdot (f\omega(Y)) - f\omega(\nabla_X Y)$$
$$= (X \cdot f)\omega(Y) + f(X \cdot (\omega(Y)) - \omega(\nabla_X Y))$$
$$= ((X \cdot f)\omega + f\nabla_X \omega)(Y).$$

(c) In these coordinates we have

$$\nabla_X \omega(Y) = X \cdot (\omega(Y)) - \omega(\nabla_X Y)$$

$$= X \cdot \left(\sum_{i=1}^{n} \omega_i dx^i(Y) \right) - \omega \left(\sum_{i=1}^{n} \left(X \cdot Y^i + \sum_{j,k=1}^{n} \Gamma^i_{jk} X^j Y^k \right) \frac{\partial}{\partial x^i} \right)$$

$$= X \cdot \left(\sum_{i=1}^{n} \omega_i Y^i \right) - \sum_{i=1}^{n} \omega_i (X \cdot Y^i) - \sum_{i,j,k=1}^{n} \Gamma^i_{jk} X^j Y^k \omega_i$$

$$= \sum_{i=1}^{n} (X \cdot \omega_i) Y^i - \sum_{i=1}^{n} \left(\sum_{j,k=1}^{n} \Gamma^k_{ji} X^j \omega_k \right) Y^i$$

$$= \left(\sum_{i=1}^{n} \left(X \cdot \omega_i - \sum_{j,k=1}^{n} \Gamma^k_{ji} X^j \omega_k \right) dx^i \right) (Y).$$

(d) Let X be a vector field and T a (k, m)-tensor field. Then we define

$$\nabla_X T(X_1, \ldots, X_k, \theta^1, \ldots, \theta^m) := X \cdot (T(X_1, \ldots, X_k, \theta^1, \ldots, \theta^m))$$
$$- T(\nabla_X X_1, \ldots, X_k, \theta^1, \ldots, \theta^m) - \cdots - T(X_1, \ldots, \nabla_X X_k, \theta^1, \ldots, \theta^m)$$
$$- T(X_1, \ldots, X_k, \nabla_X \theta^1, \ldots, \theta^m) - \cdots - T(X_1, \ldots, X_k, \theta^1, \ldots, \nabla_X \theta^m)$$

for all $X_1, \ldots, X_k \in \mathfrak{X}(M)$ and $\theta^1, \ldots, \theta^m \in \Omega^1(M)$. Notice that this definition generalizes the definition for 1-forms, and coincides with the usual definition in the case when T is a vector field. A similar calculation to the one for 1-forms yields

$$\nabla_X T = \sum_{i_1,\cdots,i_k,j_1,\cdots,j_m=1}^{n} \left(X \cdot T_{i_1\cdots i_k}^{j_1\cdots j_m} - \sum_{r,s=1}^{n} \Gamma_{r i_1}^{s} X^r T_{s i_2\cdots i_k}^{j_1\cdots j_m} - \cdots \right.$$

$$- \sum_{r,s=1}^{n} \Gamma_{r i_k}^{s} X^r T_{i_1\cdots i_{k-1}s}^{j_1\cdots j_m} + \sum_{r,s=1}^{n} \Gamma_{rs}^{j_1} X^r T_{i_1\cdots i_k}^{s j_2\cdots j_m} + \cdots$$

$$\left. + \sum_{r,s=1}^{n} \Gamma_{rs}^{j_m} X^r T_{i_1\cdots i_k}^{j_1\cdots j_{k-1}s} \right) dx^{i_1} \otimes \cdots \otimes dx^{i_k} \otimes \frac{\partial}{\partial x^{j_1}} \otimes \cdots \otimes \frac{\partial}{\partial x^{j_m}}.$$

7.3.3 Section 3.3

(3) (a) Using the fact that f is an isometry and using the Koszul formula we have

$$2\langle f_* \nabla_X Y, f_* Z \rangle_{f(p)} = 2\langle \nabla_X Y, Z \rangle_p = X_p \cdot \langle Y, Z \rangle + Y_p \cdot \langle Z, X \rangle$$
$$- Z_p \cdot \langle X, Y \rangle - \langle [Y, Z], X \rangle_p - \langle [X, Z], Y \rangle_p - \langle [Y, X], Z \rangle_p.$$

On the other hand,

$$2\langle \widetilde{\nabla}_{f_* X} f_* Y, f_* Z \rangle_{f(p)} = (f_* X)_{f(p)} \cdot \langle f_* Y, f_* Z \rangle + (f_* Y)_{f(p)} \cdot \langle f_* Z, f_* X \rangle$$
$$- (f_* Z)_{f(p)} \cdot \langle f_* X, f_* Y \rangle - \langle [f_* Y, f_* Z], f_* X \rangle_{f(p)}$$
$$- \langle [f_* X, f_* Z], f_* Y \rangle_{f(p)} - \langle [f_* Y, f_* X], f_* Z \rangle_{f(p)}$$
$$= X_p \cdot (\langle f_* Y, f_* Z \rangle \circ f) + Y_p \cdot (\langle f_* Z, f_* X \rangle \circ f)$$
$$- Z_p \cdot (\langle f_* X, f_* Y \rangle \circ f) - \langle f_*[Y, Z], f_* X \rangle_{f(p)}$$
$$- \langle f_*[X, Z], f_* Y \rangle_{f(p)} - \langle f_*[Y, X], f_* Z \rangle_{f(p)}$$
$$= X_p \cdot \langle Y, Z \rangle + Y_p \cdot \langle Z, X \rangle - Z_p \cdot \langle X, Y \rangle - \langle [Y, Z], X \rangle_p$$
$$- \langle [X, Z], Y \rangle_p - \langle [Y, X], Z \rangle_p = 2\langle f_* \nabla_X Y, f_* Z \rangle_{f(p)}.$$

(b) Let $c : I \to M$ be a geodesic and consider the map $\widetilde{c} := f \circ c : I \to N$. Then, since

$$\dot{\widetilde{c}}(t) = \frac{d(f \circ c)}{dt} = (df)_{c(t)} \dot{c},$$

we have

$$\widetilde{\nabla}_{\dot{\widetilde{c}}} \dot{\widetilde{c}} = \widetilde{\nabla}_{(df)_{c(t)}\dot{c}} \left((df)_{c(t)}\dot{c} \right) \overset{(a)}{=} (df)_{c(t)} (\nabla_{\dot{c}}\dot{c}) = 0,$$

and we conclude that $\widetilde{c}$ is a geodesic in N.

(4) (a) Since $\frac{\partial}{\partial\theta}$ is the tangent vector to the curve in S^2 obtained by varying θ while holding φ constant, we have

$$\frac{\partial}{\partial\theta} = \frac{\partial\phi}{\partial\theta} = (\cos\theta\cos\varphi, \cos\theta\sin\varphi, -\sin\theta)$$

and therefore

$$g_{\theta\theta} = \left\langle \frac{\partial}{\partial\theta}, \frac{\partial}{\partial\theta} \right\rangle = 1.$$

Similarly,

$$\frac{\partial}{\partial\varphi} = \frac{\partial\phi}{\partial\varphi} = (-\sin\theta\sin\varphi, \sin\theta\cos\varphi, 0)$$

and hence

$$g_{\varphi\varphi} = \left\langle \frac{\partial}{\partial\varphi}, \frac{\partial}{\partial\varphi} \right\rangle = \sin^2\theta;$$

$$g_{\theta\varphi} = g_{\varphi\theta} = \left\langle \frac{\partial}{\partial\theta}, \frac{\partial}{\partial\varphi} \right\rangle = 0.$$

We conclude that the metric induced on S^2 by the Euclidean metric of $\mathbb{R}^3$ is given by

$$g = d\theta \otimes d\theta + \sin^2\theta d\varphi \otimes d\varphi.$$

(b) We have

$$(g_{ij}) = \begin{pmatrix} g_{\theta\theta} & g_{\theta\varphi} \\ g_{\varphi\theta} & g_{\varphi\varphi} \end{pmatrix} = \begin{pmatrix} 1 & 0 \\ 0 & \sin^2\theta \end{pmatrix}$$

and hence

$$(g^{ij}) = (g_{ij})^{-1} = \begin{pmatrix} 1 & 0 \\ 0 & \frac{1}{\sin^2\theta} \end{pmatrix}.$$

The Christoffel symbols can be easily computed from these matrices. For instance

$$\Gamma^\theta_{\varphi\varphi} = \frac{1}{2}\sum_{i=1}^{2} g^{\theta i} \left(\frac{\partial g_{\varphi i}}{\partial\varphi} + \frac{\partial g_{\varphi i}}{\partial\varphi} - \frac{\partial g_{\varphi\varphi}}{\partial x^i} \right)$$

$$= \frac{1}{2}g^{\theta\theta}\left(0 + 0 - \frac{\partial\left(\sin^2\theta\right)}{\partial\theta} \right) = -\sin\theta\cos\theta.$$

Only three of the eight Christoffel symbols are nonzero: the one computed above and

$$\Gamma^\varphi_{\theta\varphi} = \Gamma^\varphi_{\varphi\theta} = \cot\theta.$$

(c) The geodesic equations are

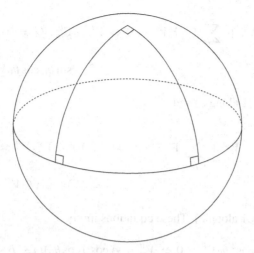

Fig. 7.3 Geodesic triangle on S^2 with three right angles

$$\ddot{\theta} + \sum_{i,j=1}^{2} \Gamma^{\theta}_{ij}\dot{x}^i\dot{x}^j = 0 \Leftrightarrow \ddot{\theta} - \sin\theta\cos\theta\,\dot{\varphi}^2 = 0;$$

$$\ddot{\varphi} + \sum_{i,j=1}^{2} \Gamma^{\varphi}_{ij}\dot{x}^i\dot{x}^j = 0 \Leftrightarrow \ddot{\varphi} + 2\cot\theta\,\dot{\theta}\dot{\varphi} = 0.$$

The curve c given in coordinates by $\hat{c}(t) = (\theta(t), \varphi(t)) = \left(\frac{\pi}{2}, t\right)$ is clearly a solution of these equations. Therefore the equator $\theta = \frac{\pi}{2}$ is the image of a geodesic.

(d) Any rotation about an axis through the origin in $\mathbb{R}^3$ is an isometry of $\mathbb{R}^3$ which preserves S^2. Since we are considering the metric in S^2 induced by the Euclidean metric on $\mathbb{R}^3$, it is clear that such a rotation will determine an isometry of S^2.

(e) Given a point $p \in S^2$ and a vector $v \in T_pS^2$, there exists a rotation $R : \mathbb{R}^3 \to \mathbb{R}^3$ such that $R(p) = (1, 0, 0)$ and $R(v) = (0, 1, 0)$. The geodesic with these initial conditions is clearly the curve c given in coordinates by $\hat{c}(t) = (\theta(t), \varphi(t)) = \left(\frac{\pi}{2}, t\right)$, whose image is the equator. By Exercise 3.3(3), the geodesic with initial condition $v \in T_pS^2$ must be $R^{-1} \circ c$. Since the image of c is the intersection of S^2 with the plane $z = 0$, the image of $R^{-1} \circ c$ is the intersection of S^2 with some plane through the origin, i.e. a great circle.

(f) For example the triangle with vertices $(1, 0, 0)$, $(0, 1, 0)$ and $(0, 0, 1)$ (cf. Fig. 7.3).

(g) The equations for parallel transport are

$$\dot{V}^\theta + \sum_{i,j=1}^{2} \Gamma_{ij}^\theta \dot{x}^i V^j = 0 \Leftrightarrow \dot{V}^\theta + \Gamma_{\varphi\varphi}^\theta V^\varphi = 0$$

$$\Leftrightarrow \dot{V}^\theta - \sin\theta_0 \cos\theta_0 V^\varphi = 0$$

(since $\dot{\varphi} = 1$ along c) and

$$\dot{V}^\varphi + \sum_{i,j=1}^{2} \Gamma_{ij}^\varphi \dot{x}^i V^j = 0 \Leftrightarrow \dot{V}^\varphi + \Gamma_{\varphi\theta}^\varphi V^\theta = 0$$

$$\Leftrightarrow \dot{V}^\varphi + \cot\theta_0 V^\theta = 0$$

(since $\dot{\varphi} = 1$ along c). These equations imply

$$\ddot{V}^\theta + \cos^2\theta_0 V^\theta = 0 \Leftrightarrow V^\theta = A\cos((\cos\theta_0)t) + B\sin((\cos\theta_0)t)$$

where $A, B \in \mathbb{R}$ are constants, and hence

$$V^\varphi = \frac{1}{\sin\theta_0 \cos\theta_0}\dot{V}^\theta = -\frac{A}{\sin\theta_0}\sin((\cos\theta_0)t) + \frac{B}{\sin\theta_0}\cos((\cos\theta_0)t).$$

The initial condition is $V^\theta(0) = 1$, $V^\varphi(0) = 0$, implying $A = 1$, $B = 0$, and thus

$$V^\theta = \cos((\cos\theta_0)t);$$

$$V^\varphi = -\frac{1}{\sin\theta_0}\sin((\cos\theta_0)t).$$

Note that in particular

$$\langle V(t), V(t)\rangle = (V^\theta)^2 + \sin^2\theta_0(V^\varphi)^2 = 1.$$

Thus the angle α between $V(0)$ and $V(2\pi)$ is given by

$$\cos\alpha = \langle V(0), V(2\pi)\rangle = V^\theta(2\pi) = \cos(2\pi\cos\theta_0),$$

that is

$$\alpha = 2\pi\cos\theta_0 \quad \text{or} \quad \alpha = 2\pi(1 - \cos\theta_0)$$

(depending on which angle one chooses to measure).

(h) Using the fact that any point on S^2 can be carried to $(0, 0, 1)$ by an appropriate
     isometry, we just have to show that no open neighborhood $U \subset S^2$ of
     $(1, 0, 0)$ is isometric to an open set $V \subset \mathbb{R}^2$ with the Euclidean metric. Now
     any such neighborhood contains the image of a curve $c(t)$ as given in (g)

(for $\theta_0 > 0$ sufficiently small). If U were isometric to W, the Levi–Civita connection on U would be the trivial connection, and hence the parallel vector field $V(t)$ in (g) would satisfy $V(0) = V(2\pi)$. Since this is not true for any $\theta_0 \in (0, \frac{\pi}{2})$, U cannot be isometric to W.

(i) The parallel postulate does not hold in S^2 because the images of any two geodesics either coincide or intersect in exactly two points.

(5) (a) As we saw in Exercise 1.10(4), we must have

$$\langle v, w \rangle_{(x,y)} = \left\langle \left(dL_{(x,y)^{-1}}\right)_{(x,y)} v, \left(dL_{(x,y)^{-1}}\right)_{(x,y)} w \right\rangle_e$$

for all $(x, y) \in H$ and $v, w \in T_{(x,y)}H$. From Exercise 7.17(3) in Chap. 1 we know that

$$\left(dL_{(x,y)}\right)_{(z,w)} = \begin{pmatrix} y & 0 \\ 0 & y \end{pmatrix}$$

and

$$(x, y)^{-1} = \left(-\frac{x}{y}, \frac{1}{y}\right).$$

Therefore we have

$$\left\langle v^x \frac{\partial}{\partial x} + v^y \frac{\partial}{\partial y}, w^x \frac{\partial}{\partial x} + w^y \frac{\partial}{\partial y} \right\rangle_{(x,y)}$$

$$= \left\langle \frac{1}{y} v^x \frac{\partial}{\partial x} + \frac{1}{y} v^y \frac{\partial}{\partial y}, \frac{1}{y} w^x \frac{\partial}{\partial x} + \frac{1}{y} w^y \frac{\partial}{\partial y} \right\rangle_{(0,1)}$$

$$= \frac{1}{y^2} \left\langle v^x \frac{\partial}{\partial x} + v^y \frac{\partial}{\partial y}, w^x \frac{\partial}{\partial x} + w^y \frac{\partial}{\partial y} \right\rangle_{(0,1)}$$

$$= \frac{1}{y^2} \left(v^x w^x + v^y w^y\right),$$

that is

$$g = \frac{1}{y^2}(dx \otimes dx + dy \otimes dy).$$

(b) We have

$$(g_{ij}) = \begin{pmatrix} g_{xx} & g_{xy} \\ g_{yx} & g_{yy} \end{pmatrix} = \begin{pmatrix} \frac{1}{y^2} & 0 \\ 0 & \frac{1}{y^2} \end{pmatrix}$$

and hence

$$(g^{ij}) = (g_{ij})^{-1} = \begin{pmatrix} y^2 & 0 \\ 0 & y^2 \end{pmatrix}.$$

The Christoffel symbols can be easily computed from these matrices. For instance

$$\Gamma_{xy}^x = \frac{1}{2} \sum_{i=1}^{2} g^{xi} \left(\frac{\partial g_{yi}}{\partial x} + \frac{\partial g_{xi}}{\partial y} - \frac{\partial g_{xy}}{\partial x^i} \right)$$

$$= \frac{1}{2} g^{xx} \left(0 + \frac{\partial}{\partial y} \left(\frac{1}{y^2} \right) - 0 \right) = -\frac{1}{y}.$$

Only four of the eight Christoffel symbols are nonvanishing: the one calculated above and

$$\Gamma_{yx}^x = -\Gamma_{xx}^y = \Gamma_{yy}^y = -\frac{1}{y}.$$

(c) The geodesic equations are

$$\ddot{x} + \sum_{i,j=1}^{2} \Gamma_{ij}^x \dot{x}^i \dot{x}^j = 0 \Leftrightarrow \ddot{x} - \frac{2}{y} \dot{x} \dot{y} = 0;$$

$$\ddot{y} + \sum_{i,j=1}^{2} \Gamma_{ij}^y \dot{x}^i \dot{x}^j = 0 \Leftrightarrow \ddot{y} + \frac{1}{y} \dot{x}^2 - \frac{1}{y} \dot{y}^2 = 0.$$

One can check that the curves α and β satisfy these equations by direct substitution. Also, it is clear that $\alpha(\mathbb{R})$ is the positive y-axis. Since

$$\tanh^2 t + \frac{1}{\cosh^2 t} = 1,$$

we see that $\beta(\mathbb{R})$ is the intersection of the unit circle with H.

(d) By Exercise 3.3(3), isometries carry images of geodesics to images of geodesics. Since the metric on H is left-invariant, any left translation is an isometry. In particular,

$$L_{(a,1)}(x, y) = (x + a, y)$$

and

$$L_{(0,b)}(x, y) = (bx, by)$$

are isometries for all $a \in \mathbb{R}$ and $b > 0$. We conclude that all vertical half-lines and all semicircles centered on the x-axis are images of geodesics of H (cf. Fig. 7.4). On the other hand, given $p \in H$ and $v \in T_p H$, v is always tangent to one of these. Indeed, if v is vertical then it is tangent to the vertical half-line through p. If v is not vertical, it is tangent to the semicircle centered at the intersection of the x-axis with the line orthogonal to v at p (cf. Fig. 7.4). Therefore the image of the geodesic with initial condition $v \in T_p M$ is either a vertical half-line or a semicircle centered on the x-axis.

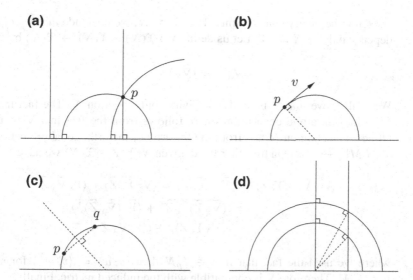

Fig. 7.4 **a** Geodesics of the hyperbolic plane and violation of the parallel postulate. **b** Geodesic tangent to a vector. **c** Geodesic through two points. **d** Internal angles of a geodesic quadrilateral

(e) If p and q are on the same vertical half-line then there is no semicircle centered on the x-axis containing p and q. If p and q are not on the same vertical half-line then there exists a unique semicircle centered on the x-axis containing p and q, whose center is the intersection of the x-axis with the perpendicular bisector of the line segment with endpoints p and q (cf. Fig. 7.4).

(f) On the sphere S^2 there are infinite geodesics which are not reparameterizations of each other going through any two antipodal points. On $\mathbb{R}^2 \setminus \{0\}$ with the usual Euclidean metric there is no geodesic connecting the points $(1, 0)$ and $(-1, 0)$.

(g) Since the metric for the hyperbolic plane is a function times the Euclidean metric, the angles measured using both metrics are equal. Consider the geodesic quadrilateral formed by two vertical half-lines and two distinct semicircles centered at the intersection of one of the half-lines with the x-axis (cf. Fig. 7.4). It is easy to check that the internal angles of this quadrilateral add up to less than 2π. Now every open set $U \subset H$ contains one such quadrilateral. If U were isometric to $V \subset \mathbb{R}^2$ then the internal angles of the quadrilateral would have to add up to exactly 2π. We conclude that U cannot be isometric to V.

(h) The parallel postulate does not hold in the hyperbolic plane. Instead, it is easy to see that given a geodesic $c : \mathbb{R} \to H$ and a point $p \notin c(\mathbb{R})$ there exist an infinite number of geodesics (up to reparameterization) $\tilde{c} : \mathbb{R} \to H$ such that $p \in \tilde{c}(\mathbb{R})$ and $c(\mathbb{R}) \cap \tilde{c}(\mathbb{R}) = \varnothing$ (cf. Fig. 7.4).

(6) (a) We start by noticing that if $p \in N$ then $(\tilde{\nabla}_{\tilde{X}} \tilde{Y})_p$ depends only on $\tilde{X}_p = X_p$ and on the values of $\tilde{Y}$ along a curve tangent to X_p, which may therefore be

chosen to be a curve on N. Since $\widetilde{Y} = Y$ on N, we conclude that $\widetilde{\nabla}_{\widetilde{X}}\widetilde{Y}|_N$ depends only on X and Y. Let us define $\nabla : \mathfrak{X}(N) \times \mathfrak{X}(N) \to \mathfrak{X}(N)$ by

$$\nabla_X Y = \left(\widetilde{\nabla}_{\widetilde{X}}\widetilde{Y}\right)^{\top}.$$

We shall prove that ∇ is the Levi–Civita connection on N. The fact that ∇ defines an affine connection on N follows from the fact that $\widetilde{\nabla}$ is an affine connection on M and from the linearity of the orthogonal projection $^{\top} : TM|_N \to TN$. On the other hand, given $X, Y, Z \in \mathfrak{X}(N)$ we have

$$\begin{aligned}
X \cdot \langle\langle Y, Z\rangle\rangle &= \widetilde{X} \cdot \langle\widetilde{Y}, \widetilde{Z}\rangle = \langle\widetilde{\nabla}_{\widetilde{X}}\widetilde{Y}, \widetilde{Z}\rangle + \langle\widetilde{Y}, \widetilde{\nabla}_{\widetilde{X}}\widetilde{Z}\rangle \\
&= \langle(\widetilde{\nabla}_{\widetilde{X}}\widetilde{Y})^{\top}, \widetilde{Z}\rangle + \langle\widetilde{Y}, (\widetilde{\nabla}_{\widetilde{X}}\widetilde{Z})^{\top}\rangle \\
&= \langle\langle\nabla_X Y, Z\rangle\rangle + \langle\langle Y, \nabla_X Z\rangle\rangle,
\end{aligned}$$

where we used the fact that if $v \in T_p N$ then $\langle v, w\rangle = \langle v, w^{\top}\rangle$ for all $w \in T_p M$. Therefore ∇ is compatible with the induced metric. Finally,

$$\begin{aligned}
\nabla_X Y - \nabla_Y X &= \left(\widetilde{\nabla}_{\widetilde{X}}\widetilde{Y}\right)^{\top} - \left(\widetilde{\nabla}_{\widetilde{Y}}\widetilde{X}\right)^{\top} \\
&= \left(\widetilde{\nabla}_{\widetilde{X}}\widetilde{Y} - \widetilde{\nabla}_{\widetilde{Y}}\widetilde{X}\right)^{\top} = [\widetilde{X}, \widetilde{Y}]^{\top} = [X, Y],
\end{aligned}$$

where we used the fact that $[\widetilde{X}, \widetilde{Y}] = [X, Y]$ on N [cf. Exercise 6.11(7) in Chap. 1]. Therefore ∇ is symmetric, and hence it is the Levi–Civita connection on $\langle\langle\cdot, \cdot\rangle\rangle$.

(b) If $(M, \langle\cdot, \cdot\rangle)$ is $\mathbb{R}^3$ with the Euclidean metric, $N \subset \mathbb{R}^3$ is a surface and $c : I \to N$ is a curve we have

$$\nabla_{\dot{c}}\dot{c} = \left(\widetilde{\nabla}_{\dot{c}}\dot{c}\right)^{\top} = \ddot{c}^{\top}.$$

Therefore c is a geodesic of N if and only if its acceleration is orthogonal to N. Assume that c is also a curve on a plane L which is orthogonal to N (i.e. $T_p L$ is orthogonal to $T_p N$ for each $p \in L \cap N$). Then both $\dot{c}(t)$ and $\ddot{c}(t)$ are contained in $T_{c(t)}L$. Since $T_{c(t)}L \cap T_{c(t)}N = \mathbb{R}\dot{c}(t)$, we see that in this case $\ddot{c}^{\top}$ is a multiple of $\dot{c}$. Finally, if we parameterize c by the arclength, we have

$$\langle\dot{c}, \dot{c}\rangle = 1 \Rightarrow \langle\ddot{c}, \dot{c}\rangle = 0 \Rightarrow \ddot{c}^{\top} = 0.$$

We conclude that any curve c, parameterized by the arclength, with image on $L \cap N$ is a geodesic of N. Similarly, it is not difficult to show that if L is not orthogonal to N and $\ddot{c}$ does not vanish then c is not a geodesic of N. These considerations make the following results obvious.

(i) On the sphere S^2, any great circle (i.e. any intersection of S^2 with a plane containing the origin) is the image of a geodesic. Any circle which is not a great circle (i.e. any intersection of S^2 with a plane which does not contain the origin) is not the image of a geodesic.

(ii) Any intersection of the torus of revolution with a plane of symmetry (i.e. a plane such that reflection with respect to that plane leaves the torus invariant) is the image of a geodesic [see also Exercise 4.8(5)]. Any intersection of the torus of revolution with a plane which is not a plane of symmetry is not the image of a geodesic.

(iii) The generators of the cone are images of geodesics (this could also be seen from the fact that they are already geodesics of $\mathbb{R}^3$). The circles obtained by intersecting the cone with planes perpendicular to its axis are not images of geodesics.

(iv) The intersections of any surface of revolution with any plane containing the axis of revolution are images of geodesics. Notice that this includes all cases above.

(c) This is immediate from the fact that

$$\nabla_{\dot{c}} V = \left(\widetilde{\nabla}_{\dot{c}} V\right)^{\top} = \dot{V}^{\top}$$

depends only on the operator $^{\top}$, which will be the same for both surfaces if they are tangent along the curve c.

(d) A parameterization of the cone is $\varphi : (0, +\infty) \times (0, 2\pi) \to \mathbb{R}^3$ given by

$$\varphi(r, \theta) = (r \cos \theta, r \sin \theta, r \cot \alpha)$$

where α is the angle between the generators and the axis. The induced metric can then be found to be

$$g = \frac{1}{\sin^2 \alpha} dr \otimes dr + r^2 d\theta \otimes d\theta.$$

Defining new coordinates

$$\begin{cases} r' := \frac{r}{\sin \alpha} \\ \theta' := \theta \sin \alpha \end{cases}$$

we have

$$g = dr' \otimes dr' + r'^2 d\theta' \otimes d\theta',$$

implying that the cone minus a generator is isometric to the open set of $\mathbb{R}^2$ given by $\theta' \in (0, 2\pi \sin \alpha)$. In particular, as suggested in the figure, parallel transport once around the cone will lead to an angle $2\pi(1 - \sin \alpha)$ between the initial and the final vectors. Now if the circle on the sphere is parameterized by

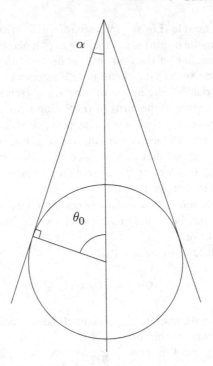

Fig. 7.5 Relation between α and θ_0

$$c(t) = (\sin \theta_0 \cos t, \sin \theta_0 \sin t, \cos \theta_0)$$

then by elementary geometry the cone tangent to the sphere along c satisfies $\alpha = \frac{\pi}{2} - \theta_0$ (cf. Fig. 7.5). Therefore the angle between the initial and the final vectors is $2\pi(1 - \cos \theta_0)$ [cf. Exercise 3.3(4)].

7.3.4 Section 3.4

(3) (a) Let $v \in \mathfrak{g} = T_e G$. Then,

$$(di)_e v = \frac{d}{dt} i(\exp(tv))_{|t=0} = \frac{d}{dt} (\exp(tv))^{-1}{}_{|t=0}$$

$$= \frac{d}{dt} (\exp(-tv))_{|t=0} = -v$$

(where exp is the Lie group exponential map), and so $(di)_e = -\,\mathrm{id}$. Moreover, for $g \in G$,

$$i(g) = g^{-1} = (h^{-1}g)^{-1}h^{-1} = (R_{h^{-1}} \circ i \circ L_{h^{-1}})(g),$$

implying that

$$(di)_g = (dR_{h^{-1}})_{(h^{-1}g)^{-1}} (di)_{h^{-1}g} (dL_{h^{-1}})_g$$

for every $h \in G$. Taking $h = g$ we obtain

$$(di)_g = (dR_{g^{-1}})_e (di)_e (dL_{g^{-1}})_g.$$

Hence, for $v, w \in T_g G$,

$$\begin{aligned}
\langle (di)_g v, (di)_g w \rangle &= \langle (dR_{g^{-1}})_e (di)_e (dL_{g^{-1}})_g v, (dR_{g^{-1}})_e (di)_e (dL_{g^{-1}})_g w \rangle \\
&= \langle (di)_e (dL_{g^{-1}})_g v, (di)_e (dL_{g^{-1}})_g w \rangle \\
&= \langle -(dL_{g^{-1}})_g v, -(dL_{g^{-1}})_g w \rangle = \langle v, w \rangle
\end{aligned}$$

where we used the fact that $R_{g^{-1}}$ and $L_{g^{-1}}$ are isometries.

(b) Let $\widetilde{c} := (c_v)^{-1} = i \circ c_v$. Then, from (a), we have $\widetilde{c}(0) = e$ and $\dot{\widetilde{c}}(0) = -v$, implying that $(c_v)^{-1}(t) = c_{-v}(t)$ is the geodesic through e with initial velocity $-v$. On the other hand, if $\gamma(t) := c_v(-t)$, we have $\gamma(0) = c_v(0) = e$ and $\dot{\gamma}(0) = -v$, and so $\gamma(t) = c_{-v}(t)$. We conclude that $(c_v)^{-1}(t) = c_v(-t)$. Now let $I := (a, b)$ be the maximal open interval (such that $0 \in I$) where c_v is defined, and consider a fixed $t_0 \in I$ such that $0 < t_0 < \varepsilon$, where $\varepsilon > 0$ is such that $\exp_e(B_\varepsilon(0))$ is a normal ball. Then $\gamma(t) := c_v(t_0) c_v(t)$ is also a geodesic (since $L_{c_v(t_0)}$ is an isometry) defined in I, satisfying

$$\gamma(-t_0) = c_v(t_0) c_v(-t_0) = c_v(t_0) (c_v(t_0))^{-1} = e$$

and $\gamma(0) = c_v(t_0)$. Since e and $c_v(t_0)$ are joined by two geodesics of length $t_0 \|v\| < \varepsilon$, these two geodesics must coincide, and so $\gamma(t) = c_v(t_0 + t)$ for all $t \in I$ where both sides are defined. If $b < +\infty$ then γ would extend c_v outside I, which is impossible. Moreover, since $c_v(-t) = (c_v(t))^{-1}$, we conclude that $a = -\infty$, and so c_v is defined for all $t \in \mathbb{R}$. As

$$c_v(t_0 + t) = \gamma(t) = c_v(t_0) c_v(t)$$

for every $t, t_0 \in \mathbb{R}$ such that $|t_0| < \varepsilon$, given $s \in \mathbb{R}$ and choosing $m > 0$ such that $|\frac{s}{m}| < \varepsilon$, we have

$$c_v(s + t) = c_v\left(\frac{m}{m} s + t\right) = c_v\left(\frac{s}{m}\right) c_v\left(\frac{m-1}{m} s + t\right)$$

$$= \cdots = c_v\left(\frac{s}{m}\right) \cdots c_v\left(\frac{s}{m}\right) c_v(t) = c_v(s) c_v(t).$$

(c) Let $v \in \mathfrak{g}$ and consider the geodesics $c_v : I \to G$ such that $c_v(0) = e$. Let X be a left-invariant vector field such that $X_e = v$. Then, since $c_v(s + t) = c_v(s) c_v(t)$, we have

$$\dot{c}_v(s) = \frac{d}{dt}c_v(s+t)|_{t=0} = \frac{d}{dt}(c_v(s)c_v(t))|_{t=0}$$
$$= (dL_{c_v(s)})_{c_v(0)}\dot{c}_v(0) = (dL_{c_v(s)})_e v = X_{c_v(s)}.$$

We conclude that c_v is an integral curve of X with $c_v(0) = e$. On the other hand, since L_g is an isometry, all geodesics of G are the images by left translations of geodesics through e. Moreover, since

$$(L_g \circ c_v)(0) = gc_v(0) = g$$

and

$$\frac{d}{dt}(L_g \circ c_v)(t)|_{t=0} = (dL_g)_e \dot{c}_v(0) = (dL_g)_e v = X_g,$$

we conclude that all geodesics are integral curves of left-invariant vector fields. Finally, we have

$$\exp_e(v) = c_v(1) = \exp(v).$$

(d) Let X be a left-invariant vector field. Then $\nabla_X X = 0$ since its integral curves are geodesics. Hence, if X and Y are two left-invariant vector fields, we have

$$0 = \nabla_{X+Y}(X+Y) = \nabla_X Y + \nabla_Y X = 2\nabla_X Y - [X,Y],$$

where we used the symmetry of the Levi–Civita connection.

(e) To check that the standard metric in $S^3 \cong SU(2)$ is also right-invariant, we notice that if $v \in SU(2)$ is a unit quaternion, that is, $v^* v = 1$, then

$$\langle q_1 v, q_2 v \rangle = \mathrm{Re}(q_1 v v^* q_2^*) = \mathrm{Re}(q_1 q_2^*) = \langle q_1, q_2 \rangle.$$

To check that the metric induced on $O(n)$ by the Euclidean metric of $\mathcal{M}_{n \times n} \cong \mathbb{R}^{n^2}$ is also right-invariant, we notice that if $S \in O(n)$ then

$$\langle AS, BS \rangle = \mathrm{tr}(ASS^t B^t) = \mathrm{tr}(AB^t) = \langle A, B \rangle.$$

(f) Let G be an n-dimensional compact Lie group and let $\langle \cdot, \cdot \rangle$ be a left-invariant metric on G. Given $v, w \in T_g G$ define

$$\langle\langle v, w \rangle\rangle = \int_G f,$$

where the integral is taken with respect to the (left-invariant) Riemannian volume element and $f : G \to \mathbb{R}$ is the function

$$f(h) = \langle (dR_h)_g v, (dR_h)_g w \rangle.$$

It is immediate to show that $\langle\langle \cdot, \cdot \rangle\rangle$ defines a Riemannian metric on G (positivity follows from $f > 0$ when $v = w \neq 0$). On the other hand, it is left-invariant, since

$$\begin{aligned}
\langle (dR_h)_{kg}(dL_k)_g v, &(dR_h)_{kg}(dL_k)_g w \rangle \\
&= \langle (d(R_h \circ L_k))_g v, (d(R_h \circ L_k))_g w \rangle \\
&= \langle (d(L_k \circ R_h))_g v, (d(L_k \circ R_h))_g w \rangle \\
&= \langle (dL_k)_{gh}(dR_h)_g v, (dL_k)_{gh}(dR_h)_g w \rangle \\
&= \langle (dR_h)_g v, (dR_h)_g w \rangle,
\end{aligned}$$

and so by integration

$$\langle\langle (dL_k)_g v, (dL_k)_g w \rangle\rangle = \langle\langle v, w \rangle\rangle.$$

Finally, it is right-invariant because

$$\begin{aligned}
\langle (dR_h)_{gk}(dR_k)_g v, &(dR_h)_{gk}(dR_k)_g w \rangle \\
&= \langle (d(R_h \circ R_k))_g v, (d(R_h \circ R_k))_g w \rangle \\
&= \langle (dR_{kh})_g v, (dR_{kh})_g w \rangle = f(kh),
\end{aligned}$$

and so

$$\langle\langle (dR_k)_g v, (dR_k)_g w \rangle\rangle = \int_G f \circ L_k = \int_G f = \langle\langle v, w \rangle\rangle$$

(as the volume element is left-invariant).

(6) (a) We have
$$R(X, Y_i)X = \nabla_X \nabla_{Y_i} X - \nabla_{Y_i} \nabla_X X - \nabla_{[X, Y_i]} X.$$

Since X and Y_i satisfy $\nabla_X X = 0$ and $[X, Y_i] = 0$, we obtain

$$R(X, Y_i)X = \nabla_X \nabla_{Y_i} X = \nabla_X ([Y_i, X] + \nabla_X Y_i) = \nabla_X \nabla_X Y_i.$$

(b) Let
$$Y(t) := \frac{\partial}{\partial s} \exp_p(tv(s))\Big|_{s=0}$$

with $v : (-\varepsilon, \varepsilon) \to T_p M$ satisfying $v(0) = \dot{c}(0)$. Since

$$\exp_p(tv(0)) = \exp_p(t\dot{c}(0)) = c(t),$$

we see that Y is a vector field along c. Moreover, we have $\exp_p(0v(s)) = p$, implying that $Y(0) = 0$. If $\{\dot{c}(t_0), Y(t_0)\}$ is linearly independent then we can extend (t, s) to a system of local coordinates in a neighborhood of $c(t_0)$ and use (a) to show that Y satisfies the Jacobi equation along c in a neighborhood of t_0. We will now show that either the set $D \subset \mathbb{R}$ of points t_0 where $\{\dot{c}(t_0), Y(t_0)\}$ is linearly dependent has empty interior (and then Y satisfies the Jacobi equation along c by continuity) or $Y(t) = \beta t \dot{c}(t)$ for some $\beta \in \mathbb{R}$ (which is also a solution of the Jacobi equation, as $R(X, X)X = 0$). Indeed, assume that D has nonempty interior, and let (a, b) be a connected component of the interior. Then we have $Y(t) = f(t)\dot{c}(t)$ for some function $f : (a, b) \to \mathbb{R}$, implying that $\frac{DY}{dt}(t) = \dot{f}(t)\dot{c}(t)$. Since D is closed, if b were finite then Y would be a solution of the Jacobi equation along c on some an open interval with infimum b, satisfying $Y(b) = \alpha \dot{c}(b)$ and $\frac{DY}{dt}(b) = \beta \dot{c}(b)$ (with $\alpha := f(b)$, $\beta := \dot{f}(b)$). Since

$$\widetilde{Y}(t) := (\alpha + \beta(t - b))\dot{c}(t)$$

is also a solution with $\widetilde{Y}(b) = \alpha \dot{c}(b)$, $\frac{D\widetilde{Y}}{dt}(b) = \beta \dot{c}(b)$, the Picard–Lindelöf theorem would imply $Y(t) = \widetilde{Y}(t)$ on this interval, and hence b would not be the supremum of the connected component (because $\widetilde{Y}(t)$ and $\dot{c}(t)$ are linearly dependent). We conclude that $b = +\infty$. Similarly we have $a = -\infty$, i.e. $D = \mathbb{R}$. Now the formula for $Y(t)$ can be written as

$$Y(t) = (d \exp_p)_{t\dot{c}(0)}(t\dot{v}(0)),$$

and so

$$\dot{v}(0) = \lim_{t \to 0} \frac{1}{t} Y(t) = \lim_{t \to 0} \frac{f(t)}{t} \dot{c}(t) = \beta \dot{c}(0) = \beta v(0),$$

where

$$\beta = \lim_{t \to 0} \frac{f(t)}{t}.$$

Also, notice that $Y(t)$ depends only on $v(0)$ and $\dot{v}(0)$. Therefore we may choose for instance $v(s) = (1 + s\beta)v(0)$, in which case

$$\exp_p(tv(s)) = \exp_p((1 + s\beta)tv(0)) = c((1 + s\beta)t)$$

and hence $Y(t) = \beta t \dot{c}(t)$.
Conversely, given a solution $Y(t)$ of the Jacobi equation along c with $Y(0) = 0$, choose

$$\widetilde{Y}(t) = \frac{\partial}{\partial s} \exp_p(t(v_0 + sw_0))\Big|_{s=0},$$

where $v_0 = \dot{c}(0)$ and $w_0 = \frac{DY}{dt}(0)$. Then $\widetilde{Y}(0) = 0$. Moreover, using the fact that the Christoffel symbols for normal coordinates vanish at the origin, it is easily seen that

$$\frac{D\widetilde{Y}}{dt}(0) = w_0 = \frac{DY}{dt}(0).$$

Since, as we saw above, $\widetilde{Y}(t)$ is also a solution of the Jacobi equation along t, by the Picard–Lindelöf theorem we conclude that $\widetilde{Y}(t) = Y(t)$ for all $t \in \mathbb{R}$.

(c) This is immediate from the formula

$$Y(t) = (d \exp_p)_{t\dot{c}(0)}(t\dot{v}(0)).$$

for a Jacobi field.

(d) From the Jacobi equation we obtain

$$
\begin{aligned}
\frac{d^2}{dt^2}\left(\|Y(t)\|^2\right) &= 2\frac{d}{dt}\left\langle \frac{DY}{dt}(t), Y(t)\right\rangle \\
&= 2\left\langle \frac{D^2Y}{dt^2}(t), Y(t)\right\rangle + 2\left\langle \frac{DY}{dt}(t), \frac{DY}{dt}(t)\right\rangle \\
&= 2\langle R(\dot{c}(t), Y(t))\dot{c}(t), Y(t)\rangle + 2\left\|\frac{DY}{dt}(t)\right\|^2 \geq 0.
\end{aligned}
$$

Moreover, it is clear that this derivative is strictly positive in a neighborhood of $t = 0$, since $\frac{DY}{dt}(0) \neq 0$ (otherwise $Y \equiv 0$). Therefore $Y(t)$ can only have one zero (for $t = 0$).

(e) (i) Regarding T^n as $[0, 1]^n$ with the usual identifications on the $(n-1)$-dimensional faces, the cut locus of the point $p = (\frac{1}{2}, \ldots, \frac{1}{2})$ is exactly the union of the $(n-1)$-dimensional faces. Each point in these faces can be reached from p by at least two geodesics with the same length but different images.

(ii) The cut locus of $p \in S^n$ is formed by the antipodal point $-p$, which is clearly conjugate to p.

(iii) Regarding $\mathbb{R}P^n$ as the northern hemisphere of S^n with antipodal identification of the equator, the cut locus of the north pole is exactly the equator. Each point on the equator can be reached from the north pole by exactly two geodesics with the same length but different images.

7.3.5 Section 3.5

(1) To prove Proposition 5.4 we start by checking that d is a distance:

(i) Since $d(p, q)$ is the infimum of a set of nonnegative numbers, it is clear that $d(p, q) \geq 0$. Also, it is immediate to check that $d(p, p) = 0$. Let $B_\varepsilon(p)$ be a normal ball and let $q \neq p$. If $q \notin B_\varepsilon(p)$ then $d(p, q) \geq \varepsilon > 0$. If $q \in B_\varepsilon(p)$ then $d(p, q) = \|\exp_p{}^{-1}(q)\| > 0$. Therefore $d(p, q) = 0$ if and only if $p = q$.

(ii) If $\gamma : [a, b] \to M$ is a piecewise differentiable curve connecting p to q then $\widetilde{\gamma} : [a, b] \to M$ defined by $\widetilde{\gamma}(t) = \gamma(a + b - t)$ is a piecewise differentiable curve connecting q to p, with the same length. Therefore $d(p, q) = d(q, p)$.

(iii) To each pair of piecewise differentiable curves γ_1 and γ_2 connecting p to q and q to r we can easily associate a piecewise differentiable curve γ connecting p to r such that $l(\gamma) = l(\gamma_1) + l(\gamma_2)$. Taking the infimum we obtain $d(p, r) \leq d(p, q) + d(q, r)$.

To see that the metric topology induced on M is the usual topology we notice that the normal balls are a basis for the metric space topology, formed by open sets of the usual topology. On the other hand, let $U \subset M$ be an open set of the usual topology and let $p \in U$. If $B_\varepsilon(p)$ is a normal ball, then $\exp_p{}^{-1}(U \cap B_\varepsilon(p))$ is an open set of $T_p M$ containing the origin, and hence it contains $B_\delta(0)$ for some $\delta > 0$, implying that $B_\delta(p) \subset U$. This shows that the normal balls are also a basis for the usual topology, and hence the two topologies coincide.

(2) The set $B_7(0, 4)$ is represented in Fig. 7.6. The larger circle has radius 7 and the smaller circles have radius 2.

(5) (a) Since f is a local isometry, it is a local diffeomorphism. To show that it is a covering map we just have to show that it is surjective and that for each point $q \in N$ there exists a neighborhood $U \ni q$ such that $f^{-1}(U)$ is a disjoint union of open sets diffeomorphic by f to U. Because (M, g) is complete and f takes geodesics to geodesics, it is clear that (N, h) is complete and that f is surjective. Let $q \in N$ be an arbitrary point and choose $U := B_\varepsilon(q)$ to be a normal ball (for $\varepsilon > 0$ sufficiently small). Then

$$f^{-1}(U) = \cup_{p \in f^{-1}(q)} B_\varepsilon(p),$$

and f is clearly a diffeomorphism when restricted to $B_\varepsilon(p)$ (because f takes geodesics through p to geodesics through q, and so in normal coordinates it is just the identity). Finally, if $p_1, p_2 \in f^{-1}(q)$ with $p_1 \neq p_2$ then $B_\varepsilon(p_1) \cap B_\varepsilon(p_2) = \varnothing$, because the intersection points have to be equidistant from p_1 and p_2 (and so cannot form an open set).

(b) The map $f : (0, +\infty) \to S^1$ given by $f(t) = e^{it}$ is a surjective local isometry (for the usual metrics) which is not a covering map: the point $1 \in S^1$ does not admit any neighborhood U such that $f^{-1}(U)$ is a disjoint union of open sets diffeomorphic by f to U.

(c) If (M, g) has nonpositive curvature then by Exercise 4.8(6) the exponential map $\exp_p : T_p M \to M$ has no critical points, that is, it is a local diffeomorphism. Consider the Riemannian metric $h := \exp_p{}^* g$ on $T_p M$. This choice makes $\exp_p$ a local isometry which takes lines through the origin to

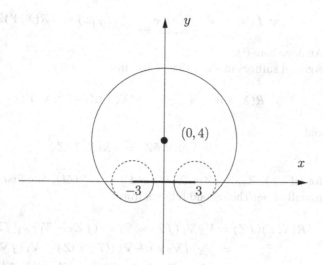

Fig. 7.6 $B_7(0, 4)$ on $\mathbb{R}^2 \backslash \{(x, 0) \mid -3 \leq x \leq 3\}$

geodesics through p. Since these geodesics are defined for all values of the parameter, we conclude that $\exp_0 : T_0 T_p M \rightarrow T_p M$ is well defined for all $v \in T_0 T_p M$, and so $(T_p M, h)$ is complete. By (a), $\exp_p$ is then a covering map.

7.4 Chapter 4

7.4.1 Section 4.1

(1) (a) (i) It suffices to prove that R satisfies

$$R(X_1 + X_2, Y)Z = R(X_1, Y)Z + R(X_2, Y)Z$$

and

$$R(fX, Y)Z = fR(X, Y)Z$$

for all $X, X_1, X_2, Y, Z \in \mathfrak{X}(M)$ and $f \in C^\infty(M)$. The first identity is trivially true. To prove the second, we notice that $[fX, Y] = f[X, Y] - (Y \cdot f)X$, and hence

$$
\begin{aligned}
R(fX, Y)Z &= \nabla_{fX}\nabla_Y Z - \nabla_Y \nabla_{fX} Z - \nabla_{[fX,Y]}Z \\
&= f\nabla_X \nabla_Y Z - \nabla_Y(f\nabla_X Z) - \nabla_{f[X,Y]-(Y \cdot f)X}Z \\
&= f\nabla_X \nabla_Y Z - f\nabla_Y \nabla_X Z - (Y \cdot f)\nabla_X Z \\
&\quad - f\nabla_{[X,Y]}Z + (Y \cdot f)\nabla_X Z
\end{aligned}
$$

$$= f(\nabla_X \nabla_Y Z - \nabla_Y \nabla_X Z - \nabla_{[X,Y]} Z) = f R(X, Y) Z.$$

(ii) Analogous to (i).

(iii) Again it suffices to show that R satisfies

$$R(X, Y)(Z_1 + Z_2) = R(X, Y)Z_1 + R(X, Y)Z_2$$

and

$$R(X, Y)(fZ) = f R(X, Y)Z$$

for all $X, Y, Z_1, Z_2, Z \in \mathfrak{X}(M)$ and $f \in C^\infty(M)$. The first identity is trivially true. The second follows from

$$
\begin{aligned}
R(X, Y)(fZ) &= \nabla_X \nabla_Y (fZ) - \nabla_Y \nabla_X (fZ) - \nabla_{[X,Y]}(fZ) \\
&= \nabla_X (f \nabla_Y Z) + \nabla_X ((Y \cdot f)Z) - \nabla_Y (f \nabla_X Z) \\
&\quad - \nabla_Y ((X \cdot f)Z) - f \nabla_{[X,Y]} Z - ([X, Y] \cdot f)Z \\
&= f \nabla_X \nabla_Y Z + (X \cdot f)\nabla_Y Z + (Y \cdot f)\nabla_X Z \\
&\quad + (X \cdot (Y \cdot f))Z - f \nabla_Y \nabla_X Z - (Y \cdot f)\nabla_X Z \\
&\quad - (X \cdot f)\nabla_Y Z - (Y \cdot (X \cdot f))Z - f \nabla_{[X,Y]} Z \\
&\quad - ([X, Y] \cdot f)Z \\
&= f(\nabla_X \nabla_Y Z - \nabla_Y \nabla_X Z - \nabla_{[X,Y]} Z) \\
&\quad + (X \cdot (Y \cdot f) - Y \cdot (X \cdot f))Z - ([X, Y] \cdot f)Z \\
&= f R(X, Y) Z.
\end{aligned}
$$

(b) If $x : V \to \mathbb{R}^n$ are local coordinates around $p \in M$, we have

$$
\begin{aligned}
R(X, Y)Z &= R\left(\sum_{i=1}^n X^i \frac{\partial}{\partial x^i}, \sum_{j=1}^n Y^j \frac{\partial}{\partial x^j} \right)\left(\sum_{k=1}^n Z^k \frac{\partial}{\partial x^k} \right) \\
&= \sum_{i,j,k=1}^n X^i Y^j Z^k R\left(\frac{\partial}{\partial x^i}, \frac{\partial}{\partial x^j} \right)\frac{\partial}{\partial x^k} = \sum_{i,j,k,l=1}^n X^i Y^j Z^k R_{ijk}{}^l \frac{\partial}{\partial x^l}.
\end{aligned}
$$

Therefore

$$(R(X, Y)Z)_p = \sum_{i,j,k,j=1}^n X^i(p)Y^j(p)Z^k(p)R_{ijk}{}^l(p)\left(\frac{\partial}{\partial x^l} \right)_p$$

depends only on X_p, Y_p, Z_p. Moreover, this dependence is linear, and hence R defines a $(3, 1)$-tensor.

(4) Let (X_p^1, X_p^2) and (Y_p^1, Y_p^2) be the components of X_p and Y_p on an orthonormal basis of the plane generated by these two vectors. Then the square of the area of

the parallelogram spanned by X_p, Y_p is

$$\begin{vmatrix} X_p^1 & Y_p^1 \\ X_p^2 & Y_p^2 \end{vmatrix}^2 = (X_p^1 Y_p^2 - X_p^2 Y_p^1)^2$$
$$= (X_p^1)^2 (Y_p^2)^2 + (X_p^2)^2 (Y_p^1)^2 - 2 X_p^1 X_p^2 Y_p^1 Y_p^2.$$

On the other hand, we have

$$\|X_p\|^2 \|Y_p\|^2 - \langle X_p, Y_p \rangle^2$$
$$= \left((X_p^1)^2 + (X_p^2)^2 \right) \left((Y_p^1)^2 + (Y_p^2)^2 \right) - \left(X_p^1 Y_p^1 + X_p^2 Y_p^2 \right)^2$$
$$= (X_p^1)^2 (Y_p^2)^2 + (X_p^2)^2 (Y_p^1)^2 - 2 X_p^1 X_p^2 Y_p^1 Y_p^2.$$

If Z_p, W_p is another basis for the plane generated by X_p, Y_p, their components on the orthonormal basis satisfy

$$\begin{pmatrix} Z_p^1 & W_p^1 \\ Z_p^2 & W_p^2 \end{pmatrix} = \begin{pmatrix} X_p^1 & Y_p^1 \\ X_p^2 & Y_p^2 \end{pmatrix} S,$$

where S is the change of basis matrix, and therefore

$$\begin{vmatrix} Z_p^1 & W_p^1 \\ Z_p^2 & W_p^2 \end{vmatrix}^2 = (\det S)^2 \begin{vmatrix} X_p^1 & Y_p^1 \\ X_p^2 & Y_p^2 \end{vmatrix}^2.$$

Finally, we have

$$R(Z_p, W_p, Z_p, W_p) = R(S_{11} X_p + S_{21} Y_p, S_{12} X_p + S_{22} Y_p, S_{11} X_p$$
$$+ S_{21} Y_p, S_{12} X_p + S_{22} Y_p)$$
$$= (S_{11} S_{22} S_{11} S_{22} - S_{11} S_{22} S_{21} S_{12} - S_{21} S_{12} S_{11} S_{22}$$
$$+ S_{21} S_{12} S_{21} S_{12}) R(X_p, Y_p, X_p, Y_p)$$
$$= (S_{11} S_{22} - S_{12} S_{21})^2 R(X_p, Y_p, X_p, Y_p)$$
$$= (\det S)^2 R(X_p, Y_p, X_p, Y_p).$$

(8) (a) First we note that the corresponding Levi–Civita connections, ∇^1 and ∇^2, coincide. Indeed, by the Koszul formula,

$$2 \langle \nabla_X^1 Y, Z \rangle_1 = X \cdot \langle Y, Z \rangle_1 + Y \cdot \langle X, Z \rangle_1 - Z \cdot \langle X, Y \rangle_1$$
$$- \langle [X, Z], Y \rangle_1 - \langle [Y, Z], X \rangle_1 + \langle [X, Y], Z \rangle_1$$
$$= 2 \rho \langle \nabla_X^2 Y, Z \rangle_2,$$

and so $2\rho\langle\nabla_X^1 Y, Z\rangle_2 = 2\rho\langle\nabla_X^2 Y, Z\rangle_2$ for every vector fields $X, Y, Z \in \mathfrak{X}(M)$, implying that $\nabla_X^1 Y = \nabla_X^2 Y$. Hence,

$$R_1(X, Y, Z, W) = \langle R_1(X, Y)Z, W\rangle_1 = \langle R_2(X, Y)Z, W\rangle_1$$
$$= \rho\langle R_2(X, Y)Z, W\rangle_2 = \rho R_2(X, Y, Z, W)$$

and so for any 2-dimensional section Π of a tangent space of M we have,

$$K_1(\Pi) = -\frac{R_1(X, Y, X, Y)}{\|X\|_1^2\|Y\|_1^2 - \langle X, Y\rangle_1^2}$$
$$= -\frac{\rho R_2(X, Y, X, Y)}{\rho^2(\|X\|_2^2\|Y\|_2^2 - \langle X, Y\rangle_2^2)} = \rho^{-1}K_2(\Pi).$$

(b) We have

$$(R_1)_{ij} = \sum_{k=1}^n (R_1)_{kij}{}^k = \sum_{k=1}^n (R_2)_{kij}{}^k = (R_2)_{ij}$$

and so $Ric_1 = Ric_2$.

(c) Since $(g_1)_{ij} = \rho(g_2)_{ij}$ implies $(g_1)^{ij} = \rho^{-1}(g_2)^{ij}$, we have

$$S_1(p) = \sum_{i,j=1}^n (R_1)_{ij}(g_1)^{ij} = \rho^{-1}\sum_{i,j=1}^n (R_2)_{ij}(g_2)^{ij} = \rho^{-1}S_2(p).$$

7.4.2 Section 4.2

(3) (a) Let us take the usual local coordinates on $S^2 \subset \mathbb{R}^3$ defined by the parameterization $\phi : (0, \pi) \times (0, 2\pi) \to \mathbb{R}^3$ given by

$$\phi(\theta, \varphi) = (\sin\theta\cos\varphi, \sin\theta\sin\varphi, \cos\theta).$$

Consider the field of frames $\{X_1, X_2\}$ where

$$X_1 := \frac{\partial}{\partial\theta} \quad \text{and} \quad X_2 := \frac{\partial}{\partial\varphi}.$$

Then

$$\langle X_1, X_1\rangle = 1, \quad \langle X_1, X_2\rangle = 0 \quad \text{and} \quad \langle X_2, X_2\rangle = \sin^2\theta,$$

and so a field of orthonormal frames $\{E_1, E_2\}$ is given by $E_1 := X_1$ and $E_2 := \frac{1}{\sin\theta} X_2$, and $\{\omega^1, \omega^2\}$, with $\omega^1 := d\theta$ and $\omega^2 := \sin\theta\, d\varphi$, is its associated field of dual coframes. Moreover,

$$dw^1 = 0 \quad \text{and} \quad dw^2 = \cos\theta\, d\theta \wedge d\varphi = \cot\theta\, \omega^1 \wedge \omega^2.$$

The first Cartan structure equations,

$$dw^1 = \omega^2 \wedge \omega_2^1 \quad \text{and} \quad dw^2 = \omega^1 \wedge \omega_1^2,$$

imply that

$$dw^1(E_1, E_2) = -\omega_2^1(E_1) = \omega_1^2(E_1)$$

and

$$dw^2(E_1, E_2) = \omega_1^2(E_2).$$

Therefore the connection form ω_1^2 is given by

$$\omega_1^2 = dw^1(E_1, E_2)\omega^1 + dw^2(E_1, E_2)\omega^2 = \cot\theta\, \omega^2 = \cos\theta\, d\varphi.$$

Hence $dw_1^2 = -\sin\theta\, d\theta \wedge d\varphi = -\omega^1 \wedge \omega^2$, and we conclude from Proposition 2.6 that the Gauss curvature of S^2 is $K = 1$.

(b) Let us consider on H the field of frames $\{X_1, X_2\}$ where

$$X_1 := \frac{\partial}{\partial x} \quad \text{and} \quad X_2 := \frac{\partial}{\partial y}.$$

Then

$$\langle X_1, X_1 \rangle = \frac{1}{y^2}, \quad \langle X_1, X_2 \rangle = 0 \quad \text{and} \quad \langle X_2, X_2 \rangle = \frac{1}{y^2},$$

and so $\{E_1, E_2\}$ given by $E_1 := y X_1$ and $E_2 := y X_2$ is a field of orthonormal frames and $\{\omega^1, \omega^2\}$, with $\omega^1 := \frac{1}{y}dx$ and $\omega^2 := \frac{1}{y}dy$, is the associated field of dual coframes. Moreover,

$$dw^1 = \frac{1}{y^2}dx \wedge dy = \omega^1 \wedge \omega^2 \quad \text{and} \quad dw^2 = 0,$$

and so the connection form ω_1^2 is given by

$$\omega_1^2 = dw^1(E_1, E_2)\omega^1 + dw^2(E_1, E_2)\omega^2 = \omega^1.$$

Hence $dw_1^2 = \frac{1}{y^2} dx \wedge dy = \omega^1 \wedge \omega^2$, and we conclude from Proposition 2.6 that the Gauss curvature of H is $K = -1$.

(6) (a) An orthonormal coframe is given by

$$\omega^r = A(r)dr, \quad \omega^\theta = rd\theta \quad \text{and} \quad \omega^\varphi = r\sin\theta d\varphi.$$

From the Cartan structure equations we obtain

$$d\omega^r = 0 = \omega^\theta \wedge \omega^r_\theta + \omega^\varphi \wedge \omega^r_\varphi;$$

$$d\omega^\theta = dr \wedge d\theta = \omega^r \wedge \omega^\theta_r + \omega^\varphi \wedge \omega^\theta_\varphi;$$

$$d\omega^\varphi = \sin\theta \, dr \wedge d\varphi + r\cos\theta \, d\theta \wedge d\varphi = \omega^r \wedge \omega^\varphi_r + \omega^\theta \wedge \omega^\varphi_\theta,$$

from which one readily guesses that

$$\omega^\theta_r = -\omega^r_\theta = \frac{1}{A}d\theta;$$

$$\omega^\varphi_r = -\omega^r_\varphi = \frac{\sin\theta}{A}d\varphi;$$

$$\omega^\varphi_\theta = -\omega^\theta_\varphi = \cos\theta d\varphi.$$

The curvature forms are given by

$$\Omega^\theta_r = d\omega^\theta_r - \omega^\varphi_r \wedge \omega^\theta_\varphi = -\frac{A'}{A^2}dr \wedge d\theta = -\frac{A'}{rA^3}\omega^r \wedge \omega^\theta;$$

$$\Omega^\varphi_r = d\omega^\varphi_r - \omega^\theta_r \wedge \omega^\varphi_\theta = -\frac{A'\sin\theta}{A^2}dr \wedge d\varphi = -\frac{A'}{rA^3}\omega^r \wedge \omega^\varphi;$$

$$\Omega^\varphi_\theta = d\omega^\varphi_\theta - \omega^r_\theta \wedge \omega^\varphi_r = \left(\frac{\sin\theta}{A^2} - \sin\theta\right)d\theta \wedge d\varphi = \frac{1}{r^2}\left(\frac{1}{A^2} - 1\right)\omega^\theta \wedge \omega^\varphi,$$

and hence the nonvanishing components of the curvature tensor on this orthonormal frame are

$$R_{r\theta r}{}^\theta = \Omega^\theta_r(E_r, E_\theta) = -\frac{A'}{rA^3};$$

$$R_{r\varphi r}{}^\varphi = \Omega^\varphi_r(E_r, E_\varphi) = -\frac{A'}{rA^3};$$

$$R_{\theta\varphi\theta}{}^\varphi = \Omega^\varphi_\theta(E_\theta, E_\varphi) = \frac{1}{r^2}\left(\frac{1}{A^2} - 1\right)$$

(plus the components related to these by symmetries). We conclude that the components of the Ricci tensor on this orthonormal frame are

$$R_{rr} = R_{\theta rr}{}^{\theta} + R_{\varphi rr}{}^{\varphi} = \frac{2A'}{rA^3};$$

$$R_{\theta\theta} = R_{r\theta\theta}{}^{r} + R_{\varphi\theta\theta}{}^{\varphi} = \frac{A'}{rA^3} - \frac{1}{r^2}\left(\frac{1}{A^2} - 1\right);$$

$$R_{\varphi\varphi} = R_{r\varphi\varphi}{}^{r} + R_{\theta\varphi\varphi}{}^{\theta} = \frac{A'}{rA^3} - \frac{1}{r^2}\left(\frac{1}{A^2} - 1\right).$$

The scalar curvature is then

$$S = R_{rr} + R_{\theta\theta} + R_{\varphi\varphi} = \frac{4A'}{rA^3} - \frac{2}{r^2}\left(\frac{1}{A^2} - 1\right).$$

(b) In this case we have $R_{rr} = R_{\theta\theta} = R_{\varphi\varphi} = 2$, i.e. $Ric = 2g$, and hence $S = 6$.

(c) In this case we have $R_{rr} = R_{\theta\theta} = R_{\varphi\varphi} = -2$, i.e. $Ric = -2g$, and hence $S = -6$.

(d) We have to solve the differential equation

$$\frac{4A'}{rA^3} - \frac{2}{r^2}\left(\frac{1}{A^2} - 1\right) = S$$

for constant S. This equation can be written as

$$-\frac{2A'r}{A^3} + \frac{1}{A^2} = 1 - \frac{Sr^2}{2} \Leftrightarrow \left(\frac{r}{A^2}\right)' = 1 - \frac{Sr^2}{2},$$

which has the immediate solution

$$A(r) = \left(1 - \frac{Sr^2}{6} + \frac{C}{r}\right)^{-\frac{1}{2}}$$

(where $C \in \mathbb{R}$ is an integration constant).

(7) Defining fields of positive orthonormal frames $\{E_1, E_2\}$ and $\{F_1, F_2\}$ such that $\{E_1, E_2\}$ is well defined on $\overline{D}$ and $F_1 = X$, we have by Proposition 2.7

$$\Delta\theta = \int_{\partial D} \sigma = \int_{\partial D} \overline{\omega}_1^2 - \omega_1^2.$$

Since X is parallel-transported along ∂D, we have

$$\nabla_{\dot{c}} F_1 = 0 \Leftrightarrow \overline{\omega}_1^2(\dot{c}) = 0 \Leftrightarrow c^* \overline{\omega}_1^2 = 0.$$

We conclude that

$$\Delta\theta = -\int_{\partial D} \omega_1^2 = -\int_D d\omega_1^2 = \int_D K\omega^1 \wedge \omega^2 = \int_D K,$$

where we have used the Stokes theorem. The formula for $K(p)$ as a limit can be obtained by standard arguments.

(9) Considering two fields of orthonormal frames $\{E_1, E_2\}$ and $\{F_1, F_2\}$ positively oriented such that $E_1 = \frac{X}{\|X\|}$ and $F_1 = \dot{c}$, we have

$$\frac{d\theta}{ds}(s) = d\theta(\dot{c}(s)) = \overline{\omega}_1^2(\dot{c}(s)) - \omega_1^2(\dot{c}(s)) = \overline{\omega}_1^2(F_1) = k_g(s),$$

where we used the fact that

$$\nabla_{\dot{c}} E_1 = 0 \Leftrightarrow \omega_1^2(\dot{c}(s)) = 0.$$

7.4.3 Section 4.3

(1) Clearly $\langle \cdot, \cdot \rangle_t$ is bilinear, symmetric and satisfies $\langle v, v \rangle_t \geq 0$. Moreover, $\langle v, v \rangle_t = 0$ if and only if

$$(1-t)\|v\|_0^2 + t\|v\|_1^2 = 0,$$

that is, if and only if $\|v\|_0^2 = \|v\|_1^2 = 0$, and so $v = 0$. To see that the function $I_p(t)$ is continuous we consider the positive orthonormal frame $\{F_{1,t}, F_{2,t}\}$ with respect to the metric $\langle \cdot, \cdot \rangle_t$ such that

$$F_{1,t} = \frac{X}{\|X\|_t}.$$

Then

$$(\overline{\omega}^1)_t := \frac{\|X\|_t}{\|X\|_1}(\overline{\omega}^1)_1$$

is continuous with respect to t. Since $F_{2,t}$, $d(\overline{\omega}^1)_t$ and $d(\overline{\omega}^2)_t$ change continuously with t, so does

$$(\overline{\omega}_1^2)_t = d(\overline{\omega}^1)_t(F_{1,t}, F_{2,t})(\overline{\omega}^1)_t + d(\overline{\omega}^2)_t(F_{1,t}, F_{2,t})(\overline{\omega}^2)_t,$$

and consequently $I_p(t)$.

(5) (a) Let $x = (x^1, \ldots, x^n)$ be local coordinates centered at p and $\hat{f} := f \circ x^{-1}$ the expression of f in these local coordinates. Since p is a critical point we have

$$\frac{\partial \hat{f}}{\partial x^i}(0, \ldots, 0) = 0.$$

Now

$$\frac{\partial}{\partial s}(f \circ \gamma) = \sum_{i=1}^{n} \frac{\partial \hat{f}}{\partial x^i} \frac{\partial \gamma^i}{\partial s}$$

(where $\gamma^i := x^i \circ \gamma$), and hence

$$\frac{\partial^2}{\partial t \partial s}(f \circ \gamma) = \sum_{i,j=1}^{n} \frac{\partial^2 \hat{f}}{\partial x^i \partial x^j} \frac{\partial \gamma^i}{\partial s} \frac{\partial \gamma^j}{\partial t} + \sum_{i=1}^{n} \frac{\partial \hat{f}}{\partial x^i} \frac{\partial^2 \gamma^i}{\partial t \partial s}.$$

Setting $(s, t) = (0, 0)$ we obtain

$$(Hf)_p(v, w) = \sum_{i,j=1}^{n} \frac{\partial^2 \hat{f}}{\partial x^i x^j}(0, \ldots, 0) v^i w^j,$$

which shows that $(Hf)_p$ is indeed a symmetric 2-tensor. Since it depends only on the components of v and w, we see that it is well defined (i.e. independent of the choice of the map γ).

(b) In local coordinates $x = (x^1, \ldots, x^n)$ centered at a particular critical point $p \in M$ we have

$$\hat{f}(x^1, \ldots, x^n) = f(p) + \sum_{i,j=1}^{n} \frac{1}{2} H_{ij} x^i x^j + o(\|x\|^2)$$

where $\hat{f} := f \circ x^{-1}$ and

$$H_{ij} := \frac{\partial^2 \hat{f}}{\partial x^i x^j}(0, \ldots, 0)$$

are the components of the Hessian, and so

$$\frac{\partial \hat{f}}{\partial x^i}(x^1, \ldots, x^n) = \sum_{j=1}^{n} H_{ij} x^j + o(\|x\|)$$

Since the Hessian is nondegenerate, the matrix (H_{ij}) is invertible, and hence there exists a neighborhood of p where p is the only critical point of f. We conclude that the critical points of f are isolated, and since M is compact, there can only be a finite number of them (otherwise they would accumulate on a non-isolated critical point). Since (H_{ij}) is symmetric and nondegenerate, there exists a linear change of coordinates which reduces it to a diagonal matrix of the form $\mathrm{diag}(1, \ldots, 1, -1, \ldots, -1)$. If M is 2-dimensional then the possibilities are $\mathrm{diag}(1, 1)$ (in which case p is a minimum), $\mathrm{diag}(-1, -1)$ (in which case p is a maximum), and $\mathrm{diag}(1, -1)$ (in which case p is a saddle point). Choosing a Riemannian metric on M which coincides with the Euclidean metric in these coordinates in a neighborhood of each critical point, we see that close to p

$$\text{grad } f = (x^1 + o(\|x\|))\frac{\partial}{\partial x^1} + (x^2 + o(\|x\|))\frac{\partial}{\partial x^2}$$

if p is a minimum,

$$\text{grad } f = (-x^1 + o(\|x\|))\frac{\partial}{\partial x^1} + (-x^2 + o(\|x\|))\frac{\partial}{\partial x^2}$$

if p is a maximum, and

$$\text{grad } f = (x^1 + o(\|x\|))\frac{\partial}{\partial x^1} + (-x^2 + o(\|x\|))\frac{\partial}{\partial x^2}$$

if p is a saddle point. From Example 3.2 we see that grad f has index 1 at p if p is a maximum or a minimum and index -1 if p is a saddle point. Since the zeros of grad f are precisely the critical points of f, we obtain

$$\chi(M) = m - s + n.$$

(6) Note that although $\partial\Delta$ is not a smooth manifold we can approximate it by a sequence of smooth manifolds by "rounding the corners" and then take the limit. In what follows we shall therefore treat $\partial\Delta$ as if it were a smooth manifold.

(a) Let us consider a vector field V parallel along $\partial\Delta$. Since the edges of Δ are geodesics, the tangent vector to $\partial\Delta$ rotates with respect to V only at the vertices, by a total amount of

$$(\pi - \alpha) + (\pi - \beta) + (\pi - \gamma).$$

This must be equal to 2π minus the angle by which V rotates. Therefore we have

$$3\pi - \alpha - \beta - \gamma = 2\pi - \int_\Delta K.$$

(b) We just saw that the tangent vector to $\partial\Delta$ rotates with respect to a parallel vector V by a total amount of

$$\int_{\partial\Delta} k_g = (\pi - \alpha) + (\pi - \beta) + (\pi - \gamma).$$

By the Gauss–Bonnet theorem for manifolds with boundary, we have

$$3\pi - \alpha - \beta - \gamma + \int_\Delta K = 2\pi\chi(\Delta).$$

Since Δ is homeomorphic to a disk, we have $\chi(\Delta) = 1$, which proves the result.

7.4.4 Section 4.4

(6) (a) The group of isometries of $\mathbb{R}^2$ is

$$Iso(\mathbb{R}^2) = \{Ax + b \mid A \in O(2), \, b \in \mathbb{R}^2\}.$$

Identifying $\mathbb{R}^2$ with the complex plane $\mathbb{C}$, we can write any orientation-preserving element $f \in Iso(\mathbb{R}^2)$ as $f(z) = e^{i\theta}z + b$, with $\theta \in \mathbb{R}$ and $b \in \mathbb{C}$. The fixed points of f are given by

$$f(z) = z \Leftrightarrow \left(1 - e^{i\theta}\right)z = b,$$

and hence f has no fixed points if and only if $e^{i\theta} = 1$, i.e. if and only if f is a translation. On the other hand, we can write any orientation-reversing element $g \in Iso(\mathbb{R}^2)$ as $g(z) = e^{i\theta}\bar{z} + b$, with $\theta \in \mathbb{R}$ and $b \in \mathbb{C}$. For instance, a gliding reflection along the real axis is given by $h(z) = \bar{z} + \xi$ (with $\xi \in \mathbb{R}$). Let $r(z) = e^{i\frac{\theta}{2}}z$ be the rotation by $\frac{\theta}{2}$ and let $t(z) = z + i\eta$ (with $\eta \in \mathbb{R}$) be the translation by $i\eta \in i\mathbb{R}$. Then

$$(r \circ h \circ t \circ r^{-1})(z) = e^{i\frac{\theta}{2}}\left(\overline{e^{-i\frac{\theta}{2}}z + i\eta + \xi}\right) = e^{i\theta}\bar{z} + e^{i\frac{\theta}{2}}(\xi - i\eta)$$

is a gliding reflection (with axis of slope $\tan\frac{\theta}{2}$ at a distance $|\eta|$ from the origin). Since this map is equal to g for ξ, η satisfying

$$\xi + i\eta = e^{i\frac{\theta}{2}}\bar{b},$$

we conclude that any orientation-reversing isometry of $\mathbb{R}^2$ is a gliding reflection. These obviously do not have fixed points as long as the translation along the reflection axis is nonzero, that is, as long as $e^{-i\frac{\theta}{2}}b \notin i\mathbb{R}$.

(b) Let Γ be a discrete group of $Iso(\mathbb{R}^2)$ acting properly and freely on $\mathbb{R}^2$. Hence, Γ can only contain translations and gliding reflections.

Suppose first that Γ only contains translations and let $t_1 \in \Gamma$ be a translation in Γ such that $t_1(0)$ has minimum length (it exists since the action of

Γ is proper). Then the group $\Gamma_1 := \langle t_1 \rangle \subset \Gamma$ generated by t_1 contains all translations of Γ with the same direction as t_1. Indeed, if t is a translation in $\Gamma \backslash \Gamma_1$ with the same direction as t_1 then, taking $m \in \mathbb{Z}$ for which $t_1^m(0)$ is the element of Γ_1 with $t_1^m(0)$ closest to $t(0)$, the length of $t^{-1} t_1^m(0)$ is smaller than the length of $t_1(0)$, contradicting our initial assumption.

If $\Gamma_1 \neq \Gamma$ then let $t_2 \in \Gamma \backslash \Gamma_1$ be such that $t_2(0)$ has minimum length. Then t_1 and t_2 generate a lattice in $\mathbb{R}^2$ and $\Gamma = \langle t_1, t_2 \rangle$. Indeed, if there were an element t in $\Gamma \backslash \langle t_1, t_2 \rangle$ then, taking $m, n \in \mathbb{Z}$ for which $t_1^m t_2^n(0)$ is closest to $t(0)$, the length of $t^{-1} t_1^m t_2^n(0)$ would either be smaller than the length of $t_1(0)$ or the length of $t_2(0)$.

If Γ contains gliding reflections then let $g \in \Gamma$ be a gliding reflection such that $g^2(0)$ has minimum length (it exists since the action of Γ is proper). If $\Gamma_1 := \langle g \rangle \neq \Gamma$ then $\Gamma \backslash \Gamma_1$ contains translations (if $g_1 \in \Gamma \backslash \Gamma_1$ is a gliding reflection then $g^{-1} g_1$ is a translation in $\Gamma \backslash \Gamma_1$). Let $t \in \Gamma \backslash \Gamma_1$ be a translation such that $t(0)$ has minimal length (it exists since the action of Γ is proper). By a suitable choice of coordinates we can assume that $g(z) = \bar{z} + \xi$ with $\xi \in \mathbb{R}$. Let $t(z) = z + b$ with $b = \alpha + i\beta \in \mathbb{C}$ (where $\alpha, \beta \in \mathbb{R}$). Then

$$(g^{-1} \circ t \circ g)(z) = g^{-1}(\bar{z} + \xi + b) = z + \bar{b},$$

and we conclude that Γ contains the translations by b and $\bar{b}$, and hence by 2α and $2i\beta$. Since $g^2(0) = 2\xi$ has minimal length, it is easy to check that 2α must be an integer multiple of ξ, as otherwise it would be possible to construct a gliding reflection $\tilde{g}$ with $|\tilde{g}^2(0)| < |g^2(0)|$. Since $t(0)$ has minimal length, we see that α must be either 0, $\pm\frac{\xi}{2}$ or $\pm\xi$; however, it cannot be $\pm\frac{\xi}{2}$, as in that case Γ would contain the reflection $z \mapsto \bar{z}$, and it cannot be $\pm\xi$, as in that case Γ would contain the reflection $z \mapsto \bar{z} + i\beta$. We conclude that t is the translation by $i\beta$, and hence the orbit of $0 \in \mathbb{R}^2$ under $\langle g, t \rangle$ is the same as its orbit by the group generated by the translations by ξ and $i\beta$. A similar argument to the one above shows that $\Gamma = \langle g, t \rangle$.

(7) To determine $\mathbb{R}^2 / \langle f \rangle$ we start by noticing that each point in the open half-plane $\{(x, y) \in \mathbb{R}^2 \mid x < 0\}$ is equivalent to a point in the open half-plane $\{(x, y) \in \mathbb{R}^2 \mid x > 0\}$. On the other hand, we have

$$f^2(x, y) = f(f(x, y)) = f(-x, y + 1) = (x, y + 2).$$

Therefore any point in $\mathbb{R}^2$ is equivalent to a point in the strip

$$S = \{(x, y) \in \mathbb{R}^2 \mid x \geq 0 \text{ and } 0 \leq y \leq 2\},$$

and hence $\mathbb{R}^2 / \langle f \rangle$ is homeomorphic to $S / \sim$, where the equivalence relation $\sim$ is defined on S by

$$(x, 0) \sim (x, 2) \quad \text{and} \quad (0, y) \sim (0, y + 1).$$

Now $S/\sim$ is clearly homeomorphic to the semi-infinite cylinder

$$C = \{(x, y, z) \in \mathbb{R}^3 \mid x^2 + y^2 = 1 \text{ and } z \geq 0\}$$

quotiented by the identification $(x, y, 0) \sim (-x, -y, 0)$ on the boundary (cf. Fig. 7.7). This, in turn, is clearly homeomorphic to a projective plane minus a closed disk, which is homeomorphic to a Möbius band (without the boundary).

To determine $\mathbb{R}^2/\langle f, g \rangle$, we notice that any point of $\mathbb{R}^2$ will be equivalent to a point in the rectangle $R = [0, 1] \times [0, 2]$. Moreover, since

$$g(f(x, y)) = g(-x, y + 1) = (1 - x, y + 1),$$

we see that actually any point in $\mathbb{R}^2$ is equivalent to a point in the square $Q = [0, 1] \times [0, 1]$. Therefore $\mathbb{R}^2/\langle f, g \rangle$ is homeomorphic to $Q/\sim$, where the equivalence relation $\sim$ is defined on Q by

$$(0, y) \sim (1, y) \quad \text{and} \quad (x, 0) \sim (1 - x, 1).$$

This is precisely a Klein bottle.

(8) (a) First note that the map

$$f(z) := \frac{az + b}{cz + d}$$

satisfies $f'(z) = \frac{1}{(cz+d)^2}$ and $\text{Im}(f(z)) = \frac{y}{|cz+d|^2}$, where $z = x + iy$. Then, we easily check that g maps H^2 onto itself and that, for $f(z) = v(x, y) + iv(x, y)$,

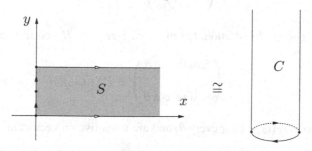

Fig. 7.7 $\mathbb{R}^2/\langle f \rangle$

$$(f^*g)_z\left(\frac{\partial}{\partial x}, \frac{\partial}{\partial x}\right) = g_{f(z)}\left(f_*\frac{\partial}{\partial x}, f_*\frac{\partial}{\partial x}\right)$$

$$= g_{f(z)}\left(\frac{\partial v}{\partial x}\frac{\partial}{\partial x} + \frac{\partial v}{\partial x}\frac{\partial}{\partial y}, \frac{\partial v}{\partial x}\frac{\partial}{\partial x} + \frac{\partial v}{\partial x}\frac{\partial}{\partial y}\right)$$

$$= g_{f(z)}\left(\frac{\partial v}{\partial x}\frac{\partial}{\partial x} - \frac{\partial v}{\partial y}\frac{\partial}{\partial y}, \frac{\partial v}{\partial x}\frac{\partial}{\partial x} - \frac{\partial v}{\partial y}\frac{\partial}{\partial y}\right)$$

$$= \frac{1}{v^2}\left(\left(\frac{\partial v}{\partial x}\right)^2 + \left(\frac{\partial v}{\partial y}\right)^2\right) = \frac{1}{v^2}|f'(z)|^2$$

$$= \frac{1}{y^2} = g_p\left(\frac{\partial}{\partial x}, \frac{\partial}{\partial x}\right),$$

where we used the Cauchy–Riemann equations for f (see for instance [Ahl79]). Similarly, we can see that

$$(f^*g)_p\left(\frac{\partial}{\partial x}, \frac{\partial}{\partial y}\right) = 0 \quad \text{and} \quad (f^*g)_p\left(\frac{\partial}{\partial y}, \frac{\partial}{\partial y}\right) = \frac{1}{y^2}$$

and so f is an isometry of H^2. Moreover, $\det(df) = |f'(z)|^2 > 0$ and so f is orientation-preserving.

(b) Recall that an isometry maps geodesics to geodesics. Hence, if we prove that given two points $p, q \in H^2$ and two unit vectors v, w respectively at p and q there exists $g \in PSL(2, \mathbb{R})$ such that $g(p) = q$ and $(dg)_p v = w$ we are done (here we also denoted by g the map given by $g(z) = g \cdot z$). For that we first see that the orbit of i is all of H^2. Indeed, given any $z_0 = x_0 + iy_0$ with $y_0 > 0$, the map $f_{z_0}(z) := y_0 z + x_0$, corresponding to the matrix

$$\begin{pmatrix} \sqrt{y_0} & \frac{x_0}{\sqrt{y_0}} \\ 0 & \frac{1}{\sqrt{y_0}} \end{pmatrix} \in PSL(2, \mathbb{R}),$$

takes i to z_0. In addition, the maps $r_\theta : H^2 \to H^2$ corresponding to the matrices

$$\begin{pmatrix} \cos\theta & \sin\theta \\ -\sin\theta & \cos\theta \end{pmatrix} \in PSL(2, \mathbb{R})$$

fix i (i.e. $r_\theta(i) = i$ for every θ) and are transitive on vectors at i, since

$$(dr_\theta)_i = \begin{pmatrix} \cos(2\theta) & -\sin(2\theta) \\ \sin(2\theta) & \cos(2\theta) \end{pmatrix}$$

(they act as $SO(2)$ on $T_i H^2$). Hence the map $g = f_q \circ r_\theta \circ f_p^{-1}$, where θ is such that $(dr_\theta)_i (df_p^{-1})_p v = (df_q^{-1})_q w$, takes p to q and is such that $(dg)_p v = w$.

(c) This is an immediate consequence of (b).

(d) Let $f : H^2 \to H^2$ be an orientation-preserving isometry. If f has two fixed points then it must fix the geodesic through them. Let z be a point in this geodesic and choose a positive orthonormal frame $\{E_1, E_2\} \in T_z H^2$ with E_1 tangent to the geodesic. Then $(df)_z E_1 = E_1$, and, since f is an orientation-preserving isometry, $(df)_z E_2 = E_2$. We conclude that $(df)_z$ must be the identity, and so f fixes all geodesics through z, that is, f is the identity map.

Now let $f : H^2 \to H^2$ be any orientation-preserving isometry, choose two points $z_1, z_2 \in H^2$ and let g be the element of $PSL(2, \mathbb{R})$ such that $g(z_1) = f(z_1)$ and $g(z_2) = f(z_2)$, given by (c). Then $f \circ g^{-1}$ is an orientation-preserving isometry with two fixed points, and so it must be the identity, that is, $f = g$.

7.4.5 Section 4.5

(1) (a) This is clear from the fact that $\left(\widetilde{\nabla}_{\widetilde{X}} \widetilde{Y}\right)_{|N}$ depends only on $\widetilde{X}_{|N} = X$ and $\widetilde{Y}_{|N} = Y$.

(b) This is also immediate from the fact that

$$B(X, Y) = \widetilde{\nabla}_{\widetilde{X}} \widetilde{Y} - \left(\widetilde{\nabla}_{\widetilde{X}} \widetilde{Y}\right)^{\mathsf{T}} = \left(\widetilde{\nabla}_{\widetilde{X}} \widetilde{Y}\right)^{\perp}.$$

(c) By Exercise 6.11(7) in Chap. 1 we know that $[\widetilde{X}, \widetilde{Y}]$ is tangent to N. Therefore

$$B(X, Y) - B(Y, X) = \left(\widetilde{\nabla}_{\widetilde{X}} \widetilde{Y}\right)^{\perp} - \left(\widetilde{\nabla}_{\widetilde{Y}} \widetilde{X}\right)^{\perp} = [\widetilde{X}, \widetilde{Y}]^{\perp} = 0.$$

(d) Let f be a function on N and let $\widetilde{f}$ be any extension of f to M. Then

$$B(fX, Y) = \left(\widetilde{\nabla}_{\widetilde{f}\widetilde{X}} \widetilde{Y}\right)^{\perp} = \widetilde{f} \left(\widetilde{\nabla}_{\widetilde{X}} \widetilde{Y}\right)^{\perp} = f B(X, Y),$$

and, by symmetry,

$$B(X, fY) = f B(X, Y).$$

Moreover, it is clear that

$$B(X + Y, Z) = B(X, Z) + B(Y, Z)$$

and
$$B(X, Y + Z) = B(X, Y) + B(X, Z).$$

(e) If $x : V \to \mathbb{R}^n$ are local coordinates around $p \in M$, we have

$$B(X, Y) = B\left(\sum_{i=1}^{n} X^i \frac{\partial}{\partial x^i}, \sum_{j=1}^{n} Y^j \frac{\partial}{\partial x^j}\right) = \sum_{i,j=1}^{n} X^i Y^j B\left(\frac{\partial}{\partial x^i}, \frac{\partial}{\partial x^j}\right).$$

Therefore

$$B(X, Y)_p = \sum_{i,j=1}^{n} X^i(p) Y^j(p) B\left(\frac{\partial}{\partial x^i}, \frac{\partial}{\partial x^j}\right)_p$$

depends only on X_p, Y_p.

(f) From Exercise 6.11(7) in Chap. 1 it is clear that $[\widetilde{X}, \widetilde{Y}]$ is an extension of $[X, Y]$. Therefore

$$\widetilde{\nabla}_{[\widetilde{X}, \widetilde{Y}]} \widetilde{X} - \nabla_{[X,Y]} X = B([X, Y], X)$$

is orthogonal to N.

(4) Let $g : D \subset N \to S^n$ be the Gauss map defined on a neighborhood D of p. Since $\det(dg)_p = (-1)^n \det(S_{n_p}) = (-1)^n K(p) \neq 0$, we may assume that g is a diffeomorphism on D. If ω is the standard volume form of S^n, we have

$$\mathrm{vol}(g(D)) = \int_{g(D)} \omega = \left|\int_{D} g^* \omega\right|$$

(as g may be orientation reversing). If $\{v_1, \ldots, v_n\}$ is an orthonormal basis for $T_q N \cong T_{g(q)} S^n$ with $q \in D$ then

$$g^* \omega(v_1, \ldots, v_n) = \omega((dg)_q v_1, \ldots, (dg)_q v_n)$$
$$= \det(dg)_q \, \omega(v_1, \ldots, v_n) = \det(dg)_q,$$

and hence $g^* \omega = \det(dg) \omega_N$ on D (where ω_N is the volume element of N). We conclude that

$$\mathrm{vol}(g(D)) = \left|\int_{D} \det(dg)\right| = \int_{D} |\det(dg)| = \int_{D} |K|$$

(as $\det(dg)$ does not change sign on D). The result then follows by the mean value theorem:

$$|K(p)| = \lim_{D \to p} \frac{\int_D |K|}{\mathrm{vol}(D)} = \lim_{D \to p} \frac{\mathrm{vol}(g(D))}{\mathrm{vol}(D)}.$$

(5) (a) The ball $B_\varepsilon(p)$ is covered by geodesics of length ε that start at p, defined by $c_v(t) := \exp_p(tv)$, with $0 \leq t \leq \varepsilon$ and $\|v\| = 1$. Hence, $N_p := \exp_p(B_\varepsilon \cap \Pi)$ is formed by the geodesics in $B_\varepsilon(p)$ that are tangent to Π. Let us choose an orthonormal basis $\{(E_1)_p, \ldots, (E_n)_p\}$ of $T_p M$ such that $\{(E_1)_p, (E_2)_p\}$ is a basis of Π. Choosing normal coordinates

$$\varphi(x^1, \ldots, x^n) = \exp_p\left(\sum_{i=1}^n x^i (E_i)_p\right),$$

we have

$$N_p = \{q \in B_\varepsilon(p) \mid x^3(q) = \cdots = x^n(q) = 0\}.$$

We conclude from Exercise 5.9(3) in Chap. 1 that N_p is a 2-dimensional submanifold of M.

(b) From Exercise 4.8(2) in Chap. 3 we know that normal coordinates satisfy

$$\left(\tilde{\nabla}_{\frac{\partial}{\partial x^i}} \frac{\partial}{\partial x^j}\right)_p = 0$$

for $i, j = 1, \ldots, n$, where $\tilde{\nabla}$ is the Levi–Civita connection of M. Consequently,

$$B\left(\left(\frac{\partial}{\partial x^i}\right)_p, \left(\frac{\partial}{\partial x^j}\right)_p\right) = \left(\tilde{\nabla}_{\frac{\partial}{\partial x^i}} \frac{\partial}{\partial x^j}\right)_p^{\perp} = 0,$$

for $i, j = 1, 2$, and hence the second fundamental form of N_p vanishes at p. From Proposition 5.3 we then have

$$K^{N_p}(\Pi) = K^M(\Pi).$$

(6) Let $c : I \to N$ be a geodesic in N parameterized by arc length and tangent at a point $c(s)$ to the principal direction $(E_i)_{c(s)}$ (a unit eigenvector of $S_{n_{c(s)}}$). Then the geodesic curvature (in M) of c is given by

$$k_g(s) = \|\tilde{\nabla}_{\dot{c}(s)} \dot{c}(s)\| = \|B(\dot{c}(s), \dot{c}(s)) - \nabla_{\dot{c}(s)} \dot{c}(s)\|$$
$$= \|B(\dot{c}(s), \dot{c}(s))\| = |\langle\langle S_{n_{c(s)}}(\dot{c}(s)), \dot{c}(s)\rangle\rangle|$$
$$= |\langle\langle S_{n_{c(s)}}((E_i)_{c(s)}), (E_i)_{c(s)}\rangle\rangle| = |\lambda_i|.$$

(7) (a) A parameterization of the paraboloid is, for instance, the map $\varphi : \mathbb{R}^2 \to \mathbb{R}^3$ given by

$$\varphi(v, v) = \left(v, v, \frac{1}{2}(v^2 + v^2)\right).$$

The tangent space to the paraboloid at $\varphi(v, v)$ is generated by

$$\frac{\partial \varphi}{\partial v} = (1, 0, v) \quad \text{and} \quad \frac{\partial \varphi}{\partial v} = (0, 1, v),$$

and a normal vector is

$$\frac{\partial \varphi}{\partial v} \times \frac{\partial \varphi}{\partial v} = (-v, -v, 1).$$

A possible choice for the Gauss map is therefore

$$g(v, v) = \frac{(v, v, -1)}{\sqrt{1 + v^2 + v^2}}.$$

Notice that the image of the paraboloid by this map is contained in the southern hemisphere of S^2, where we can again use the two first coordinate functions of $\mathbb{R}^3$ as coordinates. In these coordinates, the Gauss map is simply written

$$\hat{g}(v, v) = \frac{(v, v)}{\sqrt{1 + v^2 + v^2}},$$

and its derivative is represented by the Jacobian matrix

$$d\hat{g} = \begin{pmatrix} (1 + v^2)(1 + v^2 + v^2)^{-\frac{3}{2}} & -vv(1 + v^2 + v^2)^{-\frac{3}{2}} \\ -vv(1 + v^2 + v^2)^{-\frac{3}{2}} & (1 + v^2)(1 + v^2 + v^2)^{-\frac{3}{2}} \end{pmatrix}.$$

We conclude that the Gauss curvature of the paraboloid is

$$K = \det(dg) = (1 + v^2 + v^2)^{-2}.$$

(b) A parameterization of the saddle surface is, for instance, the map $\varphi : \mathbb{R}^2 \to \mathbb{R}^2$ given by

$$\varphi(v, v) = (v, v, vv).$$

The tangent space to the saddle surface at $\varphi(v, v)$ is generated by

$$\frac{\partial \varphi}{\partial v} = (1, 0, v) \quad \text{and} \quad \frac{\partial \varphi}{\partial v} = (0, 1, v),$$

and a normal vector is

$$\frac{\partial \varphi}{\partial v} \times \frac{\partial \varphi}{\partial v} = (-v, -v, 1).$$

A possible choice for the Gauss map is therefore

$$g(v, v) = \frac{(v, v, -1)}{\sqrt{1 + v^2 + v^2}}.$$

Notice that the image of the saddle surface by this map is contained in the southern hemisphere of S^2, where we can again use the two first coordinate functions of $\mathbb{R}^3$ as coordinates. In these coordinates, the Gauss map is simply written

$$\hat{g}(v, v) = \frac{(v, v)}{\sqrt{1 + v^2 + v^2}},$$

and its derivative is represented by the Jacobian matrix

$$d\hat{g} = \begin{pmatrix} -vv(1 + v^2 + v^2)^{-\frac{3}{2}} & (1 + v^2)(1 + v^2 + v^2)^{-\frac{3}{2}} \\ (1 + v^2)(1 + v^2 + v^2)^{-\frac{3}{2}} & -vv(1 + v^2 + v^2)^{-\frac{3}{2}} \end{pmatrix}.$$

We conclude that the Gauss curvature of the saddle surface is

$$K = \det(dg) = -(1 + v^2 + v^2)^{-2}.$$

7.5 Chapter 5

7.5.1 Section 5.1

(5) The Koszul formula yields

$$2\langle\langle\tilde{\nabla}_X Y, Z\rangle\rangle = X \cdot \langle\langle Y, Z\rangle\rangle + Y \cdot \langle\langle X, Z\rangle\rangle - Z \cdot \langle\langle X, Y\rangle\rangle \\ - \langle\langle[X, Z], Y\rangle\rangle - \langle\langle[Y, Z], X\rangle\rangle + \langle\langle[X, Y], Z\rangle\rangle.$$

Noting that for instance

$$X \cdot \langle\langle Y, Z\rangle\rangle = X \cdot \left(e^{2\rho}\langle Y, Z\rangle\right) = e^{2\rho} X \cdot \langle Y, Z\rangle + 2d\rho(X)e^{2\rho}\langle Y, Z\rangle,$$

it should be clear that

$$2\langle\langle\tilde{\nabla}_X Y, Z\rangle\rangle = 2e^{2\rho}\langle\nabla_X Y, Z\rangle + 2d\rho(X)e^{2\rho}\langle Y, Z\rangle \\ + 2d\rho(Y)e^{2\rho}\langle X, Z\rangle - 2d\rho(Z)e^{2\rho}\langle X, Y\rangle \\ = 2\langle\langle\nabla_X Y, Z\rangle\rangle + 2\langle\langle d\rho(X)Y, Z\rangle\rangle \\ + 2\langle\langle d\rho(Y)X, Z\rangle\rangle - 2e^{2\rho}\langle\text{grad } \rho, Z\rangle\langle X, Y\rangle.$$

Since

$$e^{2\rho}\langle \operatorname{grad} \rho, Z\rangle = \langle\langle \operatorname{grad} \rho, Z\rangle\rangle$$

and Z is arbitrary we obtain

$$\widetilde{\nabla}_X Y = \nabla_X Y + d\rho(X)Y + d\rho(Y)X - \langle X, Y\rangle \operatorname{grad} \rho.$$

(6) Let $s : I \to J$ be a diffeomorphism and $\gamma : J \to M$ the reparameterization of $c : I \to M$ defined by

$$c(t) := \gamma(s(t)).$$

We have

$$\dot{c}(t) = \dot{\gamma}(s(t))\frac{ds}{dt}(t) \tag{7.29}$$

and consequently

$$\nabla_{\dot{c}}\dot{c} = \nabla_{\dot{c}}\left(\frac{ds}{dt}\dot{\gamma}\right) = \left(\frac{d^2s}{dt^2}\right)\dot{\gamma} + \left(\frac{ds}{dt}\right)\nabla_{\frac{ds}{dt}\dot{\gamma}}\dot{\gamma}$$

$$= \frac{d}{dt}\left(\log\left|\frac{ds}{dt}\right|\right)\dot{c} + \left(\frac{ds}{dt}\right)^2 \nabla_{\dot{\gamma}}\dot{\gamma}$$

Therefore, if c is a reparameterization of a geodesic γ, then c satisfies

$$\nabla_{\dot{c}}\dot{c} = f(t)\dot{c}$$

with

$$f(t) = \frac{d}{dt}\left(\log\left|\frac{ds}{dt}\right|\right).$$

On the other hand, if c satisfies (7.29), then the reparameterization γ of c determined by

$$s(t) = \int \exp\left(\int f(t)dt\right) dt$$

is a geodesic.

(8) Recall that the local coordinates $(x^1, \ldots, x^n, v^1, \ldots, v^n)$ on TM parameterize the vector

$$\sum_{i=1}^{n} v^i \frac{\partial}{\partial x^i}$$

which is tangent to M at the point with coordinates $(x^1, \ldots, x^n)$. Therefore, we have

$$K(x^1, \ldots, x^n, v^1, \ldots, v^n) = \frac{1}{2} \sum_{i,j=1}^{n} g_{ij}(x^1, \ldots, x^n) v^i v^j,$$

where

$$g_{ij} = \left\langle \frac{\partial}{\partial x^i}, \frac{\partial}{\partial x^j} \right\rangle$$

are the components of the metric in this coordinate system. Consequently,

$$\frac{\partial K}{\partial v^i} = \sum_{j=1}^{n} g_{ij} v^j$$

and hence

$$\frac{\partial K}{\partial v^i}(x(t), \dot{x}(t)) = \sum_{j=1}^{n} g_{ij}(x(t)) \dot{x}^j(t),$$

leading to

$$\frac{d}{dt}\left(\frac{\partial K}{\partial v^i}(x(t), \dot{x}(t))\right) = \sum_{j=1}^{n} g_{ij}(x(t)) \ddot{x}^j(t) + \sum_{j,k=1}^{n} \frac{\partial g_{ij}}{\partial x^k}(x(t)) \dot{x}^k(t) \dot{x}^j(t).$$

Moreover,

$$\frac{\partial K}{\partial x^i} = \frac{1}{2} \sum_{j,k=1}^{n} \frac{\partial g_{jk}}{\partial x^i} v^j v^k,$$

and hence

$$\frac{\partial K}{\partial x^i}(x(t), \dot{x}(t)) = \frac{1}{2} \sum_{j,k=1}^{n} \frac{\partial g_{jk}}{\partial x^i}(x(t)) \dot{x}^j(t) \dot{x}^k(t).$$

We conclude that

$$\frac{d}{dt}\left(\frac{\partial K}{\partial v^i}(x(t), \dot{x}(t))\right) - \frac{\partial K}{\partial x^i}(x(t), \dot{x}(t)) = \sum_{j=1}^{n} g_{ij}(x(t)) \ddot{x}^j(t)$$

$$+ \sum_{j,k=1}^{n} \left(\frac{\partial g_{ij}}{\partial x^k}(x(t)) - \frac{1}{2}\frac{\partial g_{jk}}{\partial x^i}(x(t))\right) \dot{x}^j(t) \dot{x}^k(t).$$

On the other hand, if $v, w \in T_p M$ are written as

$$v = \sum_{i=1}^{n} v^i \frac{\partial}{\partial x^i}, \quad w = \sum_{i=1}^{n} w^i \frac{\partial}{\partial x^i}$$

then we have

$$\mu(v)(w) = \sum_{i,j=1}^{n} g_{ij} v^i w^j = \sum_{i,j=1}^{n} g_{ij} v^i dx^j(w),$$

and hence

$$\mu(v) = \sum_{i,j=1}^{n} g_{ij} v^i dx^j = \sum_{i,j=1}^{n} g_{ij} v^j dx^i.$$

Therefore

$$\mu\left(\frac{D\dot{c}}{dt}(t)\right) = \sum_{i,j=1}^{n} g_{ij}(x(t)) \left(\ddot{x}^j(t) + \sum_{k,l=1}^{n} \Gamma_{kl}^j(x(t))\dot{x}^k(t)\dot{x}^l(t)\right) dx^i.$$

Since

$$\sum_{j=1}^{n} g_{ij} \Gamma_{kl}^j = \frac{1}{2} \sum_{j,m=1}^{n} g_{ij} g^{jm} \left(\frac{\partial g_{ml}}{\partial x^k} + \frac{\partial g_{mk}}{\partial x^l} - \frac{\partial g_{kl}}{\partial x^m}\right)$$

$$= \frac{1}{2} \left(\frac{\partial g_{il}}{\partial x^k} + \frac{\partial g_{ik}}{\partial x^l} - \frac{\partial g_{kl}}{\partial x^i}\right),$$

we have

$$\sum_{j,k,l=1}^{n} g_{ij}(x(t)) \Gamma_{kl}^j(x(t))\dot{x}^k(t)\dot{x}^l(t)$$

$$= \frac{1}{2} \sum_{k,l=1}^{n} \left(\frac{\partial g_{il}}{\partial x^k}(x(t)) + \frac{\partial g_{ik}}{\partial x^l}(x(t)) - \frac{\partial g_{kl}}{\partial x^i}(x(t))\right) \dot{x}^k(t)\dot{x}^l(t)$$

$$= \sum_{j,k=1}^{n} \left(\frac{\partial g_{ij}}{\partial x^k}(x(t)) - \frac{1}{2}\frac{\partial g_{jk}}{\partial x^i}(x(t))\right) \dot{x}^j(t)\dot{x}^k(t),$$

which completes the proof.

7.5.2 Section 5.2

(4) (a) The position of the particle m_1 is

$$(x_1, y_1) = l_1(\sin\theta, -\cos\theta).$$

Its velocity is therefore

$$(\dot{x}_1, \dot{y}_1) = l_1\dot{\theta}(\cos\theta, \sin\theta),$$

yielding the kinetic energy

$$K_1 = \frac{1}{2}m_1l_1{}^2\dot{\theta}^2.$$

Analogously, the position of the particle m_2 is

$$(x_2, y_2) = l_1(\sin\theta, -\cos\theta) + l_2(\sin\varphi, -\cos\varphi),$$

and its velocity is

$$(\dot{x}_2, \dot{y}_2) = l_1\dot{\theta}(\cos\theta, \sin\theta) + l_2\dot{\varphi}(\cos\varphi, \sin\varphi),$$

corresponding to the kinetic energy

$$K_2 = \frac{1}{2}m_2l_1{}^2\dot{\theta}^2 + \frac{1}{2}m_2l_2{}^2\dot{\varphi}^2 + m_2l_1l_2\cos(\theta - \varphi)\dot{\theta}\dot{\varphi}.$$

The kinetic energy map $K : TN \to \mathbb{R}$ is then given in local coordinates by

$$K = \frac{1}{2}(m_1 + m_2)l_1{}^2(v^\theta)^2 + \frac{1}{2}m_2l_2{}^2(v^\varphi)^2 + m_2l_1l_2\cos(\theta - \varphi)v^\theta v^\varphi.$$

Since the potential energy is clearly

$$U = m_1gy_1 + m_2gy_2 = -(m_1 + m_2)gl_1\cos\theta - m_2gl_2\cos\varphi,$$

the equations of motion are

$$(m_1 + m_2)l_1{}^2\ddot{\theta} + m_2l_1l_2\cos(\theta - \varphi)\ddot{\varphi}$$
$$+ m_2l_1l_2\sin(\theta - \varphi)\dot{\varphi}^2 = -(m_1 + m_2)gl_1\sin\theta$$

and

$$m_2l_2{}^2\ddot{\varphi} + m_2l_1l_2\cos(\theta - \varphi)\ddot{\theta} - m_2l_1l_2\sin(\theta - \varphi)\dot{\theta}^2$$
$$= -m_2gl_2\sin\varphi.$$

(b) The linearized equations are

$$(m_1 + m_2)l_1{}^2\ddot{\theta} + m_2l_1l_2\ddot{\varphi} = -(m_1 + m_2)gl_1\theta$$

and

$$m_2 l_2{}^2 \ddot{\varphi} + m_2 l_1 l_2 \ddot{\theta} = -m_2 g l_2 \varphi.$$

Solutions of these equations satisfying $\varphi = k\theta$ must simultaneously solve

$$\left(m_1 l_1{}^2 + m_2 l_1{}^2 + k m_2 l_1 l_2 \right) \ddot{\theta} = -(m_1 + m_2) g l_1 \theta$$

and

$$\left(k m_2 l_2{}^2 + m_2 l_1 l_2 \right) \ddot{\theta} = -k m_2 g l_2 \theta.$$

Therefore k must satisfy

$$\begin{vmatrix} (m_1 + m_2) l_1{}^2 + k m_2 l_1 l_2 & (m_1 + m_2) g l_1 \\ k m_2 l_2{}^2 + m_2 l_1 l_2 & k m_2 g l_2 \end{vmatrix} = 0,$$

that is

$$k = \frac{l_2 - l_1 \pm \sqrt{(l_2 - l_1)^2 + 4 \mu l_1 l_2}}{2 \mu l_2},$$

where

$$\mu = \frac{m_2}{m_1 + m_2} \in (0, 1).$$

Notice that k has two possible values, one positive and one negative, corresponding to the two pendulums oscillating in phase or in opposition of phase.

From the first linearized equation of motion it is clear that the period of the oscillations is

$$2\pi \sqrt{\frac{l_1 + k \mu l_2}{g}} = 2\pi \sqrt{\frac{l_1 + l_2 \pm \sqrt{(l_2 - l_1)^2 + 4 \mu l_1 l_2}}{2g}}.$$

Notice that the period is longer when the two pendulums oscillate in phase, and shorter when they oscillate out of phase.

7.5.3 Section 5.3

(3) Assume without loss of generality that m is supported in the plane $\xi^3 = 0$, and consider the measure
$$m_\varepsilon := m + \varepsilon \delta_{e_3} + \varepsilon \delta_{-e_3}.$$

Since this measure is not supported on a plane, the covariant acceleration $\nabla^\varepsilon_{\dot{S}} \dot{S}$ determined by the left-invariant metric

$$\langle\langle V, W\rangle\rangle_\varepsilon := \int_{\mathbb{R}^3} \langle V\xi, W\xi\rangle \, dm_\varepsilon$$

satisfies

$$\langle\langle \nabla_{\dot{S}}^\varepsilon \dot{S}, V\rangle\rangle_\varepsilon = \int_{\mathbb{R}^3} \langle \ddot{S}\xi, V\xi\rangle \, dm_\varepsilon.$$

Now

$$\langle\langle V, W\rangle\rangle_\varepsilon = \int_{\mathbb{R}^3} \langle V\xi, W\xi\rangle \, dm + 2\varepsilon\langle Ve_3, We_3\rangle$$

converges to the left-invariant metric

$$\langle\langle V, W\rangle\rangle = \int_{\mathbb{R}^3} \langle V\xi, W\xi\rangle \, dm$$

as $\varepsilon \to 0$, and therefore $\nabla_{\dot{S}}^\varepsilon \dot{S}$ converges to the covariant acceleration $\nabla_{\dot{S}} \dot{S}$ determined by $\langle\langle \cdot, \cdot\rangle\rangle$. Hence

$$\langle\langle \nabla_{\dot{S}} \dot{S}, V\rangle\rangle = \lim_{\varepsilon \to 0} \langle\langle \nabla_{\dot{S}}^\varepsilon \dot{S}, V\rangle\rangle_\varepsilon = \lim_{\varepsilon \to 0} \int_{\mathbb{R}^3} \langle \ddot{S}\xi, V\xi\rangle \, dm_\varepsilon$$

$$= \lim_{\varepsilon \to 0} \left(\int_{\mathbb{R}^3} \langle \ddot{S}\xi, V\xi\rangle \, dm + 2\varepsilon\langle \ddot{S}e_3, Ve_3\rangle \right) = \int_{\mathbb{R}^3} \langle \ddot{S}\xi, V\xi\rangle \, dm.$$

(4) Since $\mathfrak{so}(3) = \{A \in \mathfrak{gl}(3) \mid A + A^t = 0\}$ is the space of 3×3 skew-symmetric matrices, we can define the map $\Omega : \mathfrak{so}(3) \to \mathbb{R}^3$ that, given

$$A = \begin{pmatrix} 0 & -a_3 & a_2 \\ a_3 & 0 & -a_1 \\ -a_2 & a_1 & 0 \end{pmatrix} \in \mathfrak{so}(3),$$

yields

$$\Omega(A) = (a_1, a_2, a_3).$$

The map Ω is clearly a linear isomorphism and it is easy to check that the Lie bracket on $\mathfrak{so}(3)$ is identified with the exterior product, i.e.

$$\Omega([A, B]) = \Omega(AB - BA) = \Omega(A) \times \Omega(B).$$

Moreover, given $\xi \in \mathbb{R}^3$, we have

$$A\xi = \begin{pmatrix} 0 & -a_3 & a_2 \\ a_3 & 0 & -a_1 \\ -a_2 & a_1 & 0 \end{pmatrix} \begin{pmatrix} \xi^1 \\ \xi^2 \\ \xi^3 \end{pmatrix} = \begin{pmatrix} a_2\xi^3 - a_3\xi^2 \\ a_3\xi^1 - a_1\xi^3 \\ a_1\xi^2 - a_2\xi^1 \end{pmatrix}$$

$$= (a_1, a_2, a_3) \times \xi = \Omega(A) \times \xi.$$

(7) (a) We have

$$(SIS^t)v = S(I(S^tv)) = S \int_{\mathbb{R}^3} [\xi \times (S^t v \times \xi)]dm$$

$$= \int_{\mathbb{R}^3} S[\xi \times (S^t v \times \xi)]dm$$

$$= \int_{\mathbb{R}^3} [(S\xi) \times (v \times (S\xi))]dm$$

$$= \int_{\mathbb{R}^3} [\xi \times (v \times \xi))]dm = Iv,$$

where we used the fact that S preserves the mass distribution to change variables.

(b) Let v be a nonzero vector orthogonal to the reflection plane. Then $Sv = -v$. Moreover,

$$S(Iv) = SI(S^t S)v = (SIS^t)Sv = ISv = -Iv,$$

implying that Iv is parallel to v and so there exists a principal axis with the direction of v.

(c) Let v be a vector with the direction of the rotation axis. Then $Sv = S^t v = v$. Moreover,

$$S^t(Iv) = S^t(SIS^t)v = IS^tv = Iv,$$

implying that Iv is also fixed by S^t and consequently by S. Hence Iv has the same direction as v and so the rotation axis is principal.

(d) Let v be again a vector with the direction of the rotation axis. Then, if w is a vector perpendicular to v, we know that $S^t w$ is also perpendicular to v. Moreover, if w is an eigenvector of I with eigenvalue α (i.e. if $Iw = \alpha w$) we have

$$I(S^t w) = (S^t S)IS^t w = S^t(SIS^t)w = S^t I w = \alpha S^t w$$

and so $S^t w$ is also an eigenvector of I associated to α. Considering a basis $\{e_1, e_2, e_3\}$ of $\mathbb{R}^3$ formed by principal axes so that e_1 is parallel to v, we have that if S is not a rotation by π then $S^t e_2$ is also an eigenvector for the eigenvalue I_2, independent of e_2 and perpendicular to e_1. Hence, the I_2-eigenspace contains span $\{e_2, e_3\}$. Consequently, every vector orthogonal to v is an I_2-eigenvector and so all axes orthogonal to the axis of rotation are principal.

(13) Since the kinetic energy will not depend on φ or ψ, we can assume without loss of generality that $\varphi = \psi = 0$. For this choice, the velocity of a curve on $SO(3)$ is given in terms of the Euler angles by

$$\dot{S} = \dot{\varphi} \begin{pmatrix} 0 & -1 & 0 \\ 1 & 0 & 0 \\ 0 & 0 & 0 \end{pmatrix} \begin{pmatrix} 1 & 0 & 0 \\ 0 & \cos\theta & -\sin\theta \\ 0 & \sin\theta & \cos\theta \end{pmatrix} + \dot{\theta} \begin{pmatrix} 0 & 0 & 0 \\ 0 & -\sin\theta & -\cos\theta \\ 0 & \cos\theta & -\sin\theta \end{pmatrix}$$
$$+ \dot{\psi} \begin{pmatrix} 1 & 0 & 0 \\ 0 & \cos\theta & -\sin\theta \\ 0 & \sin\theta & \cos\theta \end{pmatrix} \begin{pmatrix} 0 & -1 & 0 \\ 1 & 0 & 0 \\ 0 & 0 & 0 \end{pmatrix},$$

and hence

$$A = S^{-1}\dot{S} = \dot{\varphi} \begin{pmatrix} 0 & -\cos\theta & \sin\theta \\ \cos\theta & 0 & 0 \\ -\sin\theta & 0 & 0 \end{pmatrix} + \dot{\theta} \begin{pmatrix} 0 & 0 & 0 \\ 0 & 0 & -1 \\ 0 & 1 & 0 \end{pmatrix} + \dot{\psi} \begin{pmatrix} 0 & -1 & 0 \\ 1 & 0 & 0 \\ 0 & 0 & 0 \end{pmatrix},$$

corresponding to

$$\Omega = \dot{\varphi}(\cos\theta e_3 + \sin\theta e_2) + \dot{\theta} e_1 + \dot{\psi} e_3 = \dot{\theta} e_1 + \dot{\varphi}\sin\theta e_2 + (\dot{\varphi}\cos\theta + \dot{\psi})e_3.$$

The kinetic energy is therefore

$$K = \frac{1}{2}\langle I\Omega, \Omega \rangle = \frac{I_1}{2}\left(\dot{\theta}^2 + \dot{\varphi}^2 \sin^2\theta\right) + \frac{I_3}{2}\left(\dot{\varphi}\cos\theta + \dot{\psi}\right)^2.$$

(14) (a) Using $U = Mgl\cos\theta$ one readily obtains the equations of motion:

$$\begin{cases} \dfrac{d}{dt}\left(I_1\dot{\theta}\right) - I_1\sin\theta\cos\theta\dot{\varphi}^2 + I_3\sin\theta\dot{\varphi}\left(\dot{\varphi}\cos\theta + \dot{\psi}\right) = Mgl\sin\theta \\[2mm] \dfrac{d}{dt}\left(I_1\dot{\varphi}\sin^2\theta + I_3\cos\theta\left(\dot{\varphi}\cos\theta + \dot{\psi}\right)\right) = 0 \\[2mm] \dfrac{d}{dt}\left(I_3\left(\dot{\varphi}\cos\theta + \dot{\psi}\right)\right) = 0 \end{cases}.$$

The equilibrium points are obtained by setting $\dot{\theta} \equiv \dot{\varphi} \equiv \dot{\psi} \equiv 0$ in the equations, and are given by the condition $\sin\theta = 0$ (assuming $l \neq 0$). These correspond to the top being at rest in a vertical position (possibly upside-down).

(b) It is easy to see that one can solve the equations of motion by setting θ, $\dot{\varphi}$
and $\dot{\psi}$ constant, provided that these constants satisfy

$$-I_1 \sin\theta \cos\theta \dot{\varphi}^2 + I_3 \sin\theta \dot{\varphi} \left(\dot{\varphi}\cos\theta + \dot{\psi}\right) = Mgl \sin\theta.$$

If $|\dot{\varphi}| \ll |\dot{\psi}|$, these conditions can approximately be written as

$$I_3 \sin\theta \dot{\varphi}\dot{\psi} \simeq Mgl \sin\theta \Leftrightarrow \dot{\varphi} \simeq \frac{Mgl}{I_3\dot{\psi}}.$$

7.5.4 Section 5.4

(3) Let Σ be an integrable distribution. Then for every $p \in M$ there exists an open
set U around p and local coordinates $(x^1, \ldots, x^n) : U \to \mathbb{R}^n$ such that the
connected components of the intersection of the leaves with U are the level sets
of $(x^{m+1}, \ldots, x^n) : U \to \mathbb{R}^{n-m}$. Hence, if $X, Y \in \mathfrak{X}(\Sigma)$ then

$$X = \sum_{i=1}^{m} X^i \frac{\partial}{\partial x^i}, \qquad Y = \sum_{i=1}^{m} Y^i \frac{\partial}{\partial x^i}$$

on U. Consequently,

$$[X, Y] = \sum_{i=1}^{m} (X \cdot Y^i - Y \cdot X^i) \frac{\partial}{\partial x^i}$$

on U. Since U is arbitrary, we see that $[X, Y] \in \mathfrak{X}(U)$.

(5) Locally it is always possible to complete $\{\omega^1, \ldots, \omega^{n-m}\}$ to a local coframe
$\{\theta^1, \ldots, \theta^m, \omega^1, \ldots, \omega^{n-m}\}$. Let $\{X_1, \ldots, X_m, Y_1, \ldots, Y_{n-m}\}$ be the dual local
frame, so that Σ is locally given by $\{X_1, \ldots, X_m\}$. We have

$$dω^i (X_j, X_k) = X_j \cdot \omega^i (X_k) - X_k \cdot \omega^i (X_j) - \omega^i ([X_j, X_k])$$
$$= -\omega^i ([X_j, X_k]),$$

and therefore the distribution will be integrable if and only if

$$dω^i (X_j, X_k) = 0$$

for all $i = 1, \ldots, n - m$ and $j, k = 1, \ldots, m$. Writing

$$dω^i = \sum_{j,k=1}^{m} a^i_{jk}\theta^j \wedge \theta^k + \sum_{j=1}^{m}\sum_{k=1}^{n-m} b^i_{jk}\theta^j \wedge \omega^k + \sum_{j,k=1}^{n-m} c^i_{jk}\omega^j \wedge \omega^k,$$

we have

$$dw^i(X_j, X_k) = a^i_{jk},$$

and hence the integrability condition is equivalent to requiring that the functions a^i_{jk} vanish. Since

$$dw^i \wedge w^1 \wedge \cdots \wedge w^{n-m} = \sum_{j,k=1}^{m} a^i_{jk}\theta^j \wedge \theta^k \wedge w^1 \wedge \ldots \wedge w^{n-m},$$

this is the same as requiring that

$$dw^i \wedge w^1 \wedge \cdots \wedge w^{n-m} = 0$$

(assuming $m \geq 2$; for $m = 1$ both this condition and the integrability conditions are trivially satisfied).

(7) We have

$$\frac{dE}{dt} = \frac{d}{dt}\left(\frac{1}{2}\langle \dot{c}(t), \dot{c}(t)\rangle + U(c(t))\right) = \left\langle \frac{D\dot{c}}{dt}(t), \dot{c}(t)\right\rangle + (dU)_{c(t)}(\dot{c}(t))$$

$$= \mu\left(\frac{D\dot{c}}{dt}\right)(\dot{c}) - \mathcal{F}(\dot{c})(\dot{c}) = \mathcal{R}(\dot{c})(\dot{c}) = 0,$$

since for a perfect reaction force

$$\mathcal{R}(\dot{c})(\dot{c}) = \left\langle \mu^{-1}(\mathcal{R}(\dot{c})), \dot{c}\right\rangle = 0.$$

(8) (a) Given two points $p = (x_0, y_0, \theta_0)$ and $q = (x_1, y_1, \theta_1)$ in $\mathbb{R}^2 \times S^1$ consider the curve $c : [0, 1] \to \mathbb{R}^2 \times S^1$ given by

$$c(t) := \begin{cases} (x_0, y_0, 3(\theta_L - \theta_0)\, t + \theta_0), & \text{if } t \in [0, \tfrac{1}{3}] \\[2mm] (x_0 + (3t - 1)(x_1 - x_0),\, y_0 + (3t - 1)(y_1 - y_0), \theta_L), & \text{if } t \in [\tfrac{1}{3}, \tfrac{2}{3}] \\[2mm] (x_1, y_1, (3t - 2)(\theta_1 - \theta_L) + \theta_L), & \text{if } t \in [\tfrac{2}{3}, 1], \end{cases}$$

where

$$(\cos\theta_L, \sin\theta_L) = \frac{(x_1 - x_0,\, y_1 - y_0)}{\sqrt{(x_1 - x_0)^2 + (y_1 - y_0)^2}}$$

if $(x_0, y_0) \neq (x_1, y_1)$, and $\theta_L = \theta_0$ otherwise. Clearly c is continuous, piecewise smooth, $c(0) = p$ and $c(1) = q$. Moreover,

$$\dot{c}(t) = \begin{cases} 3(\theta_L - \theta_0)Y, & \text{if } t \in (0, \tfrac{1}{3}) \\ 3\sqrt{(x_1 - x_0)^2 + (y_1 - y_0)^2}\, X, & \text{if } t \in (\tfrac{1}{3}, \tfrac{2}{3}) \\ 3(\theta_1 - \theta_L)Y, & \text{if } t \in (\tfrac{2}{3}, 1) \end{cases}$$

with

$$X = \cos\theta_L \frac{\partial}{\partial x} + \sin\theta_L \frac{\partial}{\partial y} \quad \text{and} \quad Y = \frac{\partial}{\partial\theta},$$

and so c is compatible with Σ. The set of points accessible from p by a compatible curve is therefore $\mathbb{R}^2 \times S^1$, and so Σ cannot be integrable (if Σ were integrable any compatible curve would be restricted to an integral submanifold).

(b) Since the kinetic energy is given by

$$K = \frac{M}{2}\left((v^x)^2 + (v^y)^2\right) + \frac{I}{2}(v^\theta)^2$$

we have

$$\frac{\partial K}{\partial x} = \frac{\partial K}{\partial y} = \frac{\partial K}{\partial\theta} = 0,$$

$$\frac{\partial K}{\partial v^x} = Mv^x, \quad \frac{\partial K}{\partial v^y} = Mv^y, \quad \frac{\partial K}{\partial v^\theta} = Iv^\theta,$$

and so

$$\mu\left(\frac{D\dot{c}}{dt}\right) = M\ddot{x}\,dx + M\ddot{y}\,dy + I\ddot{\theta}\,d\theta.$$

Moreover, since the reaction force is perfect, we have

$$\mathcal{R} = \lambda\omega = -\lambda\sin\theta\,dx + \lambda\cos\theta\,dy.$$

The motion of the ice skate is then given by a solution of the equation of motion

$$\mu\left(\frac{D\dot{c}}{dt}\right) = \mathcal{R}(\dot{c})$$

which also satisfies the constraint that $(\dot{x}, \dot{y})$ is proportional to $(\cos\theta, \sin\theta)$, i.e. it is a solution of the system of ODEs

$$\begin{cases} M\ddot{x} = -\lambda\sin\theta \\ M\ddot{y} = \lambda\cos\theta \\ \ddot{\theta} = 0 \\ \dot{y}\cos\theta - \dot{x}\sin\theta = 0. \end{cases}$$

Hence $\theta(t) = \theta_0 + kt$ for some constant $k \in \mathbb{R}$.

If $k \neq 0$, differentiating $\dot{y} \cos \theta - \dot{x} \sin \theta = 0$ yields

$$\ddot{y} = \ddot{x} \tan \theta + \frac{k}{\cos^2 \theta} \dot{x} \Leftrightarrow \frac{\lambda}{M} \cos \theta = \ddot{x} \tan \theta + \frac{k}{\cos^2 \theta} \dot{x}$$

$$\Leftrightarrow -\frac{\cos \theta}{\sin \theta} \ddot{x} = \ddot{x} \tan \theta + \frac{k}{\cos^2 \theta} \dot{x} \Leftrightarrow \frac{\ddot{x}}{\dot{x}} = -k \tan \theta.$$

Then $\log |\dot{x}| = \log |\cos \theta| + \text{constant}$, yielding

$$\begin{cases} x(t) = r \sin(\theta_0 + kt) + A_0 \\[2mm] y(t) = -r \cos(\theta_0 + kt) + B_0 \\[2mm] \theta(t) = \theta_0 + kt \\[2mm] \lambda = Mk^2 r, \end{cases}$$

where r, A_0, B_0, θ_0, k are integration constants. Notice that

$$(x(t) - A_0)^2 + (y(t) - B_0)^2 = r^2$$

and so $c(t)$ traces out a circle of center $(A_0, B_0) \in \mathbb{R}^2$ and radius $|r|$ with constant speed $|kr|$. The reaction force can be interpreted as a friction force which does not allow the ice skate to slide sideways, forcing its trajectory to curve.

If $k = 0$, differentiating $\dot{y} \cos \theta - \dot{x} \sin \theta = 0$ yields

$$\ddot{y} \cos \theta_0 - \ddot{x} \sin \theta_0 = 0 \Leftrightarrow \frac{\lambda}{M} = 0,$$

and so

$$\begin{cases} x(t) = l \cos \theta_0\, t + x_0 \\[2mm] y(t) = l \sin \theta_0\, t + y_0 \\[2mm] \theta(t) = \theta_0 \\[2mm] \lambda = 0, \end{cases}$$

where x_0, y_0, θ_0, l are integration constants. Notice that in this case $c(t)$ traces out a straight line through (x_0, y_0) of slope $\tan \theta_0$ with constant speed $|l|$. Since the ice skate is sliding along its length, the reaction force vanishes in this case.

(c) The equation of motion is now

$$\mu\left(\frac{D\dot{c}}{dt}\right) = -dU + \mathcal{R}(\dot{c})$$

$$\Leftrightarrow M\ddot{x}dx + M\ddot{y}dy + I\ddot{\theta}d\theta = -Mg\sin\alpha dx - \lambda\sin\theta dx + \lambda\cos\theta dy.$$

The motion of the ice skate is then given by a solution of this equation that also satisfies the constraint equation, i.e. a solution of the system of ODEs

$$\begin{cases} M\ddot{x} = -Mg\sin\alpha - \lambda\sin\theta \\[2mm] M\ddot{y} = \lambda\cos\theta \\[2mm] \ddot{\theta} = 0 \\[2mm] \dot{y}\cos\theta - \dot{x}\sin\theta = 0. \end{cases}$$

Hence $\theta(t) = \theta_0 + kt$ for some constant $k \in \mathbb{R}$. If $k \neq 0$, differentiating $\dot{y}\cos\theta - \dot{x}\sin\theta = 0$ yields

$$\ddot{y} = \ddot{x}\tan\theta + \frac{k}{\cos^2\theta}\dot{x} \Leftrightarrow \frac{\lambda}{M}\cos\theta = \ddot{x}\tan\theta + \frac{k}{\cos^2\theta}\dot{x}.$$

From the first equation of the system of ODEs we obtain

$$\frac{\lambda}{M} = -g\frac{\sin\alpha}{\sin\theta} - \frac{1}{\sin\theta}\ddot{x},$$

and so, substituting above,

$$-\frac{\cos\theta}{\sin\theta}(g\sin\alpha + \ddot{x}) = \ddot{x}\tan\theta + \frac{k}{\cos^2\theta}\dot{x},$$

implying that

$$\frac{\ddot{x}}{\cos\theta} + k\frac{\sin\theta}{\cos^2\theta}\dot{x} = -g\sin\alpha\cos\theta$$

or, equivalently, that

$$\frac{d}{dt}\left(\frac{\dot{x}}{\cos\theta}\right) = -g\sin\alpha\cos\theta.$$

Hence

$$\dot{x} = -\frac{g}{k}\sin\alpha\sin\theta\cos\theta + l\cos\theta$$

for some integration constant $l \in \mathbb{R}$, and

$$\dot{y} = -\frac{g}{k} \sin \alpha \sin^2 \theta + l \sin \theta.$$

Differentiating this last equation yields

$$\ddot{y} = (kl - 2g \sin \alpha \sin \theta) \cos \theta$$

and so

$$\lambda = M(kl - 2g \sin \alpha \sin \theta).$$

We then obtain

$$\begin{cases} x(t) = \frac{g}{4k^2} \sin \alpha \cos(2(\theta_0 + kt)) + \frac{l}{k} \sin(\theta_0 + kt) + A_0 \\[2mm] y(t) = -\frac{g}{2k} \sin \alpha \left(t - \frac{1}{2k} \sin(2(\theta + kt))\right) - \frac{l}{k} \cos(\theta_0 + kt) + B_0 \\[2mm] \theta(t) = \theta_0 + kt \\[2mm] \lambda = M (kl - 2g \sin \alpha \sin(\theta_0 + kt)), \end{cases}$$

where θ_0, k, l, A_0, B_0 are integration constants. It is interesting to notice that unlike what one might expect $x(t)$ remains bounded, whereas $y(t)$ grows linearly.

If $k = 0$ then, again differentiating $\dot{y} \cos \theta - \dot{x} \sin \theta = 0$, we obtain

$$\begin{cases} x(t) = -\frac{g}{2} \sin \alpha \cos^2 \theta_0 \, t^2 + l \cos \theta_0 \, t + x_0 \\[2mm] y(t) = -\frac{g}{2} \sin \alpha \sin \theta_0 \cos \theta_0 \, t^2 + l \sin \theta_0 t + y_0 \\[2mm] \theta(t) = \theta_0 \\[2mm] \lambda = -Mg \sin \alpha \sin \theta_0, \end{cases}$$

where θ_0, l, x_0, y_0 are integration constants. As one would expect, the motion in this case is uniformly accelerated with acceleration $g \sin \alpha \cos \theta_0$.

7.5.5 Section 5.5

(1) If $c \in \mathcal{C}$ is a critical point of the action, $(U, x^1, \ldots, x^n)$ is a local chart and $t \in (a, b)$ is such that $c(t) \in U$ then we can find $\varepsilon > 0$ such that

$c([t - \varepsilon, t + \varepsilon]) \subset U$. Considering variations which coincide with c outside $[t - \varepsilon, t + \varepsilon]$, we conclude that c must satisfy the Euler–Lagrange equations on this local chart in the time interval $(t - \varepsilon, t + \varepsilon)$. Hence any critical point of the action must satisfy the Euler–Lagrange equations on the local chart $(U, x^1, \ldots, x^n)$ for all $t \in (a, b)$ such that $c(t) \in U$.

Conversely, suppose that $c \in C$ satisfies the Euler–Lagrange equations on any local chart. We introduce an auxiliary Riemannian metric on M and consider normal balls with center at the points of $c([a, b])$. Because $c([a, b])$ is compact, we can choose such balls such that the infimum of their radii is positive (consider an open cover of $c([a, b])$ by totally normal neighborhoods). Using the fact that the length of c is necessarily finite, we can choose a finite number of these balls (which are coordinate charts for the normal coordinates), $B_1, \ldots, B_N$, and points $t_1, \ldots, t_{N-1}$, with $a < t_1 < \cdots < t_{N-1} < b$, such that $c(a) \in B_1$, $c(b) \in B_N$ and

$$c(t_i) \in B_i \cap B_{i+1} \quad (i = 1, \ldots, N - 1).$$

For an arbitrary variation γ given by $\widetilde{\gamma} : (-\varepsilon, \varepsilon) \times [a, b] \to M$, we have, repeating the calculation in the proof of Theorem 5.3,

$$\frac{d}{ds}\bigg|_{s=0} \int_{t_{i-1}}^{t_i} L(\gamma(s))dt = (\mathbb{F}L)_{\dot{c}(t_i)}\left(\frac{\partial\widetilde{\gamma}}{\partial s}(0, t_i)\right) - (\mathbb{F}L)_{\dot{c}(t_{i-1})}\left(\frac{\partial\widetilde{\gamma}}{\partial s}(0, t_{i-1})\right)$$

for $i = 2, \ldots, N - 1$, where we used the fact that c satisfies the Euler–Lagrange equations on B_i. Analogously,

$$\frac{d}{ds}\bigg|_{s=0} \int_{a}^{t_1} L(\gamma(s))dt = (\mathbb{F}L)_{\dot{c}(t_1)}\left(\frac{\partial\widetilde{\gamma}}{\partial s}(0, t_1)\right)$$

and

$$\frac{d}{ds}\bigg|_{s=0} \int_{t_{N-1}}^{b} L(\gamma(s))dt = -(\mathbb{F}L)_{\dot{c}(t_{N-1})}\left(\frac{\partial\widetilde{\gamma}}{\partial s}(0, t_{N-1})\right).$$

Adding these formulae we finally obtain

$$\frac{d}{ds}\bigg|_{s=0} \int_{a}^{b} L(\gamma(s))dt = 0.$$

(5) (a) If we identify $\mathbb{R}^2$ with the $z = 0$ plane in $\mathbb{R}^3$ then the rotating frame's angular velocity is $\Omega = e_z$. According to Exercise 3.20(11), the third particle's equations of motion are

$$m(\ddot{x}, \ddot{y}, \ddot{z}) = (F_x, F_y, F_z) - m\Omega \times (\Omega \times (x, y, z))$$
$$- 2m\Omega \times (\dot{x}, \dot{y}, \dot{z}) - m\dot{\Omega} \times (x, y, z)$$
$$= (F_x, F_y, F_z) + m(x, y, 0) - 2m(-\dot{y}, \dot{x}, 0).$$

The first two components of this equation are precisely

$$\begin{cases} \ddot{x} = \dfrac{F_x}{m} + x + 2\dot{y} \\[2mm] \ddot{y} = \dfrac{F_y}{m} + y - 2\dot{x} \end{cases}$$

(The third component is $m\ddot{z} = F_z$, and thus requires $F_z = 0$ for a particle moving in the orbital plane).

(b) First note that

$$r_1 = \left((x - 1 + \mu)^2 + y^2\right)^{\frac{1}{2}} \quad \text{and} \quad r_2 = \left((x + \mu)^2 + y^2\right)^{\frac{1}{2}}.$$

Hence,

$$\frac{\partial L}{\partial x} = v^y + x - \frac{\mu}{r_1{}^3}(x - 1 + \mu) - \frac{1 - \mu}{r_2{}^3}(x + \mu) \quad \text{and}$$
$$\frac{\partial L}{\partial y} = -v^x + y - \frac{\mu}{r_1{}^3}y - \frac{1 - \mu}{r_2{}^3}y.$$

Moreover,

$$\frac{\partial L}{\partial v^x} = v^x - y \quad \text{and} \quad \frac{\partial L}{\partial v^y} = v^y + x,$$

and so

$$\frac{d}{dt}\left(\frac{\partial L}{\partial v^x}(x, y, \dot{x}, \dot{y})\right) = \ddot{x} - \dot{y} \quad \text{and}$$
$$\frac{d}{dt}\left(\frac{\partial L}{\partial v^y}(x, y, \dot{x}, \dot{y})\right) = \ddot{y} + \dot{x}.$$

Hence the Euler–Lagrange equations are

$$\begin{cases} \ddot{x} - 2\dot{y} = x - \dfrac{\mu}{r_1{}^3}(x - 1 + \mu) - \dfrac{1 - \mu}{r_2{}^3}(x + \mu) \\[3mm] \ddot{y} + 2\dot{x} = y - \dfrac{\mu}{r_1{}^3}y - \dfrac{1 - \mu}{r_2{}^3}y \end{cases},$$

which are exactly the equations of motion in the rotating frame.
(c) The Hamiltonian function is

$$H = v^x \frac{\partial L}{\partial v^x} + v^y \frac{\partial L}{\partial v^y} - L$$

$$= \frac{1}{2}\left((v^x)^2 + (v^y)^2\right) - \frac{1}{2}(x^2 + y^2) - \frac{\mu}{r_1} - \frac{1-\mu}{r_2}.$$

(d) Let us now find the equilibrium points, i.e. constant solutions of the equations of motion. Since in this case we have

$$\dot{x} = \ddot{x} = \dot{y} = \ddot{y} = 0,$$

we obtain

$$\begin{cases} x - \dfrac{\mu}{r_1{}^3}(x - 1 + \mu) - \dfrac{1-\mu}{r_2{}^3}(x + \mu) = 0 \\[4mm] y\left(1 - \dfrac{\mu}{r_1{}^3} - \dfrac{1-\mu}{r_2{}^3}\right) = 0 \end{cases}$$

If $y \neq 0$ we obtain from the second equation that

$$1 - \frac{\mu}{r_1{}^3} - \frac{1-\mu}{r_2{}^3} = 0.$$

Rewriting the first equation as

$$x\left(1 - \frac{\mu}{r_1{}^3} - \frac{1-\mu}{r_2{}^3}\right) + \mu(1 - \mu)\left(\frac{1}{r_1{}^3} - \frac{1}{r_2{}^3}\right) = 0$$

and using the second equation we get that $r_1 = r_2 = 1$ at the equilibrium point, which in turn satisfies the first equation. Hence we have an equilibrium point $(\frac{1}{2} - \mu, \frac{\sqrt{3}}{2})$ with $y > 0$ and another equilibrium point $(\frac{1}{2} - \mu, -\frac{\sqrt{3}}{2})$ with $y < 0$. Note that these two points are equidistant from the two massive particles.
If $y = 0$ then the equilibrium points are given by the critical points of the function

$$U(x) = \frac{1}{2}x^2 + \frac{\mu}{|x + 1 - \mu|} + \frac{1-\mu}{|x + \mu|}.$$

Since

$$U''(x) = 1 + \frac{2\mu}{|x + 1 - \mu|^3} + \frac{2(1-\mu)}{|x + \mu|^3} > 0,$$

and

$$\lim_{x \to \pm\infty} U(x) = \lim_{x \to \mu} U(x) = \lim_{x \to \mu-1} U(x) = +\infty,$$

we see that U has exactly three critical points, which are local minima, one in each interval $(-\infty, 1 - \mu)$, $(1 - \mu, \mu)$ and $(\mu, +\infty)$.

(e) To linearize the system we make the substitution

$$\begin{cases} x = \frac{1}{2} - \mu + \xi \\ y = \pm\frac{\sqrt{3}}{2} + \eta \end{cases}$$

and notice that at the equilibrium points

$$\frac{\partial r_1}{\partial x} = \frac{x - 1 + \mu}{r_1} = -\frac{1}{2}; \qquad \frac{\partial r_1}{\partial y} = \frac{y}{r_1} = \pm\frac{\sqrt{3}}{2};$$

$$\frac{\partial r_2}{\partial x} = \frac{x + \mu}{r_2} = \frac{1}{2}; \qquad \frac{\partial r_2}{\partial y} = \frac{y}{r_2} = \pm\frac{\sqrt{3}}{2},$$

so that

$$\frac{1}{r_1{}^3} = 1 + \frac{3}{2}\xi \mp \frac{3\sqrt{3}}{2}\eta + \cdots$$

and

$$\frac{1}{r_2{}^3} = 1 - \frac{3}{2}\xi \mp \frac{3\sqrt{3}}{2}\eta + \cdots.$$

Substituting on the equations of motion yields the linearized system

$$\begin{cases} \ddot{\xi} - 2\dot{\eta} = -\mu\left(\frac{3}{2}\xi \mp \frac{3\sqrt{3}}{2}\eta\right)\left(-\frac{1}{2}\right) - (1 - \mu)\left(-\frac{3}{2}\xi \mp \frac{3\sqrt{3}}{2}\eta\right)\left(\frac{1}{2}\right) \\ \\ \ddot{\eta} + 2\dot{\xi} = -\mu\left(\frac{3}{2}\xi \mp \frac{3\sqrt{3}}{2}\eta\right)\left(\pm\frac{\sqrt{3}}{2}\right) - (1 - \mu)\left(-\frac{3}{2}\xi \mp \frac{3\sqrt{3}}{2}\eta\right)\left(\pm\frac{\sqrt{3}}{2}\right) \end{cases}$$

or, equivalently,

$$\begin{cases} \ddot{\xi} - 2\dot{\eta} = \frac{3}{4}\xi \pm \frac{3\sqrt{3}}{4}(1 - 2\mu)\eta \\ \\ \ddot{\eta} + 2\dot{\xi} = \pm\frac{3\sqrt{3}}{4}(1 - 2\mu)\xi + \frac{9}{4}\eta \end{cases}$$

The matrix of corresponding linear first-order system is

$$A = \begin{bmatrix} 0 & 0 & 1 & 0 \\ 0 & 0 & 0 & 1 \\ \frac{3}{4} & \pm\frac{3\sqrt{3}}{4}(1-2\mu) & 0 & 2 \\ \pm\frac{3\sqrt{3}}{4}(1-2\mu) & \frac{9}{4} & -2 & 0 \end{bmatrix},$$

and has characteristic polynomial

$$\det(A - \lambda I) = \lambda^4 + \lambda^2 + \frac{27}{4}\mu(1-\mu).$$

The roots of this polynomial satisfy

$$\lambda^2 = \frac{-1 \pm \sqrt{1 - 27\mu(1-\mu)}}{2},$$

and so at least one will have a positive real part unless they are all pure imaginary. So the equilibrium point is unstable exactly when

$$1 - 27\mu(1-\mu) < 0 \Leftrightarrow \frac{1 - \frac{\sqrt{69}}{9}}{2} < \mu < \frac{1 + \frac{\sqrt{69}}{9}}{2}.$$

(8) (a) Given $S \in SO(3)$ we have

$$\frac{d}{dt}\Big|_{t=0} \exp(tB)S = BS = (dR_S)_I\, B = \left(X^B\right)_S,$$

where X^B is the right-invariant vector field determined by B.

(b) The Lagrangian for the free rigid body is

$$L(V) = \frac{1}{2}\langle\langle V, V \rangle\rangle$$

and is clearly $SO(3)$-invariant (because $\langle\langle \cdot, \cdot \rangle\rangle$ is). Since

$$(\mathbb{F}L)_V(W) = \langle\langle V, W \rangle\rangle,$$

The Noether theorem guarantees that the quantity

$$J^B = (\mathbb{F}L)_{\dot{S}}(X^B) = \langle\langle \dot{S}, BS \rangle\rangle = \langle\langle SA, BS \rangle\rangle$$

is conserved along the motions of the system for any $B \in \mathfrak{so}(3)$, where as usual we have written $\dot{S} = SA$. Setting $\Omega = \Omega(A)$ and $\Sigma = \Omega(B)$, we have

$$J^B = \int_{\mathbb{R}^3} \langle SA\xi, BS\xi \rangle\, dm = \int_{\mathbb{R}^3} \langle S(\Omega \times \xi), \Sigma \times (S\xi) \rangle\, dm$$

$$= \int_{\mathbb{R}^3} \langle \Omega \times \xi, (S^t \Sigma) \times \xi \rangle \, dm = \int_{\mathbb{R}^3} \langle (S^t \Sigma), \xi \times (\Omega \times \xi) \rangle \, dm$$

$$= \left\langle S^t \Sigma, \int_{\mathbb{R}^3} [\xi \times (\Omega \times \xi)] \, dm \right\rangle = \langle S^t \Sigma, P \rangle = \langle \Sigma, SP \rangle.$$

Since B, and thus Σ, is arbitrary, we conclude that the vector $p = SP$ is conserved.

7.5.6 Section 5.6

(1) Let us consider the canonical symplectic form on T^*M given in local coordinates $(x^1, \ldots, x^n, p_1, \ldots, p_n)$ by

$$\omega = \sum_{j=1}^{n} dp_j \wedge dx^j.$$

This form is clearly closed. To show that it is nondegenerate at $\alpha \in T^*M$ let us consider a vector $v \in T_\alpha(T^*M)$ such that $\omega(v, w) = 0$ for every $w \in T_\alpha(T^*M)$. Then, writing

$$v = \sum_{j=1}^{n} a^j \frac{\partial}{\partial x^j} + b_j \frac{\partial}{\partial p_j}$$

and setting $w := \frac{\partial}{\partial p_i}$ for $i \in \{1, \ldots, n\}$, we obtain

$$0 = \omega(v, w) = \sum_{j=1}^{n} (dp_j \otimes dx^j - dx^j \otimes dp_j) \left(v, \frac{\partial}{\partial p_i} \right) = -a^i.$$

If, instead, we use $w := \frac{\partial}{\partial x^i}$ for $i \in \{1, \ldots, n\}$, we get

$$0 = \omega(v, w) = \sum_{j=1}^{n} (dp_j \otimes dx^j - dx^j \otimes dp_j) \left(v, \frac{\partial}{\partial x^i} \right) = b_i.$$

We conclude that $v = 0$, and hence ω is nondegenerate. Finally, the form

$$\omega^n = \omega \wedge \cdots \wedge \omega = \left(\sum_{j=1}^{n} dp_j \wedge dx^j \right) \wedge \cdots \wedge \left(\sum_{j=1}^{n} dp_j \wedge dx^j \right)$$

$$= n! \, dp_1 \wedge dx^1 \wedge \cdots \wedge dp_n \wedge dx^n =$$

$$= n! \, (-1)^{\frac{n(n-1)}{2}} dp_1 \wedge \cdots \wedge dp_n \wedge dx^1 \wedge \cdots \wedge dx^n \neq 0$$

is a volume form on T^*M.

(2) (a) In local coordinates we have

$$L(x^1, \ldots, x^n, v^1, \ldots, v^n) = \frac{1}{2} \sum_{i,j=1}^{n} g_{ij}(x^1, \ldots, x^n) v^i v^j$$

$$+ \sum_{i=1}^{n} \alpha_i(x^1, \ldots, x^n) v^i - U(x^1, \ldots, x^n) = K + C - U$$

where $\alpha = \sum_{i=1}^{n} \alpha_i dx^i$ and $C = \sum_{i=1}^{n} \alpha_i v^i$. Now we know that

$$\mu\left(\frac{D\dot{c}}{dt}(t)\right) + (dU)_{c(t)}$$

is given in local coordinates by

$$\sum_{i=1}^{n} \left[\frac{d}{dt}\left(\frac{\partial K}{\partial v^i}(x(t), \dot{x}(t))\right) - \frac{\partial K}{\partial x^i}(x(t), \dot{x}(t)) \right] dx^i + \sum_{i=1}^{n} \frac{\partial U}{\partial x^i}(x(t)) dx^i.$$

Moreover,

$$\sum_{i=1}^{n} \left[\frac{d}{dt}\left(\frac{\partial C}{\partial v^i}(x(t), \dot{x}(t))\right) - \frac{\partial C}{\partial x^i}(x(t), \dot{x}(t)) \right] dx^i$$

$$= \sum_{i=1}^{n} \left[\frac{d}{dt}(\alpha_i(x(t))) - \sum_{j=1}^{n} \frac{\partial \alpha_j}{\partial x^i}(x(t)) \dot{x}^j(t) \right] dx^i$$

$$= \sum_{i=1}^{n} \left[\sum_{j=1}^{n} \left(\frac{\partial \alpha_i}{\partial x^j}(x(t)) - \frac{\partial \alpha_j}{\partial x^i}(x(t)) \right) \dot{x}^j(t) \right] dx^i,$$

and so $\iota(\dot{c}(t)) d\alpha$ is given in local coordinates by

$$\sum_{i=1}^{n} \left[\frac{d}{dt}\left(\frac{\partial C}{\partial v^i}(x(t), \dot{x}(t))\right) - \frac{\partial C}{\partial x^i}(x(t), \dot{x}(t)) \right] dx^i.$$

Then the Euler–Lagrange equations are equivalent to

$$\sum_{i=1}^{n}\left[\frac{d}{dt}\left(\frac{\partial L}{\partial v^i}(x(t),\dot{x}(t))\right) - \frac{\partial L}{\partial x^i}(x(t),\dot{x}(t))\right]dx^i = 0$$

$$\Leftrightarrow \sum_{i=1}^{n}\left[\frac{d}{dt}\left(\frac{\partial(K+C)}{\partial v^i}(x(t),\dot{x}(t))\right)\right.$$

$$\left. - \frac{\partial(K+C)}{\partial x^i}(x(t),\dot{x}(t)) + \frac{\partial U}{\partial x^i}(x(t))\right]dx^i = 0$$

$$\Leftrightarrow \mu\left(\frac{D\dot{c}}{dt}(t)\right) + \iota(\dot{c}(t)) + d\alpha(dU)_{c(t)} = 0,$$

and the result follows.

(b) We have

$$\frac{dE}{dt}(t) = \frac{d}{dt}\left(\frac{1}{2}\langle\dot{c}(t),\dot{c}(t)\rangle + U(c(t))\right) = \left\langle\frac{D\dot{c}}{dt}(t),\dot{c}(t)\right\rangle + (dU)_{c(t)}(\dot{c}(t))$$

$$= \mu\left(\frac{D\dot{c}}{dt}\right)(\dot{c}) + dU(\dot{c}) = -\iota(\dot{c})d\alpha(\dot{c}) = -d\alpha(\dot{c},\dot{c}) = 0.$$

(c) In local coordinates, the Legendre transformation is given by

$$p_i = \frac{\partial L}{\partial v^i} = \sum_{j=1}^{n} g_{ij}v^j + \alpha_i \quad (i = 1,\ldots,n),$$

and can be readily inverted:

$$v^i = \sum_{j=1}^{n} g^{ij}(p_j - \alpha_j) \quad (i = 1,\ldots,n).$$

This shows that the Lagrangian is hyper-regular.

(d) As a function on the tangent bundle, the Hamiltonian is

$$H = \sum_{i=1}^{n} v^i\frac{\partial L}{\partial v^i} - L = \sum_{i,j=1}^{n} g_{ij}v^iv^j + \sum_{i=1}^{n}\alpha_iv^i - L = \frac{1}{2}\sum_{i,j=1}^{n} g_{ij}v^iv^j + U.$$

Therefore, as a function on the cotangent bundle, it is given by

$$H = \frac{1}{2}\sum_{i,j,k,l=1}^{n} g_{ij}g^{ik}g^{jl}(p_k - \alpha_k)(p_l - \alpha_l) + U$$

$$= \frac{1}{2}\sum_{i,j=1}^{n} g^{ij}(p_i - \alpha_i)(p_j - \alpha_j) + U,$$

and the Hamilton equations are

$$
\begin{cases}
\dot{x}^i = \displaystyle\sum_{j=1}^{n} g^{ij}(p_j - \alpha_j) \\[2em]
\dot{p}_i = -\dfrac{1}{2} \displaystyle\sum_{k,l=1}^{n} \dfrac{\partial g^{kl}}{\partial x^i}(p_k - \alpha_k)(p_l - \alpha_l) + \displaystyle\sum_{k,l=1}^{n} g^{kl}\dfrac{\partial \alpha_k}{\partial x^i}(p_l - \alpha_l) - \dfrac{\partial U}{\partial x^i}
\end{cases}.
$$

(4) If the set of points $p \in U$ such that $\psi_t(p) \in U$ for some $t \geq T$ were not dense in U then there would exist an open set $V \subset U$ such that $\psi_t(V) \cap U = \varnothing$ for all $t \geq T$. But the Poincaré recurrence theorem applied to V would guarantee the existence of a point $p \in V$ such that $\psi_t(p) \in V \subset U$ for some $t \geq T$, and hence $p \in \psi_t(V) \cap U$.

(5) The geodesic flow corresponds to the Hamiltonian given in local coordinates by

$$
H(x^1, \ldots, x^n, p_1, \ldots, p_n) = \frac{1}{2} \sum_{i,j=1}^{n} g^{ij}(x^1, \ldots, x^n) p_i p_j.
$$

It is easily seen that the (conserved) value of H is just $\frac{1}{2}\|\dot{c}(t)\|^2$ for each geodesic $c : \mathbb{R} \to M$ given by the flow. Since M is compact, the set $K = H^{-1}([\frac{1}{8}, \frac{1}{2}])$ is also compact, as it is diffeomorphic to

$$
\left\{ v \in TM \mid \frac{1}{2} \leq \|v\| \leq 1 \right\}.
$$

Consider the open set

$$
U = \left\{ \alpha \in T^*M \mid \frac{1}{8} < H(\alpha) < \frac{1}{2} \text{ and } \pi(\alpha) \in B \right\} \subset K.
$$

By the Poincaré recurrence theorem there exists $\alpha \in U$ such that $\psi_{t_1}(\alpha) \in U$ for $t_1 \geq T$. Now the projection of $\psi_t(\alpha)$ on M is a geodesic $\gamma(t)$ such that $\gamma(0) \in B$ and $\gamma(t_1) \in B$. Moreover, $\|\dot{\gamma}(t)\| = k$ for some $\frac{1}{2} < k < 1$. Therefore $c(t) := \gamma(t/k)$ is a geodesic with $\|\dot{c}(t)\| = 1$ which satisfies $c(0) \in B$ and $c(t_1/k) \in B$, i.e. $c(t) \in B$ for some $t > T$.

7.5.7 Section 5.7

(2) Let $F_1, \ldots, F_m \in C^\infty(T^*M)$ be in involution and independent at some point $\alpha \in T^*M$ and let $X_{F_1}, \ldots, X_{F_m}$ be the corresponding Hamiltonian vector fields. Then, at α, the vectors $(X_{F_i})_\alpha \in T_\alpha(T^*M)$ are linearly independent. Indeed, if

$$\sum_{i=1}^{m} a_i \, (X_{F_i})_\alpha = 0$$

for some $a_1, \ldots, a_m \in \mathbb{R}$, then

$$0 = \iota \left(\sum_{i=1}^{m} a_i \, (X_{F_i})_\alpha \right) \omega_\alpha = \sum_{i=1}^{m} a_i \, \iota \left((X_{F_i})_\alpha \right) \omega_\alpha = - \sum_{i=1}^{m} a_i \, (dF_i)_\alpha$$

and so $a_1 = \cdots = a_m = 0$ since $(dF_1)_\alpha, \ldots, (dF_m)_\alpha$ are linearly independent. On the other hand, we have

$$\omega_\alpha \left((X_{F_i})_\alpha, (X_{F_j})_\alpha \right) = \{F_i, F_j\}(\alpha) = 0. \tag{7.30}$$

Let us take the m-dimensional space $W = \mathrm{span}\{(X_{F_1})_\alpha, \ldots, (X_{F_m})_\alpha\}$ and its symplectic orthogonal

$$W^\omega = \{w \in T_\alpha(T^*M) \mid \omega_\alpha(v, w) = 0 \text{ for all } v \in W\}.$$

Then

$$2n = \dim T_\alpha(T^*M) = \dim W + \dim W^\omega.$$

Indeed, if we consider the map

$$\Phi : T_\alpha(T^*M) \to W^*$$
$$v \mapsto (\iota(v)\omega_\alpha)_{|W}$$

we can easily see that $\ker \Phi = W^\omega$ and $\mathrm{im}\, \Phi = W^*$ (since any element in W^* can be extended to an element in $T^*_\alpha(T^*M)$ and the map $v \mapsto \iota(v)\omega_\alpha$ is an isomorphism between $T_\alpha(T^*M)$ and $T^*_\alpha(T^*M)$). Hence,

$$2n = \dim T_\alpha(T^*M) = \dim W^\omega + \dim W^* = \dim W^\omega + \dim W.$$

Since, on the other hand, we clearly have $W \subset W^\omega$ by (7.30), we conclude that

$$2n = \dim W^\omega + \dim W \geq 2 \dim W = 2m$$

and so $n \geq m$.

(3) (a) We know that the geodesics of M are the critical points of the action determined by $L = \frac{1}{2}\langle v, v \rangle$, where $\langle \cdot, \cdot \rangle$ is the metric induced in M by the Euclidean metric of $\mathbb{R}^3$. If $i : M \to \mathbb{R}^3$ is the standard inclusion, we have

$$i^*g = i^*(dx \otimes dx + dy \otimes dy + dz \otimes dz)$$
$$= (f(z))^2 d\theta \otimes d\theta + ((f'(z))^2 + 1)dz \otimes dz,$$

since

$$x = f(z) \cos \theta \quad \text{and} \quad y = f(z) \sin \theta,$$

and so

$$i^* dx = d(f(z) \cos \theta) = -f(z) \sin \theta d\theta + f'(z) \cos \theta dz$$

and

$$i^* dy = d(f(z) \sin \theta) = f(z) \cos \theta d\theta + f'(z) \sin \theta dz.$$

Therefore the Lagrangian for the geodesics of M, given by the kinetic energy, is

$$L(\theta, z, v^\theta, v^z) = \frac{1}{2} \left((f(z))^2 (v^\theta)^2 + \left((f'(z))^2 + 1 \right) (v^z)^2 \right).$$

(b) Since

$$\frac{\partial L}{\partial v^\theta} = (f(z))^2 v^\theta, \quad \frac{\partial L}{\partial v^z} = ((f'(z))^2 + 1) v^z,$$

$$\frac{\partial L}{\partial \theta} = 0, \quad \frac{\partial L}{\partial z} = f(z) f'(z) (v^\theta)^2 + f'(z) f''(z) (v^z)^2,$$

the Euler–Lagrange equations are

$$\begin{cases} f(z) \ddot{\theta} + 2 f'(z) \dot{\theta} \dot{z} = 0 \\[2mm] ((f'(z))^2 + 1) \ddot{z} + f'(z) f''(z) \dot{z}^2 - f(z) f'(z) \dot{\theta}^2 = 0. \end{cases}$$

If $\dot{\theta} = 0$ then the first Euler–Lagrange equation is trivially satisfied. Moreover, the second equation becomes

$$\frac{\ddot{z}}{\dot{z}} + \frac{f'(z) f''(z)}{(f'(z))^2 + 1} \dot{z} = 0$$

and then

$$\frac{d}{dt} \left(\log \dot{z} + \frac{1}{2} \log((f'(z))^2 + 1) \right) = 0.$$

Hence,

$$((f'(z))^2 + 1)^{1/2} \dot{z} = k$$

for some positive integration constant k and so

$$\frac{d}{dt} \left(\int_{z_0}^{z} ((f'(s))^2 + 1)^{1/2} \, ds \right) = k.$$

Noting that

$$G(z) = \int_{z_0}^{z} ((f'(s))^2 + 1)^{1/2} \, ds$$

is an increasing function of z (since $((f'(s))^2 + 1)^{1/2}$ is positive), and hence injective, we obtain the trajectory given in local coordinates by

$$\begin{cases} \theta(t) = \theta_0, \\ z(t) = G^{-1}(kt) \end{cases}$$

for some constant $\theta_0 \in \mathbb{R}$.

On the other hand if the trajectory satisfies $f'(z(t)) = 0$ the Euler–Lagrange equations become

$$\ddot{\theta} = 0 \quad \text{and} \quad \ddot{z} = 0.$$

Hence, $\theta = \theta_0 + kt$ and $z(t) = z_0 + lt$ for some integration constants k, l. Since we need $f'(z(t)) = 0$, if f is not a constant function we get $l = 0$ and $z(t) = z_0$, where z_0 is a critical point of f, and we obtain trajectories

$$\theta = \theta_0 + kt \quad \text{and} \quad z(t) = z_0.$$

If f is a constant function (that is if S is a cylinder) then any trajectory satisfying $\theta = \theta_0 + kt$ and $z(t) = z_0 + lt$ is a solution.

(c) The Legendre transformation is given in these coordinates by

$$\begin{cases} p_\theta = \dfrac{\partial L}{\partial v^\theta} = (f(z))^2 \, v^\theta \\[2mm] p_z = \dfrac{\partial L}{\partial v^z} = ((f'(z))^2 + 1) \, v^z \end{cases} \Leftrightarrow \begin{cases} v^\theta = \dfrac{p_\theta}{(f(z))^2} \\[2mm] v^z = \dfrac{p_z}{(f'(z))^2 + 1}. \end{cases}$$

Since it is clearly invertible, L is hyper-regular. The Hamiltonian function is then

$$H(\theta, z, p_\theta, p_z) = p_\theta v^\theta + p_z v^z - L = \frac{p_\theta^2}{(f(z))^2} + \frac{p_z^2}{(f'(z))^2 + 1} - L$$

$$= \frac{p_\theta^2}{2(f(z))^2} + \frac{p_z^2}{2((f'(z))^2 + 1)}.$$

(d) By the Hamilton equations,

$$\dot{p}_\theta = -\frac{\partial H}{\partial \theta} = 0$$

and hence p_θ is a first integral. Now

$$dH = \frac{p_\theta}{(f(z))^2}\,dp_\theta + \frac{p_z}{(f'(z))^2 + 1}\,dp_z$$
$$- \left[\frac{f'(z)p_\theta^2}{(f(z))^3} + \frac{f'(z)f''(z)p_z^2}{((f'(z))^2 + 1)^2}\right]dz$$

and hence dH and dp_θ are linearly independent on the dense open set of T^*M formed by the points whose coordinates (z, θ) are well defined and do not satisfy

$$p_z = f'(z)\left[\frac{p_\theta^2}{(f(z))^3} + \frac{f''(z)p_z^2}{((f'(z))^2 + 1)^2}\right] = 0.$$

Thus they are independent for instance whenever $p_z \neq 0$, i.e. outside a 3-dimensional submanifold, and so H and p_θ are independent on a dense open set.

(e) The equations for this level set are $p_\theta = l$ and

$$H = E \Leftrightarrow \frac{l^2}{(f(z))^2} + \frac{p_z^2}{(f'(z))^2 + 1} = 2E$$
$$\Leftrightarrow \frac{p_z^2}{(f'(z))^2 + 1} = 2E - \frac{l^2}{(f(z))^2}.$$

These can be solved for p_z on the set of points for which the right-hand side is nonnegative, i.e. for

$$f(z) \geq \frac{l}{\sqrt{2E}}.$$

If f has a strict local maximum at $z = z_0$ then the projections of invariant level sets $L_{(E,l)}$ close to the geodesic with image $z = z_0$ will be sets of the form $z_{min} \leq z \leq z_{max}$, with z_{min}, z_{max} close to z_0 and satisfying $z_{min} < z_0 < z_{max}$. Thus geodesics with initial condition close to a vector tangent to $z = z_0$ will remain close to $z = z_0$, meaning that this geodesic is stable.

7.5.8 Section 5.8

(2) (a) If $\omega = \sum_{i=1}^{n} dp_i \wedge dx^i$ then we saw in Exercise 6.15(6) that $\{x^i, x^j\} = \{p_i, p_j\} = 0$ and $\{p_i, x^j\} = \delta_{ij}$ for $i, j = 1, \ldots, n$. On the other hand, if this latter condition holds then

$$X_{x^i} = \sum_{i=1}^{n} \left(\{x^i, x^j\} \frac{\partial}{\partial x^j} + \{x^i, p_j\} \frac{\partial}{\partial p_j} \right) = -\frac{\partial}{\partial p_i}$$

and

$$X_{p_i} = \sum_{i=1}^{n} \left(\{p_i, x^j\} \frac{\partial}{\partial x^j} + \{p_i, p_j\} \frac{\partial}{\partial p_j} \right) = \frac{\partial}{\partial x^i},$$

implying that

$$\omega \left(\frac{\partial}{\partial x^i}, \frac{\partial}{\partial x^j} \right) = \omega(X_{p_i}, X_{p_j}) = \{p_i, p_j\} = 0;$$

$$\omega \left(\frac{\partial}{\partial p_i}, \frac{\partial}{\partial p_j} \right) = \omega(X_{x^i}, X_{x^j}) = \{x^i, x^j\} = 0;$$

$$\omega \left(\frac{\partial}{\partial p_i}, \frac{\partial}{\partial x^j} \right) = -\omega(X_{x^i}, X_{p_j}) = -\{x^i, p_j\} = \delta_{ij},$$

and so $\omega = \sum_{i=1}^{n} dp_i \wedge dx^i$.

(b) It is immediate from the Darboux theorem and Exercise 6.15(1) that $\omega^n = \omega \wedge \cdots \wedge \omega$ is a volume form on S.

(c) Assume that S is compact. If $\omega = d\theta$ then $\omega^n = d(\theta \wedge \omega \wedge \cdots \wedge \omega)$ (as $d\omega = 0$), and so by the Stokes theorem we have

$$\int_S \omega^n = \int_S d(\theta \wedge \omega \wedge \cdots \wedge \omega) = 0,$$

which is a contradiction.

(4) (a) We have

$$B^{ij} = B(dx^i, dx^j) = \{x^i, x^j\}$$

for $i, j = 1, \ldots, n$.

(b) If $F, G \in C^{\infty}(M)$ then

$$X_F \cdot G = \{F, G\} = B(dF, dG) = \sum_{i,j=1}^{n} B^{ij} \frac{\partial F}{\partial x^i} \frac{\partial G}{\partial x^j}$$

$$= \left(\sum_{i,j=1}^{n} B^{ij} \frac{\partial F}{\partial x^i} \frac{\partial}{\partial x^j} \right) \cdot G.$$

(c) From the Jacobi identity we have

$$\{x^i, \{x^j, x^k\}\} + \{x^j, \{x^k, x^i\}\} + \{x^k, \{x^i, x^j\}\} = 0$$
$$\Leftrightarrow \{x^i, B^{jk}\} + \{x^j, B^{ki}\} + \{x^k, B^{ij}\} = 0$$
$$\Leftrightarrow X_{x^i} \cdot B^{jk} + X_{x^j} \cdot B^{ki} + X_{x^k} \cdot B^{ij} = 0$$

for $i, j, k = 1, \ldots, n$. Noticing that

$$X_{x^i} = \sum_{l=1}^{n} B^{il} \frac{\partial}{\partial x^l}$$

we obtain

$$\sum_{l=1}^{n} \left(B^{il} \frac{\partial B^{jk}}{\partial x^l} + B^{jl} \frac{\partial B^{ki}}{\partial x^l} + B^{kl} \frac{\partial B^{ij}}{\partial x^l} \right) = 0.$$

(d) The definition of the Hamiltonian vector field X_F on a symplectic manifold (M, ω) can be written in local coordinates as

$$\iota(X_F)\omega = -dF \Leftrightarrow \sum_{i,j=1}^{n} X_F^i \omega_{ij} dx^i \otimes dx^j = -\sum_{i=1}^{n} \frac{\partial F}{\partial x^i} dx^i.$$

Using the antisymmetry of (ω_{ij}) we then have

$$\sum_{j=1}^{n} \omega_{ij} X_F^j = \frac{\partial F}{\partial x^i} \Leftrightarrow X_F^i = \sum_{j=1}^{n} \omega^{ij} \frac{\partial F}{\partial x^j}$$

where $(\omega^{ij}) := (\omega^{ij})^{-1}$. Since by (b)

$$X_F^i = \sum_{j=1}^{n} B^{ji} \frac{\partial F}{\partial x^j} = -\sum_{j=1}^{n} B^{ij} \frac{\partial F}{\partial x^j}$$

we must have $(B^{ij}) = -(\omega^{ij}) = -(\omega_{ij})^{-1}$.

(e) If B is nondegenerate then we can define a linear isomorphism $\Phi : T_p^* M \to T_p M$ for each $p \in M$ through

$$\Phi(\omega)(\eta) = B(\omega, \eta)$$

for all $\omega, \eta \in T_p^* M$. In local coordinates, we have

$$\Phi(\omega)(\eta) = \sum_{i,j=1}^{n} B^{ij} \omega_i \eta_j = \sum_{i,j=1}^{n} B^{ij} \omega_i \frac{\partial}{\partial x^j}(\eta),$$

that is,

$$\Phi(\omega) = \sum_{i,j=1}^{n} B^{ij} \omega_i \frac{\partial}{\partial x^j}.$$

We can then define a 2-form $\omega \in \Omega^2(M)$ through

$$\omega(v, w) = B(\Phi^{-1}(v), \Phi^{-1}(w))$$

for all $v, w \in T_p M$. Setting $(B_{ij}) := (B^{ij})^{-1}$, we can write ω in local coordinates as

$$\omega(v, w) = \sum_{i,j,k,l=1}^{n} B^{ij}(-B_{ik}v^k)(-B_{jl}v^l)$$

$$= \sum_{k,l=1}^{n} B_{lk}v^k v^l = - \sum_{i,j=1}^{n} B_{ij}v^i v^j,$$

that is, $(\omega_{ij}) = -(B_{ij})^{-1}$. This shows that ω determines the same Poisson bracket as B and is nondegenerate. To show that it is closed we notice that

$$\sum_{i,j,k,l=1}^{n} B_{pi} B_{qj} B_{rk} B^{il} \frac{\partial B^{jk}}{\partial x^l} = - \sum_{j,k=1}^{n} B_{qj} \frac{\partial B^{jk}}{\partial x^p} B_{kr} = \frac{\partial B_{qr}}{\partial x^p}$$

for $p, q, r = 1, \ldots, n$, and so, multiplying the formula in (c) by $B_{pi} B_{qj} B_{rk}$ and summing over i, j, k we have

$$\frac{\partial B_{qr}}{\partial x^p} + \frac{\partial B_{rp}}{\partial x^q} + \frac{\partial B_{pq}}{\partial x^r} = 0.$$

This is equivalent to

$$\sum_{i,j,k=1}^{n} \left(\frac{\partial \omega_{jk}}{\partial x^i} + \frac{\partial \omega_{ki}}{\partial x^j} + \frac{\partial \omega_{ij}}{\partial x^k} \right) dx^i \otimes dx^j \otimes dx^k = 0,$$

or, noticing that the expression in brackets is antisymmetric in each pair of indices, to

$$\sum_{i,j,k=1}^{n} \frac{\partial \omega_{jk}}{\partial x^i} dx^i \wedge dx^j \wedge dx^k = 0 \Leftrightarrow d\omega = 0.$$

(9) (a) From the expression of the group operation it is clear that

$$(x, y)^{-1} = \left(-\frac{x}{y}, \frac{1}{y}\right),$$

and so

$$L_{(a,b)^{-1}}(x, y) = \left(\frac{x}{b} - \frac{a}{b}, \frac{y}{b}\right).$$

Therefore, by Example 5.4, the lift of the action of H on itself to T^*H is given by

$$(a, b) \cdot (p_x dx + p_y dy) = \left(L_{(a,b)^{-1}}\right)^* (p_x dx + p_y dy)$$
$$= \frac{p_x}{b} dx + \frac{p_y}{b} dy,$$

which can be written in local coordinates as

$$(a, b) \cdot (x, y, p_x, p_y) = \left(bx + a, by, \frac{p_x}{b}, \frac{p_y}{b}\right).$$

Since

$$K\left(bx + a, by, \frac{p_x}{b}, \frac{p_y}{b}\right) = \frac{b^2 y^2}{2}\left(\frac{p_x^2}{b^2} + \frac{p_y^2}{b^2}\right) = K(x, y, p_x, p_y),$$

we see that K is H-invariant.

(b) The functions F and G are H-invariant as

$$F\left(bx + a, by, \frac{p_x}{b}, \frac{p_y}{b}\right) = by\frac{p_x}{b} = y p_x = F(x, y, p_x, p_y)$$

and

$$G\left(bx + a, by, \frac{p_x}{b}, \frac{p_y}{b}\right) = by\frac{p_y}{b} = y p_y = G(x, y, p_x, p_y).$$

These functions are coordinates on the quotient manifold T^*H/H (they are the components on a left-invariant basis), and so the Poisson structure of the quotient is determined by

$$\{F, G\} = X_F \cdot G = \frac{\partial F}{\partial p_x}\frac{\partial G}{\partial x} + \frac{\partial F}{\partial p_y}\frac{\partial G}{\partial y} - \frac{\partial F}{\partial x}\frac{\partial G}{\partial p_x} - \frac{\partial F}{\partial y}\frac{\partial G}{\partial p_y}$$

$$= -p_x y = -F.$$

The Poisson bivector on the quotient is therefore

$$B = \{F, G\}\frac{\partial}{\partial F} \otimes \frac{\partial}{\partial G} + \{G, F\}\frac{\partial}{\partial G} \otimes \frac{\partial}{\partial F}$$

$$= -F\frac{\partial}{\partial F} \otimes \frac{\partial}{\partial G} + F\frac{\partial}{\partial G} \otimes \frac{\partial}{\partial F}.$$

Since B vanishes for $F = 0$, the quotient T^*H/H is not a symplectic manifold.

(c) Differentiating the expression

$$L_{(a,b)}(x, y) = (bx + a, by)$$

along a curve $(a(t), b(t))$ through the identity $e = (0, 1)$, it is readily seen that the infinitesimal action of $V = \alpha\frac{\partial}{\partial x} + \beta\frac{\partial}{\partial y} \in \mathfrak{h}$ is

$$X^V = (\alpha + \beta x)\frac{\partial}{\partial x} + \beta y\frac{\partial}{\partial y}.$$

From Example 5.4, the momentum map for the action of H on T^*H is the map $J : T^*H \to \mathfrak{h}^*$ given by

$$J(p_x dx + p_y dy)(V) = (p_x dx + p_y dy)(X^V) = (\alpha + \beta x)p_x + \beta y p_y.$$

Since K is H-invariant, J is constant along the Hamiltonian flow of K, and so, choosing $\alpha = 0$ and $\beta = 1$, we obtain the nontrivial first integral

$$I(x, y, p_x, p_y) = xp_x + yp_y$$

for the Hamiltonian flow of K (in addition to the obvious first integrals K and p_x). A geodesic for which $K = E$, $p_x = l$ and $I = m$ then satisfies

$$y^2\left(p_x{}^2 + p_y^2\right) = 2E \Leftrightarrow y^2 l^2 + (m - xl)^2 = 2E,$$

which for $l \neq 0$ is the equation of a circle centered on the x-axis.

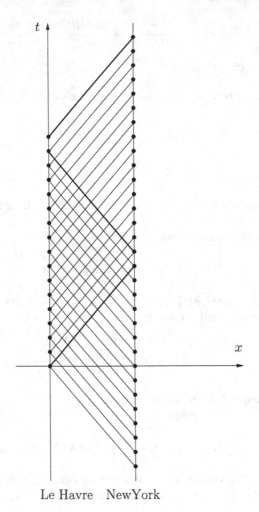

Fig. 7.8 Spacetime diagram for the Lucas problem

7.6 Chapter 6

7.6.1 Section 6.1

(1) The solution becomes trivial when one represents the transatlantic ships' motions as curves in a 2-dimensional Galileo spacetime (cf. Fig. 7.8). Thus, each transatlantic ship would meet 13 others when at sea, at midnight and at noon of every day of its voyage. Allowing one day at the arrival port for unloading, refueling and reloading, it would be possible to run this service with 15 transatlantic ships.

(3) By composing with an appropriate translation we can assume without loss of generality that f maps the origin to the origin. Therefore we just have to prove that f is linear.

We start by noticing that f maps 2-planes to 2-planes bijectively. Indeed, take any 2-plane $\Pi \subset \mathbb{R}^n$ and consider 3 straight lines on Π which intersect pairwise. Then their images must also be straight lines which intersect pairwise, and hence define a 2-plane Π'. Any straight line contained on Π intersects at least 2 of the 3 pairwise intersecting lines, and hence so does its image, which is therefore contained in Π'. We conclude that $f(\Pi) \subset \Pi'$. The same argument shows that $f^{-1}(\Pi') \subset \Pi$, and hence f is a surjection of Π onto Π'. Since it must be injective, it is a bijection.

Consider the restriction of f to a 2-plane Π. Since it is bijective, it must take parallel lines to parallel lines. Therefore it takes parallelograms to parallelograms. Since f maps the origin to the origin, we see that $f(v+w) = f(v) + f(w)$ for any two vectors $v, w \in \mathbb{R}^n$.

Finally, consider two parallel lines on Π and draw the parallel line which is equidistant from both. Any parallelogram with two sides on the two initial lines will have diagonals which intersect on a point of the third line. Because f is a bijection and preserves parallel lines, the same will be true for the image. We conclude that f maps equidistant parallel lines to equidistant parallel lines. We can easily use this fact to show that f is continuous on each 2-plane. Since it is additive, it must be linear.

7.6.2 Section 6.2

(1) Since $\langle \cdot, \cdot \rangle$ is nondegenerate there exist vectors $v, w \in V$ such that $\langle v, w \rangle \neq 0$. Moreover, there is a vector $\tilde{v} \in V$ such that $\langle \tilde{v}, \tilde{v} \rangle \neq 0$. Indeed, even if $\langle v, v \rangle = \langle w, w \rangle = 0$ we can take $\tilde{v} := v + w$ and then

$$\langle v + w, v + w \rangle = 2\langle v, w \rangle \neq 0.$$

We will now show the existence of an orthonormal basis by induction in n, the dimension of V. If $n = 1$ we take $w \in V$ such that $\langle w, w \rangle \neq 0$ and define

$$v_1 := \frac{w}{|w|},$$

where $|w| := |\langle w, w \rangle|^{\frac{1}{2}}$. Clearly $\langle v_1, v_1 \rangle = \pm 1$ and $\{v_1\}$ is the required orthonormal basis.

If $n > 1$ we again take $w \in V$ such that $\langle w, w \rangle \neq 0$ and let

$$v_1 := \frac{w}{|w|}.$$

If W is the orthogonal complement in V of the space spanned by v_1 then dim $W = n - 1$, $v_1 \notin W$ and the restriction of $\langle \cdot, \cdot \rangle$ to W is symmetric and nondegenerate. We can apply the induction hypothesis and obtain a basis $\{v_2, \ldots, v_n\}$ of W such that $\langle v_i, v_j \rangle = 0$ if $i \neq j$ and $\langle v_i, v_i \rangle = \pm 1$ for $i = 2, \ldots, n$. Therefore $\{v_1, \ldots, v_n\}$ is the desired basis of V.

To show that the signature of $\langle \cdot, \cdot \rangle$ does not depend on the choice of orthonormal basis we note that it can be invariantly defined as the dimension of a maximal subspace of V where $\langle \cdot, \cdot \rangle$ is positive definite minus the dimension of a maximal subspace of V where $\langle \cdot, \cdot \rangle$ is negative definite.

(2) Fix inertial coordinates (x^0, x^1, x^2, x^3). Then

$$v = \sum_{i=0}^{3} v^i \frac{\partial}{\partial x^i} \quad \text{and} \quad w = \sum_{i=0}^{3} w^i \frac{\partial}{\partial x^i}.$$

(a) (i) Since v is timelike and future-pointing we have

$$\langle v, v \rangle = -(v^0)^2 + \sum_{i=1}^{3}(v^i)^2 < 0 \quad \text{and} \quad v^0 > 0.$$

Similarly, since w is timelike or null and future-pointing we have

$$\langle w, w \rangle = -(w^0)^2 + \sum_{i=1}^{3}(w^i)^2 \leq 0 \quad \text{and} \quad w^0 > 0.$$

Then by the Cauchy–Schwarz inequality

$$\langle v, w \rangle = -v^0 w^0 + \sum_{i=1}^{3} v^i w^i$$

$$\leq -v^0 w^0 + \left(\sum_{i=1}^{3}(v^i)^2\right)^{\frac{1}{2}} \left(\sum_{i=1}^{3}(w^i)^2\right)^{\frac{1}{2}}$$

$$< -v^0 w^0 + |v^0||w^0| = 0.$$

(ii) Since

$$\langle v + w, v + w \rangle = \langle v, v \rangle + 2\langle v, w \rangle + \langle w, w \rangle$$

and $\langle v, v \rangle < 0$, $\langle w, w \rangle \leq 0$ and $\langle v, w \rangle < 0$ (from (i)), we conclude that $\langle v + w, v + w \rangle < 0$. Moreover,

$$(v + w)^0 = v^0 + w^0 > 0$$

and so $v + w$ is also future-pointing.

(iii) From (i) we conclude that if $\langle v, w \rangle = 0$ then w cannot be timelike nor null (except for the zero vector).

(b) (i) Since v is null and future-pointing we have $(v^0)^2 = \sum_{i=1}^{3}(v^i)^2$ and $v^0 > 0$. Moreover, since w is timelike or null and future-pointing, we have $(\sum_{i=1}^{3}(w^i)^2)^{\frac{1}{2}} \leq w^0$ and $w^0 > 0$. Then by the Cauchy–Schwarz inequality

$$\langle v, w \rangle = -v^0 w^0 + \sum_{i=1}^{3} v^i w^i$$

$$\leq -v^0 w^0 + \left(\sum_{i=1}^{3}(v^i)^2 \right)^{\frac{1}{2}} \left(\sum_{i=1}^{3}(w^i)^2 \right)^{\frac{1}{2}}$$

$$\leq -v^0 w^0 + |v^0||w^0| = 0.$$

Moreover, equality holds if and only if

$$\sum_{i=1}^{3} v^i w^i = \left(\sum_{i=1}^{3}(v^i)^2 \right)^{\frac{1}{2}} \left(\sum_{i=1}^{3}(w^i)^2 \right)^{\frac{1}{2}}$$

and

$$\left(\sum_{i=1}^{3}(w^i)^2 \right)^{\frac{1}{2}} = w^0.$$

Hence equality holds if and only if $w^i = \lambda v^i$ with $\lambda > 0$ and

$$w^0 = \lambda \left(\sum_{i=1}^{3}(v^i)^2 \right)^{\frac{1}{2}} = \lambda v^0.$$

(ii) Since
$$\langle v + w, v + w \rangle = 2\langle v, w \rangle + \langle w, w \rangle$$

and $\langle w, w \rangle \leq 0$ and $\langle v, w \rangle \leq 0$ (from (i)), we conclude that $\langle v + w, v + w \rangle \leq 0$. Moreover, equality holds if and only if $\langle v, w \rangle = \langle w, w \rangle = 0$, that is, if and only if $w = \lambda v$ with $\lambda > 0$. Clearly in all cases $v + w$ is future-pointing.

(iii) From (i) we conclude that if $\langle v, w \rangle = 0$ then w cannot be timelike, and if w is null then it is a multiple of v.

(c) Since v is spacelike we must have $A := \sum_{i=1}^{3}(v^i)^2 > (v^0)^2 \geq 0$. Then the vector $w := (A, v^0v^1, v^0v^2, v^0v^3)$ is timelike and is such that $\langle v, w \rangle = 0$. Moreover, one of v^1, v^2, v^3 must be different from zero. Assuming, without loss of generality, that $v^1 \neq 0$, the vector $\widetilde{w} = (0, -v^2, v^1, 0)$ is spacelike and satisfies $\langle v, \widetilde{w} \rangle = 0$. Moreover, $\langle w, \widetilde{w} \rangle = 0$, and so

$$\langle aw + b\widetilde{w}, aw + b\widetilde{w} \rangle = a^2\langle w, w \rangle + b^2\langle \widetilde{w}, \widetilde{w} \rangle.$$

Since w is timelike $\widetilde{w}$ is spacelike, this can be made to vanish for nonzero a and b.

(4) (a) Let us use years and light-years as our time and length units. On the Earth's frame, the motion of the Earth is the timelike line $x = 0$, whereas the motion of Planet X is the timelike line $x = 8$. If we choose $t = 0$ so that the departure of Bob is the event $(0, 0)$, then Bob's arrival at Planet X is the event $(10, 8)$, and the reunion of the twins the event $(20, 0)$. The motion of Bob is the broken line connecting these events, and hence the time measured by Bob between his departure and his return is

$$|(10, 8) - (0, 0)| + |(20, 0) - (10, 8)| = |(10, 8)| + |(10, -8)|$$
$$= \sqrt{10^2 - 8^2} + \sqrt{10^2 - (-8)^2} = 6 + 6 = 12.$$

Therefore Bob is $20 + 12 = 32$ years old when he meets Alice again.

(b) Although Bob can indeed claim that in his frame it is Alice who is moving, his frame is not an inertial frame, as he must accelerate at event $(10, 8)$ to reverse his velocity. Therefore one cannot use the Minkowski geometry in Bob's frame.

(c) At event $(10, 8)$, Bob is receiving light that left the Earth at $t = 2$. Therefore, in the 6 years it takes him to get to Planet X, Bob sees only 2 years of Alice's life (cf. Fig. 7.9). Consequently, he sees Alice moving in slow motion, 3 slower than normal. In the 6 years of the return trip, Bob will see the remaining 18 years experienced by Alice until they meet again, and hence he will see her moving in fast motion, 3 times faster than normal.

On the other hand, light emitted at event $(10, 8)$ doesn't reach Alice until $t = 18$ (cf. Fig. 7.9). Therefore she spends 18 years seeing the 6 years of Bob's trip towards Planet X, and hence sees him moving in slow motion, 3 times slower than normal. In the remaining 2 years, Alice will see the 6 years of the return trip, and will thus see Bob moving in fast motion, 3 times faster than normal.

(13) (a) According to its crew, the *Enterprise*'s trip lasts

$$|(13, 12)| = \sqrt{13^2 - 12^2} = \sqrt{25} = 5 \text{ years.}$$

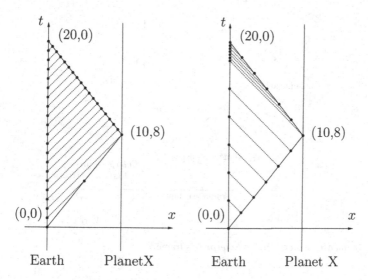

Fig. 7.9 Twin paradox

(b) The *Enterprise*'s frame moves with velocity $v = \frac{12}{13}$ with respect to the Earth. Consequently, $\sqrt{1 - v^2} = \frac{5}{13}$, and hence

$$\begin{cases} t' = \left(1 - v^2\right)^{-\frac{1}{2}} t - v \left(1 - v^2\right)^{-\frac{1}{2}} x = \frac{13}{5}t - \frac{12}{5}x \\ x' = \left(1 - v^2\right)^{-\frac{1}{2}} x - v \left(1 - v^2\right)^{-\frac{1}{2}} t = \frac{13}{5}x - \frac{12}{5}t \end{cases}.$$

Therefore in the *Enterprise*'s frame the radio signal is sent at event $(0, 0)$, the missile is launched at event $(28.6, -26.4)$, the Earth is destroyed at event $(2.4, 2.4)$ and the *Enterprise* arrives at the Earth's ruins at event $(5, 0)$ (as it had to be).

(c) Figure 7.10 shows a plot of these events on the *Enterprise*'s frame. The sequence of events is surreal: the Earth explodes without any reason at $t' = 2.4$; the faster-than-light missile jumps intact from the explosion and travels backwards in the direction of the Klingon planet, where an exact replica is being built; the two missiles vanish simultaneously at $t' = 28.6$, in the event that should be the cause of the Earth's destruction, much after the *Enterprise* has arrived at the Earth's ruins. This illustrates the absurdities that one can get if faster-than-light speeds are allowed.

7.6.3 Section 6.3

(2) From the expression $\Omega^{\mu}_{\nu} = \sum_{\alpha < \beta} R_{\alpha\beta\nu}{}^{\mu} \omega^{\alpha} \wedge \omega^{\beta}$ of the curvature forms we conclude that the nonvanishing components of the Riemann tensor are

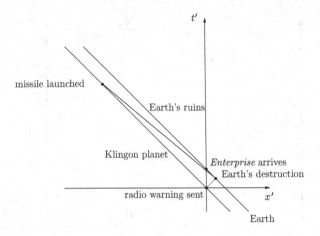

Fig. 7.10 Sequence of events on the *Enterprise*'s frame

$$R_{j00}{}^i = -R_{0j0}{}^i = \frac{\partial^2 \Phi}{\partial x^j \partial x^i}.$$

($i, j = 1, 2, 3$). Therefore the only nonzero coefficient of the Ricci curvature tensor is

$$R_{00} = \sum_{i=1}^{3} R_{i00}{}^i = \sum_{i=1}^{3} \frac{\partial^2 \Phi}{\partial (x^i)^2}.$$

(4) The nonvanishing components of the Riemann tensor for the Cartan connection are

$$R_{j00}{}^i = -R_{0j0}{}^i = \frac{\partial \Phi}{\partial x^i \partial x^j}$$

($i, j = 1, 2, 3$). If the Cartan connection were the Levi–Civita connection for a pseudo-Riemannian metric on $\mathbb{R}^4$ then the nonvanishing components of the curvature tensor would be

$$R_{j00\mu} = -R_{0j0\mu} = \sum_{i=1}^{3} g_{\mu i} \frac{\partial \Phi}{\partial x^i \partial x^j}$$

where $(g_{\mu\nu})$ is the matrix of the metric. The symmetry property $R_{j00\mu} = -R_{j0\mu 0}$, which still holds for pseudo-Riemannian metrics, would then imply that the curvature tensor is actually zero (because $R_{j0\mu 0} = 0$), meaning that the Riemann tensor is also zero. Therefore if the curvature of the Cartan connection is not zero then it cannot be the Levi–Civita connection of any pseudo-Riemannian metric.

7.6.4 Section 6.4

(1) Let p be a point in M^n. The signature of g_p is the signature of the quadratic form $Q_p(v) = g_p(v, v)$ defined on T_pM. This quadratic form is represented in local coordinates $x : V \subset M \to \mathbb{R}^n$ by a matrix $A(x) = (g_{ij}(x))$. The entries of this matrix are continuous functions of x, implying that its eigenvalues are also continuous functions of x. Indeed, they are the roots of the characteristic polynomial of $A(x)$, whose coefficients are continuous functions of the entries of $A(x)$. Since the eigenvalues of $A(x)$ can never be zero (as g is nondegenerate, implying that $A(x)$ is invertible) and M is connected, their signs cannot change and so the result follows.

(2) (a) In any given coordinate system we have

$$
Ric - \frac{S}{2}g = 8\pi E \iff R_{\mu\nu} - \frac{S}{2}g_{\mu\nu} = 8\pi E_{\mu\nu}
$$

($\mu, \nu = 0, 1, 2, 3$). Multiplying by the inverse of the metric matrix, $g^{\mu\nu}$, and summing over μ and ν yields

$$
S - \frac{S}{2} \cdot 4 = 8\pi \sum_{\mu,\nu=0}^{3} g^{\mu\nu} E_{\mu\nu} \iff S = -8\pi \sum_{\mu,\nu=0}^{3} g^{\mu\nu} E_{\mu\nu}.
$$

Therefore Einstein's field equation can be written as

$$
Ric = 8\pi E - 4\pi \left(\sum_{\mu,\nu=0}^{3} g^{\mu\nu} E_{\mu\nu} \right) g = 8\pi T.
$$

(b) We just have to notice that, since ν is timelike and unit,

$$
\sum_{\mu,\nu=0}^{3} g^{\mu\nu} E_{\mu\nu} = \rho \langle v, v \rangle = -\rho.
$$

(7) (a) This is immediate from the fact that the expression of the Euclidean metric $dx \otimes dx + dy \otimes dy + dz \otimes dz$ in cylindrical coordinates (r, θ, z) is $dr \otimes dr + r^2 d\theta \otimes d\theta + dz \otimes dz$.

(b) We just have to notice that

$$
d\theta \otimes d\theta = (d\theta' + \omega dt) \otimes (d\theta' + \omega dt)
$$
$$
= d\theta' \otimes d\theta' + \omega^2 dt \otimes dt + \omega dt \otimes d\theta' + \omega d\theta' \otimes dt.
$$

(c) For $r < \frac{1}{\omega}$ one has (in the new coordinate system (t, r, θ', z))

$$\left\langle \frac{\partial}{\partial t}, \frac{\partial}{\partial t} \right\rangle = -1 + \omega^2 r^2 < 0,$$

and so the curves of constant (r, θ', z) are timelike curves. Since along these curves r and z are constant but $\theta = \theta' + \omega t$, with θ' constant, it is clear that they correspond to observers who are rotating rigidly with angular velocity ω relative to the inertial observers of constant (r, θ, z).

(d) If we use local coordinates (t, r, θ', z) on U and (r, θ', z) on Σ then the projection map is locally represented by the map $\hat{\pi}(t, r, \theta', z) = (r, \theta', z)$. Therefore if

$$v = v^r \frac{\partial}{\partial r} + v^{\theta'} \frac{\partial}{\partial \theta'} + v^z \frac{\partial}{\partial z}$$

then

$$v^\dagger = v^0 \frac{\partial}{\partial t} + v^r \frac{\partial}{\partial r} + v^{\theta'} \frac{\partial}{\partial \theta'} + v^z \frac{\partial}{\partial z}$$

for some $v^0 \in \mathbb{R}$. Now

$$\left\langle \frac{\partial}{\partial t}, v^\dagger \right\rangle = -(1 - \omega^2 r^2) v^0 + \omega r^2 v^{\theta'},$$

and so we must have

$$v^0 = \frac{\omega r^2}{1 - \omega^2 r^2} v^{\theta'}.$$

Consequently

$$\langle v^\dagger, w^\dagger \rangle = -(1 - \omega^2 r^2) v^0 w^0 + \omega r^2 \left(v^0 w^{\theta'} + v^{\theta'} w^0 \right)$$
$$+ v^r w^r + r^2 v^{\theta'} w^{\theta'} + v^z w^z$$
$$= v^r w^r + \frac{r^2}{1 - \omega^2 r^2} v^{\theta'} w^{\theta'} + v^z w^z.$$

Notice that this does not depend on the choice of the vectors $v^\dagger$ and $w^\dagger$ projecting to v and w.

(e) As was shown in (d), the condition for a vector w to be orthogonal to $\frac{\partial}{\partial t}$ is

$$w^0 = \frac{\omega r^2}{1 - \omega^2 r^2} w^{\theta'} \Leftrightarrow dt(w) = \frac{\omega r^2}{1 - \omega^2 r^2} d\theta'(w) \Leftrightarrow \alpha(w) = 0.$$

Therefore the curve c consists of simultaneous events if and only if $\alpha(\dot{c}) = 0$. If $\gamma : [0, 1] \to \Sigma$ is a closed curve, and $c : [0, 1] \to U$ is a curve consisting of simultaneous events such that $\pi \circ c = \gamma$, then

$$t(c(1)) - t(c(0)) = \int_0^1 dt(\dot{c}) = \int_0^1 \frac{\omega r^2}{1 - \omega^2 r^2} d\theta'(\dot{c})$$

$$= \int_0^1 \frac{\omega r^2}{1 - \omega^2 r^2} d\theta'(\dot{\gamma}).$$

Since

$$d\left(\frac{\omega r^2}{1 - \omega^2 r^2} d\theta'\right) = \frac{2\omega r}{(1 - \omega^2 r^2)^2} dr \wedge d\theta' \neq 0,$$

we see that in general $t(c(1)) \neq t(c(0))$. Therefore, if the rotating observers synchronize clocks around a closed path, they will conclude that the initial clock is not synchronized with itself.

7.6.5 Section 6.5

(2) (a) Let $\{E_0, E_r, E_\theta, E_\varphi\}$ be the orthonormal frame dual to the orthonormal coframe $\{\omega^0, \omega^r, \omega^\theta, \omega^\varphi\}$, that is

$$E_0 = \frac{1}{A}\frac{\partial}{\partial t}, \qquad E_r = \frac{1}{B}\frac{\partial}{\partial r},$$
$$E_\theta = \frac{1}{r}\frac{\partial}{\partial \theta}, \qquad E_\varphi = \frac{1}{r \sin \theta}\frac{\partial}{\partial \varphi}.$$

Then, since

$$\dot{c} = \dot{t}\frac{\partial}{\partial t} + \dot{\varphi}\frac{\partial}{\partial \varphi} = \dot{t}A E_0 + \dot{\varphi}r E_\varphi$$

($\sin \theta = 1$), we have

$$\nabla_{\dot{c}}\dot{c} = 0 \Leftrightarrow \frac{d}{d\tau}(\dot{t}A) E_0 + \dot{t}A\nabla_{\dot{c}}E_0 + \frac{d}{d\tau}(\dot{\varphi}r) E_\varphi + \dot{\varphi}r\nabla_{\dot{c}}E_\varphi = 0.$$

Moreover,

$$\nabla_{\dot{c}}E_0 = \dot{t}A\nabla_{E_0}E_0 + \dot{\varphi}r\nabla_{E_\varphi}E_0$$

and

$$\nabla_{\dot{c}}E_\varphi = \dot{t}A\nabla_{E_0}E_\varphi + \dot{\varphi}r\nabla_{E_\varphi}E_\varphi.$$

Since

$$\nabla_{E_0} E_0 = \sum_{\mu=0}^{3} \omega_0^\mu (E_0) E_\mu = \omega_0^r (E_0) E_r = \frac{A'}{AB} E_r = A' E_r,$$

$$\nabla_{E_0} E_\varphi = \sum_{\mu=0}^{3} \omega_\varphi^\mu (E_0) E_\mu = 0,$$

$$\nabla_{E_\varphi} E_0 = \sum_{\mu=0}^{3} \omega_0^\mu (E_\varphi) E_\mu = 0,$$

$$\nabla_{E_\varphi} E_\varphi = \sum_{\mu=0}^{3} \omega_\varphi^\mu (E_\varphi) E_\mu = \omega_\varphi^r (E_\varphi) E_r = -\frac{1}{Br} E_r,$$

we have

$$\nabla_{\dot c} E_0 = \dot t A A' E_r \quad \text{and} \quad \nabla_{\dot c} E_\varphi = -\frac{\dot\varphi}{B} E_r,$$

and so the geodesic equation becomes

$$\ddot t A E_0 + \left(\dot t^2 A^2 A' - \frac{\dot\varphi^2 r}{B} \right) E_r + \ddot\varphi r E_\varphi = 0,$$

or, equivalently,

$$\begin{cases} \ddot t = 0 \\ \ddot\varphi = 0 \\ \dot t^2 A^2 A' = \frac{\dot\varphi^2 r}{B} \Leftrightarrow AA' \dot t^2 = \dot\varphi^2 r \Leftrightarrow \frac{m}{r^2} \dot t^2 = \dot\varphi^2 r. \end{cases}$$

Finally, since we must have $\langle \dot c, \dot c \rangle = -1$, we obtain

$$-\left(1 - \frac{2m}{r}\right) \dot t^2 + r^2 \dot\varphi^2 = -1 \Leftrightarrow \left(1 - \frac{2m}{r} - \frac{m}{r}\right) \dot t^2 = 1$$

$$\Leftrightarrow \left(1 - \frac{3m}{r}\right) \dot t^2 = 1.$$

(b) From the last equation above we see that for $r = 3m$ we have $\langle \dot c, \dot c \rangle = 0$, and we get an equatorial circular null geodesic. A stationary observer placed on this circular light ray would see it as a straight line, with infinite images of himself placed at regular spacings (equal to $2\pi r$), corresponding to light rays completing an integer number of orbits before reaching his eyes. Different images would be images of the observer at different times.

(c) Let $V = V^0 E_0 + V^r E_r + V^\theta E_\theta + V^\varphi E_\varphi$ be the angular momentum vector field of a free-falling spinning particle on a circular orbit around a pointlike mass m. Since this vector field is orthogonal to the motion we conclude that

$$-iAV^0 + \dot{\varphi}rV^\varphi = 0. \tag{7.31}$$

From the fact that this vector field is parallel-transported along its motion we have

$$\nabla_{\dot{c}}V = 0 \Leftrightarrow \frac{dV^0}{d\tau}E_0 + V^0\nabla_{\dot{c}}E_0 + \frac{dV^r}{d\tau}E_r + V^r\nabla_{\dot{c}}E_r$$
$$+ \frac{dV^\theta}{d\tau}E_\theta + V^\theta\nabla_{\dot{c}}E_\theta + \frac{dV^\varphi}{d\tau}E_\varphi + V^\varphi\nabla_{\dot{c}}E_\varphi = 0.$$

Since

$$\nabla_{E_0}E_r = \sum_{\mu=0}^{3}\omega_r^\mu(E_0)E_\mu = \omega_r^0(E_0)E_0 = A'E_0,$$

$$\nabla_{E_0}E_\theta = \sum_{\mu=0}^{3}\omega_\theta^\mu(E_0)E_\mu = 0,$$

$$\nabla_{E_\varphi}E_r = \sum_{\mu=0}^{3}\omega_r^\mu(E_\varphi)E_\mu = \omega_r^\varphi(E_\varphi)E_\varphi = \frac{1}{rB}E_\varphi,$$

$$\nabla_{E_\varphi}E_\theta = \sum_{\mu=0}^{3}\omega_\theta^\mu(E_\varphi)E_\mu = 0,$$

we have

$$\nabla_{\dot{c}}E_r = iA\nabla_{E_0}E_r + \dot{\varphi}r\nabla_{E_\varphi}E_r = iAA'E_0 + \frac{\dot{\varphi}}{B}E_\varphi,$$
$$\nabla_{\dot{c}}E_\theta = iA\nabla_{E_0}E_\theta + \dot{\varphi}r\nabla_{E_\varphi}E_\theta = 0,$$

and so we obtain

$$\frac{dV^0}{d\tau} = -iAA'V^r, \tag{7.32}$$

$$\frac{dV^r}{d\tau} = -iAA'V^0 + \frac{\dot{\varphi}}{B}V^\varphi,$$

$$\frac{dV^\theta}{d\tau} = 0, \tag{7.33}$$

$$\frac{dV^\varphi}{d\tau} = -\frac{\dot{\varphi}}{B}V^r.$$

Substituting (7.31) in (7.32) yields

$$\frac{dV^r}{d\tau} = \dot{\varphi}V^\varphi \left(\frac{1}{B} - A'r\right) = B\left(1 - \frac{3m}{r}\right)\dot{\varphi}V^\varphi, \qquad (7.34)$$

and so, from (7.33) and (7.34), we get

$$\frac{dV^r}{d\varphi} = \frac{dV^r}{d\tau}\frac{1}{\dot{\varphi}} = B\left(1 - \frac{3m}{r}\right)V^\varphi,$$

$$\frac{dV^\varphi}{d\varphi} = \frac{dV^\varphi}{d\tau}\frac{1}{\dot{\varphi}} = -\frac{V^r}{B}.$$

Hence,

$$\frac{d^2V^\varphi}{d\varphi^2} = -\frac{1}{B}\frac{dV^r}{d\varphi} = -\left(1 - \frac{3m}{r}\right)V^\varphi,$$

implying that

$$V^\varphi(\varphi) = \alpha\cos\left(\sqrt{1 - \frac{3m}{r}}\,\varphi\right) + \beta\sin\left(\sqrt{1 - \frac{3m}{r}}\,\varphi\right)$$

and

$$V^r(\varphi) = -B\frac{dV^\varphi}{d\varphi}$$

$$= -B\sqrt{1 - \frac{3m}{r}}\left(-\alpha\sin\left(\sqrt{1 - \frac{3m}{r}}\,\varphi\right) + \beta\cos\left(\sqrt{1 - \frac{3m}{r}}\,\varphi\right)\right).$$

Since the vector field V is initially aligned with the radial direction we have $V^\varphi(0) = V^\theta(0) = V^0(0) = 0$, implying that $V^\theta \equiv 0$, and

$$V^\varphi(\varphi) = \beta\sin\left(\sqrt{1 - \frac{3m}{r}}\,\varphi\right),$$

$$V^r(\varphi) = -B\beta\sqrt{1 - \frac{3m}{r}}\cos\left(\sqrt{1 - \frac{3m}{r}}\,\varphi\right),$$

$$V^0(\varphi) = \frac{\dot{\varphi}r}{\dot{t}A}V^\varphi(\varphi) = \frac{\dot{\varphi}r}{\dot{t}A}\beta\sin\left(\sqrt{1 - \frac{3m}{r}}\,\varphi\right).$$

Hence,

$$V(0) = -\beta B\sqrt{1 - \frac{3m}{r}}\,X^r,$$

$$V(2\pi) = \frac{\dot{\varphi}r}{\dot{t}A}\beta \sin\left(2\pi\sqrt{1-\frac{3m}{r}}\right)X^0$$

$$- \beta B\sqrt{1-\frac{3m}{r}}\cos\left(2\pi\sqrt{1-\frac{3m}{r}}\right)X^r + \beta\sin\left(2\pi\sqrt{1-\frac{3m}{r}}\right)X^\varphi,$$

and so, since

$$\langle V(0), V(2\pi)\rangle = \beta^2 B^2\left(1-\frac{3m}{r}\right)\cos\left(2\pi\sqrt{1-\frac{3m}{r}}\right)$$

and

$$|V(0)|^2 = \beta^2 B^2\left(1-\frac{3m}{r}\right) = |V(2\pi)|^2,$$

we get

$$\cos(\delta) = \frac{\langle V(0), V(2\pi)\rangle}{|V(0)|\,|V(2\pi)|} = \cos\left(2\pi\sqrt{1-\frac{3m}{r}}\right),$$

where δ is the angle between the two vectors $V(0)$ and $V(2\pi)$ (after one revolution). Choosing $\beta < 0$, so that initially V has the same direction as E_r, one easily checks that $\delta > 0$, and so

$$\delta = 2\pi - 2\pi\left(1-\frac{3m}{r}\right)^{\frac{1}{2}}.$$

(7) (a) From Exercise 5.1(5) we see that radial ($\dot{\theta} = \dot{\varphi} = 0$) timelike or null geodesics satisfy

$$\begin{cases} \dot{r}^2 = E^2 - \sigma\left(1-\frac{2m}{r}\right) \\ \left(1-\frac{2m}{r}\right)\dot{t} = E \end{cases}$$

where E is an integration constant, $\sigma = 1$ for timelike geodesics and $\sigma = 0$ for null geodesics. In either case, geodesics satisfying $r(0) = r_0$ cannot be extended beyond the values of the affine parameter

$$\lambda(2m) = \lambda(2m) - \lambda(r_0) = \int_{r_0}^{2m}\frac{d\lambda}{dr}\,dr = \int_{r_0}^{2m}\frac{dr}{\dot{r}}$$

or

$$\lambda(0) = \lambda(0) - \lambda(r_0) = \int_{r_0}^{0}\frac{d\lambda}{dr}\,dr = \int_{r_0}^{0}\frac{dr}{\dot{r}},$$

that is,

$$\pm \int_{r_0}^{2m} \frac{\sqrt{r}\,dr}{\sqrt{(E^2 - \sigma)r + 2m\sigma}} \quad \text{or} \quad \pm \int_{r_0}^{0} \frac{\sqrt{r}\,dr}{\sqrt{(E^2 - \sigma)r + 2m\sigma}}$$

($\pm$ is the sign of $\dot{r}$). Since both these integrals are clearly finite, the geodesics are necessarily incomplete. Notice that

$$\int_{r_0}^{2m} \frac{dt}{dr}\,dr = \int_{r_0}^{2m} \frac{\dot{t}}{\dot{r}}\,dr = \int_{r_0}^{2m} \frac{E r^{\frac{3}{2}}\,dr}{(r - 2m)\sqrt{(E^2 - \sigma)r + 2m\sigma}}$$

diverges, implying that t diverges as $r \to 2m$ for all geodesics satisfying $E \neq 0$ ($E = 0$ can only happen for the timelike geodesics in the region $\{r < 2m\}$ with constant t).

(b) We already know that $r \to 2m$ for a finite value of the affine parameter. From the definition of the Painlevé time coordinate we see that it approaches

$$\int_{r_0}^{2m} \frac{dt'}{dr}\,dr = \int_{r_0}^{2m} \frac{dt}{dr}\,dr + \int_{r_0}^{2m} \frac{\sqrt{2mr}}{r - 2m}\,dr$$

$$= \pm \int_{r_0}^{2m} \frac{E r^{\frac{3}{2}}\,dr}{(r - 2m)\sqrt{(E^2 - \sigma)r + 2m\sigma}} + \int_{r_0}^{2m} \frac{\sqrt{2mr}}{r - 2m}\,dr$$

$$= \int_{r_0}^{2m} \frac{\sqrt{2mr}\sqrt{(E^2 - \sigma)r + 2m\sigma} \pm E r^{\frac{3}{2}}}{(r - 2m)\sqrt{(E^2 - \sigma)r + 2m\sigma}}\,dr,$$

as $r \to 2m$ ($\pm$ is the sign of $\dot{r}$). Consider the geodesics in the region $\{r > 2m\}$, and choose $E > 0$, corresponding to $\dot{t} > 0$. Then the Painlevé time coordinate t' diverges for $\dot{r} > 0$, and converges for $\dot{r} < 0$. Therefore ingoing geodesics can be extended past $r = 2m$, but not outgoing ones. In other words, radial timelike and null geodesics are asymptotic to $r = 2m$ in the past (with t' diverging for a finite value of the affine parameter), but cross this hypersurface in the future.

(c) Since

$$\frac{\partial}{\partial t'} = \frac{\partial t}{\partial t'}\frac{\partial}{\partial t} + \frac{\partial r}{\partial t'}\frac{\partial}{\partial r} = \frac{\partial}{\partial t},$$

we see that $\frac{\partial}{\partial t'}$ is still a Killing vector field. Therefore the equations for a radial curve $c : \mathbb{R} \to M$ to be a future-directed timelike geodesic can be written as

$$\begin{cases} g(\frac{\partial}{\partial t'}, \dot{c}) = -E \\ g(\dot{c}, \dot{c}) = -1 \end{cases} \Leftrightarrow \begin{cases} \dot{t}' - \sqrt{\frac{2m}{r}} \left(\dot{r} + \sqrt{\frac{2m}{r}} \dot{t}' \right) = E \\ \dot{t}'^2 - \left(\dot{r} + \sqrt{\frac{2m}{r}} \dot{t}' \right)^2 = 1 \end{cases}.$$

Therefore radial curves satisfying

$$\frac{dr}{dt'} = -\sqrt{\frac{2m}{r}} \Leftrightarrow \dot{r} + \sqrt{\frac{2m}{r}} \dot{t}' = 0$$

are future-directed timelike geodesics as long as

$$\dot{t}' = E = 1.$$

In particular this implies that the Painlevé time coordinate t' coincides with the proper time along these curves.

(d) The light received by the stationary observer corresponds to outgoing null geodesics. As we saw in (b), these geodesics cannot cross the horizon, and hence accumulate along it, as shown in Fig. 7.11. Consequently, all of them intersect the curve representing the falling particle. This means that the stationary observer sees the particle forever, moving slower and slower, and increasingly redshifted.

7.6.6 Section 6.6

(3) (a) The distance between the two galaxies is clearly $d(t) = a(t)x$, where

$$x = \int_0^{r_1} \frac{dr}{\sqrt{1 - kr^2}}.$$

Therefore

$$\dot{d} = \dot{a}x = \frac{\dot{a}}{a}ax = Hd.$$

(b) The tangent vector to these curves is

$$\dot{c} = \dot{t}\frac{\partial}{\partial t} + \frac{\partial}{\partial r},$$

which is null if and only if

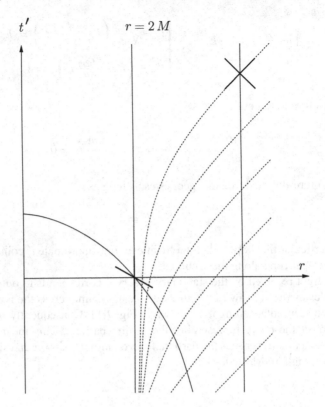

Fig. 7.11 Stationary observer watching a particle fall through the event horizon

$$-\dot{t}^2 + \frac{a^2(t)}{1 - kr^2} = 0 \Leftrightarrow \dot{t} = \pm\frac{a(t)}{\sqrt{1 - kr^2}}.$$

We must choose the positive sign for null geodesics connecting the first galaxy to the second galaxy. Therefore

$$\dot{c} = \frac{a(t)}{\sqrt{1 - kr^2}}\frac{\partial}{\partial t} + \frac{\partial}{\partial r} = \frac{a(t)}{\sqrt{1 - kr^2}}(E_0 + E_r).$$

Since we have

$$\omega_r^0 = \omega_0^r = \frac{\dot{a}}{a}\omega^r,$$

we see that

$$\nabla_{E_0+E_r}(E_0 + E_r) = \omega_0^r(E_0 + E_r)E_r + \omega_r^0(E_0 + E_r)E_0$$

$$= \frac{\dot{a}}{a}(E_0 + E_r).$$

Therefore $E_0 + E_r$ is tangent to a reparameterized geodesic, and consequently so is $\dot{c}$.

(c) Differentiating the equation for $\frac{dt}{dr}$ we obtain

$$\frac{dt'}{dr} = \frac{\dot{a}(t)t'}{\sqrt{1 - kr^2}} = \frac{\dot{a}(t)}{a(t)}\frac{dt}{dr}t',$$

where $t' = \frac{\partial t}{\partial t_0}$. Integrating yields

$$\int_0^{r_1} \frac{1}{t'}\frac{dt'}{dr}dr = \int_0^{r_1} \frac{\dot{a}(t)}{a(t)}\frac{dt}{dr}dr \Leftrightarrow \log(t'(r_1)) = \log\left(\frac{a(t_1)}{a(t_0)}\right).$$

(8) Using spherical coordinates, the metric for the 4-dimensional Minkowski spacetime is written

$$g = -dt \otimes dt + dr \otimes dr + r^2 h,$$

where h is the standard metric in S^3.

(a) The equation for the "cylinder" is $r^2 = \frac{1}{\Lambda}$, and the induced metric is therefore

$$g_{Einstein} = -dt \otimes dt + \frac{1}{\Lambda}h.$$

This is the metric for a FLRW model with $k = 1$ and $a(t) = \frac{1}{\sqrt{\Lambda}}$. In this case,

$$\frac{\dot{a}^2}{2} - \frac{\alpha}{a} - \frac{\Lambda}{6}a^2 = -\alpha\sqrt{\Lambda} - \frac{1}{6} = -\frac{1}{2}$$

if $\alpha = \frac{1}{3\sqrt{\Lambda}}$ [cf. Exercise 6.1(7)]. We conclude that this metric satisfies the Einstein equation with cosmological constant $\Lambda > 0$ for a pressureless fluid with density

$$\rho = \frac{3\alpha}{4\pi a^3} = \frac{\Lambda}{4\pi}.$$

(b) The equation for the "sphere" is $-t^2 + r^2 = \frac{3}{\Lambda}$, and is solved by

$$\begin{cases} t = \sqrt{\frac{3}{\Lambda}}\sinh\left(\sqrt{\frac{\Lambda}{3}}\tau\right) \\ r = \sqrt{\frac{3}{\Lambda}}\cosh\left(\sqrt{\frac{\Lambda}{3}}\tau\right) \end{cases}.$$

The induced metric is therefore

$$g_{deSitter} = -d\tau \otimes d\tau + \frac{3}{\Lambda} \cosh^2\left(\sqrt{\frac{\Lambda}{3}}\,\tau\right)h.$$

This is the metric for a FLRW model with $k = 1$ and $a(\tau) = \sqrt{\frac{3}{\Lambda}} \cosh\left(\sqrt{\frac{\Lambda}{3}}\,\tau\right)$.
In this case,

$$\frac{\dot{a}^2}{2} - \frac{\alpha}{a} - \frac{\Lambda}{6}a^2 = -\frac{1}{2} - \frac{\alpha}{a} = -\frac{1}{2}$$

if $\alpha = 0$. We conclude that this metric satisfies the Einstein equation with cosmological constant $\Lambda > 0$ for a pressureless fluid with density $\rho = 0$.

(9) (a) From Exercise 6.1(3) we know that the null geodesic connecting the two galaxies can be found by solving the differential equation

$$\frac{dt}{dr} = \frac{t}{\sqrt{1 + r^2}}.$$

This equation is separable, and can be integrated to give

$$\int_{t_0}^{t_1} \frac{dt}{t} = \int_0^{r_1} \frac{dr}{\sqrt{1 + r^2}}$$

$$\Leftrightarrow \log\left(\frac{t_1}{t_0}\right) = \text{arcsinh}(r_1)$$

$$\Leftrightarrow r_1 = \sinh(\log(1 + z)) = \frac{1}{2}\left(1 + z - \frac{1}{1 + z}\right).$$

Therefore

$$R = t_1 r_1 = t_1 \frac{z^2 + 2z}{2 + 2z}.$$

(b) In this case we must solve

$$\frac{dt}{dr} = e^{Ht},$$

and so

$$\int_{t_0}^{t_1} e^{-Ht} dt = \int_0^{r_1} dr \Leftrightarrow r_1 = \frac{1}{H}\left(e^{-Ht_0} - e^{-Ht_1}\right).$$

Therefore

$$R = e^{Ht_1} r_1 = \frac{z}{H}.$$

(c) In this case we must solve

$$\frac{dt}{dr} = \left(\frac{t}{t_1}\right)^{\frac{2}{3}},$$

and so

$$\int_{t_0}^{t_1} \left(\frac{t}{t_1}\right)^{-\frac{2}{3}} dt = \int_0^{r_1} dr \Leftrightarrow$$

$$r_1 = 3t_1^{\frac{2}{3}} \left(t_1^{\frac{1}{3}} - t_0^{\frac{1}{3}}\right) = 3t_1 \left(1 - (1+z)^{-\frac{1}{2}}\right).$$

Therefore

$$R = r_1 = 3t_1 \left(1 - (1+z)^{-\frac{1}{2}}\right).$$

7.6.7 Section 6.7

(2) (a) Take $q \in I^+(p)$. Then there exists a future-directed timelike curve $c :$ $[0, 1] \to M$ connecting p to q. Let V be a geodesically convex neighborhood of q as in Proposition 7.1, and take $s \in (0, 1)$ such that $c([s, 1]) \subset V$. The chronological future $I^+(r, V)$ of the point $r := c(s)$ with respect to the spacetime (V, g) is clearly an open set (image of an open set by the exponential map) satisfying $q \in I^+(r, V) \subset I^+(p)$. Therefore q is an interior point of $I^+(p)$. Since q is arbitrary, $I^+(p)$ is open.

(b) Let M be $\mathbb{R}^2$ with the point $(1, 1)$ deleted. If p is the origin then $I^+(p)$ is the Minkowski cone

$$I^+(p) = \{(t, x) \in \mathbb{R}^2 \mid t > |x|\}.$$

However, no point in the line $t = x$ with $t > 1$ can be reached from p by a causal curve. Hence,

$$J^+(p) = I^+(p) \cup \{(t, -t) \in \mathbb{R}^2 \mid t \geq 0\} \cup \{(t, t) \in \mathbb{R}^2 \mid 0 < t < 1\}$$

is not closed.

(c) Let $q \in J^+(p)$. Then there is a future-directed causal curve $c : [0, 1] \to M$ connecting p to q. Let T be a future-directed timelike vector field parallel along c, and $\gamma : [0, 1] \times (-\varepsilon, \varepsilon) \to M$ a smooth map such that $\gamma(0, t) = c(t)$ and

$$\frac{\partial \gamma}{\partial s}(0, t) = t T_{c(t)}.$$

Setting

$$\gamma' = \frac{\partial\gamma}{\partial s} = \gamma_* \frac{\partial}{\partial s} \quad \text{and} \quad \dot\gamma = \frac{\partial\gamma}{\partial t} = \gamma_* \frac{\partial}{\partial t}$$

we have

$$\nabla_{\gamma'}\dot\gamma - \nabla_{\dot\gamma}\gamma' = \gamma_* \left[\frac{\partial\gamma}{\partial s}, \frac{\partial\gamma}{\partial t} \right] = 0.$$

Therefore

$$\frac{\partial}{\partial s}\Big|_{s=0} \langle \dot\gamma, \dot\gamma \rangle = 2\langle \nabla_{\gamma'}\dot\gamma, \dot\gamma \rangle|_{s=0} = 2\langle \nabla_{\dot\gamma}\gamma', \dot\gamma \rangle|_{s=0}$$

$$= 2\langle \nabla_{\dot c}(t\, T_{c(t)}), \dot c \rangle = 2\langle T_{c(t)}, \dot c \rangle < 0,$$

and so $\gamma(s, t)$ is timelike and future-directed for small positive s. Therefore

$$q = \lim_{s\to 0^+} \gamma(s, 1) \in \overline{I^+(p)}.$$

(d) Clearly $I^+(p) \subset \text{int}\, J^+(p)$. To see that $\text{int}\, J^+(p) \subset I^+(p)$ let $q \in \text{int}\, J^+(p)$. Taking a geodesically convex neighborhood V of q with $V \subset J^+(p)$ it is easily seen that there exists a point $r \in J^+(p)$ which can be connected to q by a future-directed timelike curve. If $r_n \in I^+(p)$ is a sequence with $r_n \to r$ we have $r_n \in I^-(q)$ (hence $q \in I^+(r_n)$) for sufficiently large n. Thus $q \in I^+(p)$.

(e) It is clear that $I^+(r)$ is an open subset of $J^+(p)$. Therefore $I^+(r)$ is an open subset of $\text{int}\, J^+(p) = I^+(p)$.

(f) This is equivalent to proving that if $r \in J^-(q)$ and $p \in I^-(r)$ then $p \in I^-(q)$, which is done as above.

(g) If M is the quotient of the Minkowski 2-dimensional spacetime by the group of isometries generated by $f(t, x) = (t + 1, x)$ then $I^+(p) = M$ for any point $p \in M$.

(4) Take local coordinates around each point $p \in K := \text{svpp}\, h$. By continuity, the matrix of the components of g_ε in these coordinates, $(g_{\mu\nu} + \varepsilon h_{\mu\nu})$, has one negative and three positive eigenvalues for $\varepsilon \in (-\varepsilon_p, \varepsilon_p)$ in some neighborhood V_p of p. Since $\{V_p\}_{p\in K}$ is an open cover of K, we can take a finite subcover $\{V_{p_1}, \ldots, V_{p_N}\}$. If $\varepsilon_0 = \min\{\varepsilon_{p_1}, \ldots, \varepsilon_{p_N}\}$, then it is clear that g_ε is a Lorentzian metric for $|\varepsilon| < \varepsilon_0$.

Let $t : M \to \mathbb{R}$ be a global time function for g. Since K is compact, we have on K

$$g(\text{grad}\, t, \text{grad}\, t) \le -\delta$$

for some $\delta > 0$. Consider the map $f : (-\varepsilon_0, \varepsilon_0) \times K \to \mathbb{R}$ given by

$$f(\varepsilon, p) = g_\varepsilon \left(\left(\text{grad}_\varepsilon\, t\right)_p, \left(\text{grad}_\varepsilon\, t\right)_p \right),$$

where $\mathrm{grad}_\varepsilon$ is the gradient with respect to the metric g_ε. This map is clearly continuous and satisfies

$$f(0, p) \le -\delta$$

for all $p \in K$. Therefore there exists $\varepsilon_1 \in (o, \varepsilon_0)$ such that

$$f(\varepsilon, p) < 0$$

for $|\varepsilon| < \varepsilon_1$ and $p \in K$. We conclude that if $|\varepsilon| < \varepsilon_1$ then t is still a global time function for g_ε, and so (M, g_ε) satisfies the chronology condition.

(6) (a) Trivial.
(b) Taking for instance $S = \{0\} \times [-1, 1]$ in Minkowski 2-dimensional space-time, we have that $D^+(S)$ is the closed triangle with vertices $(0, -1)$, $(0, 1)$ and $(1, 0)$, and hence is not open.
(c) Taking for instance $S = \{0\} \times (-1, 1)$ in Minkowski 2-dimensional space-time, we have that $D^+(S)$ is the union of S with the open triangle with vertices $(0, -1)$, $(0, 1)$ and $(1, 0)$, and hence is not closed.

(8) Assume first that (Σ, h) is complete. Let $c : I \to M$ be an inextendible causal curve. Since t is clearly a time function we can assume without loss of generality that c is parameterized by t. Therefore $c(t) = (t, \gamma(t))$ with $\gamma : I \to \Sigma$. The fact that c is timelike is equivalent to

$$h(\dot\gamma, \dot\gamma) \le 1.$$

Let $t_n \in I$ be an increasing bounded sequence with limit t_0. If d is the Riemannian distance on Σ and $n > m$ then

$$d(\gamma(t_n), \gamma(t_m)) = \int_{t_m}^{t_n} (h(\dot\gamma, \dot\gamma))^{\frac{1}{2}} \, dt \le t_n - t_m.$$

Thus $\{\gamma(t_n)\}$ is a Cauchy sequence, which must converge, implying that there exists $\lim_{t \to t_0} \gamma(t)$. Since c is inextendible, we conclude that I cannot be bounded above. By a similar argument it cannot be bounded below, and hence $I = \mathbb{R}$. We conclude that every inextendible causal curve intersects every level set of t, and hence (M, g) is globally hyperbolic.

Assume now that (Σ, h) is not complete, but (M, g) is globally hyperbolic. Let $\gamma : I \to \Sigma$ be a geodesic parameterized by arclength which cannot be extended for $t \ge 0$. Then $c(t) = (t_0 + t, \gamma(t))$ is a future inextendible causal curve contained in the region where $t < t_0$. Since the integral curves of $\frac{\partial}{\partial t}$ are timelike, any Cauchy surface $S \subset M$ is a graph

$$S = \{(t, p) \in M \mid t = f(p)\}$$

of some smooth function $f : \Sigma \to \mathbb{R}$. The projection $d : I \to S$ of c on S, given by $d(t) = (f(\gamma(t)), \gamma(t))$, is necessarily spacelike, and so satisfies $\dot{f}^2 < h(\dot{\gamma}, \dot{\gamma}) = 1$ (where $f(t) := f(\gamma(t))$). This implies that the limit $f_0 := \lim_{t \to 0} f(t)$ exists. As $|\dot{f}| < 1$, we then have $f(t) > f_0 + t$ for all $t < 0$. Choosing $t_0 = f_0 - 1$, we guarantee that c and d do not intersect, that is, c does not intersect S. Therefore we reach a contradiction, meaning that (M, g) cannot be globally hyperbolic.

(10) Let $t : M \to \mathbb{R}$ be a time function whose level sets are Cauchy surfaces. Since grad t does not vanish, t cannot have maxima or minima, and so its range must be an open interval $I \subset \mathbb{R}$. If $f : I \to \mathbb{R}$ is a diffeomorphism then $f \circ t$ is also a time function whose level sets are Cauchy surfaces, and so we can assume $I = \mathbb{R}$. Define

$$X := \frac{\operatorname{grad} t}{\langle \operatorname{grad} t, \operatorname{grad} t \rangle},$$

so that

$$X \cdot t = \langle X, \operatorname{grad} t \rangle = 1.$$

Since X is timelike, its integral curves must cross all level sets of the time function t (as (M, g) is globally hyperbolic), and thus X is complete. If ψ_t is the flow of X then it is easy to check that $F : \mathbb{R} \times S \to M$ defined by $F(t, p) = \psi_t(p)$ is a diffeomorphism.

7.6.8 Section 6.8

(1) The only non-vanishing Christoffel symbols on $(\overline{M}, \overline{g})$ are

$$\Gamma^v_{vv} = -\frac{2v}{v^2 + v^2} \quad \text{and} \quad \Gamma^v_{vv} = -\frac{2v}{v^2 + v^2}.$$

Hence the geodesics equations are

$$\ddot{v} - \frac{2v}{v^2 + v^2}\dot{v}^2 = 0$$

and

$$\ddot{v} - \frac{2v}{v^2 + v^2}\dot{v}^2 = 0.$$

Taking for instance $v \equiv 0$ we obtain

$$\frac{\ddot{v}}{\dot{v}} = 2\frac{\dot{v}}{v}$$

and so

$$\dot{v} = -av^2,$$

implying that

$$v(t) = \frac{1}{at + b}$$

for some constants $a, b \in \mathbb{R}$. Therefore, the curve

$$c(t) := \left(\frac{1}{at + b}, 0 \right)$$

is a null geodesic through the point $c(0) = (\frac{1}{b}, 0)$. If for instance $b > 0$ and $a < 0$ this geodesic is defined for $t < -\frac{b}{a}$. The image of this geodesic is

$$\{(v, 0) \mid v > 0\}$$

and so this curve is inextendible in $\overline{M}$. We conclude that $\overline{M}$ (and consequently M) is not geodesically complete.

(7) Clearly, simple neighborhoods form a basis for the topology of M, and so for every open cover $\{V_\alpha\}_{\alpha \in A}$ there is a refinement $\{U_\beta\}_{\beta \in B}$ by simple neighborhoods, i.e.

$$\bigcup_{\alpha \in A} V_\alpha = \bigcup_{\beta \in B} U_\beta$$

and for each $\beta \in B$ there exists $\alpha \in A$ such that $U_\beta \subset V_\alpha$.
Let us assume first that $V := \bigcup_{\alpha \in A} V_\alpha$ is compact. Then there is a finite subcover $\{U_{\beta_i}\}_{i=1}^k$ such that

$$V = \bigcup_{i=1}^k U_{\beta_i}.$$

Clearly $\{U_{\beta_i}\}_{i=1}^k$ is a countable locally finite refinement of the open cover $\{V_\alpha\}_{\alpha \in A}$ by simple neighborhoods.

If V is not compact we can use a compact exhaustion, that is a sequence $\{K_i\}_{i \in \mathbb{N}}$ of compact subsets of V such that $K_i \subset K_{i+1}$ and $M = \bigcup_{i=1}^\infty K_i$ (see Remark 7.4 in Chap. 2). The family $\{U_\beta\}_{\beta \in B}$ is a cover of K_1 so we can consider a finite subcover of K_1

$$\{U_{\beta_1}, \ldots, U_{\beta_{k_1}}\}.$$

By induction, we obtain a finite collection of neighborhoods

$$\{U_{\beta_1^i}, \ldots, U_{\beta_{k_i}^i}\}$$

that covers $K_i \setminus \text{int } K_{i-1}$ (a compact set). Note that, by taking smaller simple neighborhoods if necessary, we can assume that

$$\bigcup_{j=1}^{k_i} U_{\beta_j^i} \subset \text{int } K_{i+1} \setminus K_{i-2}$$

and so the countable cover $\{U_{\beta_j^i}\}_{i \in \mathbb{N}, \, 1 \leq j \leq k_i}$ is locally finite and the result follows.

(8) First note that a similar argument to that of Proposition 8.6 shows that $D^-(S) \cap J^+(p)$ is compact.

  (i) Consider a sequence of points $q_n \in J^+(p)$ converging to a point $q \in M$, and let S be a Cauchy surface with $t(S) > t(q)$ (where $t : M \to \mathbb{R}$ is the global time function of M), so that $q \in D^-(S)$. Then for sufficiently large $n \in \mathbb{N}$ we have $q_n \in A := D^-(S) \cap J^+(p)$, and so, since A is compact, $q \in A \subset J^+(p)$.

  (ii) First note that (i) holds for $J^-(q)$, that is, $J^-(q)$ is closed for any $q \in M$. Hence the set $B := J^+(p) \cap J^-(q)$ is closed. Taking a Cauchy surface S such that $t(S) = t(q)$, we have that B is a closed subset of the compact set $D^-(S) \cap J^+(p)$, implying that it is itself a compact set.

(9) Let $\gamma \in C(S, p)$. Then there exists a sequence $\gamma_n \in T(S, p)$ such that $\gamma_n \to \gamma$. We begin by showing that γ intersects each level set $S_a := t^{-1}(a)$ for $0 \leq a \leq t(p)$. Indeed, if $\gamma \cap S_a = \varnothing$ then, since γ is compact and S_a closed, the distance between them would be greater than some $\varepsilon > 0$. But since each γ_n intersects S_a we would have $d_H(\gamma, \gamma_n) > \varepsilon$, a contradiction. A similar argument shows that γ cannot intersect S_a for $a < 0$ or $a > t(p)$.

It is easy to check that the map $\pi_a : C(S, p) \to C(S_a)$ given by $\pi_a(c) = c \cap S_a$ is a continuous map (here $C(S_a)$ is the set of all compact subsets of S_a with the Hausdorff metric). Therefore $\pi_a(\gamma_n) \to \pi_a(\gamma)$. Since each $\pi_a(\gamma_n)$ is a point, $\pi_a(\gamma)$ is also a point, and so γ can be thought of as a map $\gamma : [0, t(p)] \to M$. To see that γ is a causal curve we notice that if $0 \leq a < b \leq t(p)$ we have $\gamma_n(a) \to \gamma(a)$ and $\gamma_n(b) \to \gamma(b)$ with $\gamma_n(b) \in I^+(\gamma_n(a))$. If $q \in I^-(\gamma(a))$ then, for sufficiently large n, $\gamma_n(a) \in I^+(q)$, and hence $\gamma_n(b) \in I^+(q)$. It follows that $\gamma(b) \in \overline{I^+(q)} = J^+(q)$ (recall that $J^+(q) \subset \overline{I^+(q)}$ is closed), and hence $q \in J^-(\gamma(b))$. Taking a sequence of points $q_n \in I^-(\gamma(a))$ with $q_n \to \gamma(a)$, we conclude that $\gamma(a) \in J^-(\gamma(b))$.

Finally, γ must be continuous: if $a_n \in [0, t(p)]$ is an increasing sequence with limit $a \in [0, t(p)]$ then $\gamma(a_n) \to \gamma(a)$, for otherwise there would exist a subsequence b_n with $\gamma(b_n) \to q \in S_a$, and we would have $\gamma(b_n) \notin J^-(\gamma(a))$ for sufficiently large n.

(12) Let $p \in M$ and take a normal sphere $S = S_\delta(p)$. Because (M, g) is complete, we can define the map $\exp : \mathbb{R} \times S \to M$ using the outward-pointing unit normal, and $M = B_\delta(p) \cup \exp(\mathbb{R} \times S)$. Since S is compact, we have $\theta \leq \theta_0$ on S for some $\theta_0 > 0$. Exactly the same calculation as in the proof of Proposition 8.4

shows that

$$\frac{\partial \theta}{\partial t} + \frac{1}{n-1}\theta^2 \le -\varepsilon,$$

where $n = \dim M$. In particular,

$$\frac{\partial \theta}{\partial t} \le -\varepsilon,$$

and hence $\theta \le -1$ for some $t \le \frac{1+\theta_0}{\varepsilon}$ along any geodesic (note that the line $\theta = -\varepsilon t + \theta_0$ crosses the line $\theta = -1$ at $t = \frac{1+\theta_0}{\varepsilon}$). From that point on we have

$$\frac{\partial \theta}{\partial t} + \frac{1}{n-1}\theta^2 \le 0 \Rightarrow \frac{1}{\theta} \ge \frac{t}{n-1} - 1.$$

We conclude that each outward-directed geodesic orthogonal to S reaches a conjugate point in arclength at most

$$\frac{1+\theta_0}{\varepsilon} + n - 1,$$

where it ceases to be minimizing. Thus all points in M are closer to p than

$$\delta + \frac{1+\theta_0}{\varepsilon} + n - 1,$$

meaning that M is bounded and hence compact.

Notice that in the Riemannian setting no contradictions are obtained: all points on a geodesic past the conjugate point can be reached from S by a geodesic which is shorter (not longer, as in the Lorentzian case), and hence hasn't necessarily reached a conjugate point itself. Thus the proof of the singularity theorem does not work in Riemannian geometry.

(13) (a) Minkowski spacetime does not contain a Cauchy hypersurface whose expansion satisfies $\theta \le \theta_0 < 0$.
   (b) The Einstein universe does not contain a Cauchy hypersurface whose expansion satisfies $\theta \le \theta_0 < 0$
   (c) The de Sitter universe does not satisfy the strong energy condition [cf. Exercise 8.12(6)].
   (d) The 2-dimensional anti-de Sitter spacetime is not globally hyperbolic; notice, however, that any globally hyperbolic open subset of this spacetime is geodesically incomplete.

7.6.9 Section 6.9

(1) It is easily seen that

$$
(g^{\mu\nu}) = \begin{pmatrix} \alpha & -1 & \beta_2 & \beta_3 \\ -1 & 0 & 0 & 0 \\ \beta_2 & 0 & \gamma_{22} & \gamma_{23} \\ \beta_3 & 0 & \gamma_{32} & \gamma_{33} \end{pmatrix}^{-1} = \begin{pmatrix} 0 & -1 & 0 & 0 \\ -1 & \delta & \beta^2 & \beta^3 \\ 0 & \beta^2 & \gamma^{22} & \gamma^{23} \\ 0 & \beta^3 & \gamma^{32} & \gamma^{33} \end{pmatrix},
$$

where $\beta^i = \sum_{j=2}^{3} \gamma^{ij} \beta_j$ and $\delta = \sum_{i=2}^{3} \beta_i \beta^i - \alpha$. Consequently, we have for instance

$$
\Gamma_{vr}^{v} = \frac{1}{2} \sum_{\alpha=0}^{3} g^{v\alpha} \left(\frac{\partial g_{r\alpha}}{\partial v} + \frac{\partial g_{v\alpha}}{\partial r} - \frac{\partial g_{vr}}{\partial x^\alpha} \right)
$$

$$
= -\frac{1}{2} \left(\frac{\partial g_{rr}}{\partial v} + \frac{\partial g_{vr}}{\partial r} - \frac{\partial g_{vr}}{\partial r} \right) = 0,
$$

and similarly $\Gamma_{rr}^{v} = \Gamma_{ri}^{v} = \Gamma_{rr}^{r} = \Gamma_{rr}^{i} = 0$. Finally,

$$
\Gamma_{rj}^{i} = \frac{1}{2} \sum_{\alpha=0}^{3} g^{i\alpha} \left(\frac{\partial g_{j\alpha}}{\partial r} + \frac{\partial g_{r\alpha}}{\partial x^j} - \frac{\partial g_{rj}}{\partial x^\alpha} \right)
$$

$$
= \frac{1}{2} \beta^i \left(\frac{\partial g_{vj}}{\partial r} + \frac{\partial g_{vr}}{\partial x^j} - \frac{\partial g_{rj}}{\partial v} \right) + \frac{1}{2} \sum_{k=2}^{3} \gamma^{ik} \left(\frac{\partial g_{jk}}{\partial r} + \frac{\partial g_{rk}}{\partial x^j} - \frac{\partial g_{rj}}{\partial x^k} \right)
$$

$$
= \frac{1}{2} \sum_{k=2}^{3} \gamma^{ik} \frac{\partial \gamma_{jk}}{\partial r} = \sum_{k=2}^{3} \gamma^{ik} \beta_{kj}.
$$

(3) (a) Minkowski spacetime does not contain trapped surfaces.
   (b) The Einstein universe has compact Cauchy hypersurfaces.
   (c) The de Sitter universe has compact Cauchy hypersurfaces.
   (d) The 2-dimensional anti-de Sitter spacetime is not globally hyperbolic; notice however that any globally hyperbolic open subset of this spacetime is geodesically incomplete.

Reference

[Ahl79] Ahlfors, L.: Complex Analysis. McGraw-Hill, New York (1979)

Index